Fluorescent Chemosensors

Monographs in Supramolecular Chemistry

Series editors:
Philip Gale, *The University of Sydney, Australia*
Jonathan Steed, *Durham University, UK*

Titles in this series:

1: Cyclophanes
2: Calixarenes
3: Crown Ethers and Cryptands
4: Container Molecules and Their Guests
5: Membranes and Molecular Assemblies: The Synkinetic Approach
6: Calixarenes Revisited
7: Self-assembly in Supramolecular Systems
8: Anion Receptor Chemistry
9: Boronic Acids in Saccharide Recognition
10: Calixarenes: An Introduction, 2nd Edition
11: Polymeric and Self Assembled Hydrogels: From Fundamental Understanding to Applications
12: Molecular Logic-based Computation
13: Supramolecular Systems in Biomedical Fields
14: Synthetic Receptors for Biomolecules: Design Principles and Applications
15: Polyrotaxane and Slide-Ring Materials
16: Boron: Sensing, Synthesis and Supramolecular Self-Assembly
17: Porous Polymers: Design, Synthesis and Applications
18: Pillararenes
19: Supramolecular Chemistry at Surfaces
20: Aromatic Interactions: Frontiers in Knowledge and Application
21: Naphthalenediimide and its Congeners: From Molecules to Materials
22: Functional Supramolecular Materials: From Surfaces to MOFs
23: Supramolecular Amphiphiles
24: Co-crystals: Preparation, Characterization and Applications
25: Molecular Gels: Structure and Dynamics
26: Understanding Intermolecular Interactions in the Solid State: Approaches and Techniques
27: Metallomacrocycles: From Structures to Applications
28: Cucurbiturils and Related Macrocycles
29: Dendrimer Chemistry: Synthetic Approaches Towards Complex Architectures
30: Supramolecular Protein Chemistry: Assembly, Architecture and Application
31: Reactivity in Confined Spaces
32: Porphyrin-based Supramolecular Architectures: From Hierarchy to Functions
33: Supramolecular Chemistry in Biomedical Imaging
34: Fluorescent Chemosensors

How to obtain future titles on publication:
A standing order plan is available for this series. A standing order will bring delivery of each new volume immediately on publication.

For further information please contact:
Book Sales Department, Royal Society of Chemistry, Thomas Graham House, Science Park, Milton Road, Cambridge, CB4 0WF, UK
Telephone: +44 (0)1223 420066, Fax: +44 (0)1223 420247
Email: booksales@rsc.org
Visit our website at http://www.rsc.org/Shop/Books/

Fluorescent Chemosensors

Edited by

Luling Wu
University of Bath, UK
Email: wllcyl@126.com

Adam C. Sedgwick
University of Oxford, UK
Email: adam.sedgwick@chem.ox.ac.uk

Xiao-Peng He
East China University of Science and Technology, China
Email: xphe@ecust.edu.cn

and

Tony D. James
University of Bath, UK
Email: t.d.james@bath.ac.uk

ROYAL SOCIETY OF CHEMISTRY

Monographs in Supramolecular Chemistry No. 34

Print ISBN: 978-1-83916-386-9
PDF ISBN: 978-1-83916-732-4
EPUB ISBN: 978-1-83916-733-1
Print ISSN: 1368-8642
Electronic ISSN: 2041-7144

A catalogue record for this book is available from the British Library

The Royal Society of Chemistry is a charity, registered in England and Wales, Number 207890, and a company incorporated in England by Royal Charter (Registered No. RC000524), registered office: Burlington House, Piccadilly, London W1J 0BA, UK, Telephone: +44 (0) 20 7437 8656.

Visit our website at www.rsc.org/books

Printed in the United Kingdom by CPI Group (UK) Ltd, Croydon, CR0 4YY, UK

Dedication

Dedicated to the Memory of Our Close Friend John S. Fossey (29 April 1977 to 15 April 2022)

John completed his first degree in chemistry in 2000 at Cardiff University of Wales. He was awarded a PhD from Queen Mary University London with Professor Christopher J. Richards in 2004. He then relocated to Japan where he worked as a JSPS postdoctoral research fellow with Professor Shū Kobayashi at the University of Tokyo. During his time in Japan, he met his future wife Rumi Fossey (née Matsumoto) with whom he had three children (Karen, Toby, and Edwin). John then returned to the UK from Japan in 2005 where he took up a temporary position at the University of Bath and pointedly in 2008 with Tony D. James and Steven D. Bull established the CAtalysis and SEnsing (CASE) Network. The first CASE networking meeting was held in Bath in December 2008 and has since been held at ECUST (China, 2009 host Xuhong Qian), The University of Birmingham (UK, 2011 host John S. Fossey), SIOC (China, 2012 host Shu-Li You), University of Texas at Austin (USA, 2013 hosts Eric V. Anslyn and Jonathan L. Sessler), Xiamen University (China, 2014 host Yun-Bao Jiang), The Royal College of Surgeons & Trinity College (Ireland, 2015 hosts Thorri Gunnlaugsson and Donal O'Shea), Shanghai Polytechnic University (China, 2017 host Yufang Xu), and Northwest A&F University (China, 2019 hosts Leyong Wang, Zhichao Pei, Yuxin Pei, and Xiao-Yu Hu).

John then moved to the University of Birmingham in 2008 where he rose through the ranks to become a professor of synthetic chemistry. Throughout his career John received many awards and honours, including a visiting professorship at Henan Normal University and guest professorship at East China University of Science and Technology. In addition, he received a JSPS Bridge Fellowship in 2010 and was chair of the JSPS alumni executive committee. John was a former industry fellow of the Royal Society and was jointly

Monographs in Supramolecular Chemistry No. 34
Fluorescent Chemosensors
Edited by Luling Wu, Adam C. Sedgwick, Xiao-Peng He and Tony D. James

Published by the Royal Society of Chemistry, www.rsc.org

awarded the Daiwa Adrian Prize in 2013 with Seiji Shinkai, Tony D. James, Steven D. Bull, Kazuo Sakurai, and Yuji Kubo for research into 'chemonostics'. In 2016, John received the inaugural Czarnik Emerging Investigator Award for work on CAtalysis and SEnsing. While in 2018 he was awarded a CRUK Pioneer Award to support his research in establishing early detection potential from single-molecule chemosensors. More recently, John was the principal investigator of a JDRF project focused on translating boronic acid-mediated recognition to smart drug delivery for diabetes.

Throughout his career John was deeply involved with promoting Chemistry and this formed part of his lifelong commitment to promote CAtalysis and SEnsing research through international collaborations. In addition to his tireless energy for research he was also a devoted husband and father who was an enthusiastic mountain biker along with his wife and dogs.

John's endless wit and boundless vitality will be sorely missed by all who knew him. Farewell our good friend, you can rest in peace knowing you were a very special person to many people far and near.

Empty-handed I entered the world
Barefoot I leave it.
My coming, my going —
Two simple happenings
That got entangled.
—***Kozan Ichikyo*** (1283–1360)

John S. Fossey pictured with Tony D. James in Korea 2007 (Itaewon drinking a Soju Kettle), Xiamen, in 2009 (Nanputuo Temple), Egypt in 2016 (the Sphinx), Japan in 2007 (Tokyo University), and Birmingham in 2015 (John's office with view of Old Joe) just after submitting the introduction for the Royal Society of Chemistry book on *Boron: Sensing, Synthesis and Supramolecular Self-Assembly*.

Biographies

Jusung An received his BS degree from Department of Chemistry at Hallym University of Korea in 2018. He is currently pursuing his PhD degree under the guidance of Prof. Jong Seung Kim at Korea University. His current research interests are in development of small-molecular fluorescent probes for optical imaging of aberrant proteinaceous biomarkers in Alzheimer's disease and therapeutic platforms for neurodegenerative disease.

Eric V. Anslyn received his BS in Chemistry from the California State University Northridge in 1982. He performed his thesis studies under the direction of Robert Grubbs at the California Institute of Technology, receiving a PhD in 1987. Afterwards, he was an NSF postdoctoral fellow at Columbia University, working with the late Ronald Breslow. From there he started as an assistant professor of chemistry at the University of Texas at Austin in 1989. At UT Austin he rose through the ranks to currently hold the Welch Regents Chair of Chemistry, and is a University Distinguished Teaching Professor, as well

Monographs in Supramolecular Chemistry No. 34
Fluorescent Chemosensors
Edited by Luling Wu, Adam C. Sedgwick, Xiao-Peng He and Tony D. James

Published by the Royal Society of Chemistry, www.rsc.org

as a Howard Hughes Medical Institute Professor. His research is broadly in the areas of physical organic chemistry and supramolecular chemistry with a specialization on molecular sensing, mechanistic organic chemistry studies, and most recently soft materials and sequence-defined polymers. He is a co-author of the graduate-level textbook entitled *Modern Physical Organic Chemistry*.

What aspect of your research are you most excited about at the moment?
The ability to employ the tools of physical organic chemistry to a wide variety of chemical endeavors.

What is the best part of your job?
Interpersonal relationships with all my students.

What is your advice to young emerging scientists?
Being a scientist is a fun and exciting job, so keep that passion and love throughout your career.

What do you do in your spare time?
Gardening, car racing, and cooking.

By the time I'm 100 I would like to have...
...seen all my previous co-workers achieve satisfaction in their careers.

Michał Bartkowski is a researcher specializing in the application of carbon nanomaterials as nanocarriers for cancer therapy. He received his BSc in Pharmaceutical Chemistry from Dublin City University in 2019. Subsequently, he worked on a targeted drug delivery system project involving carbon nano-onions under the direction of Prof. Silvia Giordani. Michał is currently pursuing a PhD at the School of Chemistry at Dublin City University, investigating the 'Engineering of Nanomaterials for Cancer Therapy (ENACT)' under Prof. Giordani's supervision. His research interests include the preparation and characterization of carbon nanomaterials, surface functionalization of nanomaterials, and the development of novel bioimaging and drug delivery systems utilizing carbon nanomaterials.

What aspect of your research are you most excited about at the moment?
The use of carbon nanomaterials as drug carriers is a rapidly evolving area of research that has the potential to transform the way we treat diseases, and improve the quality of life for patients. The challenge of working with a problem that has yet to be solved is both daunting and inspiring. It provides an opportunity to push the boundaries of current knowledge and to develop

innovative solutions that could have a significant impact. The prospect of exploring emerging technologies and developing novel solutions is both intellectually stimulating and rewarding.

What is your advice to young emerging scientists?
Practice optimizing your processes. By streamlining repetitive tasks, you can free up space for more engaging work. Additionally, optimization can lead to greater efficiency, consistency, and higher-quality results.

What do you do in your spare time?
As someone with ADHD, I often find myself driven by curiosity and starting new little projects that capture my interest, as they keep me energized and engaged. While I rarely finish everything I start, that's not what drives me. As a researcher, I'm motivated by the pursuit of knowledge and understanding. These little projects enable my learning, and once I've gained the knowledge I seek, I move on to the next project.

What is the best part of your job?
Colleagues who are much more intelligent and experienced than I am—working with whom has provided me with a wealth of knowledge and expertise and expanded my understanding of my work, challenged me to grow both personally and professionally, and enriched my overall experience—are surrounding the best part of my job.

By the time I'm 100 I would like to have...
By the time I'm 100, I would like to have made significant contributions to the fields that I'm passionate about. Throughout my career, I hope to have made a positive impact and left a lasting legacy in these areas. I aspire to have continued to learn, grow and develop my skills, and to have mentored and inspired others to pursue their own goals. Ultimately, I want to have led a fulfilling life where I've made a meaningful difference in the world, and left it a little better than I found it.

Amy Bowyer completed her BSc degree (Hons I, University Medal) in Chemistry and Biology at the University of Sydney in 2018. She is currently undertaking her PhD under the supervision of Prof. Elizabeth New. Her research interests include developing selective fluorescent probes for heavy metal ions, and cross-reactive sensing arrays for monitoring metal contamination in environmental water sources.

What aspect of your research are you most excited about at the moment?
In my research, I develop fluorescent probes and sensing arrays for toxic metal detection. I'm

excited about the potential to use these systems in real-world environmental monitoring applications as an alternative strategy to spectrometry-based techniques.

What is the best part of your job?
My favourite part about working in the New research group is the people. Everyone in the group is very enthusiastic and friendly: we have great, collaborative workplace.

What is your advice to young emerging scientists?
My advice would be to study widely early on to find the area of research that you are most passionate about, and also to say yes to any opportunities that are presented to you.

What do you do in your spare time?
My favourite things to do in my spare time are hiking around Sydney, reading and knitting.

By the time I'm 100 I would like to have...
Travelled all around Australia to see the many beautiful landscapes first-hand.

Samuel Bradberry studied for his MChem between 2008 and 2012 at the University of Southampton, UK where his research projects focused on supramolecular chemistry and particularly anion transporter compounds under the supervision of Prof. Philip A. Gale. He then moved to Trinity College Dublin in 2012 with Prof. Thorri Gunnlaugsson to carry out doctoral and post-doctoral research in physical organic chemistry working on self-assembly and photophysical properties of lanthanide complexes as responsive probes in solution and materials. Since 2021, he has worked in industry specializing in scientific data analysis and data science tools for the biotechnology and life sciences sectors.

What aspect of your research are you most excited about at the moment?
The power of lab automation and data science in accelerating the discovery process for so many different interest areas in biotechnology and pharmaceutical development.

What is the best part of your job?
Collaborating with colleagues who are experts in different scientific disciplines from molecular and computational biology to cell biology and bioinformatics to biochemistry and chemistry.

What is your advice to young emerging scientists?
Don't be afraid. Don't be afraid of learning a new field. Don't be afraid of a difficult path to finding a solution. Don't be afraid to take a risk for the role

you want. Intellectual risk and perseverance are integral parts of what pioneers discovery and invention.

What do you do in your spare time?
Learn languages.

By the time I'm 100 I would like to have...
Learned to read fluently in as many languages as possible to be able to read my favourite international authors in their original form.

Stephen M. Butler completed his undergraduate (2015) and PhD (2020) studies with Prof. Jonathan Morris at UNSW Sydney, investigating the development of small-molecule probes for biologically relevant kinases and phosphatases. He is currently employed in a postdoctoral role working with Prof. Kate Jolliffe at the University of Sydney, investigating macrocycles for the detection of anions. He is the second of his name to work with Prof. Jolliffe, which at times is about as confusing as you could imagine!

What aspect of your research are you most excited about at the moment?
I'm loving exploring the world of fluorescence and some of the responsive properties of different fluorophores. The many ways in which fluorescent probes can be used to visualize molecular-scale changes and demystify cellular processes continues to amaze me!

What is the best part of your job?
Teaching students and seeing them take on their projects and discover more. It's fantastic being able to discuss new ideas and see them come up with things I would never have imagined.

What is your advice to young emerging scientists?
I'm not sure I'm qualified to answer this one! If my young nieces and nephews count as emerging scientists it would be to never stop asking "why?".

What do you do in your spare time?
I enjoy absorbing science-fiction and fantasy to escape to different worlds, listening to film soundtracks, playing French horn (in a group, so I can't hear myself too much), and exploring the hidden corners of Sydney's parks and reserves.

By the time I'm 100 I would like to have...
... travelled the length and breadth of Europe, solved a David Astle cryptic crossword unassisted, seen an otter in the wild, learned to swim, grown a giant vegetable, finished The Wheel of Time books, learned taxidermy, seen Jungle live (again)...

Dr David F. Caffrey completed his Moderatorship (B.A. Hons) in Medicinal Chemistry at Trinity College Dublin and PhD in Supramolecular Chemistry under the direction of Prof. Gunnlaugsson. His postgraduate research focused on the development of novel lanthanide-directed self-assemblies from dipicolinic acid derivatives, studying the impact of different ligand substitution strategies on the complex luminescent properties. Currently, he is working as Associate Director, MS&T Bioprocess Laboratory Lead at Merck Sharp and Dohme (MSD).

What aspect of your research are you most excited about at the moment?
I don't work directly in research anymore but I still get to work on solving complex scientific problems in my day job so that always excites me.

What is the best part of your job?
The opportunity to support the professional growth and development of other scientists.

What is your advice to young emerging scientists?
Challenges and setbacks are unavoidable as a scientist when pursuing ambitious goals. Don't let it get you down, learn from it.

What do you do in your spare time?
I have two young kids so don't have any spare time.

By the time I'm 100 I would like to have....
Retired.

Christopher J. Chang (right) is the Class of 1942 Chair Professor of Chemistry and Molecular and Cell Biology at the University of California, Berkeley, and Faculty Scientist at Lawrence Berkeley National Laboratory. Chris graduated with BS/MS degrees from Caltech in 1997 and worked with Prof. Harry Gray, spent a year as a Fulbright scholar with Dr Jean-Pierre Sauvage, earned his PhD from MIT in 2002 with his thesis advisor Prof. Dan Nocera, and then after postdoctoral studies with Prof. Steve Lippard, also at MIT, joined the Berkeley faculty in 2004. Research in the Chang laboratory focuses on the study of metals and redox-active molecules in biology and energy, focusing on the development of activity-based sensing and proteomics probes and catalysts and applying them to questions in neuroscience, metabolism, and sustainable synthesis.

What aspect of your research are you most excited about at the moment?
Many activities in our lab are converging on the concept of single-atom signaling, where we are using imaging and proteomics probes to decipher how addition of a one-element unit, like a metal or oxygen atom, can regulate complex protein function.

What is the best part of your job?
Getting to work with talented and enthusiastic students and postdocs and helping them achieve their goals.

What is your advice to young emerging scientists?
Be optimistic, be determined, and be supported by people that are rooting for your success.

What do you do in your spare time?
During this COVID time, lots of walking and cooking (and eating!).

By the time I'm 100 I would like to have...
Continued health and happiness.

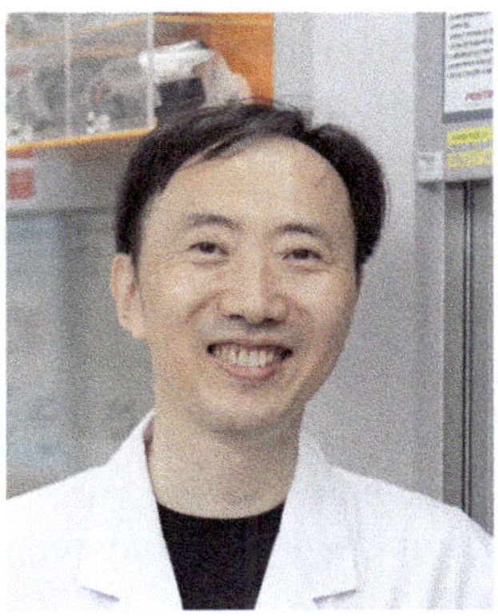

Young-Tae Chang was born in Busan, Korea, in 1968. He studied chemistry at Pohang University of Science and Technology (POSTECH), Korea and received his BS in 1991. After one and half years of army service in Korea, he started his graduate study at POSTECH and received a PhD in 1997 under the supervision of Prof. Sung-Kee Chung, working on the divergent synthesis of all possible regioisomers of myo-inositol phosphates. He did his postdoctoral work with Prof. Peter Schultz at UC Berkeley and The Scripps Research Institute. In 2000, he was appointed assistant professor at New York University and promoted to associate professor in 2005. He received the NSF Career award in 2005 and his research interests have been chemical genetics, molecular evolution, and artificial tongues. In September 2007, he moved to National University of Singapore and Singapore Bioimaging Consortium. He was a full professor of chemistry and leader of Medicinal Chemistry Program of NUS, and Lab Head of Bioimaging Probe Development at SBIC, Biopolis. In 2017, he came back to POSTECH, as a professor in the Department of Chemistry, and joined Center for Self-assembly and Complexity (CSC), Institute for Basic Science (IBS) as an Associate Director. He has published more than 400 scientific papers, 3 books and filed 60 patents so far.

What aspect of your research are you most excited about at the moment?
Discriminating old *vs.* young cells and elucidation of the novel biomarkers of aging.

What is the best part of your job?
Academic freedom and working together with young students in the lab.

What is your advice to young emerging scientists?
Select the job or work which makes your heart beat in full steam!

What do you do in your spare time?
Testing the colorful sensors and probes for many things around.

By the time I'm 100 I would like to have...
Many colourful babies, *i.e.* sensors and probes!

Zhijun Chen, born in October 1987 and received PhD degree in 2017 from the Max Planck Institute for Polymer Research, Germany, working on near-infrared sensitive materials under the supervision of Prof. Hans-Juergen Butt. Returned to China in 2017 and kicked off an independent research career at Northeast Forestry University as a full professor. Main research interest is nature-inspired design of photosensitive biomass materials. The research is funded by Natural Science Foundation of China and National Key Research Program. In 2018, awarded "Young Elite Scientist" by the China Association for Science and Technology.

What aspect of your research are you most excited about at the moment?
Now I am working on the smart sustainable materials for pollutant remediation, which could protect our mother earth and also help solve the energy crisis. I feel very excited about it.

What is the best part of your job?
The exploration of the unknown in science itself is very fascinating. Although you will face some challenges, the process of overcoming difficulties will make you grow and feel fulfilled.

What is your advice to young emerging scientists?
Don't be affected by too much external pressure, just enjoy your work. Talk with professionals in and outside the academic circle and learn as much as possible from the experience of others.

What do you do in your spare time?
During my spare time, I will do some exercises such as jogging to keep fit. Because a healthy body is the guarantee of doing scientific research well.

By the time I'm 100 I would like to have...
By the time I'm 100, I would like to have the ability to travel in space and explore the mysteries of the universe.

Ziyi Cheng received his PhD degree under the supervision of Professor Jaebum Choo from Department of Bionanotechnology, Hanyang University, Korea in 2019. Currently he is an associate professor at the Institute of Functional Materials and Molecular Imaging, Hainan Medical University. His research interests focus on surface-enhanced Raman scattering-based microfluidic devices and POCT.

Anthony W. Czarnik (born 1957) is an American chemist and inventor. He is best known for pioneering studies in the field of fluorescent chemosensors and co-founding Illumina, Inc., a biotechnology company in San Diego. Czarnik was also the founding editor of *ACS Combinatorial Science*. He currently serves as an adjunct visiting professor at the University of Nevada, Reno.

He attended the University of Wisconsin and received his BS in Biochemistry in 1977. He then studied with Nelson J. Leonard at the University of Illinois at Urbana–Champaign and earned an MS in Biochemistry in 1980 and a PhD in Chemistry in 1981 with a thesis, "Chemical studies on nucleic acid analogues". He then did postdoctoral fellowships with Ronald Breslow at Columbia University (1981–1983) as an NIH Postdoctoral Fellow.

In 1983 he joined the Chemistry Department at Ohio State University as assistant professor and was subsequently promoted to associate professor. He worked at Ohio State University until 1993, when he was offered a position as director of the BioOrganic chemistry group at Parke-Davis Pharmaceutical Research in Ann Arbor, Michigan.

He was the founding editor of *ACS Combinatorial Science* (formerly *Journal of Combinatorial Chemistry*), an academic journal published by the American Chemical Society. In April 1998, he co-founded Illumina, Inc., a biotechnology company now traded on NASDAQ and specializing in sequencing, genotyping, and gene expression, with David Walt, John Stuelpnagel, Larry Bock, and Mark Chee. He served as Illumina's chief scientific officer (CSO) until 2000.

In 2001, he was recruited by Sensors for Medicine and Science, Inc., where he served as chief scientific officer. Since 2003, he has co-founded a number of biotechnology companies, including Deuteria Pharmaceuticals LLC and Protia LLC.

He is a founder of RenoCares, a charity that provides support to alcohol and drug addicts convicted of misdemeanors in the form of financial aid for rehabilitation treatment, counseling, and psychological services. The organization is managed by the Community Foundation of Western Nevada. Since 2016, annual Czarnik Awards are given for exceptional work in the area of chemosensors at the International Conference on Molecular Sensors and Molecular Logic Gates (MSMLG). In 2007, he took part as an executive producer of *Electric Heart: Don Ellis*, a documentary about Don Ellis, an American jazz musician.

What aspect of your research are you most excited about at the moment?
After all these years, a compound I invented is in Phase II clinical trials. It is based on slowing the rate of pioglitazone racemization by placing a deuterium at the chiral center. In this way, we've been able to separate the desirable activity from the undesirable side effects. Intellectually satisfying and likely useful.

What is the best part of your job?
Completing a research project that pleases my creative side.

What is your advice to young emerging scientists?
Never allow others to make you misrepresent your research results.

What do you do in your spare time?
Increasingly, I am contacting people who I've known during my career just to catch up with them. One real advantage of having had multiple jobs: you meet one or two real friends at each, and eventually you have a great cohort of friends.

By the time I'm 100 I would like to have...
lol I won't make it to 100. Don't even want to. I'm close to having accomplished everything I'd hoped for in my life.

Hi, my name is **Bruno D'Agostino,** I was born in Milan, but I grew up in Catania (Sicily). I have a Masters degree in Organic and Bio-organic Chemistry and a Masters in Nanotechnology for Drug Delivery Systems, and I got my PhD in 2020. My research dealt with the formation of different Ln(III) cyclen based complexes and self-assembly and their applications in several fields. This includes the use of such complexes in solution and on solid surfaces (such as thin films) in the monitoring of enzymatic reactions in real-time. Presently, I am working for a multinational corporation, improving efficiency and quality within the manufacturing field, and developing new technologies, and new strategies.

What aspect of your research are you most excited about at the moment?
When I got the final molecule for the first time, I was really excited.

What is the best part of your job?
It is never boring, every day is a new challenge.

What is your advice to young emerging scientists?
Believe, even when everything looks dark, keep believing.

What do you do in your spare time?
I love sport and travel.

By the time I'm 100 I would like to have...
I want to be reminded I'm good at what I do.

Musa Dirak got his BSc degree in Chemistry from Istanbul Technical University. During his undergraduate studies, he spent a year abroad as a visiting student in Wrocław University of Science and Technology. He is now a PhD student in the Kölemen group at Koç University. His interests include organic fluorophores, chemiluminescence, activatable photosensitizer for photodynamic therapy and aggregation-induced emissive molecules.

Cidália M. G. dos Santos was born in Tondela, a small town in the center-north of Portugal. She first joined Thorri's Supramolecular Chemistry group as a visiting student in November 2000, while still an undergraduate student at the University of Coimbra. It was at this stage that her passion for lanthanide chemistry and anion sensing started to germinate. After finishing her undergraduate degree, she joined the Gunnlaugsson group in April 2003 as a postgraduate student. Her focus at the time was on the design, synthesis and photophysical evaluation of novel luminescent sensing devices. Having completed her PhD in 2008 she continued working within the research group as a post-doctoral researcher for several years. Thereafter, she moved with her family to the North East of England and for several years worked in the research group of Prof. David Parker FRS as a postdoctoral fellow.

Silvia Giordani is a Full Professor Chair of Nanomaterials and the Head of School of Chemical Sciences at Dublin City University. After receiving a "Laurea" in Chemistry and Pharmaceutical Technology from the University of Milan (Italy) in 1999, she moved to the Center for Supramolecular Science at the University of Miami (USA) where she graduated with a Masters and a PhD in Chemistry. In 2003 she moved to Trinity College Dublin (TCD), Ireland, to work on an EU-funded Marie Curie project on "Template Grown Molecular Nanomaterials" as the young researcher. She successfully applied for the Marie Curie reintegration grant to work on a research project at the University of Trieste. In 2007 she received the prestigious President of Ireland Young Researcher Award and started her independent career as Research Assistant Professor at TCD. In September 2013 she funded the "Nano Carbon Materials" research laboratory at the Istituto Italiano di Tecnologia (IIT). In December 2016 she was appointed Associate Professor in Organic Chemistry at the University of Turin, Italy, and in October 2018 Professor Chair of Nanomaterials within the School of Chemical Sciences at Dublin City University. Her main research interests are in the design, synthesis, and characterization of a wide range of nanomaterials for applications in smart and responsive biorelated nanotechnologies. She is the author/co-author of approx. 140 manuscripts, reviews, and book chapters. She is the recipient of many international prizes and honours, including the L'Oreal UNESCO for Women in Science fellowship, the William Evans visiting fellowship from the University of Otago (New Zealand), and is a Visiting Scientist to the Bio-Nano Institute at Toyo University (Japan).

What aspect of your research are you most excited about at the moment?
My aim is to design small, molecular, or nanosized, devices that can carry out a series of different functions within the context of smart and responsive medicine.

The starting point is a carbon nano-onion that is compatible with living biological systems. We attach one or more components to it in such a way that the resulting device can carry out a pre-programmed task. The different components can be added and removed in an interchangeable manner, which means that the same platform can be rapidly modified and refined depending on requirements. We are currently examining components that allow us to track and monitor the location of the device, and we are also exploring the possibilities for using this platform as a vehicle for targeted and controlled drug delivery within a biological system. Finally, we also hope to equip the device with components that will allow us to switch it on and off using different types of "remote control" (external stimuli).

What is the best part of your job?
Science simply offers unparalleled opportunities for creativity and imagination. I have always enjoyed immersing myself in new thoughts and concepts and on some level there is almost a magical beauty to the way in which a set of experimental data can suddenly come together and offer a clue or a solution to a long-standing problem. Every day is different, every day offers a new challenge, and I think I am very fortunate that I can do something together with our students, on a regular basis, that has never been done or seen before.

What is your advice to young emerging scientists?
Stay curious, ask questions, believe in yourself, and there is no substitute for hard work. Don't be afraid to speak up; your voice is worth as much as anybody else's.

What do you do in your spare time?
In my spare time I enjoy long walks in nature, swimming and gardening.

By the time I'm 100 I would like to have...
delivered knowledge and new technology that can be of real benefit to the general public.

Thorri (Thorfinnur) Gunnlaugsson MRIA is a Professor of Chemistry in the School of Chemistry, Trinity College Dublin. His main research focus is on the synthesis and formation of novel functional self-assembly structures and materials, functional luminescent materials, chemosensors and imaging agents, *etc.* and the applications of such materials. Having established his research group in the School of Chemistry at TCD in 1998, he is the author of close to 300 publications that have attracted an *h*-index of 82. He has a strong background in mentoring PhD students (over 60 to date) and early career researchers, many of who now hold independent academic positions. His work has been recognised with several awards, including membership of the Royal Irish Academy in 2011 (MRIA), and a fellowship of the Institute of Chemistry of Ireland (FICI) in 2020. He is the recipient of the 2021 Molecular Sensors and Molecular Logic Gates (MSMLG) Czarnik Award, the Institute of Chemistry of Ireland (ICI) Annual Award for Chemistry 2014 and the 2006

Bob Hay Award from the Royal Society of Chemistry, Macrocycle and Supramolecular Chemistry Interest Group.

What aspect of your research are you most excited about at the moment?
We are developing novel luminescent self-assembly structures and materials such as cages and polymers and exploring their host–guest chemistry in various ways.

What is the best part of your job?
Having the opportunity to work with people who inspire me and enrich our research work constantly!

What is your advice to young emerging scientists?
To think positively and focus on the work they love to do!

What do you do in your spare time?
Spend time with my two children and friends.

By the time I'm 100 I would like to have...
Retired.... My dad is turning 95 this year and is still working so that might be wishful thinking on my behalf!

Xiao-Peng He (Franck) is a Fellow of the Royal Society of Chemistry and a senior member of the Chinese Society of Chemistry. He received his BSc in Applied Chemistry (2006) and PhD in Pharmaceutical Engineering (2011) from East China University of Science and Technology (ECUST). Sponsored by the French Embassy in China, He completed a co-tutored doctoral program at ENS Cachan (France) from July 2008 to February 2009. Then he carried out his postdoctoral research with Prof. Kaixian Chen (SIMM, CAS) at ECUST from 2011 to 2013. In 2018, he became a professor at the Feringa Nobel Prize Scientists Joint Research Center, School of Chemistry and Molecular Engineering, ECUST, where his research interests span chemical glycobiology to fluorescent molecular probes for disease diagnostics and theranostics. Prof. He has co-authored over 200 peer-reviewed papers with a citation number of over 7700 (*h*-index 44) and is co-inventor of 10 Chinese patents. He was one of the 2020–2021 Most Cited Chinese Researchers (Shanghai Ranking, Elsevier), and was awarded the Thieme Chemistry Journal Award (2020), the Chinese Chemistry Society Prize for Young Scientists (2018) and the Chinese Chemistry Society Prize in Chemical Biology (2018). He serves as an associate editor for *Frontiers in Chemistry* (*Supramolecular Section*), *Science Bulletin* (*Chemical Biology*) and *Chemical Research in Chinese Universities* (*Chemical Biology*).

What aspect of your research are you most excited about at the moment?
To unravel new biomolecular targets for disease diagnosis and therapy using chemical biological approaches.

What is the best part of your job?
Interdisciplinary cooperation.

What is your advice to young emerging scientists?
Think in a way that researchers in other research fields do.

What do you do in your spare time?
Watching basketball games.

By the time I'm 100 I would like to have...
My own lab to do research.

Dr Tasuku Hirayama received his BS degree from Kyoto University in 2004. He also obtained his MS (2006) and PhD (2009) in chemical biology from Kyoto University under the direction of Prof. Yukio Yamamoto. He joined Prof. Christopher Chang's group at University of California, Berkeley as a postdoctoral fellow. In 2010, he joined Prof. Hideko Nagasawa's group in Gifu Pharmaceutical University as an assistant professor. He has been an associate professor in Gifu Pharmaceutical University since 2016. His research interests include development of unique chemical tools to understand biological functions of metal ions.

What aspect of your research are you most excited about at the moment?
Our group has recently succeeded in the selective detection of heme, and I'm now excited about cellular imaging of heme in cells by the small-molecule probes.

What is the best part of your job?
Creating molecules that no one has ever made before and being able to witness the moment when they work in living cells is the most exciting part of my job.

What is your advice to young emerging scientists?
I think that I'm still one of the young scientists, but my advice to the younger generation is to cherish your curiosity.

What do you do in your spare time?
Camping with my family.

By the time I'm 100 I would like to have...
I would like to have 100 PhD students graduating from my lab.

James Howard was born and raised in Spring Hill, Tennessee. James earned his BS degree in Chemistry from the University of Alabama. He is currently a 4th year PhD student in the Anslyn research group at the University of Texas at Austin. His current research is focused on high-throughput experimentation, automation, and chemistry-based computer programming.

What aspect of your research are you most excited about at the moment?
I'm excited to improve my computer programming skills and diversify my skillset.

What is the best part of your job?
Teaching undergraduate students how to conduct research.

What is your advice to young emerging scientists?
Don't get tunnel vision and think you can only do one thing! Your research and collaborations may take you from chromatography to writing code as mine did. Embrace these new paths.

What do you do in your spare time?
I enjoy baking, playing video games, and tinkering with hobby electronics.

By the time I'm 100 I would like to have...
...a boat and retained the ability to water ski!

Tony D. James is a Professor at The University of Bath and Fellow of the Royal Society of Chemistry. He received his BSc from the University of East Anglia (1986), PhD from the University of Victoria (1991), and was a Postdoctoral Research Fellow in Japan with Seiji Shinkai (1991–95). He was awarded a Royal Society University Research Fellowship at the University of Birmingham (1995–2000) before moving to the University of Bath in 2001. He has been a visiting professor at Tsukuba, Osaka, Kyushu and Sophia Universities, an AMADEus invited professor at the University of Bordeaux and is a guest Professor at East China University of Science and Technology, Xiamen University, Shandong Normal University, Nanjing University, Shaanxi University of Science and Technology, Changzhou University, Zhejiang University, Qufu Normal University, Beijing University of Chemical Technology, Shanghai Normal University, Ewha Womans University, Henan Normal University and is a Hai-Tian (Sea-Sky) Scholar at Dalian University of Technology. He was awarded the Daiwa-Adrian Prize (2013), Inaugural CASE Prize (2015), MSMLG Czarnik Award (2018) Frontiers in Chemistry Diversity Award (2020) and Royal Society Wolfson Research Merit Award (2017–2022) and listed by Clarivate as a Highly Cited Researcher (2022). He has an *h*-index of 83 (ORCID 0000-0002-4095-2191).

What aspect of your research are you most excited about at the moment?
I am always excited when fundamental research can lead to something with practical utility.

What is the best part of your job?
Acting as a mentor for the next generation of researchers.

What is your advice to young emerging scientists?
Work hard and play hard – but above all keep in mind that research should be fun.

What do you do in your spare time?
Walking and watching movies.

By the time I'm 100 I would like to have...
Enough energy to sit on a beach with a cold beer.

Anal Jana is currently pursuing a PhD at the Department of Chemistry in Shiv Nadar University (SNU), Delhi NCR, India, under the supervision of Dr Animesh Samanta. His doctoral thesis focused on developing activity-based fluorescent probes to detect reactive carbonyl species (RCSs) in living cells and environmental matrices. Before joining SNU in 2017, he obtained his Master's degree in Applied Chemistry from Birla Institute of Technology, Mesra, and a Bachelor's degree in Chemistry from University of Calcutta. His future research interest mainly lies in the conjunction of chemistry and biology and focused on monitoring the fate of the cellular microenvironment by the action of various xenobiotic species.

Paramesh Jangili received his PhD in 2016 from Jawaharlal Nehru Technological University, Hyderabad, India. He subsequently joined as a research professor in the research group of Prof. Dae-Kee-Kim, Ewha Womans University, South Korea and in 2017 he joined as a research professor in the research group of Prof. Jong Seung Kim, Korea University, South Korea. His research interests include preparation of organic molecules for drug delivery and imaging of important biomarkers in different diseases.

Shang Jia (left) received his BS in chemistry in 2014 from Peking University, China. He then completed his PhD in 2019 at the University of California, Berkeley in the laboratory of Prof. Chris Chang, developing fluorescent sensors for detection of intracellular Cu(I) and bioconjugation methods for labeling of histidine and methionine residues. Shang is currently a postdoctoral scholar with Prof. Ellen Sletten at UCLA, developing molecular probes for *in vivo* imaging.

What aspect of your research are you most excited about at the moment?
The ability of using optical imaging to look at processes happening in cells and animals.

What is the best part of your job?
Working on projects that I'm interested in and being able to (slightly) expand the boundary of our knowledge in the process.

What is your advice to young emerging scientists?
If you like it, then go for it.

What do you do in your spare time?
Exploring software, especially those for embedded electronics.

By the time I'm 100 I would like to have...
Health and an up-to-date mind with contemporary culture.

Jianfeng Jin received his PhD degree under the supervision of Professor Xu Gao from Department of Biochemistry, Harbin Medical University, China in 2013. Currently he is an associate professor at the Institute of Functional Materials and Molecular Imaging, Hainan Medical University. His research interests focus on molecular mechanism of cancer research.

Katrina (Kate) Jolliffe received her BSc (Hons) in 1993 and PhD in 1997 from the University of New South Wales. She then held positions at Twente University, The Netherlands; the University of Nottingham, UK and the Australian National University, before taking up an Australian Research Council QEII fellowship at The University of Sydney in 2002. In 2007 she became a senior lecturer at the same institution and was promoted to associate professor in 2008 and to full professor in 2009, and was head of school from 2013–2016. She currently holds the position of Payne-Scott Professor at The University of Sydney. She is a Fellow of the Australian Academy of Science and has been awarded the Beckwith (2004), Biota (2006), A. J. Birch (2017), H. G. Smith (2018) and Margaret Shiel (2021) medals of the Royal Australian Chemical Institute. Her research interests are in the areas of supramolecular, peptide and organic chemistry, with a focus on the design and synthesis of functional molecules, such as molecular sensors capable of detecting anions in biological environments.

What aspect of your research are you most excited about at the moment?
We're currently working on developing a suite of fluorescent probes to sense different cell membrane phospholipids. If we can do this, it will help unravel a lot of biological questions, which is very exciting!

What is the best part of your job?
Working with other people. My favourite thing is to watch PhD students and early career researchers develop into outstanding scientists who are way smarter than I am!

What is your advice to young emerging scientists?
Work on developing your time management skills so you work smarter, not longer, hours.

What do you do in your spare time?
Read trashy detective novels (and sometimes actual literature), preferably while lying on a beach. I also love to cook (and eat!), attempt to grow vegetables in my small Sydney courtyard and try to squeeze in some exercise to clear my mind – mostly involving Pilates, Yoga, swimming or some high intensity skipping.

By the time I'm 100 I would like to have...
...been retired for quite a while so I can enjoy all the other things in life that aren't work related.

Vinayak Juvekar is Research Professor at the Department of Energy Systems Research, Ajou University. He received his BS (2009) and MS (2011) degrees from Karnataka University, India. After finishing his PhD (2017) degree under the guidance of Prof. Young Dae Gong from Dongguk University, South Korea, he joined Prof. Hwan Myung Kim's lab as postdoctoral fellow at Ajou University in 2017. His research interest is development of small-molecule fluorescent probes for biomedical applications.

Eda Kilic received her BSc degree from Istanbul Technical University, Department of Chemistry (2018). She is currently doing her MSc studies at Koç University, Department of Chemistry under the supervision of Dr Kölemen. She is mainly working on activatable fluorophores and photosensitizers for imaging and therapy.

Hwan Myung Kim is a Professor at the Department of Chemistry and Department of Energy Systems Research, Ajou University. He received his BS degree from Chungnam National University in 2001 and his MS (2003) and PhD (2008) degrees from Korea University under the guidance of Prof. Bong Rae Cho. He joined the faculty at Ajou University in September 2008. Since 2012, he has been a principal investigator of National Leading Research Laboratory. He has received the Young Organic Chemist Award (2013) from the Korean Chemical Society. His research interest is the development of organic nonlinear optical materials and two-photon probes for biological imaging and therapy. He resides in Suwon city in Korea with wife and two children (Si Woo and Si Won).

What aspect of your research are you most excited about at the moment?
Diversity of molecular probes that can be used for basic research and biomedical applications.

What is the best part of your job?
Given freedom to design exciting projects.

What do you do in your spare time?
Exercising, walking, listening to music and enjoying wine.

By the time I'm 100 I would like to have...
A house with a beautiful garden. Time to relax and have a cup of tea with my wife.

Jong Seung Kim received his PhD from the Department of Chemistry and Biochemistry at Texas Tech University in 1993. Currently he is a full professor in the Department of Chemistry at Korea University in Seoul. His main research interests are application of organic chemistry to drug delivery and diagnosis of various pathologies, including Alzheimer's disease and malignant neoplasm and their super-resolution imaging. He has been selected as 1% HCR for last 8 years. His *h*-index records 105 (Google Scholar).

What aspect of your research are you most excited about at the moment?

Presently, my focus is on developing new discoveries, which could be helpful to diagnose and therapy of many diseases that the world is mostly suffering in recent years. A number of diagnosing techniques and drugs were available to diagnose/treat certain diseases; however, they are still facing many impediments in their useful applications. We mainly focus on resolving those impediments. For a number of years in organic chemistry, we focused on organic systems that could be useful in successfully delivering many commercially available fluorophores (as diagnosing agents) and sensitizers/drugs (therapeutics) to the diseased tissues to eliminate them permanently. In addition, for the same purpose, we also focus on the useful developments based on nanosystems, including different nanoparticles, porous organic polymers (POPs), and covalent organic frameworks (COFs).

What is the best part of your job?

The best part of my job is teaching students and training them to learn many new experimental techniques. Also, build knowledge and create interest in them to passionate about research, which could be helpful to today's society. In addition, making new useful discoveries and working collaboratively with many outstanding professional scientists.

What is your advice to young emerging scientists?

In my vision, young emerging scientists are sufficiently strong enough to reach their goals in the field of their research. If I had to give some advice, I would like to suggest they update themselves, find the important problems, and make the research goals what the world really wanted. Make many fruitful collaborations and get scientific advice from colleagues to successfully reach their big challenges.

What do you do in your spare time?

I always try to be up to date with learning new emerging technologies. From time to time, I used to play tennis and jogging around my apartment.

By the time I'm 100 I would like to have...

I want to fulfill all my dreams in the field of my research area and have a peaceful life at an older age.

Jungryun Kim received her BS degree from the Department of Chemistry at the Sookmyung Women's University of Korea in 2020. She is currently pursuing her PhD degree under the guidance of Prof. Jong Seung Kim at Korea University. Her current research interests are in development of a drug-delivery system with prodrug for drug resistance cancers, including breast cancers.

Dr Safacan Kolemen graduated from the Department of Chemistry at Bilkent University, Turkey in 2008, and then completed his MSc studies in 2010 at the same department. He received his PhD from the Institute of Material Science and Nanotechnology (UNAM) at Bilkent University in 2014 under the supervision of Prof. Engin Akkaya, where he worked on designing photosensitizers for photodynamic therapy of cancer. Then, he joined to Prof. Christopher Chang's research group in the Department of Chemistry at University of California, Berkeley, USA, as a postdoctoral researcher, where he mainly worked on the design of fluorescent probes for live cell imaging.

Since September 2017, he has been working as a faculty member in the Department of Chemistry at Koç University, Istanbul, Turkey. His research lies in interdisciplinary approaches within the field of synthetic organic chemistry and chemical biology. He particularly focuses on development of small-molecule-based fluorescent/chemiluminescent probes for bioimaging, design of activatable photosensitizers for phototherapy applications as well as preparation of drug-delivery and theranostic (therapy + diagnosis) systems. He received "Outstanding Teaching Award" given at Koç University, Faculty of Sciences in 2018 and recently the prestigious "Outstanding Young Scientist-BAGEP 2021" award, which is given by the Science Academy of Turkey. Dr Kölemen is among the editorial advisory board members of ACS Bio&Med Chem Au journal.

What aspect of your research are you most excited about at the moment?
Observing that our molecules are doing their job in tumors.

What is the best part of your job?
Every work day is different from the other.

What is your advice to young emerging scientists?
Never give up and be patient.

What do you do in your spare time?
Spending time with my dog (Rita).

By the time I'm 100 I would like to have...
a magnificent party to celebrate this big achievement.

Oxana Kotova is a researcher in the group of Professor T. Gunnlaugsson at the School of Chemistry, Trinity College Dublin, Ireland. Her research focuses on lanthanide based supramolecular systems including those containing chiral centres for imaging and sensing applications. Oxana graduated with a Masters degree in chemistry and materials science as well as a PhD in inorganic chemistry from Lomonosov Moscow State University working on luminescent lanthanide complexes as electroluminescent materials for organic light emitting diodes.

What aspect of your research are you most excited about at the moment?
At the moment I am most excited about characterisation of chiral Eu(III) complexes with helical structures.

What is the best part of your job?
The best part of my job is the possibility to create and learn.

What is your advice to young emerging scientists?
My advice would be to keep exploring as the one who searches will always find.

What do you do in your spare time?
In my spare time I enjoy time with my family, exploring new places and reading.

By the time I'm 100 I would like to have...
A clear mind and keep an interest in the natural world as it is full of research projects.

Toshiyuki Kowada received his PhD degree from Kyoto University in 2010 under the supervision of Prof. Kouichi Ohe. After one year of postdoctoral work with Prof. Kazuya Kikuchi at Immunology Frontier Research Center (IFReC) in Osaka, he became a specially appointed assistant professor. After postdoctoral study with Prof. Jianghong Rao at Stanford University (2014–2016), he was appointed as an assistant professor at Tohoku University. His research interests include the development of small-molecule-based functional probes for biological applications.

What aspect of your research are you most excited about at the moment?
I am so excited when I look into a fluorescence microscope and see the fluorescent probe that I designed and synthesized shining brightly by detecting target substances in a cell.

What is the best part of your job?
The real thrill is that I can create the molecules I envision in my mind as a reality and polish them to be useful to the world.

What is your advice to young emerging scientists?
I'm still a young researcher, but I think it's important to talk with other researchers around you, sometimes from different fields, in order to carve out your own path, instead of being immersed in your world.

What do you do in your spare time?
Recently, I've been playing exclusively with my little kids. I'm just amazed at their growth ability every day.

By the time I'm 100 I would like to have...
I would like to travel with my wife to all the prefectures in Japan, visit the famous places, including the World Heritage Sites, and enjoy the local specialties.

Meng Li now is an associate professor at the North China Electric Power University. She obtained her BSc in 2011 (East China University of Science and Technology), and was a master combined with PhD student from 2011 to 2012 (with Prof. Weihong Zhu in China) and then worked with Prof. Tony D James for her PhD degree at University of Bath since October 2012 till July 2015. Her research interests comprise many aspects of supramolecular chemistry, environmental chemistry and materials chemistry, including molecular recognition and materials fabrication.

Ping Li received her PhD degree in 2008 from Shandong Normal University. In 1998, she joined the faculty at Shandong Normal University, where she is currently a professor of College of Chemistry, Chemical Engineering and Materials Science. She is Taishan Distinguished Professor (2017), Millions of Talent Projects National candidates (2019) and the leader of the Changjiang Scholars and Innovative Research Team in University. Her research interests include the synthesis and bioimaging applications of fluorescent probes for biologically active molecules.

What aspect of your research are you most excited about at the moment?
I have been most excited about how to accurate detect biological molecules that regulate diseases.

What is the best part of your job?
When we figure out a quite exciting and original research subjects which other researchers haven't thought of, I feel really great.

What is your advice to young emerging scientists?
Be diligent and think deeply, that's what I suggest young emerging scientists to do.

What do you do in your spare time?
I love reading, watching American TV drama and traveling.

By the time I'm 100 I would like to have...
I would like to still be alive with sober-mind, trying to write memoirs. At the same time, I would hope not to disturb young people.

Xiaoning Li is a PhD at the North China Electric Power University. She obtained her bachelor's degree from Hebei University of Engineering in 2016 and majored in Water Supply and Drainage. From 2016 to now, her main research interests are in the development of synthesis and application of cellulose-based aggregation-induced emission luminogens.

Jongdoo Lim was born in Chonan, South Korea. He studied chemistry at Korea University where he obtained his bachelor's and master's degrees. He then earned his PhD in Chemistry at Texas A&M University under the supervision of Professor Eric Simanek, working on developing drug-delivery systems based on triazine dendrimers. His current projects in the Anslyn group at the University of Texas at Austin are to synthesize sequence-defined unnatural polymers *via* ribosome-mediated polymerization and develop rapid optical methods to determine enantiomeric excess (ee), diastereomeric excess (de), and reaction yield, using a variety of techniques such as CD (circular dichroism), UV, and dynamic molecular assembly.

What aspect of your research are you most excited about at the moment?
Work collaboratively in a team, exchanging ideas and learning new technologies.

What is the best part of your job?
There is always an opportunity to learn something new.

What is your advice to young emerging scientists?
Learn broadly spending some time in other areas of research.

What do you do in your spare time?
I read books and play board games.

By the time I'm 100 I would like to have...
...traveled around the world.

Heng Liu received his PhD degree from University of Chinese Academy of Sciences in 2014. Then, he started to work at Hubei University. During 2017–2018, he joined Prof. Ming Xian's group as a visiting scholar at Washington State University. In 2019, he moved to Hainan Medical University as a full professor. His research interests included development of fluorescent probes and advanced molecular imaging agents.

Shin Mizukami received his PhD in 2002 from the University of Tokyo, where he worked in the laboratory of Professor Tetsuo Nagano. After his postdoctoral studies at the National Institute of Advanced Science and Technology (AIST) and Stanford University (Prof. Eric T. Kool lab), he was appointed as an Assistant Professor in the Graduate School of Engineering at Osaka University (Prof. Kazuya Kikuchi lab) in 2005. He was promoted to Associate Professor at Osaka University in 2009 and to Professor at Tohoku University in 2016. His research interests are in developing molecule-based technologies to solve unsolved problems in biology and medicine.

What aspect of your research are you most excited about at the moment?
We have developed molecules that function in cells so far by intuition, but the molecular design has become more rational with the recent advances in computational science and informatics.

What is the best part of your job?
Since it is a boundary area of study between chemistry and biology, our research can receive professional evaluations and criticisms from both sides.

What is your advice to young emerging scientists?
Both broadening the base and raising the top of research are necessary because they are complementary and yield synergistic effects.

What do you do in your spare time?
Studying information technology skills, relaxing (watching YouTube and playing online games), and sometimes thinking about the future of our country.

By the time I'm 100 I would like to have...
I would like to have an observable contribution to our country and human society beyond basic study.

Sarah Moor is a 5th year PhD student in the Anslyn research group at the University of Texas at Austin where she also obtained her BS in Chemistry as well as a Certificate in Computer Science in 2017. Her research has primarily focused on the development of high-throughput experimentation techniques and oligourethane-based sequencing platform for information technologies.

What aspect of your research are you most excited about at the moment?
I am most excited to continue working at the interface of chemistry and computer science.

What is the best part of your job?
Getting to work with people who are passionate about their research.

What is your advice to young emerging scientists?
Graduate school is a great place to acquire additional skills (computer science, statistics, business skills, *etc.*) to help you in your future career so take advantage of what universities have to offer.

What do you do in your spare time?
I like to play video games and hang out with my dog.

By the time I'm 100 I would like to have...
...seen every species of owl in North America.

Subrata Munan completed his Bachelors of Science in Chemistry (Honours) from Midnapore College, West Bengal, India. He received a Master's degree in Chemistry from Vidyasagar University, West Bengal, in 2018. Currently, he is a doctoral student under the supervision of Dr Animesh Samanta in the School of Natural Sciences, Department of Chemistry, Shiv Nadar University, Delhi NCR. His research interests include designing and synthesizing new functional fluorescent probes and their biological activity towards reactive species inside the cellular microenvironment. During his bachelors's and master's degrees, he received a prestigious INSPIRE Scholarship (Innovation in Science Pursuit for Inspired Research) from the Department of Science and Technology, Govt. of India. He loves to read scientific books and discusses with others for learning and teaching.

Elizabeth (Liz) J. New undertook her undergraduate and masters studies at the University of Sydney, working with Prof. Trevor Hambley on fluorescently tagged platinum complexes. She then completed her PhD in 2010 at the University of Durham (UK) with Prof. David Parker studying luminescent lanthanoid complexes for use as cellular probes. Liz was a Royal Commission for the Exhibition of 1851 Postdoctoral Fellowship at the University of California, Berkeley, where she worked with Prof. Chris Chang on fluorescent copper sensors. In 2012, she returned to the University of Sydney to establish her independent research career, holding an ARC DECRA fellowship (2012–2014) and Westpac Research Fellowship (2016–2019). Her research focusses on the development of small-molecule fluorescent sensors for biological and environmental applications. She is passionate about university teaching, and has also served on a number of national committees that aim to improve opportunities for future generations of scientists. Her awards for research and teaching including the 2019 Malcolm McIntosh Prize for Physical Scientist of the Year, 2018 Eureka 3M Emerging Leader Prize, and the 2016 RACI Chemistry Educator of the Year.

What aspect of your research are you most excited about at the moment?
We're starting to translate our work into clinical and environmental industrial applications, and it's great to be working towards impacts outside research.

What is the best part of your job?
The people I work with – my talented and enthusiastic group members, and all of my collaborators who let us explore so many exciting new avenues.

What is your advice to young emerging scientists?
Spend time building networks: with your peers who will become your colleagues, and with more senior researchers who will become your mentors.

What do you do in your spare time?
At the moment, my spare time is consumed by two energetic young children! When I get a moment, I love reading biographies and playing the viola.

By the time I'm 100 I would like to have...
Played a part in revolutionising the academic sector to ensure that researchers can flourish and innovate.

Yuvin Noh received her BS degree from the Department of Chemistry at the Sookmyung Women's University of Korea in 2020. She then received her MS degree under the guidance of Prof. Jong Seung Kim at Korea University. Her current research interest is harnessing the activity-based fluorescent chemosensors to illuminate a novel biomarker toward pathologies, including senescence, and Alzheimer's disease.

Animesh Samanta received his MSc in Chemistry at Indian Institute of Technology (IIT) Madras in 2007. He obtained his PhD in the interface of chemistry and biology under the supervision of Prof. Young-Tae Chang in 2012 at National University of Singapore (NUS). He continued his research career as a postdoctoral fellow at Singapore Bioimaging Consortium (SBIC) in Prof. Chang's lab until 2015. Later he moved to National University of Singapore, Department of Biomedical Engineering, to develop his research skills in a microfluidic device for diagnosis. He came back to India with the prestigious Ramanujan Fellowship in 2016 and started an independent research career at CSIR – National Institute for Interdisciplinary Science and Technology (NIIST). He is currently an Assistant Professor in the Department of Chemistry, Shiv Nadar University, Delhi NCR, since 2019. His research interests are creating molecular entities of fluorophores, discovering new optical phenomena, and exploring non-invasive optical imaging to study unknown phenomena in life at the molecular level leading to a better understanding of cancer and neurodegenerative diseases.

What aspect of your research are you most excited about at the moment?
The research engages you in creating new knowledge and makes you think differently from the traditional concepts. The challenging research may give you unexpected results, opening new research areas that always excite me. These new findings of science offer you new windows and stretch your mind to search for new ideas and new possibilities.

What is the best part of your job?
Solving global challenges together as a family member of the scientific community and educating the students is the best part of my job.

What is your advice to young emerging scientists?
As science is an art of invention from the imagination, young emerging scientists should solve the emerging problems from the creative mind.

What do you do in your spare time?
I love to read inspirational quotes from books, the internet, and different apps and find the meaning of a happy life.

By the time I'm 100 I would like to have...
In my 100s, I would like to have a few books that can guide living happily.

Adam C. Sedgwick is a Glasstone Research Fellow at the University of Oxford and Junior Research Fellow at Jesus College. He received his MChem in Drug Discovery (2014) and PhD in Organic Chemistry (2014) from the University of Bath. In 2018, he moved to Austin, Texas where he worked under the supervision of Prof. Jonathan L. Sessler (2018–2021). His research interests include chemical sensing, theranostics and macrocyclic chemistry. To date, he has co-authored over 70 peer-reviewed papers with over 3500 citations (*h*-index 26) and is co-inventor on 2 US patents.

What aspect of your research are you most excited about at the moment?
Developing tumour-localising near-infrared platforms for chemical sensing and theranostic applications.

What is the best part of your job?
Interdisciplinary collaboration and building research networks.

What is your advice to young emerging scientists?
To enjoy research and follow your passions.

What do you do in your spare time?
Have beers with my friends.

By the time I'm 100 I would like to have...
My own lab to do research.

Jinwoo Shin received his BS degree (Summa cum laude) from the Department of Chemistry at the Hallym University of Korea in 2015. He then received a PhD in organic chemistry from the Department of Chemistry at Korea University in 2021 (Supervisor: Professor Jong Seung Kim). He stayed in the group as a postdoctoral fellow before moving to Stanford University in 2023 (Supervisor: Professor Eric T. Kool). His research interests include multi-functional theranostic agents, advanced fluorescent imaging agents, and DNA repair systems.

Bo Tang is a professor of chemistry at Shandong Normal University. He received his PhD degree in 1994 from Nankai University. He began his independent career as a professor of chemistry at Shandong Normal University in 1994. He won National Fund for Outstanding Young Scientists in 2007, Chief scientist for 973 Program in 2012. He has contributed more than 300 journal articles, as well as 21 invited book chapters and reviews, and obtained 39 granted patents. His research interests include the synthesis of molecular and nano probes and their application in biological imaging, green chemical production, synthesis of fluorescent materials, and solar chemical conversion and storage.

What aspect of your research are you most excited about at the moment?
Visualizing the dynamic modification of proteins.

What is the best part of your job?
Discussion about research subjects with my students is the best part of my job.

What is your advice to young emerging scientists?
Try to find an interesting and exciting subject and then to pursue it during your whole life.

What do you do in your spare time?
Play ping pong.

By the time I'm 100 I would like to have...
Continued research and published papers.

Rui Wang received his PhD degree under the supervision of Professor Jaebum Choo from Department of Bionanotechnology, Hanyang University, Korea in February 2018. Then he joined Hainan Medical University in March 2018. Currently he is an associate professor at the Institute of Functional Materials and Molecular Imaging, Hainan Medical University. His research interests focus on surface-enhanced Raman scattering (SERS) imaging and functional bionanomaterials.

Zhenkai Wang is a PhD candidate at the Institute of Functional Materials and Molecular Imaging, Hainan Medical University. His main research direction is the design and application of small-molecule fluorescent probes for clinical medical detection.

Luling Wu obtained a PhD in 2021 supported by scholarships from the China Scholarship Council and the University of Bath under the supervision of Professor Tony D. James. He is a referee for peer-reviewed journals including *JACS Au*, *Chemical Science* and *Chemical Reviews*. His research focuses on fluorescent probes/prodrugs and imaging. His *h*-index is 17.

What aspect of your research are you most excited about at the moment?
Making fluorescent probes commercially available and help to diagnose diseases.

What is the best part of your job?
I have the opportunities to collaborate with people who are experienced in different research fields.

What is your advice to young emerging scientists?
Paying attention to experimental details.

What do you do in your spare time?
Travelling and meeting friends.

By the time I'm 100 I would like to have...
More time to spend with my family.

Yanlong Xing received her PhD degree from the Technical University of Berlin (TU Berlin), Germany in 2017. After her postdoctoral research in the Leibniz Institute of Analytical Science (ISAS Berlin), she joined the Hainan Medical University as a professor in August 2018. Her current research interests focus on extracellular vesicle-based liquid biopsy, microfluidic technology and the development of functional nanomaterials for biosensors.

Fabiao Yu is now as a full professor in Key Laboratory of Hainan Functional Materials and Molecular Imaging, Hainan Medical University. He was awarded Hainan Hundred Talents and Hainan Leading Talents. He received his MS degree from Shandong Normal University in 2009. He then further received his PhD degree from Dalian University of Technology and Dalian Institute of Chemical Physics, Chinese Academy of Sciences in 2013. After four years' work at Yantai Institute of Coastal Zone Research, Chinese Academy of Sciences, he joined Hainan Medical University from 2018. His research interests focus on functional biomaterials, optical molecular imaging, and disease point-of-care testing.

What aspect of your research are you most excited about at the moment?
The most interested research field for me is optical molecular imaging, and I am now committed to transforming visual detection technology to clinical applications.

What is the best part of your job?
Design, synthesis and bio-application of organic small-molecular fluorescent probes.

What is your advice to young emerging scientists?
Faced with busy experiments, difficult writing and publishing papers, and fierce job competition, young scientists may be afraid of future career development. To become a successful scientist, the most important thing is to cultivate curiosity about science. The better way is to learn from the experience of others, rather than let yourself make all the mistakes.

What do you do in your spare time?
Read classical literature, appreciate Chinese opera and grow flowers.

By the time I'm 100 I would like to have...
First of all, I have to live to be 100 years old, and then I hope I can walk around freely, read literature and sunbathe.

Feifei Yu is an organic chemistry engineer in college of pharmacy, Hainan Medical University. She received her BS degree from Shandong Normal University in 2006 and worked as a chemistry teacher until 2018. Then she joined Hainan Medical University. Her research interests focus on organic synthesis of functional biomaterials and biological analysis.

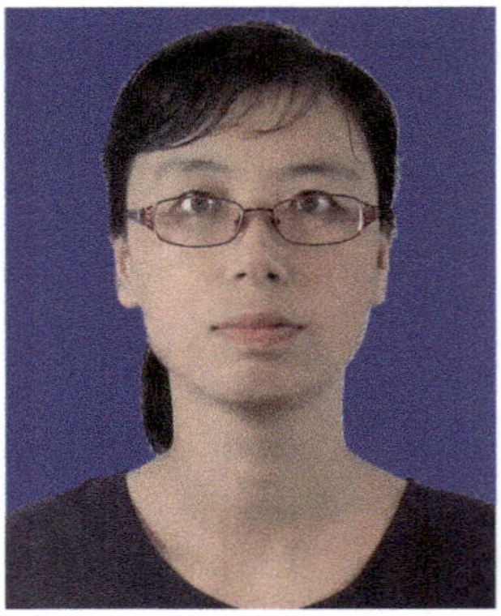

Linlu Zhao received her PhD under the supervision of Prof. Junqiu Liu in the State Key Laboratory of Supramolecular Structure and Materials, College of Chemistry, Jilin University in 2018. Currently she is an associate professor at the Institute of Functional Materials and Molecular Imaging, Hainan Medical University. Her main research interests include the study of detection and sensing of small-molecule fluorescent probes, and development of self-assembled nanomaterials.

Contents

Monographs in Supramolecular Chemistry No. 34
Fluorescent Chemosensors
Edited by Luling Wu, Adam C. Sedgwick, Xiao-Peng He and Tony D. James

Published by the Royal Society of Chemistry, www.rsc.org

CHAPTER 1

Welcome to Fluorescent Chemosensors

LULING WU[a], ADAM C. SEDGWICK[b], XIAO-PENG HE[c] AND TONY D. JAMES*[a,d]

[a]Department of Chemistry, University of Bath, Bath, BA2 7AY, UK; [b]Chemistry Research Laboratory, University of Oxford, Mansfield Road, Oxford, OX1 3TA, UK; [c]Key Laboratory for Advanced Materials and Joint International Research Laboratory of Precision Chemistry and Molecular Engineering, Feringa Nobel Prize Scientist Joint Research Center, Frontiers Center for Materiobiology and Dynamic Chemistry, School of Chemistry and Molecular Engineering, East China University of Science and Technology, 130 Meilong Road, Shanghai 200237, P. R. China; [d]School of Chemistry and Chemical Engineering, Henan Normal University, Xinxiang 453007, China
*E-mail: wllcyl@126.com, t.d.james@bath.ac.uk, adam.sedgwick@chem.ox.ac.uk, xphe@ecust.edu.cn

> A fluorescent chemosensor is a compound of abiotic origin that complexes to an analyte reversibly with concomitant fluorescence signal transduction.
>
> A. W. Czarnik (1993)

Monographs in Supramolecular Chemistry No. 34
Fluorescent Chemosensors
Edited by Luling Wu, Adam C. Sedgwick, Xiao-Peng He and Tony D. James

Published by the Royal Society of Chemistry, www.rsc.org

1.1 Introduction

This book, entitled *Fluorescent Chemosensors*, is part of the Monographs in Supramolecular Chemistry series edited by Philip Gale and Jonathan Steed and follows in the footsteps of the seminal book on *Fluorescent Chemosensors for Ion and Molecule Recognition* published as part of the ACS Symposium Series and edited by Anthony W. Czarnik in 1993. That book provided access to a collection of groundbreaking chapters on fluorescent chemosensors by pioneers of the field, including Bernard Valeur, A. Prasanna de Silva, Roger Y. Tsien and Joseph R. Lakowicz.

With this book on fluorescent chemosensors, the editorial team of Wu, Sedgwick, He and James have assembled a diverse collection of chapters from world-leading experts involved in fluorescent chemosensors, including Anthony Czarnik; Christopher Chang; Meng Li; Young Tae Chang; Hwan Myung Kim; Jong Seung Kim; Eric Anslyn; Silvia Giordani; Kate Jolliffe; Bo Tang; Shin Mizukami; Fabiao Yu; Tasuku Hirayama; Safacan Kölemen, Elizabeth New and Thorfinnur Gunnlaugsson.

Together the chapters take us on a journey from receptor-based to reaction-based sensing systems using traditional synthetic design and diversity-oriented synthesis for detecting chirality, phosphates, reactive oxygen, zinc ions (Zn^{2+}), sulfane sulfur/selenium species and other transition metals. The systems described are both molecular-based and nanoparticle-based and have uses in disparate fields, including medical diagnostics, theranostics and environmental sensing. Specific techniques discussed include two-photon fluorescence, ratiometric and aggregation-induced emission (AIE) (Figure 1.1).

Whether you are visiting for the first time or as a regular visitor, we are delighted to welcome you to the fluorescent chemosensor realm, we hope you enjoy your stay.

1.2 Summary of Contents

Chapter 2 (Anthony W. Czarnik)

Tony Czarnik presents his original introductory chapter for the ACS Symposium Series book on *Fluorescent Chemosensors for Ion and Molecule Recognition*. While the chapter was written to coincide with the original chemosensors collection, the content remains pertinent and as fresh today as it did almost 30 years ago.

Chapter 3 (Christopher J. Chang)

Chris Chang provides an excellent overview and introduction to activity-based sensing (ABS) for biological applications. As an emerging area of sensor research, ABS leverages the unique chemical reactivity of a given analyte of interest, rather than traditional binding-based approaches, which rely on

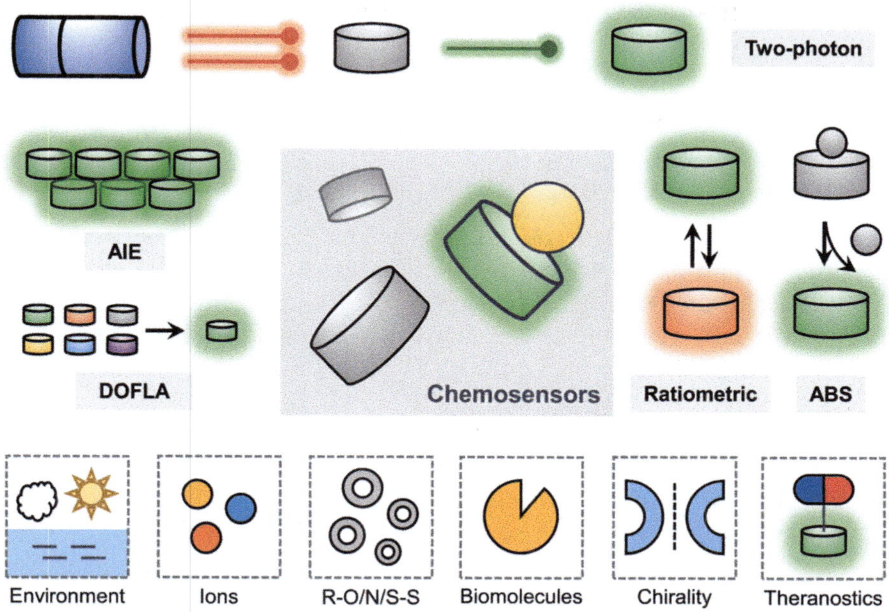

Figure 1.1 Schematic describing the chemosensor systems covered in this book. AIE, aggregation-induced emission; ABS, activity-based sensing; DOFLA, diversity-oriented fluorescent library approach, R-O/N/S-S, reactive oxygen, nitrogen, and sulfur species.

lock-and-key molecular recognition, to achieve selectivity in complex biological environments. The chapter covers design principles for ABS and provides examples of activity-based imaging probes, highlighting the applicability of the ABS approach to identify and characterize new biology.

Chapter 4 (Meng Li)

The chapter by Meng Li provides the latest research achievements on AIE-based fluorescent systems. AIE is a relatively recently discovered approach towards the development of fluorescent chemosensors, where molecules in an aggregated state exhibit enhanced fluorescence emission. The chapter covers the developments of AIE over the past decade, including organic luminescent molecules and clusteroluminogens.

Chapter 5 (Animesh Samanta and Young-Tae Chang)

The chapter by Animesh Samanta and Young-Tae Chang summarizes the recent developments of chemical probes/sensors and chemosensors using a diversity-oriented fluorescent library approach (DOFLA) and their biological and environmental applications. Consisting of six sections, including the concept of diversity-oriented fluorescent library, mechanism of signal transduction, screening strategies and reviews of recent examples from *in vitro* spectra-based screening, cell-based screening and *in vivo* whole organelle-based screening.

Chapter 6 (Hwan Myung Kim)

The chapter by Hwan Myung Kim describes the development of two-photon (TP) probes with improved sensitivities for measuring target analytes in biological specimens. Coverage includes photo-physiological properties, basic TP probe design strategies and representative TP sensors for biological systems.

Chapter 7 (Paramesh Jangili and Jong Seung Kim)

The chapter by Paramesh Jangili and Jong Seung Kim discusses how ratiometric chemosensors can be developed to detect biological/environmentally important analytes. Highlighting the basic principles and design strategies of ratiometric fluorescent chemosensors, including photophysical/chemical sensing mechanisms based on different molecular types (*i.e.*, small molecule and nanoparticle), with appropriate examples.

Chapter 8 (Eric V. Anslyn)

The chapter by Eric Anslyn covers both historical and modern examples of chirality sensing *via* UV-visible, fluorescence and circular dichroism spectroscopy. Each technique possesses unique advantages, which are leveraged to tailor sensing schemes to a particular application. These optical signaling methods are well suited for high-throughput screening platforms as they can be performed in parallel with multiwell plates. This allows for rapid detection at low cost, which is advantageous compared to intrinsically serial conventional chromatographic methods.

Chapter 9 (Silvia Giordani)

The chapter by Silvia Giordani provides an overview of fluorescent carbon nanoparticles, supported by selected literature examples, namely on carbon nano-onions, fluorescent carbon dots and detonation nanodiamonds. Significantly, the fluorescent and sensing properties of these types of nanoparticles can be easily modulated, affording chemical tools with wide applicability in imaging and sensing.

Chapter 10 (Katrina Jolliffe)

The chapter by Kate Jolliffe outlines the development of small-molecule fluorescent chemosensors for phosphate species. Phosphate species are ubiquitous in nature, play diverse roles in biological systems and display extensive variation of types of functionalized phosphate species (*e.g.*, ATP, DNA). Key examples of interaction types, varied approaches to chemosensor design and fluorescence response mechanisms are highlighted.

Chapter 11 (Ping Li and Bo Tang)

The chapter by Ping Li and Bo Tang covers sensors for reactive oxygen species (ROS) that are widely involved in signal transduction of physiological processes. Non-invasive fluorescence imaging techniques have gradually matured enabling the tracking of active molecules such as ROS. This chapter summarizes the development of fluorescent sensors for ROS in living cells and *in vivo*.

Chapter 12 (Shin Mizukami)

The chapter by Shin Mizukami introduces the fluorescence-based imaging of zinc, which is important since disruption of zinc homeostasis is associated with various diseases, such as Alzheimer's disease and diabetes. Therefore, investigating its functions and physiological roles is important. The chapter highlights organelle-targeting small-molecule probes and protein tag–small-molecule hybrid probes, which are expected to have a higher potential for accurately visualizing and quantifying labile Zn^{2+} ions at the organelle level.

Chapter 13 (Linlu Zhao and Fabiao Yu)

The chapter by Linlu Zhao and Fabiao Yu covers the development of selective sensors for sulfane sulfur and reactive selenium species. These species are important antioxidant biomarkers for aging and disease development. Since the intracellular levels and distribution of sulfane sulfur and reactive selenium species can directly evidence the dynamic state of oxidative stress, they may reveal the difference between physiological and pathological processes. Fluorescence bioimaging has the advantages of high temporal and spatial resolution, low invasiveness and fast response, and has become a powerful tool for intracellular detection. This chapter outlines the design strategy and development of fluorescent probes for the detection of sulfane sulfur and reactive selenium species.

Chapter 14 (Tasuku Hirayama)

The chapter by Tasuku Hirayama covers examples of activity-based probes for transition metal ion detection. Such metal ions are involved in various biological processes, and their essential roles and pathological involvement have provided a motive for the development of new synthetic chemical tools for the detection and monitoring of metal ions in cells and in living organisms. Activity-based probes provide new possibilities for the detection of biological metals and avoid many disadvantages associated with chelation-based probes (*e.g.*, fluorescence quenching effect of *d* unsaturated metal ions).

Chapter 15 (Safacan Kolemen)

The chapter by Safacan Kolemen introduces photodynamic therapy, which is a clinically approved treatment modality for a wide range of medical conditions, including malignant cancers. The technique employs cytotoxic reactive oxygen species (ROS), particularly singlet oxygen ($^1O_2$), to kill diseased cells. Spatiotemporal control of ROS generation and consequent cancer cell selectivity is one of the most important characteristics of photosensitizers to minimize severe adverse effects, as well as to enhance therapeutic outcomes. As such, activatable photosensitizers that are OFF prior to activation with various tumor-associated intracellular stimuli are essential.

Chapter 16 (Elizabeth J. New)

The chapter by Liz New covers fluorescent sensors developed for environmental monitoring of chemical pollutants and other substances that can cause harm to society and the natural world. Fluorescent-sensing platforms provide high sensitivity using much cheaper and smaller instrumentation than traditional analytical techniques. The chapter highlights fluorescent platforms developed for environmental sensing, with a particular focus on examples that have been successfully applied in the field.

Chapter 17 (Thorfinnur Gunnlaugsson)

The chapter by Thori Gunnlaugsson discusses the construction of lanthanide-based sensors/probes and their use in sensing and imaging. Lanthanides (*e.g.*, Eu(III) and Tb(III)) exhibit many advantageous optical properties, including long excited-state lifetimes, improved signal-to-noise ratios and long emission wavelengths. Design strategies towards lanthanide-based luminescence systems have been discussed and representative examples provided which illustrate the versatility and wide-ranging applications of these systems.

CHAPTER 2

Supramolecular Chemistry, Fluorescence, and Sensing†

ANTHONY W. CZARNIK*[a]

[a]Department of Chemistry, University of Nevada, Reno, Reno, NV 89557, USA
*E-mail: Aczarnik@unr.edu

A ***sensor*** is a device that interacts with matter or energy and reversibly yields a measurable signal in response. Many scientists to date have considered sensors as macroscopic devices, such as the pH electrode (an H^+ sensor) or the bimetallic strip (a temperature sensor). However, with the increasing recognition of nanotechnology there remains little reason not to view suitably engineered individual molecules as devices capable of performing useful work on other molecules. A catalyst is therefore a construction device able to fabricate new compounds. A transport protein is a transportation device able to move species through barriers.

Likewise, phenolphthalein is a sensing device able to signal the concentration of hydrogen ion. As such, phenolphthalein is an abiotic molecule that senses an abiotic analyte. Because living organisms have provided such a rich array of both analytes worth sensing and receptors for many of them, biotic sensors have received considerable attention. As we begin a discussion of fluorescent chemosensors, a definition of terms is in order. ***Chemical sensor***:

---

†This chapter has been reproduced from *Fluorescent Chemosensors for Ion and Molecule Recognition*, ed. A. Czarnik, ACS Symposium Series, 1993, vol. 538 with permission from American Chemical Society [ISBN: 0841227284].

Monographs in Supramolecular Chemistry No. 34
Fluorescent Chemosensors
Edited by Luling Wu, Adam C. Sedgwick, Xiao-Peng He and Tony D. James

Published by the Royal Society of Chemistry, www.rsc.org

a micro- or macroscopic device that interacts reversibly with a chemical analyte with signal transduction. ***Chemosensor***: a molecule of abiotic origin that signals the presence of matter or energy. ***Biological sensor***: a micro- or macroscopic device that interacts reversibly with a biological analyte with signal transduction. ***Biosensor***: a molecule of biotic origin that signals the presence of matter or energy. ***Intrinsic fluorescence probe***: a chemosensor in which the mechanism for signal transduction involves interaction of the analyte with a ligand that is part of the fluorophore π-system. ***Conjugate fluorescence probe***: a chemosensor in which the mechanism for signal transduction involves interaction of the analyte with a ligand electronically insulated from the fluorophore π-system.

Thus, phenolphthalein is a chemosensor. In the presence of a sufficient concentration of OH^-, phenolphthalein is deprotonated with visible light signaling at 552 nm. ***Signal transduction*** is the mechanism by which an interaction of sensor with analyte yields a measurable form of energy, which may reside in wavelength regimes characterized by various spectroscopies (*e.g.*, UV, visible, NMR) or may yield electrochemical responses. In fact, each mode of signal transduction occupies an entire literature whose summary is beyond the intent of this chapter. Rather, the function of this symposium was to survey contemporary research into the synthesis of abiotic molecules that sense analyte concentrations with fluorescent signal transduction: ***fluorescent chemosensors***.

2.1 Statement of the Research Topic

The requisite research issues essential to the creation of fluorescent chemosensors are: (1) *how can one bind a molecular entity with selectivity (preferably from water)*, (2) *what molecular changes result in fluorescence changes, and* (3) *what mechanisms for binding and fluorescence signal transduction intersect.*

2.1.1 Binding a Molecular Entity with Selectivity

While much of organic chemistry focusses on the selective creation of covalent bonds, such irreversible associations of potential analytes yield chemodosimeters (designed for cumulative assay) in contrast to chemosensors (designed for real-time assay). Thus, sensing applications require the existence of receptors (hosts) that associate with analytes (guests) selectively *and reversibly*.

One approach to the fabrication of such hosts is to utilize the many such receptors available from biological sources. Selective binding of species is demonstrated by enzymes for their substrates (and cofactors), by transport proteins for their transportees, and by antibodies for their antigens. In particular, because antibodies can be raised to many antigens, the biotic creation of hosts proves a powerful method for binding analytes with selectivity. Indeed, it is fair to say that biotic receptors have realized far greater practical utility to date than have synthetic, abiotic receptors. Biosensors are sensors derived from such biotic receptors.

Having said this, there is a strong case to be made for the complementary, abiotic approach to receptor formulation. Except for antibodies, the analytes amenable to biotic receptor fabrication are limited to those provided by nature; there are many analytes of interest, including small molecules and metal ions, against which high affinity antibodies cannot be raised. For example, it is not at present possible to raise a high affinity antibody against glucose. There is a sizable class of analytes for which the biotic approach to selective binding is not likely to be successful even in principle. Additionally, biotic receptors often have limited stabilities that make their translation from the benchtop to the field difficult. And sometimes they are not readily amenable to scaled-up production, which makes them expensive.

An alternative approach to the fabrication of selective receptors for small molecules is to design and synthesize one from scratch. This field of chemistry has been in existence for perhaps 30 years, and the subfield has been referred to as *supramolecular chemistry*.[1] Lessons learned from inspection of biotic host–guest interactions provide some guidance for the engineering of synthetic receptors. The non-peptidic frameworks used in the synthesis of abiotic receptors typically engender enhanced stability. However, the efforts required to chemically synthesize molecules with either cleft-like or cavity-like binding sites is often substantial and rate-limiting to their study. Additionally, not all of the synthetic receptors evaluated to date are water soluble or compete effectively with water for guest complexation. However, the abiotic approach to receptor fabrication provides at present the only intellectually-viable approach to the design of chemosensors for many analytes of interest. While synthetic schemes may be long and some steps low yielding, the synthesis of even a few milligrams of a fluorescent chemosensor might well be sufficient for thousands of analyses.

2.2 Molecular Changes that Result in Fluorescence Changes

2.2.1 A Brief Introduction to Fluorescence

In 1565 the Spanish physician and botanist Nicolas Monardes made the first recorded observation of fluorescence when he noted a strange blue glimmer from water contained in a cup made from a specific wood (*Ligirium nephiticiem*). In the 17th century this phenomenon was more extensively described by Robert Boyle and Isaac Newton. It was not until 1845 that the first crude fluorescence emission spectrum was obtained for quinine by John Herschel. Great strides were made toward the understanding of fluorescence in the late 19'th century by George Stokes. He established the technique of observing fluorescence with two different colored filters, one for the excitation beam, and one through which to observe the emission. In 1852 Stokes became the first to determine and report that emission was at a longer wavelength than the excitation. This characteristic of fluorescence is now known as Stokes' Law. Stokes also described the relationship between fluorescence intensity and concentration. He outlined the quenching of fluorescence at high concentration and by the presence of foreign substances and used this information to propose the use of fluorescence for the detection of organic substances. In 1867 F. Goppelsroder developed a method for the determination of non-fluorescent Al(III) by forming a strongly fluorescent morin chelate, which was the first analysis based on fluorescence.

The contemporary explanation for the fluorescence phenomenon may be summarized as follows. The absorption of a quantum of light by a molecule results in the elevation of an electron from the molecule's ground electronic state (S_0) to one of several vibrational levels in the an electronic excited state. In solution, the excited state molecule rapidly relaxes to the lowest vibrational level of the lowest electronic state (S_1). The energy thus stored in the excited state may be released in several ways. The electron may return to the electronic ground state with only the release of heat (radiationless relaxation). After relaxing thermally to the lowest vibrational level of the S_1 state, the electron may return to the S_0 state with light emission (fluorescence). Or, if the molecule is sufficiently long-lived in the S_1 state, it may cross into a lower energy triplet state (T_1). Relaxation from the T_1 state to the S_0 state can also occur with light emission in solids (phosphorescence), by the release of energy (radiationless transition), or by chemical reaction.

2.2.2 Fluorescence as a Signal Transduction Mechanism

There are many reasons for which fluorescence might be identified as the optimal signal transduction mechanism in potential sensing applications. Fluorescence is an enormously sensitive technique, in large part because the observing wavelength is always longer than that of the exciting wavelength. Thus, given appropriate choice of fluorophores, a signal may be read *versus* zero or near-zero background. In the extreme, the fluorescence of even single molecules has been observed. In some applications (*e.g.*, water quality monitoring), background fluorescence is not a particular problem. However, the autofluorescence of most biological samples generates background signal up to about 650 nm. Long wavelength fluorophores (*i.e.*, near-IR) generally avoid these interferences, although sometimes at the cost of severely narrowing the signal transduction mechanisms that function in such compounds. Fluorescence signaling also permits the monitoring of both excitation and emission wavelengths. The emission signal may be observed in the form of intensity, intensity ratio, or lifetime measurements. We possess a rudimentary ability to engineer molecules with predictable responses to the first two signal types, if not yet the third. Fluorescence is usually nondestructive.

Electrochemical methods benefit from Nemstian behavior permitting the signaling of analytes over very large concentration ranges (*e.g.*, pH electrodes), while spectroscopic signalling is in general limited in range to perhaps two concentration decades. This has the effect of requiring that the K_D for a given fluorescent chemosensor be engineered near to the actual concentration of the analyte of interest. However, light can travel through and egress from environments without an absolute requirement for a physical waveguide; electrochemical sensors rely on wires. This proves enormously powerful in, for example, the simultaneous visualization of concentrations in all regions of a living cell. Waveguides *can* be used advantageously, however, in conjunction with fluorescence signalling. While the laser, fiber optic, and detector technologies are quite well established to permit a vast array of remote sensing applications, *what is lacking is the scientific basis upon which to design and synthesize fluorescent chemosensors with appropriate selectivities that translate to immobilization on the tips of fiber optics.*

In listing the criteria required for fluorescent chemosensors that would yield functional, real-time fiber optic sensors, we delineate the broad scientific questions to be addressed.

(a) The binding domain must have sufficient selectivity for the analyte of interest as compared to others present.
(b) The analyte must be capable of dissociating from the binding domain in real-time.
(c) There must exist a signal transduction mechanism between the binding and the fluorescing domains.
(d) The ideal system will incorporate a method for internal calibration.
(e) The excitation wavelength should correspond to that of an inexpensive, portable laser source.
(f) The emission wavelength should be consistent with inexpensive detectors and removed from that of adventitious fluorescent impurities in the media to be monitored.
(g) The conjugate fluoroionophore (*i.e.*, both binding and fluorescing domains) should prove stable to both hydrolysis and oxidation reactions.
(h) The fluorescence signal being measured should not vary with pH in the range encountered in the environment.
(i) The signal being measured should not be substantially modulated by adventitious fluorescence quenchers encountered in the environment.
(j) The above properties of the chemosensor should remain largely intact upon immobilization onto the waveguide material.

Even in doing so, it must be acknowledged that no chemosensors are yet known that meet all these criteria. Fluorescence is the ideal signal transduction mechanism for fiber optic remote sensing applications. In the near future, the applications that await chemosensor synthesis will serve as a major driving force for basic research into the organic chemistry of fluorescent chemosensors.

2.2.3 Chemical Mechanisms for Fluorescence Modulation

While UV/visible signalling almost always results from the ionization of a conjugated substituent, there exists a plethora of mechanisms by which fluorescence signal transduction may be engendered. It is useful to categorize the mechanisms for fluorescence modulation described to date *via* the type of measurement that is made. These are: *intensity, intensity-ratio,* and *lifetime.* A pictoral summary of each is found in Figure 1 of the chapter by Szmacinski and Lakowicz.

2.2.4 Intensity Measurement

By far the most common mode of fluorescence modulation described to date is the increase or decrease of fluorescence intensity at a single emission wavelength upon analyte binding. Methods for halide ion and molecular oxygen analyses rely on non-chelative (collisional) quenching by these species. There is a substantial literature describing chelative metal ion detection in water using intrinsic fluorescent chemosensors, in which donor atoms of the ligand are a part of the fluorophore's structure. The complexation of a metal ion results in either enhanced fluorescence of the chemosensor ('chelation-enhanced fluorescence', or **CHEF**), or in decreased fluorescence ('chelation-enhanced quenching', or **CHEQ**). In such intrinsic fluorescent chemosensors, **CHEF** typically results upon chelation of one of the ions that is not inherently quenching (non-redox active, closed-shell, *e.g.*, Zn(II), Cd(II), Al(III)) and CHEQ is observed on chelation of an inherently quenching ion (*e.g.*, Cu(II), Hg(II), Ni(II)). The chapter by Bell describes the use of intrinsic sensors with intensity signalling. It is this author's conclusion that there is no single mechanistic rationale that can be used to explain **CHEF** in intrinsic fluorescent chemosensors.

In 1977, the application of supramolecular chemistry to fluorescent chemosensor design took root with the publication of an article by Sousa describing naphthalene-crown ether compounds that bind alkali-metal ions with changes in emission intensity (either **CHEF** or **CHEQ**, depending on the chemosensor).[2] Because no donor atom was a part of or conjugated to the fluorophore 1t-system, these are examples of conjugate fluorescent chemosensors. The observed changes were attributed to a heavy atom effect (for Cs^+ and Rb^+) and/or a complexation induced change in triplet energy relative to ground and excited singlet state energies. The chapter by Sousa describes other examples of intensity changes arising *via* these mechanisms.

Interestingly, a chemosensor for calcium that fits the definition of a conjugate fluorescent chemosensor was reported in 1956 by Diehl and Ellingboe.[3] *Calcein can* be viewed as two iminodiacetate ligands conjugated to fluorescein by methylene spacers. Basic solutions of calcein undergo a color change from brown to yellow-green upon addition of calcium ions. The development of calcein is best considered an isolated incident in that the fluorescent properties of the molecule and its metal complexes were not studied in detail (in fact, the authors caution against carrying out titrations of calcium in the presence of strong or fluorescent light).

Intensity modulations also result when photoinduced electron transfer (**PET**) from a donor atom in a conjuage sensor is enhanced or inhibited. For example, triethylamine quenches the fluorescence of anthracene, with greater quenching efficiency at higher concentrations. When the amine is an integral, but nonconjugated, part of the fluorophore structure its effective concentration can greatly exceed an actual concentration achievable in solution. Benzylic amines appear uniquely effective at efficient quenching of anthracene (but not of many longer wavelength fluorophores). Thus, intramolecular **PET** can be very efficient in such systems. Protonation or metal ion complexation of the amine effectively prevents the **PET** mechanism, resulting most often in **CHEF**. The chapters by de Silva, Czarnik, and Masilamani elaborate on sensing systems utilizing this mechanism for signal transduction. (If fluoroionophores undergoing photoinduced intramolecular charge transfer exhibit a shift of their emission and/or excitation spectra upon complexation, intensity-ratio measurements are possible.)

Fluorescence quantum yield (intensity) can be strongly influenced by the polarity of the microenvironment. With creative molecule engineering, the microenvironment in which a fluorophore resides can be linked physically to the presence of absence of a binding analyte. The chapter by Ueno provides examples.

2.2.5 Intensity-ratio Measurements

Despite the desirable flexibility in the design of intensity sensors, there are drawbacks inherent to intensity-based signalling schemes. The most important drawback is that fluorescence intensity can also vary in complex samples for reasons other than analyte concentration, among them: pH changes, photobleaching (decomposition of the fluorophore mediated by the light source), compartmentalization within cells that changes microenvironment, and light scattering. This proves a particular problem in fiber optic remote sensing applications, where constant change in the chemosensor concentration requires frequent (and inconvenient) recalibration.

One approach that ameliorates many of these problems is to take intensity measurements at two excitation or emission wavelengths at which intensity responds differently; for example, an emission band at 500 nm might decrease at the same time a different band at 400 nm is increasing. Once calibrated, the ratio of intensities at these two wavelengths yields analyte concentrations independent of the absolute concentration of the chemosensor (assuming the analyte is present in large excess). Mechanistically, intensity-ratio measurements can be made using chemosensors in which both bound and unbound states are fluorescent, but with different emission maxima because a donor atom becomes deconjugated from the fluorophore π-system upon analyte chelation. Mechanisms leading to ratiometric measurements of calcium and sodium ions have been pioneered by Tsien, and examples are found in the chapters by Tsien and by Kuhn.

Nonradiative relaxation from S_1 states can be affected *via* fluorescence resonance energy transfer. The rate of energy transfer in donor–acceptor pairs is directly related to their distance, which can of course be affected by analyte complexation. Excitation energy transfer leads to quenching of the donor, but also to enhancement of the acceptor fluorescence; therefore, the ratio of fluorescence intensities at donor and acceptor emission wavelengths provides information on complexation *via* changes in donor–acceptor distance. The chapters by Valeur, Bonas-Laurent, and Krafft describe uses of the energy transfer mechanism.

A less well explored mechanism leading to ratiometry is that of ligand-induced excimer formation or dissociation between chemosensor containing two fluorophores. For example, anthracene forms an intermolecular excimer (an excited state dimer) at sufficiently high concentrations; excimer formation has the net effect of yielding ratiometric signalling (see Figure 1 of the chapter by Bouas-Laurent). Chemosensors posessing two fluorophores in their structures can form intramolecular excimers even at very low concentration, whose formation may be influenced by the binding of an analyte. The chapter by Bonas-Laurent describes examples of guest-influenced excimer formation. Also, Tsien's chapter discusses dissociation of two fluorophores induced by cyclic AMP.

One might mention also that the covalent interaction between a guest and the fluorophore substructure might also give rise to a new fluorescing species; if the new species displayed an emission maximum different from that of the unbound chemosensor, ratiometric signalling will result One example of such a mechanism is found in the chapter by Czarnik, relating to chelatoselective Cd(II) indication.

2.2.6 Lifetime Measurement

An alternative and newer approach to overcoming some of the shortcomings of intensity measurements is the use of signalling *via* fluorescent lifetime measurements. The accumulation of lifetime data requires more sophisticated equipment than that for intensity or intensity-ratio data, but there appear to be advantages. The measured lifetime is independent of the chemosensor concentration, which as previously described is of value. Furthermore, changes that affect a fluorophore's emission intensity often also affect its fluorescence lifetime; thus, known intensity sensors may find new utilities as lifetime sensors. The chapter by Szmacinski and Lakowicz describes their studies into the use of lifetime modulations to signal guest binding.

2.2.7 What Mechanisms for Binding and Signal Transduction Intersect?

The collection of successful intersection strategies was the primary goal of our symposium. The invention of solutions requires an unusual degree of creativity from chemists willing to place one foot firmly in each of two research arenas. Such solutions comprise the body of this collection of contributions by researchers in the fluorescent chemosensor field.

A Bibliography of Reviews Relevant to these Topics

Ion and Molecule Recognition:

M. L. Bender and M. Komiyama, *Cyclodextrin Chemistry*, Springer-Verlag, New York, 1977.
C. D. Gutsche, *Calixarenes*, Royal Society of Chemistry, Cambridge, 1989.
G. W. Gokel, *Crown Ethers & Cryptands*, Royal Society of Chemistry, Cambridge, 1991.
F. Diederich, *Cyclophanes*, Royal Society of Chemistry, Cambridge, 1991.
F. Vogtle, *Supramolecular Chemistry*, VCH, Weinhein, FRG, 1991.

Fluorescence:

G. G. Guilbault, *Practical Fluorescence*, Marcel Dekker, Inc, New York, 1973.
G. G. Guilbault, *Practical Fluorescence*, Marcel Dekker, Inc, New York, 1990 (this second edition does not contain all the references of the first).
B. M. Krasovitskii and B. M. Bolotin, *Organic Luminescent Materials*, VCH, Weinheim, FRG, 1988.
J. R. Lakowicz, *Principles of Fluorescence Spectroscopy*, Plenum, New York, 1983.
Biophysical and Biochemical Aspects of Fluorescence Spectroscopy, ed. T. G. Dewey, Plenum, 1991.
M. Zander, *Fluorimetrie*, Springer, 1981.

Reviews of Fluorescent Chemosensor Research:

G. Schwarzenbach and H. Flaschka, *Complexometric Titrations*, (translation by H. Irving), Metheun, London, 1969.
T. S. West, *Complexometry with EDTA and Related Reagents*, Broglia Press Ltd, Bournemouth, England, 1969.
Indicators, ed. E. Bishop, Pergamon, New York, 1972.
A. Fernandez-Gutierrez and A. Munoz de la Pena, in *Molecular Luminescence Spectroscopy: Methods and Applications. Part I*, ed. S. G. Schulman, Wiley, New York, 1985, p. 371.

Fiber Optic Chemical Sensors and Biosensors, ed. O. S. Wolfbeis, CRC Press, Boca Raton, Florida, 1991, vol. I and II.
R. A. Bissell, A. P. de Silva, H. Q. N. Gunaratne, P. L. Lynch, G. E. M. Maguire and K. R. A. S. Sandanayake, *Chem. Soc. Rev.*, 1992, **21**, 187.
R. Y. Tsien, *Methods in Cell Biology*, 1989, **30**, 127.
R. Y. Tsien and A. Waggoner, Fluorophores for Confocal Microscopy: Photophysics and Photochemistry, in *Handbook for Biological Confocal Microscopy*, ed. J. B. Pawley, Plenum, 1990.

References

1. A.k.a. *host–guest chemistry, biomimetic chemistry, or molecular recognition.*
2. L. R. Sousa and J. M. Larson, *J. Am. Chem. Soc.*, 1977, **99**, 307.
3. H. Diehl and J. L. Ellingboe, *Anal. Chem.*, 1956, **28**, 882.

CHAPTER 3

Activity-based Sensing: Principles and Probes for Selective Bioimaging

SHANG JIA[a] AND CHRISTOPHER J. CHANG*[a,b]

[a]Department of Chemistry, University of California, Berkeley, Berkeley, California 94720, USA; [b]Department of Molecular and Cell Biology, University of California, Berkeley, Berkeley, California 94720, USA
*E-mail: chrischang@berkeley.edu

3.1 Introduction

Molecular imaging technologies provide powerful platforms for studying biological processes in their native contexts. These tools often rely on building functional chemical probes that can enable selective detection of chemical entities in live or fixed biological specimens, with high spatial and temporal resolution, to help decipher their physiological and/or pathological contributions. By incorporating biochemically responsive moieties onto imaging contrast agents, a myriad of molecular probes have been developed for modalities spanning fluorescence,[1–3] bioluminescence,[4,5] magnetic resonance imaging (MRI),[6,7] and photoacoustic detection.[8] These reagents have found utility in the detection of a diverse array of important biological analytes, including transient small molecules (*e.g.*, reactive oxygen, nitrogen, and carbon species)[9–12] and metal ions.[13–17] In particular, the explosive increase in

Monographs in Supramolecular Chemistry No. 34
Fluorescent Chemosensors
Edited by Luling Wu, Adam C. Sedgwick, Xiao-Peng He and Tony D. James

Published by the Royal Society of Chemistry, www.rsc.org

available fluorescent sensors has enriched basic research and clinical diagnostics in biology and medicine.[18]

In this context, two general mechanisms are employed for the design of responsive imaging probes for biological analyte detection: binding-based sensing (BBS) and activity-based sensing (ABS). In both cases, a recognition moiety is appended to an imaging agent to elicit a change in signal upon interaction with the analyte of interest. BBS achieves specificity by relying on molecular recognition and lock-and-key binding of the analyte. For example, binding-based designs have enabled success in detection of endogenous labile metal ion pools,[13–17] including calcium,[19] zinc,[20,21] copper,[22–25] and nickel.[26] However, weakly binding metal ions and small, transient reactive molecules present challenges to this design strategy. To address these limitations, ABS provides a complementary approach that exploits the intrinsic chemical reactivity of an organic or inorganic analyte to achieve chemoselectivity. The ABS concept has origins in promiscuous chemodosimeters for use in organic solvents and related abiotic environments; for example, thiophilic heavy metals, such as mercury and silver, can be leveraged to produce a signal change in a fluorescent dye in response to hydrolysis.[27] Building on this early work, advances in selectivity and biological compatibility have given rise of a robust field of ABS for imaging of biological analytes,[3,28–31] including detection of transition metals,[16,17,32] reactive nitrogen species (RNS),[33,34] reactive oxygen species (ROS),[9] reactive sulfur species (RSS),[10] and reactive carbon species (RCS).[11,35]

In this chapter, we provide an overview of ABS and its bioimaging applications. We start with a discussion on foundational design principles of ABS probes, followed by a survey of representative examples of reaction classes that exploit this approach for selective detection, focusing on work from our laboratory on reactive oxygen, sulfur, and carbon species in biological models spanning cells to animals. Because such species are small, transient, and are similar in shape and size to other competing analytes in biological environments, these examples highlight the broad utility of ABS as a platform to help decipher the physiological and/or pathological roles of these important biological molecules. We then close by discussing current and future opportunities of ABS in imaging and other applications. We refer the reader to a more comprehensive treatment of ABS for further reading.[3]

3.2 Principles of Activity-based Sensing

The design of ABS probes relies on exploiting the unique reactivity of the analyte as a chemical substrate, which goes beyond the molecular recognition mode used for BBS (Figure 3.1). Drawing inspiration from the areas of supramolecular chemistry[36] and bio-orthogonal chemistry,[37] as well as activity-based protein profiling (ABPP),[38,39] ABS is built upon the development of chemoselective and biocompatible reaction triggers that react with analytes in their native biological environments. In this way, ABS provides a dose-dependent readout of the analyte substrate and is generally applicable for a broad range of small and large organic and inorganic analytes.

Because ABS is inherently a synthetic method, one can use the tools of retrosynthetic analysis to construct these types of probes. In general, a reaction trigger is incorporated into the imaging agent to maintain a physical and/or chemical property of the probe molecule in its "off" state. Among reversible interactions with competing molecules in complex biological milieus, only the targeted analyte will promote a chemical transformation upon reacting with the ABS trigger, due to its intrinsic chemoselectivity. This tandem binding/activity scheme can potentially augment selectivity compared to probes that rely on simple reversible binding events, particularly for analytes that are similar in shape and/or size to biological competitors. The reaction cleaves, adds to, or modifies the imaging agent to give an "on" state of the probe with a signal that can be captured by its imaging platform. The generality in this reaction-for-selectivity approach means that developing an effective reaction trigger of ABS can lead to its facile application to a variety of imaging modalities (Figure 3.1).

A variety of chemical transformations spanning the fields of synthetic methodology and bio-orthogonal chemistry has been tailored for ABS of biological analytes, with privileged reactions featuring selectivity, biocompatibility, and tunable kinetics. These reactions can be generally categorized into four broad classes: oxidation, reduction, redox-neutral, and metal-based/organometallic (Figure 3.2). The following sections discuss selected examples of ABS from our laboratory for imaging of reactive oxygen, sulfur, and carbon species, which represent oxidation, reduction, and redox-neutral reaction methods, respectively. ABS sensing methods involving metal-based/

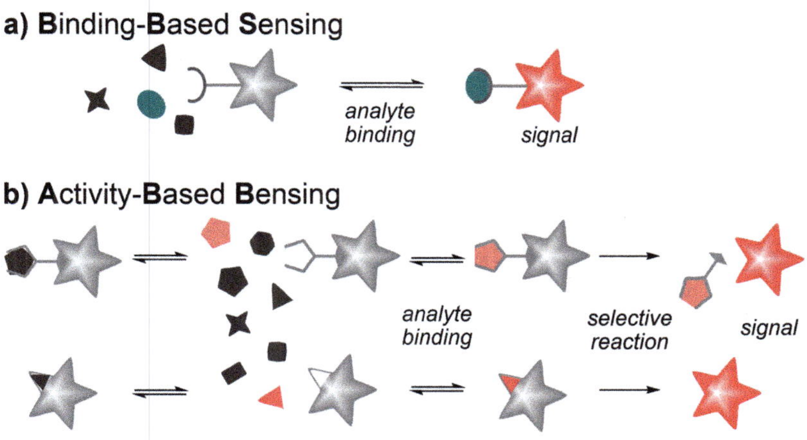

Figure 3.1 General schemes of binding-based sensing (BBS) and activity-based sensing (ABS). (a) BBS relies on the specific binding between the analyte and its ligand. (b) ABS involves interaction between the analyte and the reactive moiety to initiate a subsequent reaction that modifies the imaging agent.

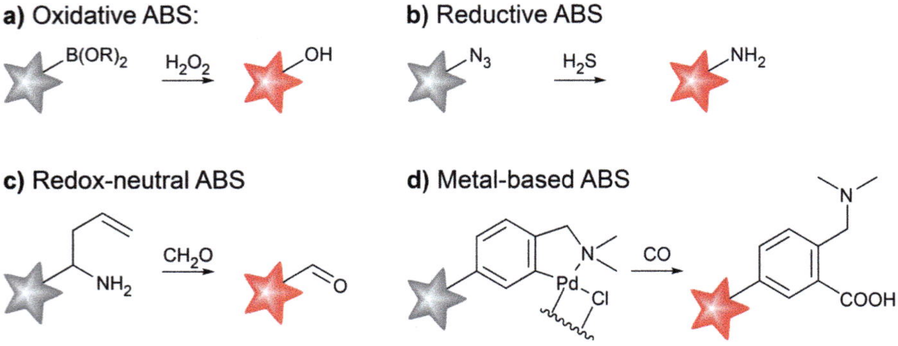

Figure 3.2 Major types of ABS reactions and representative examples for transient (a) reactive oxygen, (b) sulfur, and (c, d) carbon species detection.

organometallic reactions, including palladium-catalyzed carbonylation for CO detection[40,41] and ruthenium-catalyzed metathesis for ethylene sensing,[42] have been reviewed previously.[17] Readers can also refer to a more comprehensive review on the history of ABS and applications to bioimaging and beyond.[3]

3.3 Oxidative ABS: Boronates for Selective Hydrogen Peroxide Detection as One Type of Reactive Oxygen Species

Reactive oxygen species (ROS) are a class of transient small molecules derived from molecular oxygen through oxidative metabolism.[43–45] As by-products of cellular respiration, as well as product from dedicated enzymatic pathways, ROS is an umbrella term for a family of molecules that include superoxide ($O_2^{\bullet-}$), hydroxyl radical ($HO^{\bullet}$), hydrogen peroxide (H_2O_2), hypochlorous acid (HOCl), singlet oxygen ($^1O_2$), and lipid peroxides, among others. In addition to causing oxidative stress and damage at the cell and tissue level, ROS play diverse roles in physiology and pathology and participate in redox signalling events.[43,44,46–49] To help elucidate the biochemistry of ROS with more precise molecular resolution, it is essential to develop ways to distinguish between different members of the ROS family, to decipher contributions of individual species. In this context, because ROS are inherently similar in shape and size as well as being transient, they provide a challenge for traditional binding-based approaches for detection. Considering the high but dissimilar oxidative reactivity of these molecules, ABS provides a more fruitful approach towards chemoselective detection of each individual type of ROS. Indeed, a number of ABS probes have been reported that allow selective bioimaging of O_2^{-},[50,51] H_2O_2,[9,52–55] $HO^{\bullet}$,[56–58] and HOCl.[59,60] Among these examples, our laboratory has advanced the use of H_2O_2-mediated boronate oxidation as a selective and robust ABS reaction for developing H_2O_2 imaging sensors.[9,61]

To develop an effective activity-based method for H_2O_2 detection that achieves chemoselectivity over other biologically relevant ROS, we noted the unique ambiphilic reactivity of H_2O_2 as both a two-electron electrophilic oxidant and as a nucleophile, owing to the alpha effect of adjacent lone pairs of electrons. As such, we identified aryl boronates as a complementary ambiphile for reaction with H_2O_2.[62,63] Following the nucleophilic complexation with H_2O_2, the resulting C–B bond in the tetrahedral boron center can act as a nucleophile to reduce the peroxide by two electrons, releasing water and borate as biologically benign by-products (Figure 3.3a). In our initial report, which opened a field of ABS, we masked the two hydroxyl groups of fluorescein with boronates to afford peroxyfluor-1 (**PF1**) as a fluorescent probe for H_2O_2 (Figure 3.3b). As expected, upon reaction with H_2O_2 under physiological conditions, the closed, lactone form of the fluorophore in **PF1** is converted to the open form of fluorescein, resulting in a >1000-fold fluorescence increase. Most importantly, **PF1** is selective towards H_2O_2 over competing ROS, including $O_2^{\bullet -}$, nitric oxide, $HO^{\bullet}$ and *tert*-butyl hydroperoxide, and was applied in fluorescent imaging of H_2O_2 in living HEK293T cells.[61] Interestingly, complementary work reported that other versions of aryl boronates can react with peroxynitrite ($ONOO^-$) as well, with rates that can exceed H_2O_2 in *in vitro* assays.[12,64] However, we note that in biological contexts hydrogen peroxide is typically generated in much higher concentrations (up to mM) and has a longer half-life[46] compared to peroxynitrite, the latter having an exceedingly short lifetime and low steady-state concentration in the sub-nM range.[65] As such, we advocate for additional control experiments with boronate-based ABS probes to distinguish between peroxide- and peroxynitrite-dependent signals and to consider the biological context for their use, particularly as the former is generated in higher concentrations and has a longer biological half-life. For example, the addition of a nitric oxide synthase (NOS) inhibitor can block peroxynitrite formation, so if signal is not affected by NOS inhibition then H_2O_2 is likely the more dominant reactive species in these situations.[66,67] Catalase or ebselen addition can scavenge H_2O_2 but can also show reactivity with $OONO^-$. Chemistry specific to the probe must also be considered and can be informed by kinetic analyses with both species.[68] Indeed, the higher reactivity of $OONO^-$ relative to H_2O_2 can lead to dye reporter degradation in some situations, and loss of signal.[69,70]

We and others have further demonstrated the wide applicability of the boronate group as a selective H_2O_2 trigger for developing a suite of ABS imaging probes. The double boronate trigger for H_2O_2 was also applied to other fluorophore scaffolds, including resorufin (in peroxyresorufin-1, **PR1**), xanthone (in peroxyxanthone-1, **PX1**),[71] and naphthofluorescein (in naphtho peroxyfluorescein-1, **NPF1**),[72] each showing high selectivity for H_2O_2 over other ROS. To enhance the sensitivity of boronate-based ABS probes, we later replaced the double boronate masks with a single trigger, with Peroxy Green-1 (**PG1**) as a next-generation reagent derived from an asymmetric Tokyo Green scaffold (Figure 3.3c).[66] **PG1** provided the first direct imaging evidence for selective H_2O_2 signalling, as shown by its ability

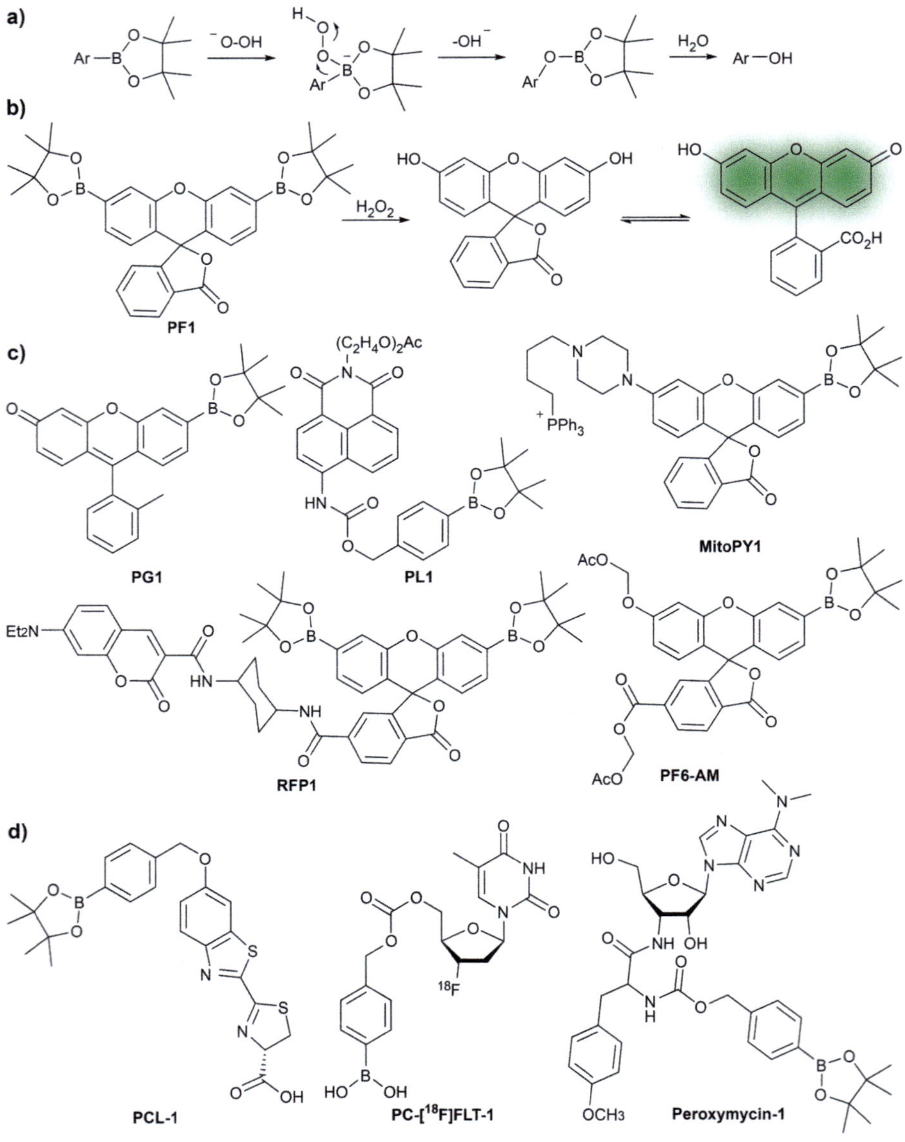

Figure 3.3 Boronate ABS probes for H_2O_2 detection and imaging. (a) Reactivity of aryl boronates with H_2O_2. (b) Peroxyfluor-1, a first-generation boronate-based probe for ABS of H_2O_2 for fluorescence imaging. (c) Selected boronate-based ABS probes for H_2O_2 fluorescent imaging bearing additional functionalities. (d) Boronate-based ABS probe for H_2O_2 detection using other imaging modalities beyond fluorescence.

to detect endogenously produced H_2O_2 during epidermal growth factor (EGF) stimulation in A431 cell line and in hippocampal neuron models. This scaffold was further expanded to a palette of H_2O_2 fluorescent dyes for imaging cellular H_2O_2 signalling events.[73]

In addition to intensity-based turn-on fluorescent probes for selective H_2O_2 detection, our laboratory also designed boronate reagents operating with dual optical signatures that enable ratiometric peroxide imaging with an internal calibration standard. For example, **PF1** was used as a fluorescence resonance energy transfer (FRET) acceptor that was covalently attached to a coumarin donor through a rigid linker to produce Ratio Peroxyfluor-1 (**RPF1**), where FRET efficiency by changes in spectral overlap can be monitored after reaction with H_2O_2 (Figure 3.3c).[74] Alternatively, we achieved ratiometric imaging by modulating the internal charge transfer (ICT) process of an aminonaphthalimide fluorophore in Peroxy Lucifer-1 (**PL1**). In this design, an electron-rich aminonapthalimide fluorophore was conjugated to an arylboronate trigger *via* a self-immolative linker that rendered the fluorophore in a blue-shifted carbamate form until its release by reaction with H_2O_2 (Figure 3.3c).[75]

To increase spatial resolution from the single-cell to subcellular level, we sought to limit the free diffusion of the fluorescent H_2O_2 probe before and after ABS. These efforts typically incorporated a second functional moiety to the fluorescent H_2O_2 sensor scaffold to afford bifunctional activity-based H_2O_2 probes with an added cellular targeting property and/or bioconjugation capability. For example, incorporation of phosphonium cation as the targeting group enables mitochondrial H_2O_2 imaging by Mitochondrial Peroxy Yellow-1 (**MitoPY1**, Figure 3.3c).[76,77] Highly sensitive H_2O_2 imaging within the cytosol can be achieved by introduction of ester-protected carboxylates that can be deprotected by intracellular esterases in Peroxy Yellow-1 methyl ester (**PY1-ME**) and Peroxyfluor-6 acetoxymethyl ester (**PF6-AM**) (Figure 3.3c),[78,79] Moreover, localized H_2O_2 detection in organelles, including plasma membrane, nucleus, mitochondria, and endoplasmic reticulum, is possible with hybrid boronate probe protein reporters, such as SNAP Peroxy Green dyes **SPG1** and **SPG2** containing a SNAP-tag for bioconjugation with organelle-specific proteins.[80] Finally, we achieved permanent marking of localized H_2O_2 fluxes with Peroxy Green-1 fluoromethyl (**PG1-FM**) featuring a fluoromethyl moiety as a masked quinone methide precursor that covalently traps nearby proteins upon activation by boronate-based H_2O_2 uncaging,[67] and microtubule-targeted imaging of H_2O_2 with Microtubule Peroxy Yellow-1 (**MT-PY1**) carrying a taxoid targeting group.[81]

By separating the detection trigger and reporter moiety into distinct units, the modularity of the ABS approach enables facile expansion of H_2O_2 sensing to other imaging modalities. For bioluminescence imaging, we incorporated the aryl boronate trigger as a blocking group for the phenolic hydroxyl group of luciferin *via* a self-immolative linker to afford peroxy-caged luciferin-1 (**PCL-1**). Upon reaction with H_2O_2, **PCL-1** releases the free luciferin substrate for reaction with firefly luciferase for bioluminescent imaging in living mice (Figure 3.3d).[82] We expanded this strategy to develop dual-analyte reporters that operate by an AND-type logic gate, by incorporating the boronate trigger on a hydroxyl cyanobenzothiazole in peroxy-caged luciferin-2 (**PCL-2**), which releases the luciferin precursor upon reaction with H_2O_2. Together with another precursor released from a D-cysteine peptide under caspase 8 activity, luciferin forms *in situ* to facilitate bioluminescent imaging of both H_2O_2 and caspase 8 activity.[83] Similarly, we incorporated the boronate trigger with a self-immolative linker to protect 5′-hydroxyl of a commonly used positron emission tomography (PET) tracer, [^{18}F]fluorothymidine-1 ([^{18}F]FLT), to afford peroxy-caged fluorothymidine-1 (**PC-[^{18}F]FLT-1**, Figure 3.3d). Upon reaction with H_2O_2, **PC-[^{18}F]FLT-1** releases [^{18}F]FLT as a PET radiotracer that metabolically accumulates in cells.[84] Additionally, we developed a histochemical probe for H_2O_2 by synthesizing the boronate-protected puromycin, **peroxymycin-1** (Figure 3.3d). H_2O_2-mediated boronate cleavage releases puromycin, which can be covalently incorporated into nascently synthesized peptides to leave a permanent mark on biological specimens, which can be visualized with commercial puromycin antibodies in a standard immunofluorescence assay.[85] Taken together, these representative examples illustrate the broad utility and impact of boronates as versatile activity-based reaction triggers for H_2O_2, enabling selective detection of this single type of ROS in a variety of cell and animal models with different classes of imaging modalities.

Activity-based probes for H_2O_2 with high selectivity and sensitivity provide a platform to reveal new biology mediated by this important ROS molecule. Prior to the advent of boronate H_2O_2 reporters, it was accepted that ROS are produced during extracellular growth factor stimulation, which subsequently mediates intracellular kinase activity. However, the identity of the ROS involved, their molecular sources, and membrane trafficking mechanism were elusive. With the high selectivity and sensitivity of the monoboronate probe **PG1**, we were able to directly visualize H_2O_2 produced upon EGF stimulation in living A431 cells. These cells have a high endogenous expression of EGF receptors (EGFRs), a classic example of receptor tyrosine kinases that mediate signal transduction in a variety of cell types. In addition, by imaging in the presence of specific pharmacological inhibitor compounds, we identified the involvement of phosphatidylinositol-3-OH kinase and the NADPH oxidase (Nox) enzyme in physiological production of H_2O_2. We also showed for the first time that H_2O_2 is endogenously produced in the brain upon EGF/Nox stimulation (Figure 3.4a).[66] In another example of new biological discovery enabled by ABS, using the cell-trappable H_2O_2 fluorescent indicator **PY1-ME**, we were able to identify certain classes of aquaporin water channels that facilitate the uptake of extracellular H_2O_2 as part of the intracellular signalling pathway (Figure 3.4b).[78] In parallel, we used **PF6-AM** to help explore the indispensable physiological roles of H_2O_2 in the growth and proliferation of neural stem cells. Intrigued by the H_2O_2 increase in adult hippocampal progenitor cells (AHPs) upon fibroblast growth factor-2 (FGF-2) stimulation, as revealed by **PF6-AM** imaging, we discovered that basal generation of this ROS is essential for AHP growth and differentiation, and identified the Nox2 enzyme and PTEN phosphatase as sources and targets of H_2O_2, respectively (Figure 3.4c).[79] More recently, our bifunctional sensor **PG1-FM** enabled direct observation of transcellular redox signalling between microglia and neurons. Neuronal cells cocultured with microglia exhibited an elevated level of H_2O_2 when microglia were selectively stimulated by lipopolysaccharide (LPS) and interferon gamma (IFN-γ, Figure 3.4d), with negligible increases in neuronal

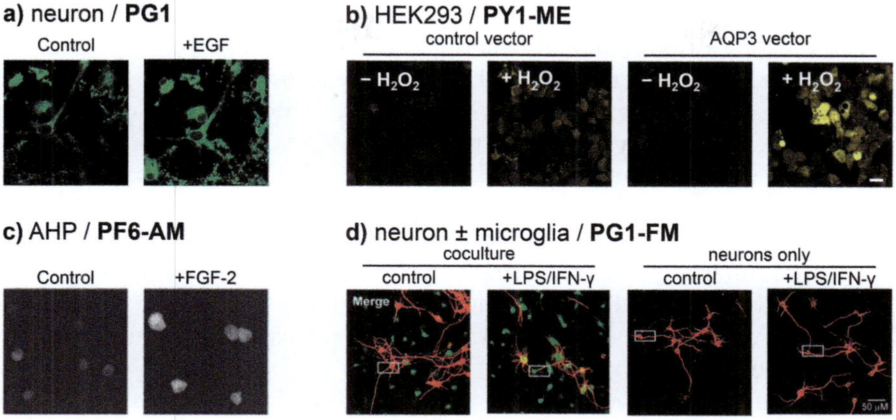

Figure 3.4 ABS boronate probes reveal new principles of biological H_2O_2 signalling. (a) **PG1** imaging of elevated H_2O_2 levels produced in live hippocampal neurons with EGF stimulation of receptor tyrosine kinase EGFR compared to controls. Reproduced from ref. 66 with permission from Springer Nature, Copyright 2007. (b) Detection of increased intracellular H_2O_2 level with **PY1-ME** in HEK293 cells transfected with aquaporin-3 (AQP3) vector in media containing H_2O_2 compared to cells with control vector, showing that this isoform and others are mediators of membrane H_2O_2 transport. Reproduced from ref. 78 with permission from National Academy of Sciences. (c) Visualization of H_2O_2 generated in adult hippocampal stem/progenitor (AHP) cells from Nox2 enzyme upon FGF-2 stimulation using **PF6-AM**, showing a physiological role for H_2O_2 signalling in the brain *via* growth and proliferation of neural stem cell populations for neurogenesis. Reproduced from ref. 79 with permission from Springer Nature, Copyright 2010. (d) Observation of intercellular H_2O_2 signalling between microglia and neurons with **PG1-FM**, where selective generation of H_2O_2 within microglia causes spillover into adjacent neurons. Red channel shows the MAP1 neuronal marker; green channel shows the **PG1-FM** signal. Reproduced from ref. 67 with permission from National Academy of Sciences.

H_2O_2 levels in the coculture resting state compared to neuronal monocultures. This phenomenon was confirmed to be a result of transcellular peroxide signalling from microglia to neurons with experiments using genetic knockout of Nox as an endogenous H_2O_2 source.[67] These examples showcase the power of boronate-based ABS methods to discover and characterize new biological principles involving H_2O_2 as a reactive oxygen signalling molecule.

3.4 Reductive ABS: Azides for Selective Hydrogen Sulfide Detection as One Type of Reactive Sulfur Species

Similar to the definition of ROS and RNS, reactive sulfur species (RSS) are a collection of sulfur chemotypes that can oxidize, reduce, or react with other molecules in living systems. The multiple oxidation states of sulfur, from −2 to +6, give rise to a myriad of members in this family, including organic and inorganic sulfides, polysulfides, thiosulfates, and sulfates, playing diverse and largely unexplored roles in biology.[86–88] Among these members, H_2S, as the smallest thiol, is a gaseous signalling molecule that regulates a variety of biological processes,[89] including neuronal activity,[90] cardiovascular protection,[91,92] inflammation,[93,94] and antioxidant defence,[89] whereas misregulation of H_2S is observed in disease states, including neurodegenerative Huntington's,[95] Parkinson's,[96] and Alzheimer's[97] diseases. The broad network of H_2S signalling pathways motivates the development of methods for selective and precise detection of this RSS in cells and in animals, to help unravel its biological contributions.

The strong reductive potential and relatively high acidity of H_2S under physiological conditions led us to develop a reductive ABS strategy for detecting this gasotransmitter molecule. Notably, H_2S has a pK_a of 7.0, suggesting that at physiological pH a considerable portion of this species exists as HS^-, a potent nucleophile, while other competing thiols in the cell, such as glutathione and protein cysteine residues, have weaker acidity (pK_a > 8) and thus remain mostly in the neutral form. In this context, we identified the reduction of organic azides to amines as a suitable reaction for ABS H_2S probes (Figure 3.5a).[98] The synthetic accessibility of organic azides, along with broad applications of organic azides in chemical biology as bio-orthogonal functional groups,[99] facilitated their development as a reductive ABS probe trigger for H_2S imaging. Our laboratory and Wang's

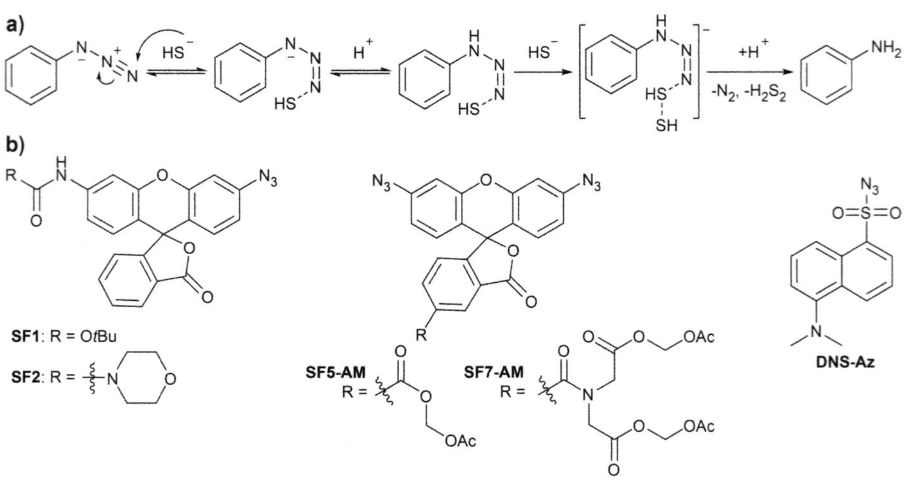

Figure 3.5 Azide-based ABS probes for H_2S detection and imaging. (a) Reductive reactivity of aryl azides with H_2S to provide anilines.[98] (b) Examples of azide-based H_2S ABS probes.

group independently developed azide-to-amine conversion as a selective H_2S trigger for ABS detection.[100,101] From our first report, Sulfidefluor-1 and Sulfidefluor-2 (**SF1** and **SF2**) are azide-functionalized rhodamine derivatives locked in a non-fluorescent spirolactone form, where H_2S-mediated reduction of the azide to amine generates the free dye, which can subsequently open to give its fluorescent form (Figure 3.5b).[100] Notably, the use of aryl azides, rather than alkyl ones, promotes H_2S-mediated reduction. In parallel, Wang and co-workers utilized dansyl azide (**DNS-Az**) as the reporter, which also exhibits a fluorescent turn-on upon reduction by H_2S to dansyl amide (Figure 3.5b).[101] We later expanded our design concept to Sulfidefluor-5 and -7 acetoxymethyl esters (**SF5-AM** and **SF7-AM**), which are bis-azide derivatives of rhodamine with minimized background and contain esterase-sensitive groups that undergo hydrolysis in cells to increase cell retention (Figure 3.5b).[102]

From this starting point, the H_2S-mediated reduction of azides has provided a general and versatile method for many others in the ABS field to develop H_2S probes for bioimaging.[10,52,103] The azide trigger has been incorporated onto a variety of fluorophore scaffolds for fluorescent imaging,[10,52,103] including coumarin,[95,104] rhodol,[105] BODIPY,[106] aminonaphthalimide,[107] and benzothiazole[108] dyes. Of note is that the azide trigger on red-shifted or two-photon fluorophores enables *in vivo* imaging of H_2S in tissue and in mouse.[109,110] Additionally, the azide functionality has been used to mask bioluminescent and chemiluminescent cores, either directly or indirectly *via* a self-immolative linker, to afford activity-based H_2S imaging agents in mice without light activation, which bypasses the photoactivation background introduced from fluorescence imaging.[111–114]

The availability of ABS probes for H_2S detection has enabled new investigations into the biological roles of this RSS. Owing to the high selectivity and sensitivity of **SF7-AM**, our laboratory developed and used this fluorescent probe to identify and characterize endogenously produced H_2S in living cells upon growth factor stimulation. Indeed, we observed elevated H_2S levels in living human umbilical vein endothelial cell (HUVEC) models with vascular endothelial growth factor (VEGF) stimulation (Figure 3.6a and b). We further studied the signalling proteins involved in this process, using biochemistry methods to modulate enzyme activity while monitoring change in H_2S level with **SF7-AM**. The result indicates that production of H_2S in the VEGF-stimulated pathway is dependent on VEGFR2 activation, cystathionine γ-lyase (CSE) activity, and Nox2-generated H_2O_2 (Figure 3.6c). Thus, in addition to identifying endogenous H_2S production and its sources, this study provided evidence for the oxidative/reductive crosstalk in this growth factor stimulation model between oxidative H_2O_2 generation and reductive H_2S generation.[102] With the continued development of H_2S ABS imaging probes, new biological principles are being revealed that add to our knowledge about this simple but common RSS.

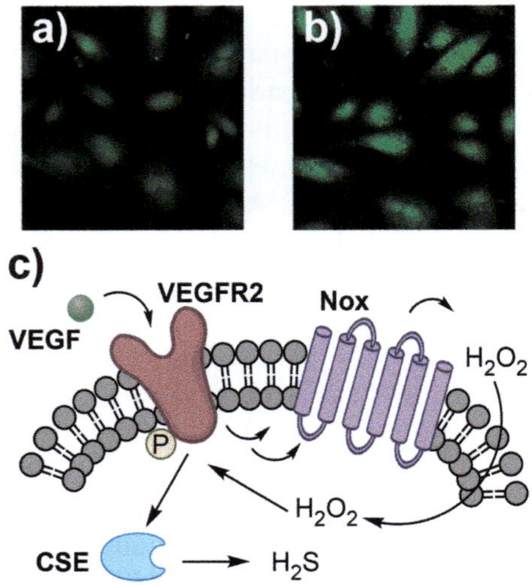

Figure 3.6 ABS enables imaging of endogenous H_2S produced in signalling pathways. (a) HUVEC cells stained with **SF7-AM**, and (b) same field of HUVEC cells treated with VEGF. Reproduced from ref. 102 with permission from National Academy of Sciences. (c) Proposed signalling pathway of H_2S generation upon growth factor stimulation.

3.5 Redox-Neutral ABS: Aza-Cope–Hardy Chemistry for Selective Formaldehyde Detection as One Type of Reactive Carbon Species

Redox-neutral reactions make up a large and growing portion of ABS designs, traditionally focusing on detection of nucleophilic biological analytes. More recently, with the development of new sensing triggers, electrophilic biomolecules, and in particular RCS, have emerged as a target of interest for ABS imaging. In this context, formaldehyde (FA) is a one-carbon electrophile that represents the smallest RCS and smallest aldehyde. FA is traditionally considered an environmental carcinogen from industrial sources, which can cause genetic damage *via* DNA-interstrand crosslinks,[115] but this one-carbon unit is also constantly being produced endogenously in the body through protein and nucleic acid demethylation and folate metabolism,[116] where it is metabolized into less toxic one-carbon fuel molecules, such as formate, *via* glutathione redox cycling.[117] The growing recognition of the broader biological significance of FA thus motivates the development of new chemical methods to monitor FA fluctuations with selectivity over other competing biological analytes, including single-carbon molecules, longer chain aldehydes, and related reactive oxygen, nitrogen, and sulfur species.

Against this backdrop, our laboratory[118] and Chan's group[119] in parallel designed an aza-Cope–Hardy rearrangement for ABS of formaldehyde.[11,35] This ABS approach relies on the reversible formation of an imine between amine and aldehyde, which can presumably occur with other biological aldehydes, but only the intermediate from formaldehyde can proceed *via* the aza-Cope–Hardy rearrangement with adjacent homo-allyl owing to its small steric hindrance (Figure 3.7a), endowing such probes with high selectivity for FA over other aldehydes, particularly over probe designs that are based on simple imine formation. Our first-generation formaldehyde detection reagent, Formaldehyde Probe-1 (**FAP-1**), embeds the amine-containing trigger within a silicon rhodamine fluorophore scaffold to favor the closed, lactone form of the silicon rhodamine. After selective reaction with FA, the aza-Cope–Hardy rearrangement and subsequent hydrolysis steps result in an aldehyde product that favors the open form of the fluorophore (Figure 3.7c).[118] Following this initial discovery, in subsequent work we introduced

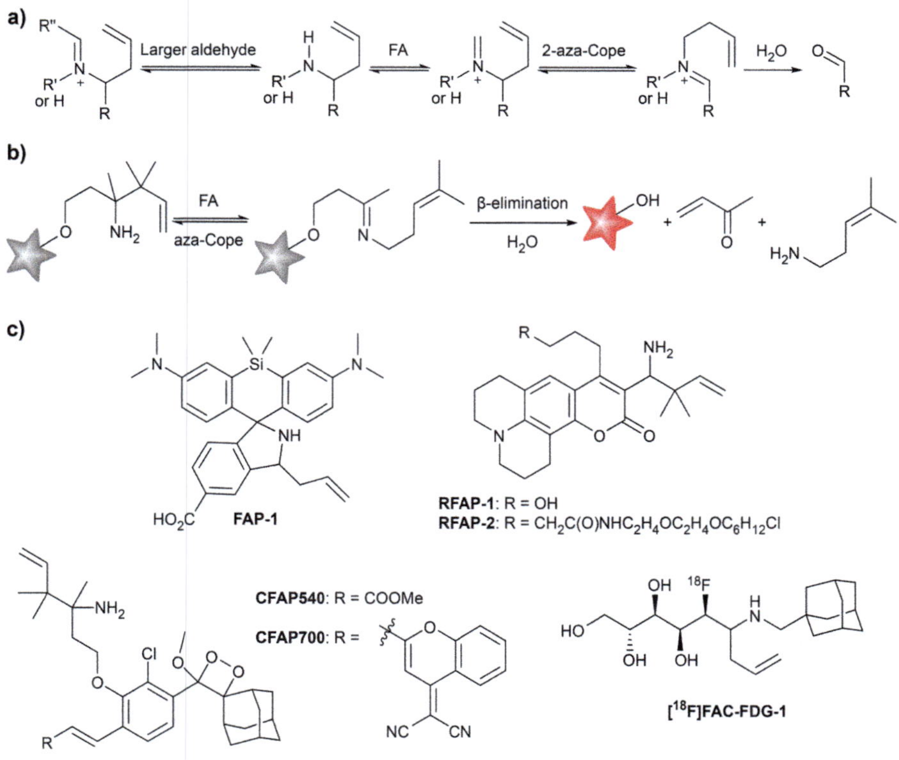

Figure 3.7 ABS detection and imaging of formaldehyde (FA) using an aza-Cope–Hardy arrangement. (a) First-generation ABS trigger for FA that forms an aldehyde upon FA reaction. (b) Modular ABS trigger for FA that releases any protected hydroxyl dye. (c) Structures of aza-Cope–Hardy probes for FA imaging in cells and in animals.

a geminal dimethyl moiety to accelerate the rearrangement, and reported Ratio Formaldehyde Probes 1 and 2 (**RFAP-1** and **RFAP-2**) as ratiometric formaldehyde reporters by modulating ICT on a coumarin scaffold before and after the aza-Cope–Hardy rearrangement (Figure 3.7c).[120] We later demonstrated the versatility of this transformation for FA detection with a modular trigger that can protect a hydroxyl group on a variety of dyes, which is unmasked upon aza-Cope–Hardy rearrangement followed by β-elimination (Figure 3.7b).[121] This modular trigger was introduced onto hydroxyl-containing fluorophores, including coumarin, rhodol, fluorescein, and resorufin, to afford **FAP385, FAP498, FAP555**, and **FAP573** as a palette of FA fluorescent probes with excitation/emission profiles covering the ultraviolet to visible spectrum.[121] Beyond fluorescence imaging, we also expanded the same design principle to the development of the PET imaging agent formaldehyde caged fluorodeoxy glucose-1 ([^{18}F]**FAC-FDG-1**, Figure 3.7c)[122] and the chemiluminescent reagents Chemiluminescent Formaldehyde Probes 540 and 700 (**CFAP540** and **CFAP700**) for FA imaging from cell to mouse models (Figure 3.7c).[123] These examples collectively showcase the modularity of the aza-Cope–Hardy trigger for ABS formaldehyde detection, which enables facile translation to different imaging modalities tailored for specific biological applications.

The aforementioned ABS sensors provide valuable tools to help unravel the physiological and pathological roles of formaldehyde as an emerging one-carbon unit in biology. With **FAP-1**, we were able to detect the endogenously produced FA in MCF7 cells, where, in this breast cancer model, the high expression of the epigenetic enzyme lysine-specific demethylase 1 (LSD1) leads to higher basal levels of FA compared to healthy cells. Upon inhibition of LSD1, **FAP-1** faithfully reflected the lowered levels of cellular FA (Figure 3.8a).[118] Likewise, Patel and co-workers also utilized **FAP-1** to support the discovery of spontaneous generation of FA from endogenous folate metabolism, with different oxidation levels of folate derivatives showing varying FA-releasing capabilities (Figure 3.8b).[116] In collaboration with the Shabat laboratory, we utilized activity-based chemiluminescent FA reporters to observe increases in FA levels in live mice upon treatment of tetrahydrofolate, confirming the FA-releasing character of folate derivatives *in vivo* (Figure 3.8c).[123] More recently, we identified and characterized the distinct mechanistic pathways of oxidative *N*-demethylation of RNA catalyzed by non-heme epigenetic enzymes FTO and ALKBH5, where ALKBH5 directly produces adenosine from methyladenosine with rapid FA release, in contrast to slow FA release by FTO. We further verified that FA produced by ALKBH5 provides a dynamic source of FA in cells with the aid of **FAP573**,[121] showing that knockdown of ALKBH5 results in a substantial decrease of basal cellular FA levels in both HEK293 and MCF7 cell lines (Figure 3.8d).[124] These examples collectively showcase the power of ABS imaging to discover and decipher new principles of FA and one-carbon signalling in living systems.

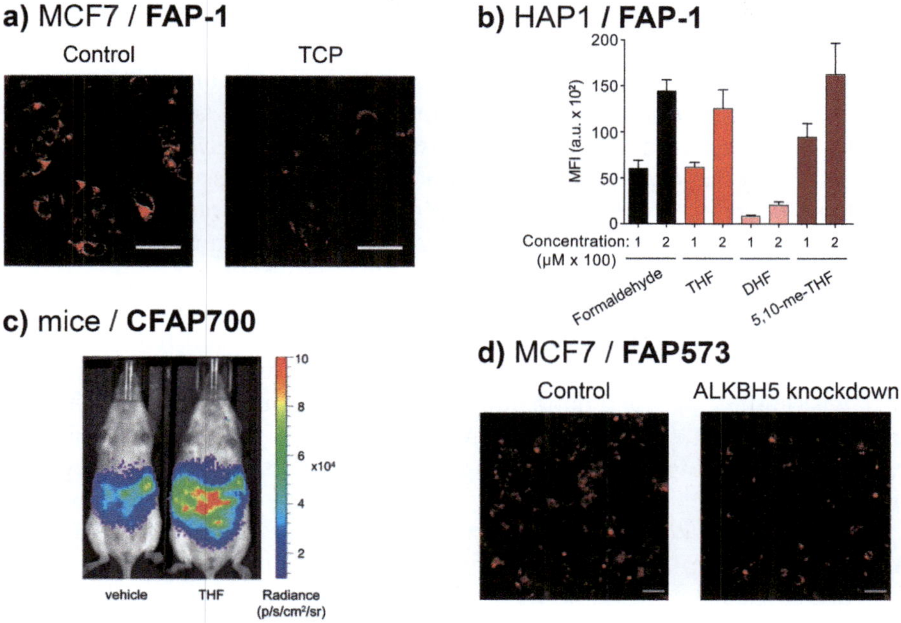

Figure 3.8 Aza-Cope–Hardy-based FA probes for biological studies. (a) Imaging of endogenously produced FA in MCF7 cells that overexpress LSD1 using **FAP-1**, showing a lowering of basal FA levels when treated with the LSD1 inhibitor tranylcypromine (TCP, right). Reproduced from ref. 118 with permission from American Chemical Society, Copyright 2015. (b) Detection of endogenous FA from spontaneous degradation of dietary and metabolically generated folate derivatives (DHF, dihydrofolate; THF, tetrahydrofolate; 5,10-me-THF, 5,10-dimethyltetrahydrofolate) in HAP1 cells sensitive to oxidative degradation. Reproduced from ref. 116 with permission from Springer Nature, Copyright 2017. (c) **CFAP700** imaging of FA produced from folate metabolism on mice exposed to tetrahydrofolate, and basal FA levels in mice with vehicle control. Reproduced from ref. 123 with permission from John Wiley & Sons, Copyright © 2018 Wiley-VCH Verlag GmbH & Co. KGaA, Weinheim. (d) Imaging of FA produced from the ALKBH5 demethylation pathway using **FAP573** in control MCF7 cells and MCF7 with ALKBH5 knockdown. Reproduced from ref. 124 with permission from National Academy of Sciences.

3.6 Concluding Remarks

As a rapidly growing area of chemistry, chemical biology, and molecular engineering, ABS provides a versatile platform to translate foundational knowledge from synthetic methodology to solving leading analytical challenges in biology. By exploiting the intrinsic chemical reactivity of biological analyte molecules, rather than simple shape and size consideration, ABS can achieve higher specificity than traditional binding-based molecular recognition approaches. Fundamental chemical properties, including hard-soft acidity and basicity, reduction and oxidation potentials, and multi-step organic and inorganic transformations, can all be employed to design probes to attain selectivity in complex biological environments. Coupling activity-based triggers with a change in the chemical and physical properties of a reporter unit before and after the sensing reaction can create a diverse array of imaging agents. Indeed, the modularity of ABS designs lend themselves well to imaging modalities that can be tailored for biological use. In this chapter, we have highlighted examples of ABS imaging platforms that operate by oxidative, reductive, and redox-neutral activity, which have been advanced by our laboratory and others to uncover new biology, particularly for reactive oxygen, sulfur, and carbon species. Motivated to expand our knowledge of biological signalling and metabolism pathways mediated by these and related molecules, and point the way to new diagnostic and therapeutic interventions, we envision the continued rapid advancement of selective and sensitive ABS platforms to meet the growing demands of real-time detection of biological analytes in living contexts and to help unravel these complex biological networks.

Against this backdrop, many directions still await further exploration in ABS. The development of new selective reaction triggers for sensing key biological analytes is a fundamental and foundational topic. While there are an ever-increasing number of reports on new biological sensors, selective and sensitive ABS probes for transition metals, transient reactive species, and small biomolecules, including signalling messengers and metabolites, are still underdeveloped compared to the infinite number of biomolecules. In parallel, more thorough applications of existing and future ABS imaging probes will continue to inspire new biological discoveries. Another avenue to explore is the expansion of imaging modalities. At this time, fluorescent imaging is the most widely applied imaging platform in ABS sensing due to its broad accessibility, focusing primarily on cells and other small, transparent model animals to address fundamental biological questions. For research on systems biology and clinical diagnostics, it is necessary to emphasize chemiluminescence, bioluminescence, and fluorescence imaging using near-infrared light for improved penetration, as well as advanced medical imaging platforms, including MRI, PET, and ultrasound imaging for detection of biological analytes and their events in animal models and humans.[125] Indeed, the modular design of ABS is well suited to this type of translational direction for biology and medicine. Another key challenge in ABS sensing lies in increasing its sensitivity and signal-to-noise ratio. Most existing ABS designs rely on stoichiometric reactions, causing consumption of original analytes during the sensing process. In this context, a growing area of ABS focuses on improvements in sensitivity by catalytic detection or regeneration of analytes, such as seminal reports on sensing FA,[126] H_2S,[127] and fluoride,[128] or by incorporation of signal amplification scaffolds, such as self-immolative linkers, to achieve multiple turnover detection reactions.[129] Reversible ABS reactions would also provide another advance in the field.

Beyond ABS detection by imaging techniques to meet the needs of biological research and clinical diagnostics, this chemical approach also offers possibilities in the development of less resource-demanding detection methods for broader applications beyond the research bench. Indeed, the design principles of ABS sensing can also be applied to analytical colorimetric and fluorometric assays, which in some cases can limit spatial and temporal resolution but in turn require only a UV-visible spectrometer or fluorometer, which are more broadly available, or naked eye and camera for readout, making it possible for broader field projects and public use. On the other hand, the chemistry behind ABS parallels the theory of ABPP, which leverages bioconjugation reactions to probe protein expression and activity as subjects of detection.[38,39] A key to ABPP is the development of labelling strategies to capture active amino acid residues for sensing using proteomics methodologies.[130] Moving forward from extensive studies on cysteine,[131] lysine,[132] and serine[133] sensing, we and others in the field have been developing labelling methods on amino acid residues, including methionine,[134–136] tyrosine,[137–139] histidine,[140] and tryptophan,[141] towards their activity-based sensing. We foresee that the interplay between ABS and ABPP and their underlying principles will offer valuable tools for chemical biologists to advance the knowledge about the molecular basis of biology and at the same time provide new concepts in building chemical probes for applications in imaging, biomolecule engineering, drug discovery, and diagnostics.

Acknowledgements

We thank the NIH (GM139465, GM79465, and ES28096), Agilent, and Novartis for supporting this work. C.J.C. is a CIFAR Fellow.

References

1. R. Y. Tsien, *Nat. Rev. Mol. Cell Biol.*, 2003, **Suppl**, SS16–SS21.
2. D. Wu, A. C. Sedgwick, T. Gunnlaugsson, E. U. Akkaya, J. Yoon and T. D. James, *Chem. Soc. Rev.*, 2017, **46**, 7105–7123.
3. K. J. Bruemmer, S. W. M. Crossley and C. J. Chang, *Angew. Chem.*, 2020, **59**, 13734–13762.
4. T. A. Su, K. J. Bruemmer and C. J. Chang, *Curr. Opin. Biotechnol.*, 2019, **60**, 198–204.
5. A. C. Love and J. A. Prescher, *Cell Chem. Biol.*, 2020, **27**, 904–920.
6. E. L. Que and C. J. Chang, *Chem. Soc. Rev.*, 2009, **39**, 51–60.
7. J. Wahsner, E. M. Gale, A. Rodríguez-Rodríguez and P. Caravan, *Chem. Rev.*, 2019, **119**, 957–1057.
8. H. J. Knox and J. Chan, *Acc. Chem. Res.*, 2018, **51**, 2897–2905.
9. A. R. Lippert, G. C. Van de Bittner and C. J. Chang, *Acc. Chem. Res.*, 2011, **44**, 793–804.

10. V. S. Lin, W. Chen, M. Xian and C. J. Chang, *Chem. Soc. Rev.*, 2015, **44**, 4596–4618.
11. J. Ohata, K. J. Bruemmer and C. J. Chang, *Acc. Chem. Res.*, 2019, **52**, 2841–2848.
12. L. Wu, A. C. Sedgwick, X. Sun, S. D. Bull, X.-P. He and T. D. James, *Acc. Chem. Res.*, 2019, **52**, 2582–2597.
13. E. L. Que, D. W. Domaille and C. J. Chang, *Chem. Rev.*, 2008, **108**, 1517–1549.
14. D. W. Domaille, E. L. Que and C. J. Chang, *Nat. Chem. Biol.*, 2008, **4**, 168–175.
15. K. P. Carter, A. M. Young and A. E. Palmer, *Chem. Rev.*, 2014, **114**, 4564–4601.
16. A. T. Aron, K. M. Ramos-Torres, J. A. Cotruvo and C. J. Chang, *Acc. Chem. Res.*, 2015, **48**, 2434–2442.
17. D. A. Iovan, S. Jia and C. J. Chang, *Inorg. Chem.*, 2019, **58**, 13546–13560.
18. T. Ueno and T. Nagano, *Nat. Methods*, 2011, **8**, 642–645.
19. A. Minta, J. P. Kao and R. Y. Tsien, *J. Biol. Chem.*, 1989, **264**, 8171–8178.
20. E. M. Nolan and S. J. Lippard, *Acc. Chem. Res.*, 2009, **42**, 193–203.
21. Z. Xu, J. Yoon and D. R. Spring, *Chem. Soc. Rev.*, 2010, **39**, 1996–2006.
22. S. C. Dodani, D. W. Domaille, C. I. Nam, E. W. Miller, L. A. Finney, S. Vogt and C. J. Chang, *Proc. Natl. Acad. Sci. U. S. A.*, 2011, **108**, 5980–5985.
23. C. J. Fahrni, *Curr. Opin. Chem. Biol.*, 2013, **17**, 656–662.
24. J. J. A. Cotruvo, A. T. Aron, K. M. Ramos-Torres and C. J. Chang, *Chem. Soc. Rev.*, 2015, **44**, 4400–4414.
25. K. M. Ramos-Torres, S. Kolemen and C. J. Chang, *Isr. J. Chem.*, 2016, **56**, 724–737.
26. S. C. Dodani, Q. He and C. J. Chang, *J. Am. Chem. Soc.*, 2009, **131**, 18020–18021.
27. M. Y. Chae and A. W. Czarnik, *J. Am. Chem. Soc.*, 1992, **114**, 9704–9705.
28. D.-G. Cho and J. L. Sessler, *Chem. Soc. Rev.*, 2009, **38**, 1647–1662.
29. J. Chan, S. C. Dodani and C. J. Chang, *Nat. Chem.*, 2012, **4**, 973–984.
30. Y. Yang, Q. Zhao, W. Feng and F. Li, *Chem. Rev.*, 2013, **113**, 192–270.
31. M. H. Lee, J. S. Kim and J. L. Sessler, *Chem. Soc. Rev.*, 2015, **44**, 4185–4191.
32. A. T. Aron, A. G. Reeves and C. J. Chang, *Curr. Opin. Chem. Biol.*, 2018, **43**, 113–118.
33. H. Kojima, N. Nakatsubo, K. Kikuchi, S. Kawahara, Y. Kirino, H. Nagoshi, Y. Hirata and T. Nagano, *Anal. Chem.*, 1998, **70**, 2446–2453.
34. H. Kojima, M. Hirotani, N. Nakatsubo, K. Kikuchi, Y. Urano, T. Higuchi, Y. Hirata and T. Nagano, *Anal. Chem.*, 2001, **73**, 1967–1973.
35. K. J. Bruemmer, T. F. Brewer and C. J. Chang, *Curr. Opin. Chem. Biol.*, 2017, **39**, 17–23.
36. J.-M. Lehn, *Angew. Chem., Int. Ed. Engl.*, 1988, **27**, 89–112.
37. E. M. Sletten and C. R. Bertozzi, *Angew. Chem., Int. Ed*, 2009, **48**, 6974–6998.
38. G. C. Adam, E. J. Sorensen and B. F. Cravatt, *Mol. Cell. Proteomics*, 2002, **1**, 781–790.

39. *Activity-Based Protein Profiling*, ed. B. F. Cravatt, K.-L. Hsu and E. Weerapana, Springer International Publishing, Cham, 2019.
40. B. W. Michel, A. R. Lippert and C. J. Chang, *J. Am. Chem. Soc.*, 2012, **134**, 15668–15671.
41. J. Morstein, D. Höfler, K. Ueno, J. W. Jurss, R. R. Walvoord, K. J. Bruemmer, S. P. Rezgui, T. F. Brewer, M. Saitoe, B. W. Michel and C. J. Chang, *J. Am. Chem. Soc.*, 2020, **142**, 15917–15930.
42. S. N. W. Toussaint, R. T. Calkins, S. Lee and B. W. Michel, *J. Am. Chem. Soc.*, 2018, **140**, 13151–13155.
43. C. C. Winterbourn, *Nat. Chem. Biol.*, 2008, **4**, 278–286.
44. B. C. Dickinson and C. J. Chang, *Nat. Chem. Biol.*, 2011, **7**, 504–511.
45. M. P. Murphy, A. Holmgren, N.-G. Larsson, B. Halliwell, C. J. Chang, B. Kalyanaraman, S. G. Rhee, P. J. Thornalley, L. Partridge, D. Gems, T. Nyström, V. Belousov, P. T. Schumacker and C. C. Winterbourn, *Cell Metab.*, 2011, **13**, 361–366.
46. B. D'Autréaux and M. B. Toledano, *Nat. Rev. Mol. Cell Biol.*, 2007, **8**, 813–824.
47. M. Schieber and N. S. Chandel, *Curr. Biol.*, 2014, **24**, R453–R462.
48. D. Reichmann, W. Voth and U. Jakob, *Mol. Cell*, 2018, **69**, 203–213.
49. H. Sies and D. P. Jones, *Nat. Rev. Mol. Cell Biol.*, 2020, **21**, 363–383.
50. H. Zhao, S. Kalivendi, H. Zhang, J. Joseph, K. Nithipatikom, J. Vásquez-Vivar and B. Kalyanaraman, *Free Radicals Biol. Med.*, 2003, **34**, 1359–1368.
51. K. M. Robinson, M. S. Janes, M. Pehar, J. S. Monette, M. F. Ross, T. M. Hagen, M. P. Murphy and J. S. Beckman, *Proc. Natl. Acad. Sci. U. S. A.*, 2006, **103**, 15038–15043.
52. V. S. Lin, B. C. Dickinson and C. J. Chang, in *Methods in Enzymology*, ed. E. Cadenas and L. Packer, Academic Press, 2013, vol. 526, pp. 19–43.
53. T. F. Brewer, F. J. Garcia, C. S. Onak, K. S. Carroll and C. J. Chang, *Annu. Rev. Biochem.*, 2015, **84**, 765–790.
54. S. Ye, J. J. Hu and D. Yang, *Angew. Chem., Int. Ed.*, 2018, **57**, 10173–10177.
55. D. Pham, U. Basu, I. Pohorilets, C. M. S. Croix, S. C. Watkins and K. Koide, *Angew. Chem.*, 2020, **59**, 17435–17441.
56. K. Setsukinai, Y. Urano, K. Kakinuma, H. J. Majima and T. Nagano, *J. Biol. Chem.*, 2003, **278**, 3170–3175.
57. K. Żamojć, M. Zdrowowicz, D. Jacewicz, D. Wyrzykowski and L. Chmurzyński, *Crit. Rev. Anal. Chem.*, 2016, **46**, 160–169.
58. J.-T. Hou, M. Zhang, Y. Liu, X. Ma, R. Duan, X. Cao, F. Yuan, Y.-X. Liao, S. Wang and W. Xiu Ren, *Coord. Chem. Rev.*, 2020, **421**, 213457.
59. S. Kenmoku, Y. Urano, H. Kojima and T. Nagano, *J. Am. Chem. Soc.*, 2007, **129**, 7313–7318.
60. Y. Koide, Y. Urano, K. Hanaoka, T. Terai and T. Nagano, *J. Am. Chem. Soc.*, 2011, **133**, 5680–5682.
61. M. C. Y. Chang, A. Pralle, E. Y. Isacoff and C. J. Chang, *J. Am. Chem. Soc.*, 2004, **126**, 15392–15393.
62. H. G. Kuivila, *J. Am. Chem. Soc.*, 1954, **76**, 870–874.
63. H. G. Kuivila and A. G. Armour, *J. Am. Chem. Soc.*, 1957, **79**, 5659–5662.

64. J. Zielonka, A. Sikora, J. Joseph and B. Kalyanaraman, *J. Biol. Chem.*, 2010, **285**, 14210–14216.
65. S. Carballal, S. Bartesaghi and R. Radi, *Biochim. Biophys. Acta*, 2014, **1840**, 768–780.
66. E. W. Miller, O. Tulyathan, E. Y. Isacoff and C. J. Chang, *Nat. Chem. Biol.*, 2007, **3**, 263–267.
67. H. Iwashita, E. Castillo, M. S. Messina, R. A. Swanson and C. J. Chang, *Proc. Natl. Acad. Sci. U. S. A.*, 2021, **118**(9), e2018513118.
68. L. Wu, J. Huang, K. Pu and T. D. James, *Nat. Rev. Chem.*, 2021, **5**, 406–421.
69. L. Wu, J. Liu, X. Tian, R. R. Groleau, S. D. Bull, P. Li, B. Tang and T. D. James, *Chem. Sci.*, 2021, **12**, 3921–3928.
70. L. Wu, J. Liu, X. Tian, R. R. Groleau, B. Feng, Y. Yang, A. C. Sedgwick, H.-H. Han, Y. Wang, H.-M. Wang, F. Huang, S. D. Bull, H. Zhang, C. Huang, Y. Zang, J. Li, X.-P. He, P. Li, B. Tang, T. D. James and J. L. Sessler, *J. Am. Chem. Soc.*, 2022, **144**(1), 174–183.
71. E. W. Miller, A. E. Albers, A. Pralle, E. Y. Isacoff and C. J. Chang, *J. Am. Chem. Soc.*, 2005, **127**, 16652–16659.
72. A. E. Albers, B. C. Dickinson, E. W. Miller and C. J. Chang, *Bioorg. Med. Chem. Lett.*, 2008, **18**, 5948–5950.
73. B. C. Dickinson, C. Huynh and C. J. Chang, *J. Am. Chem. Soc.*, 2010, **132**, 5906–5915.
74. A. E. Albers, V. S. Okreglak and C. J. Chang, *J. Am. Chem. Soc.*, 2006, **128**, 9640–9641.
75. D. Srikun, E. W. Miller, D. W. Domaille and C. J. Chang, *J. Am. Chem. Soc.*, 2008, **130**, 4596–4597.
76. B. C. Dickinson and C. J. Chang, *J. Am. Chem. Soc.*, 2008, **130**, 9638–9639.
77. B. C. Dickinson, V. S. Lin and C. J. Chang, *Nat. Protoc.*, 2013, **8**, 1249–1259.
78. E. W. Miller, B. C. Dickinson and C. J. Chang, *Proc. Natl. Acad. Sci. U. S. A.*, 2010, **107**, 15681–15686.
79. B. C. Dickinson, J. Peltier, D. Stone, D. V. Schaffer and C. J. Chang, *Nat. Chem. Biol.*, 2011, **7**, 106–112.
80. D. Srikun, A. E. Albers, C. I. Nam, A. T. Iavarone and C. J. Chang, *J. Am. Chem. Soc.*, 2010, **132**, 4455–4465.
81. S. Jia and C. J. Chang, *Bioorg. Med. Chem. Lett.*, 2021, **48**, 128252.
82. G. C. V. de Bittner, E. A. Dubikovskaya, C. R. Bertozzi and C. J. Chang, *Proc. Natl. Acad. Sci. U. S. A.*, 2010, **107**, 21316–21321.
83. G. C. Van de Bittner, C. R. Bertozzi and C. J. Chang, *J. Am. Chem. Soc.*, 2013, **135**, 1783–1795.
84. V. Carroll, B. W. Michel, J. Blecha, H. VanBrocklin, K. Keshari, D. Wilson and C. J. Chang, *J. Am. Chem. Soc.*, 2014, **136**, 14742–14745.
85. C. Yik-Sham Chung, G. A. Timblin, K. Saijo and C. J. Chang, *J. Am. Chem. Soc.*, 2018, **140**, 6109–6121.
86. G. I. Giles and C. Jacob, *Biol. Chem.*, 2002, **383**, 375–388.

87. T. V. Mishanina, M. Libiad and R. Banerjee, *Nat. Chem. Biol.*, 2015, **11**, 457–464.
88. N. Lau and M. D. Pluth, *Curr. Opin. Chem. Biol.*, 2019, **49**, 1–8.
89. H. Kimura, *Antioxid. Redox Signaling*, 2014, **20**, 783–793.
90. B. D. Paul and S. H. Snyder, *Biochem. Pharmacol.*, 2018, **149**, 101–109.
91. Y.-H. Liu, M. Lu, L.-F. Hu, P. T.-H. Wong, G. D. Webb and J.-S. Bian, *Antioxid. Redox Signaling*, 2012, **17**, 141–185.
92. L. Testai, V. Citi, A. Martelli, S. Brogi and V. Calderone, *Pharmacol. Res.*, 2020, **160**, 105125.
93. M. Whiteman and P. G. Winyard, *Expert Rev. Clin. Pharmacol.*, 2011, **4**, 13–32.
94. J. L. Wallace, J. G. P. Ferraz and M. N. Muscara, *Antioxid. Redox Signaling*, 2012, **17**, 58–67.
95. B. D. Paul, J. I. Sbodio, R. Xu, M. S. Vandiver, J. Y. Cha, A. M. Snowman and S. H. Snyder, *Nature*, 2014, **509**, 96–100.
96. L.-F. Hu, M. Lu, C. X. Tiong, G. S. Dawe, G. Hu and J.-S. Bian, *Aging Cell*, 2010, **9**, 135–146.
97. D. Giuliani, A. Ottani, D. Zaffe, M. Galantucci, F. Strinati, R. Lodi and S. Guarini, *Neurobiol. Learn. Mem.*, 2013, **104**, 82–91.
98. H. A. Henthorn and M. D. Pluth, *J. Am. Chem. Soc.*, 2015, **137**, 15330–15336.
99. N. J. Agard, J. M. Baskin, J. A. Prescher, A. Lo and C. R. Bertozzi, *ACS Chem. Biol.*, 2006, **1**, 644–648.
100. A. R. Lippert, E. J. New and C. J. Chang, *J. Am. Chem. Soc.*, 2011, **133**, 10078–10080.
101. K. Wang, H. Peng, N. Ni, C. Dai and B. Wang, *J. Fluoresc.*, 2014, **24**, 1–5.
102. V. S. Lin, A. R. Lippert and C. J. Chang, *Proc. Natl. Acad. Sci. U. S. A.*, 2013, **110**, 7131–7135.
103. V. S. Lin and C. J. Chang, *Curr. Opin. Chem. Biol.*, 2012, **16**, 595–601.
104. W. Chen, A. Pacheco, Y. Takano, J. J. Day, K. Hanaoka and M. Xian, *Angew. Chem., Int. Ed*, 2016, **55**, 9993–9996.
105. M. D. Hammers, M. J. Taormina, M. M. Cerda, L. A. Montoya, D. T. Seidenkranz, R. Parthasarathy and M. D. Pluth, *J. Am. Chem. Soc.*, 2015, **137**, 10216–10223.
106. T. Saha, D. Kand and P. Talukdar, *Org. Biomol. Chem.*, 2013, **11**, 8166–8170.
107. L. A. Montoya and M. D. Pluth, *Chem. Commun.*, 2012, **48**, 4767–4769.
108. Z. Jingyu and G. Wei, *Chem. Commun.*, 2014, **50**, 4214–4217.
109. S. K. Das, C. S. Lim, S. Y. Yang, J. H. Han and B. R. Cho, *Chem. Commun.*, 2012, **48**, 8395–8397.
110. W. Sun, J. Fan, C. Hu, J. Cao, H. Zhang, X. Xiong, J. Wang, S. Cui, S. Sun and X. Peng, *Chem. Commun.*, 2013, **49**, 3890–3892.
111. T. S. Bailey and M. D. Pluth, *J. Am. Chem. Soc.*, 2013, **135**, 16697–16704.
112. J. Cao, R. Lopez, J. M. Thacker, J. Y. Moon, C. Jiang, S. N. S. Morris, J. H. Bauer, P. Tao, R. P. Mason and A. R. Lippert, *Chem. Sci.*, 2015, **6**, 1979–1985.

113. X. Tian, Z. Li, C. Lau and J. Lu, *Anal. Chem.*, 2015, **87**, 11325–11331.
114. B. Ke, W. Wu, W. Liu, H. Liang, D. Gong, X. Hu and M. Li, *Anal. Chem.*, 2016, **88**, 592–595.
115. L. B. Pontel, I. V. Rosado, G. Burgos-Barragan, J. I. Garaycoechea, R. Yu, M. J. Arends, G. Chandrasekaran, V. Broecker, W. Wei, L. Liu, J. A. Swenberg, G. P. Crossan and K. J. Patel, *Mol. Cell*, 2015, **60**, 177–188.
116. G. Burgos-Barragan, N. Wit, J. Meiser, F. A. Dingler, M. Pietzke, L. Mulderrig, L. B. Pontel, I. V. Rosado, T. F. Brewer, R. L. Cordell, P. S. Monks, C. J. Chang, A. Vazquez and K. J. Patel, *Nature*, 2017, **548**, 549–554.
117. M. Koivusalo, M. Baumann and L. Uotila, *FEBS Lett.*, 1989, **257**, 105–109.
118. T. F. Brewer and C. J. Chang, *J. Am. Chem. Soc.*, 2015, **137**, 10886–10889.
119. A. Roth, H. Li, C. Anorma and J. Chan, *J. Am. Chem. Soc.*, 2015, **137**, 10890–10893.
120. T. F. Brewer, G. Burgos-Barragan, N. Wit, K. J. Patel and C. J. Chang, *Chem. Sci.*, 2017, **8**, 4073–4081.
121. K. J. Bruemmer, R. R. Walvoord, T. F. Brewer, G. Burgos-Barragan, N. Wit, L. B. Pontel, K. J. Patel and C. J. Chang, *J. Am. Chem. Soc.*, 2017, **139**, 5338–5350.
122. W. Liu, C. Truillet, R. R. Flavell, T. F. Brewer, M. J. Evans, D. M. Wilson and C. J. Chang, *Chem. Sci.*, 2016, **7**, 5503–5507.
123. K. J. Bruemmer, O. Green, T. A. Su, D. Shabat and C. J. Chang, *Angew. Chem., Int. Ed.*, 2018, **57**, 7508–7512.
124. J. D. W. Toh, S. W. M. Crossley, K. J. Bruemmer, E. J. Ge, D. He, D. A. Iovan and C. J. Chang, *Proc. Natl. Acad. Sci. U. S. A.*, 2020, **117**, 25284–25292.
125. T. F. Massoud and S. S. Gambhir, *Genes Dev.*, 2003, **17**, 545–580.
126. H. Xu, H. Xu, S. Ma, X. Chen, L. Huang, J. Chen, F. Gao, R. Wang, K. Lou and W. Wang, *J. Am. Chem. Soc.*, 2018, **140**, 16408–16412.
127. Y. Zhao, M. M. Cerda and M. D. Pluth, *Chem. Sci.*, 2018, **10**, 1873–1878.
128. M. S. Baker and S. T. Phillips, *Org. Biomol. Chem.*, 2012, **10**, 3595–3599.
129. J. Yan, S. Lee, A. Zhang and J. Yoon, *Chem. Soc. Rev.*, 2018, **47**, 6900–6916.
130. J. L. Counihan, B. Ford and D. K. Nomura, *Curr. Opin. Chem. Biol.*, 2016, **30**, 68–76.
131. J. M. Chalker, G. J. L. Bernardes, Y. A. Lin and B. G. Davis, *Chem. – Asian J.*, 2009, **4**, 630–640.
132. C. C. Ward, J. I. Kleinman and D. K. Nomura, *ACS Chem. Biol.*, 2017, **12**, 1478–1483.
133. G. M. Simon and B. F. Cravatt, *J. Biol. Chem.*, 2010, **285**, 11051–11055.
134. S. Lin, X. Yang, S. Jia, A. M. Weeks, M. Hornsby, P. S. Lee, R. V. Nichiporuk, A. T. Iavarone, J. A. Wells, F. D. Toste and C. J. Chang, *Science*, 2017, **355**, 597–602.
135. M. T. Taylor, J. E. Nelson, M. G. Suero and M. J. Gaunt, *Nature*, 2018, **562**, 563–568.
136. A. H. Christian, S. Jia, W. Cao, P. Zhang, A. T. Meza, M. S. Sigman, C. J. Chang and F. D. Toste, *J. Am. Chem. Soc.*, 2019, **141**, 12657–12662.

137. H. Ban, J. Gavrilyuk and C. F. Barbas, *J. Am. Chem. Soc.*, 2010, **132**, 1523–1525.
138. J. W. Brulet, A. L. Borne, K. Yuan, A. H. Libby and K.-L. Hsu, *J. Am. Chem. Soc.*, 2020, **142**, 8270–8280.
139. F. Sun, S. Suttapitugsakul and R. Wu, *Anal. Chem.*, 2021, **93**, 10334–10342.
140. S. Jia, D. He and C. J. Chang, *J. Am. Chem. Soc.*, 2019, **141**, 7294–7301.
141. J. M. Antos and M. B. Francis, *J. Am. Chem. Soc.*, 2004, **126**, 10256–10257.

CHAPTER 4

Aggregation-induced Emission-based Fluorescent Systems

MENG LI*[a], XIAONING LI[a] AND ZHIJUN CHEN[b]

[a]Hebei Key Lab of Power Plant Flue Gas Multi-Pollutants Control, Department of Environmental Science and Engineering, North China Electric Power University, Baoding 071003, PR China; [b]Key Laboratory of Bio-based Material Science and Technology of Ministry of Education, Material Science and Engineering College, Northeast Forestry University, Hexing Road 26, Harbin 150040, PR China
*E-mail: mlincepu@hotmail.com

4.1 Introduction

Fluorescent materials have aroused great interest owing to their specific photophysical properties and widespread applications in the fields of optoelectronics, sensing and bioimaging.[1–5] However, most aromatic organic luminescent molecules exhibit concentration quenching and aggregation-caused quenching (ACQ) problems.[6–10] These ACQ molecules have faced the problem of poor solubility and weak solid-state luminescence, thus greatly limiting their practical applications.[11] Therefore, it is critical to find new solutions that can answer the ACQ problem.[12–14] In 2001, Tang and co-workers presented the concept of aggregation-induced emission luminogens (AIEgens), which are highly emissive as aggregates/clusters.[15]

These AIEgens generally include organic molecules and polymeric materials. Mostly, the organic AIEgens contain bulky aromatic or conjugated units

Monographs in Supramolecular Chemistry No. 34
Fluorescent Chemosensors
Edited by Luling Wu, Adam C. Sedgwick, Xiao-Peng He and Tony D. James

Published by the Royal Society of Chemistry, www.rsc.org

with remarkable conjugation in individual molecules, exhibiting excellent AIE characteristics.[16,17] Recently, non-conventional luminogens, including synthetic polymers, nanodots and natural AIEgens derived from sustainable biomass materials, have been developed and investigated. In addition, the excellent properties of non-conventional materials, in terms of good solubility, enhanced biocompatibility and high stability, make them more suitable for biological and medical applications than synthetic AIEgens.[18,19]

The luminescence mechanism is a particularly important one to the scientific community, making the study of the AIE mechanism of great significance for a deeper understanding of the photophysical processes, and facilitating the synthesis of new high-efficiency fluorescent materials. Tang's group first proposed restriction of intramolecular rotation (RIR) to explain the mechanism for traditional organic molecules. The mechanisms currently exploited in the AIE area include RIR, restriction of intramolecular motion (RIM), J-aggregate formation and clustering-triggered emission (CTE), among others.[20–22] Herein, we will summarize recent advances in this exciting area of AIE and its luminescence mechanisms.

4.2 Organic Luminescent Molecules

Organic luminescent molecules generally consist of aromatic groups and conjugated subunits, which function as chromophoric units. Restriction of intramolecular motion can reasonably explain the appearance of AIE, which leads to enhanced structural rigidification and reduces the non-radiative decay rates.

Tang and co-workers first proposed an unusual molecule of 1-methyl-1,2,3,4,5-pentaphenylsilole (**1**) (see Scheme 4.1a).[15] Unprecedentedly, the molecules are almost non-luminescent when dissolved, but exhibit a visible fluorescence emission in the aggregated state. The emission intensity of **1** was increased by approximately 333 times when aggregated. In addition, the prepared thin film also produced a strong photoluminescence (PL) emission. A significant increase and red shift in the absorption and emission bands owing to conjugation between the periphery and the core of **1** was also found. To further confirm this phenomenon, various composite films were prepared using **1** and poly(methyl methacrylate) (PMMA). These films exhibited a blue emission with good fluorescence of ~24 000 cd m^{-2} with 90% content of **1**, suggesting that aggregation increased the emission.

Silole is an important light-emitting molecule in the design of AIE-based molecules. As shown in Scheme 4.1b, various novel silole derivatives modified with 1-phenyl-1*H*-benzo[*d*]imidazole (PBI) or 1-phenyl-1*H*-phenanthro[9,10-*d*]imidazole (PPI) groups were reported.[23] These prepared molecules are highly emissive and have high quantum yields of 49.5–62.1% in solid films, indicating that they have AIE features and great solid-state emission phenomena. In addition, they show good thermal stability and solubility. Meanwhile, these new silole derivatives also exhibit low lowest

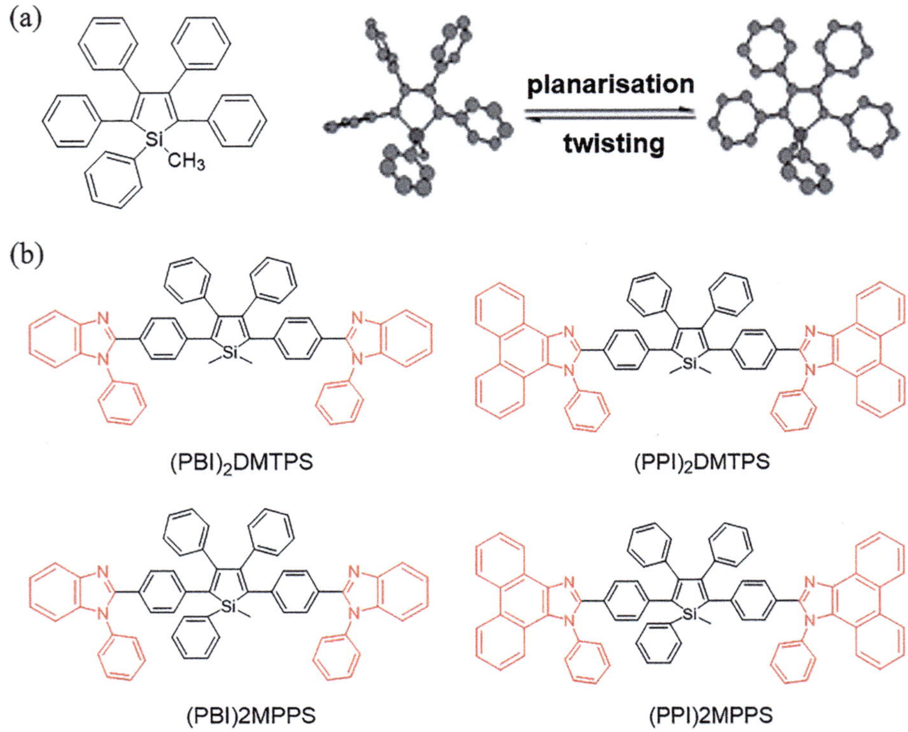

Scheme 4.1 Chemical structures of fluorescence molecules. (a) Chemical structure and conformational rotamers of 1-methyl-1,2,3,4,5-pentaphenylsilole. Reproduced from ref. 15 with permission from the Royal Society of Chemistry. (b) Chemical structures of silole derivatives.

unoccupied molecular orbital (LUMO) energy levels, which are typical π–π^* electron transitions. Based on their excellent solid-state PL properties, non-doped trilayer organic light-emitting diodes (OLEDs) of (PPI)2DMTPS show a great electroluminescence efficiency of 15.06 cd A^{-1}, 16.24 lm W^{-1} and 4.84%. For (PBI)2DMTPS, it can still demonstrate good electroluminescence efficiency (13.30 cd A^{-1}, 14.51 lm W^{-1} and 4.25%) as a double-layer OLED, which provides an efficient approach for synthesizing n-type light emitters.

Subsequently, a series of bis(2,4,5-trimethylphenyl)methane (DPM), 1,1,2,2-tetrakis-(2,4,5-trimethylphenyl)ethane (s-TPE-TM) and 1,1,2,2-tetraphenylethane (s-TPE) structures with isolated phenyl rings were investigated by Tang and co-workers.[24] It was found that all the molecules displayed typical AIE features in tetrahydrofuran (THF)/water mixture, even at longer excitation wavelengths. The s-TPE showed a longer wavelength emission at ~467 nm and an excellent fluorescence quantum yield of 70% under solid-state conditions. Such excellent fluorescence performance can be well rationalized by through-space conjugation between the phenyl rings (see Scheme 4.2a). The behavior of s-TPE-TM and DPM also proved that through-space conjugation contributes to the AIE phenomenon.

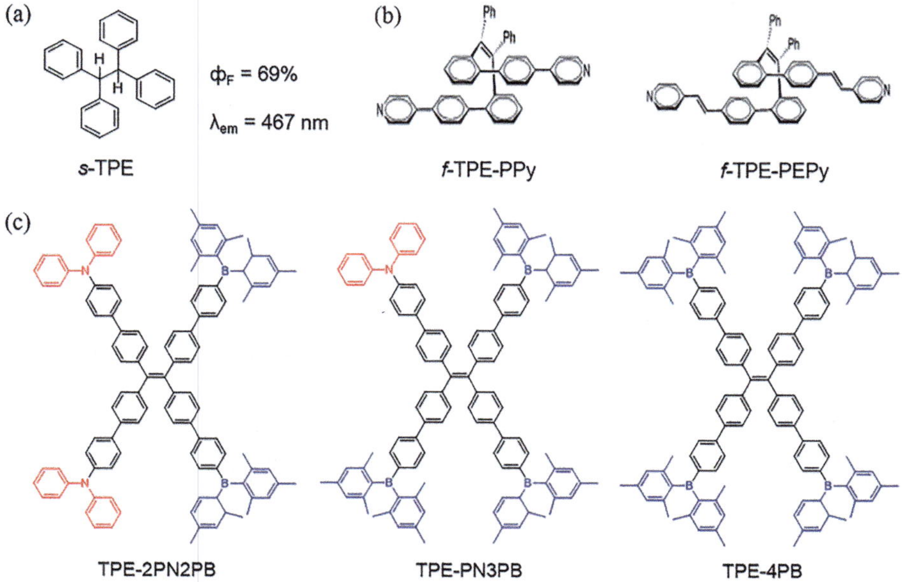

Scheme 4.2 Molecular structures of TPE derivatives. (a) The chemical structure of s-TPE. (b) Molecular structures of f-TPE-PPy and f-TPE-PEPy. Reproduced from ref. 25 with permission from John Wiley & Sons, Copyright © 2015 WILEY-VCH Verlag GmbH & Co. KGaA, Weinheim. (c) Chemical structures of TPE-2PN2PB, TPE-PN3PB and TPE-4PB.

Tang and co-workers[25] have also synthesized tetraphenylethene (TPE) derivatives (f-TPE-PPy/f-TPE-PEPy) with a folded conformation (see Scheme 4.2b). The folded molecules showed enhanced rigidity in comparison with l-TPE-PPy and l-TPE-PEPy. It was found that f-TPE-PPy and f-TPE-PEPy showed an obvious emission at 485 and 488 nm in THF solution. Also, the photoluminescence efficiency of the folded molecules can increase greatly, to 45.7% and 46.2% in solution, due to the suppression of intramolecular motion. From the scanning tunneling microscope (STM)-based break-junction technique and theoretical calculations, it can be found that there is through-bond and through-space conjugation in folded molecules, which indicates that a single-molecule junction can simultaneously contain two conducting channels.

Owing to its unique AIE properties, the tetraphenylethene (TPE) unit has been employed to prepare various tailored, star-shaped TPE derivatives (TPE-2PN2PB, TPE-PN3PB and TPE-4PB) using different proportions of diphenylamino and dimesitylboryl as donor and acceptor groups (see Scheme 4.2c).[26] Surprisingly, it is found that the fluorescence intensity of the TPE derivatives is weak initially, but highly emissive when aggregated in water. Moreover, the luminogens of TPE derivatives display excellent quantum yields of 95, 92 and 91% in the solid state, respectively, presenting a typical AIE characteristic. In addition, the thermal and morphological properties were investigated and showed good thermal stability and film-forming ability. According to these properties, these solution-processible AIE luminogens were successfully employed as luminescent materials in high-efficiency nondoped OLEDs with a good electroluminescence efficiency up to 8.3 cd A^{-1}.

While the unique AIE features of various molecules have been developed and investigated, the "always on" nature of the fluorescence would hinder *in situ* detection and long-term imaging. Motivated by these factors, Zhu's group reported quinoline-malononitrile (QM) units for the preparation of water-soluble AIE-active fluorescent nanoprobes, which can realize the extension of AIE fluorescence to red light and near-infrared (NIR) regions, and achieve a high-fidelity signal in cells. Various QM derivatives with controllable morphology, by varying the electron donor groups and thiophene π-bridge, were reported (see Figure 4.1a).[27] The results of this study showed that all QM compounds exhibited red to NIR AIE-active characteristics in a tetrahydrofuran/water (THF/H_2O) solvent system. In particular, the quantum efficiency of QM-1 and QM-2 derivatives was ~178/183-fold higher in the optimum ratio of THF/H_2O mixed solutions than in pure THF. The aggregation-induced luminescence properties may be attributed to the moderate interplanar angles between the ethylene and quinoline units, resulting in twisted conformations in the aggregated state. The characterization results indicated that QM-2 and QM-5 are typical representatives of rod-like and spherical shapes when aggregated, respectively. The spherical QM-5 was

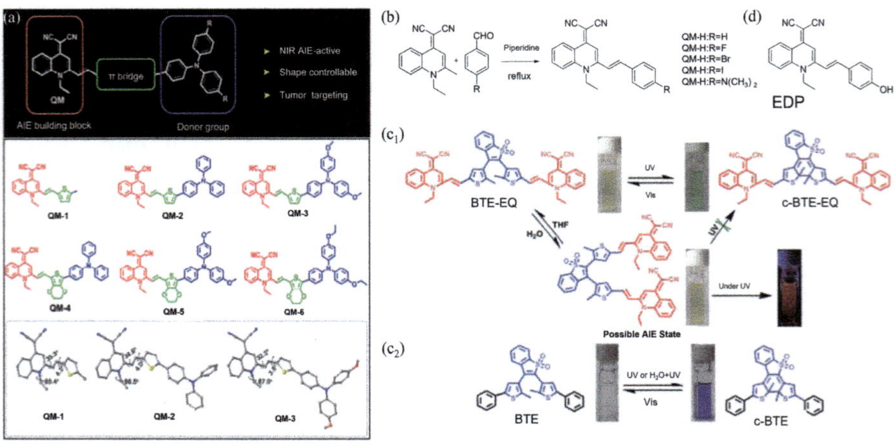

Figure 4.1 Schematic diagram of fluorescence properties and molecular structures of QM derivatives. (a) The structure of QM derivatives. Reproduced from ref. 27 with permission from John Wiley & Sons, Copyright © 2015 WILEY-VCH Verlag GmbH & Co. KGaA, Weinheim. (b) Structures of compounds QM-H, QM-F, QM-Br, QM-I and QM-N. (c_1) The images of BTE-EQ and c-BTE-EQ under UV and visible light, and (c_2) the fluorescence behavior of BTE under UV and visible light. Reproduced from ref. 29 with permission from the Royal Society of Chemistry. (d) Molecular structures of EDP.

first successfully applied to tumor-targeted bioimaging *in vivo* and provides a good platform for *in situ* and *in vivo* tumor imaging agents.

Additionally, Zhu and co-workers have reported and investigated various different substituents of quinoline-malononitrile AIE compounds (QM-H, QM-F, QM-Br, QM-I and QM-N) using dicyanomethylene-4*H*-pyran (DCM) as the precursor (see Figure 4.1b).[28] These compounds exhibit weak fluorescence in good solvents, but are highly emissive as aggregates at ~528–614 nm. In addition, they also exhibit solid-state luminescence effects in the powder state, with good fluorescence quantum efficiencies of 14.5%, 9.7%, 15.1%, 5.3% and 5.2%, respectively. It was found that the introduction of the electron-donating unit can form a donor (D)–π-acceptor (A) structure with the QM parent unit, and its enhanced push–pull electron effect extends the aggregation state wavelength and narrows the energy-level band gap.

After that, a new type of AIE-active photochromic dithienylethene derivative (BTE-EQ) was designed by connecting two quinoline-malononitrile (EQ) units to a dithienylethene core.[29] Upon addition of water, the BTE-EQ showed an increased fluorescence emission, displaying unique AIE properties (see Figure 4.1c_1 and c_2). It was also found that the BTE-EQ exhibited strong long-wavelength emission at 613 nm in the aggregated state. Specifically, Figure 4.1c demonstrates that the EQ unit was key for the AIE behavior of BTE-EQ as an AIE building block. These compounds also exhibited great solid-state enhanced red emission, which may be due to non-radiative deactivation and the suppression of close $\pi\cdots\pi$ packing of the molecules in the aggregated state. In addition, density functional theory (DFT) calculations indicated that the photochromic activity of BTE-EQ can be reversibly controlled by the aggregation state in aqueous solution.

Interestingly, Zhu and co-workers[30] have successfully synthesized novel AIE-active EDP nanoparticles (NPs) with desired sizes by a flash nanoprecipitation (FNP) method using QM as the AIE-active core (see Figure 4.1d). Under the optimized conditions of FNP, the obtained EDP NP could be adjusted to between 20 and 60 nm with ideal strong fluorescence performance. Upon addition of 70% water, the EDP showed an enhanced 60-fold fluorescence increase at 555 nm, in contrast to that in pure THF solution. What is more, with the addition of stream velocity, the EDP NPs also exhibited an obvious fluorescence enhancement while their size remained unchanged. These results could be ascribed to the fact that the EDP molecules were encapsulated in the hydrophobic moiety of the polymer, thus resulting in an enhanced fluorescence emission.

4.3 Clusteroluminogens

4.3.1 Synthetic Luminogens

Compared with organic molecules, the synthetic luminogens are usually free from aromatic components. These synthetic luminogens exhibit novel AIE features, which can be well explained by the CTE mechanism.[1]

4.3.1.1 Synthetic Polymers

Recently, various polymers have been developed and investigated. The design of luminogens is expected to display typical AIE phenomena in the aggregated state or concentrated solution. In this section, we will summarize non-aromatic polymers with simple carbonyl (C=O), cyanide (C≡N), Si–O and amine moieties.

Tang *et al.*[31] developed poly[(maleic anhydride)-*alt*-(vinyl acetate)] (PMV), a well-known non-conjugated alternating polymer with pure oxygen atoms (see Scheme 4.3). Surprisingly, PMV was shown to have excellent solvatochromism. This polymer also exhibits tunable absorption and PL emission by using different solvents. To test the mechanism, the polymerization of maleic anhydride (PMAh) and vinyl acetate (PVAc) were also studied. The PVAc was found to have no PL signal, whereas PMAh showed a strong PL emission. Moreover, there was also no color change and emission red shift throughout the whole experiment and the related experiments indicated that the VAc unit was necessary for the solvatochromism process, and the MAh cluster was the chromophore of PMV. Meanwhile, these units can interact with each other to generate a longer wavelength. These results indicate that the photoluminescence of PMV can be ascribed to the clustering of the locked carbonyl groups and the formation of a polymer/solvent complex.

Recently, in order to further investigate the CTE mechanism, Yuan and co-workers[32] reported non-aromatic polyurethanes (PUs) bearing carbamate (NHCOO) groups. It was observed that the PUs were basically non-emissive in dilute solutions, but showed a distinct blue emission in aggregated states. In addition, the prepared cast films displayed an intense luminescence with a quantum yield of 13.1%, revealing typical AIE features. Similarly, the absorption spectra showed an enhanced signal with increasing concentration. The photophysical properties might be attributed to the clustering of NHCOO groups, rigidified conformations and electronic communication between carbonyl (C=O) units, N and O atoms (see Scheme 4.4), which is consistent with the IR results. Moreover, room-temperature phosphorescence (RTP) was observed in solids and films of PUs, illustrating the potential for RTP emission from non-conventional macromolecular luminogens. Based on the luminescence properties, this polymer could be effectively used for the detection of explosives and cellular imaging.

Zhang and co-workers[33] oxidized 1-hexanamine (HA, a primary amine), *N*-ethylbutanamine (EBA, a secondary amine) and *N*,*N*-diethylethanamine (DEA, a tertiary amine), three isomeric small molecular aliphatic amines, by

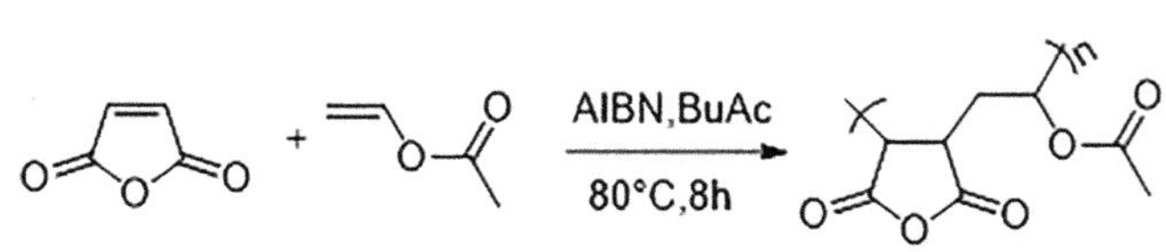

Scheme 4.3 Synthetic route to PMV.

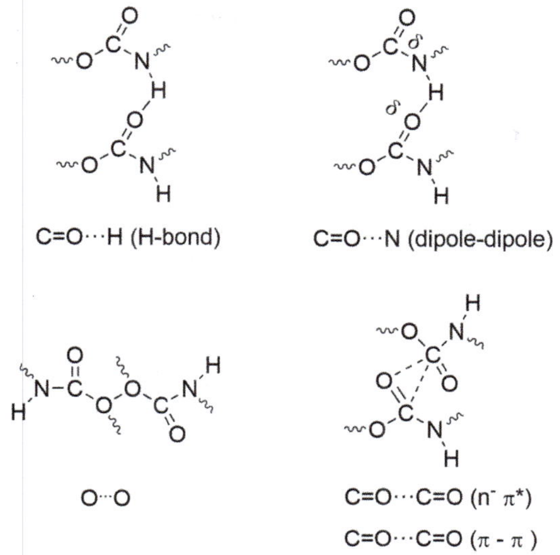

Scheme 4.4 Schematic diagram of the fluorescence mechanism of PUs.

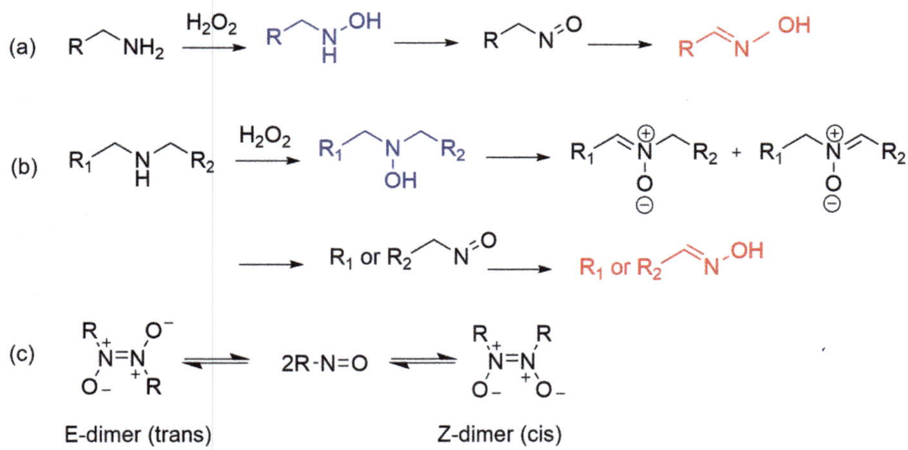

Scheme 4.5 Schematic diagram of oxidation of (a) primary amines and (b) secondary amines, and (c) preparation of *C*-nitroso compounds.

using H_2O_2 as an oxidant (see Scheme 4.5). The oximes were first verified to be the fluorescent substance of the oxidized primary and secondary amines. As expected, the hexanal oxime displayed an enhanced fluorescence emission in concentrated solutions and in the solid state. It was also found that the particle size of the aggregates positively correlated with the fluorescence intensity, exhibiting typical AIE characteristics. The single-crystal structure and theoretical calculations indicated that the intermolecular hydrogen bonding of the oxime groups helped to form extended conjugation and a rigid conformation, resulting in fluorescence enhancement in the aggregated state. Furthermore, the oximes were used to detect temperature and metal ions.

An unusual polyacrylonitrile (PAN) containing a polyethylene backbone and C≡N groups has been developed by Yuan and co-workers.[34] Upon addition of the concentration of PAN, a much brighter blue emission and an obvious fluorescence enhancement at 405 nm was observed. This molecule exhibited no fluorescence emission at 77 K in dilute solution, whereas it showed an enhanced emission in the concentrated state, indicative of the clustering of cyano units in the aggregated state. Subsequently, the fluorescence properties were investigated in mixed solutions to understand the CTE mechanism. It was found that PAN showed a remarkable fluorescence enhancement with increase of the non-solvent (DCM). Furthermore, the solid aggregates of PAN were shown to have a strong blue emission under room light, and a green, long-lasting phosphorescence at 77 K. These AIE features can be well explained by formation of cyano clusters together with through-space electronic interactions and the rigidified conformation (see Figure 4.2).

To further understand the CTE mechanism, Yuan's group developed and investigated the luminescence properties of non-aromatic

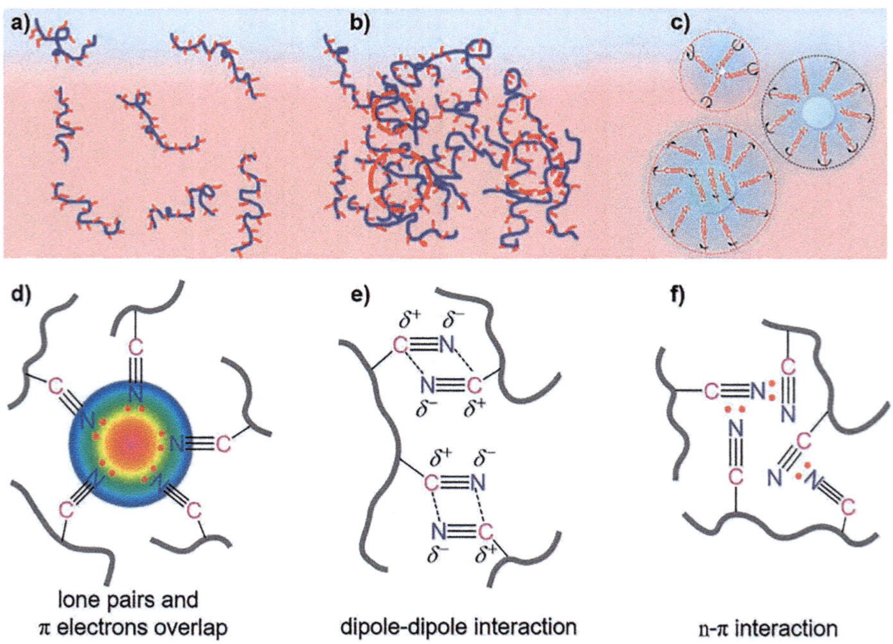

Figure 4.2 Schematic illustration of the mechanism for the PAN molecule. Pictures of PAN molecules in (a) isolated and (b) aggregated states; (c) cyano clusters in the aggregated states. Possible intra- and intermolecular interactions within cyano clusters: (d) electron overlap between lone pairs and π electrons, (e) dipole–dipole interactions, and (f) n–π interactions. Reproduced from ref. 34 with permission from John Wiley & Sons, Copyright © 2016 WILEY-VCH Verlag GmbH & Co. KGaA, Weinheim.

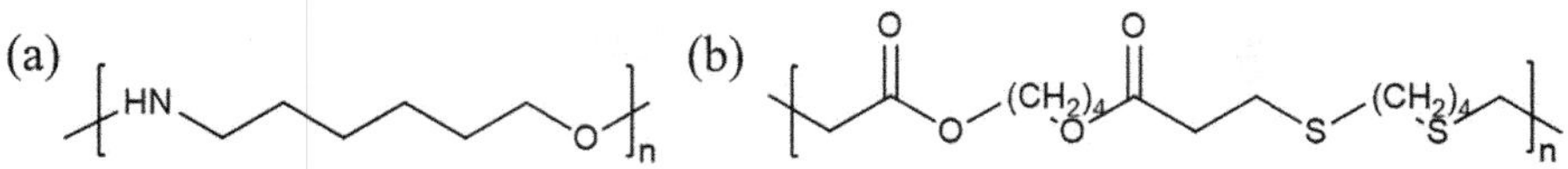

Scheme 4.6 Chemical structures of fluorescent compounds. The structures of non-aromatic polyamide-6 (PA-6) (a) and (b) sulfur-containing polythioether (P1).

polyamide-6 (PA-6) by using THF and dimethylformamide (DMF) solvents (see Scheme 4.6a).[35] The PA-6 was found to have a weak fluorescence emission in dilute solution, but showed a visible emission when the formic acid (FA) solution was increased. The prepared PA-6a/FA and PA-6b/FA displayed a bright green and blue emission in concentrated solution and solid state, respectively. The formation of variable clusters of PA-6/FA was responsible for the luminescence, which was confirmed by the dynamic light scattering (DLS) and scanning electron microscopy (SEM) results. Furthermore, the PA-6 cast film and PA-6 electrospun film with α (antiparallel molecular arrangement) and γ (parallel molecular arrangement) crystals showed a distinct blue fluorescence under UV light, which further confirmed that the luminescence is associated with the molecular arrangement. The highly regular molecular arrangement contributed to the aggregation of amide clusters and conformation rigidification, thus promoting the enhanced fluorescence emission.

A sulfur-containing polythioether (P1) was developed for studying the luminescence mechanism (see Scheme 4.6b).[36] P1 was found to exhibit an enhanced fluorescence emission and absorption spectra with increase in concentration from 5 to 100 mg L^{-1}, indicating typical AIE characteristics. It was found that the clustering of COO groups and S units might contribute to the AIE properties. To further confirm the CTE mechanism, the polysulfoxide (P2) and polysulfone (P3) were prepared and showed excellent fluorescence emission in concentrated solution and the solid state. For comparison, P2 and P3 displayed a higher quantum yield than P1. Such a result strongly demonstrated that the fluorescence enhancement was related to the conformation rigidification and the intra- and intermolecular electronic interactions in the clusters. Furthermore, P1, P2 and P3 were successfully employed for the detection of explosives, providing guidance for constructing non-aromatic polymers and expanding their potential applications.

4.3.1.2 Nano-based Luminogens

The excellent AIE fluorescence of some nano-based luminogens has been observed for a long time, due to their unique luminescence properties and great photostability. In this section, we will focus on two sorts of nano-based luminogens: nanodots and metal nanoparticles.

Nanodot-based luminogens, including Cu–In–S quantum dots (CIS QDs),**37** graphene quantum dots[38,39] and carbon dots,[40,41] are widely developed and investigated owing to their tunable photoluminescence and chemical stability. Mou and co-workers designed hydrophilic CIS QDs to investigate the AIE mechanism.[42] The transmission electron microscopy (TEM) images showed that the CIS QDs were uniformly dispersed spherical nanoparticles with an average diameter of 2.8 ± 0.4 nm in water. The CIS QDs were found to be non-emissive, whereas they displayed a remarkable fluorescence enhancement in the presence of Zn^{2+}. Compared with the original CIS QDs, the Cu–In–S QDs-Zn^{2+} displayed a higher quantum yield of 24%. In this CIS QDs-Zn^{2+} complex, the addition of Zn^{2+} can trigger the CIS QDs into three-dimensional (3D) networks *via* self-assembly, which was proved by the SEM and TEM results. The cross-linked aggregates with a 3D network structure would restrict the motion of CIS QDs and decrease the non-radiative decay, thus boosting the fluorescence emission (see Figure 4.3). According to the metal-enhanced fluorescent effect, the Cu–In–S QDs were exploited to detect the Zn^{2+} as a nanosensor. This nanosensor showed good sensitivity, selectivity and a linear dependence on Zn^{2+}, revealing its potential application in real samples.

In addition, graphene quantum dots (GQDs) have gained wide interest due to their significant fluorescence emission and non-toxicity. Single-layered graphene quantum dots (sGQDs) were reported *via* a self-exothermic system. The synthesized sGQDs showed a broad absorption band and a fluorescence emission at 463 nm. With increase in the concentration of Al^{3+}, the fluorescence intensity of sGQDs was significantly enhanced. This observation

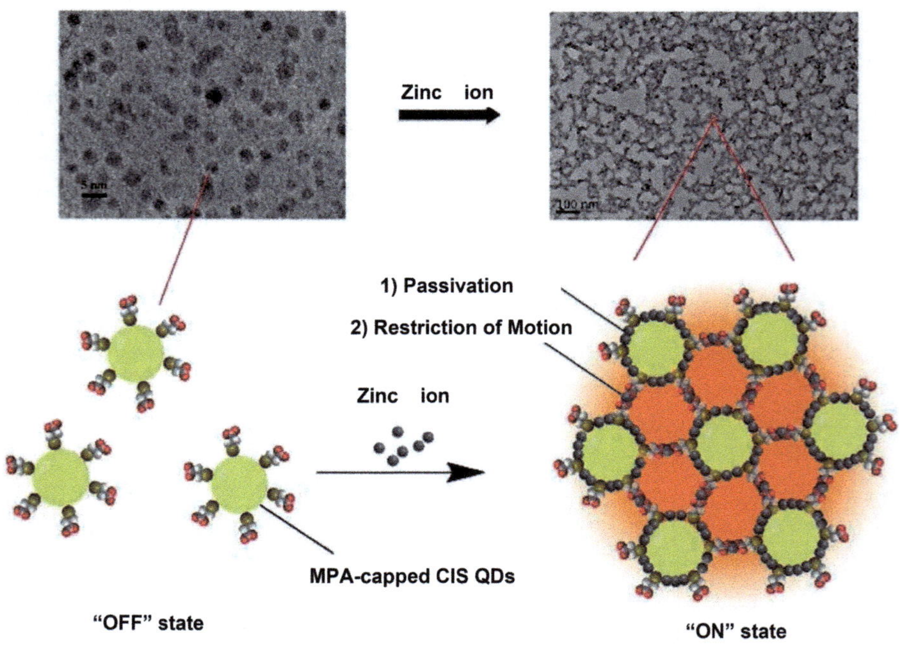

Figure 4.3 Schematic diagram of the luminescence mechanism of CIS QDs. Reproduced from ref. 42 with permission from Elsevier, Copyright 2019.

presents the opportunity for sGQDs to be used as a probe to detect Al^{3+}. Remarkably, the sGQDs displayed high sensitivity, a linear response and excellent selectivity for Al^{3+}. The AIE phenomenon is associated with the formation of sGQD aggregates, which was verified using TEM (see Figure 4.4a and b). The change of lifetime and quantum yield further revealed that the aggregation of sGQDs reduced the non-radiative rate, thus producing an AIE

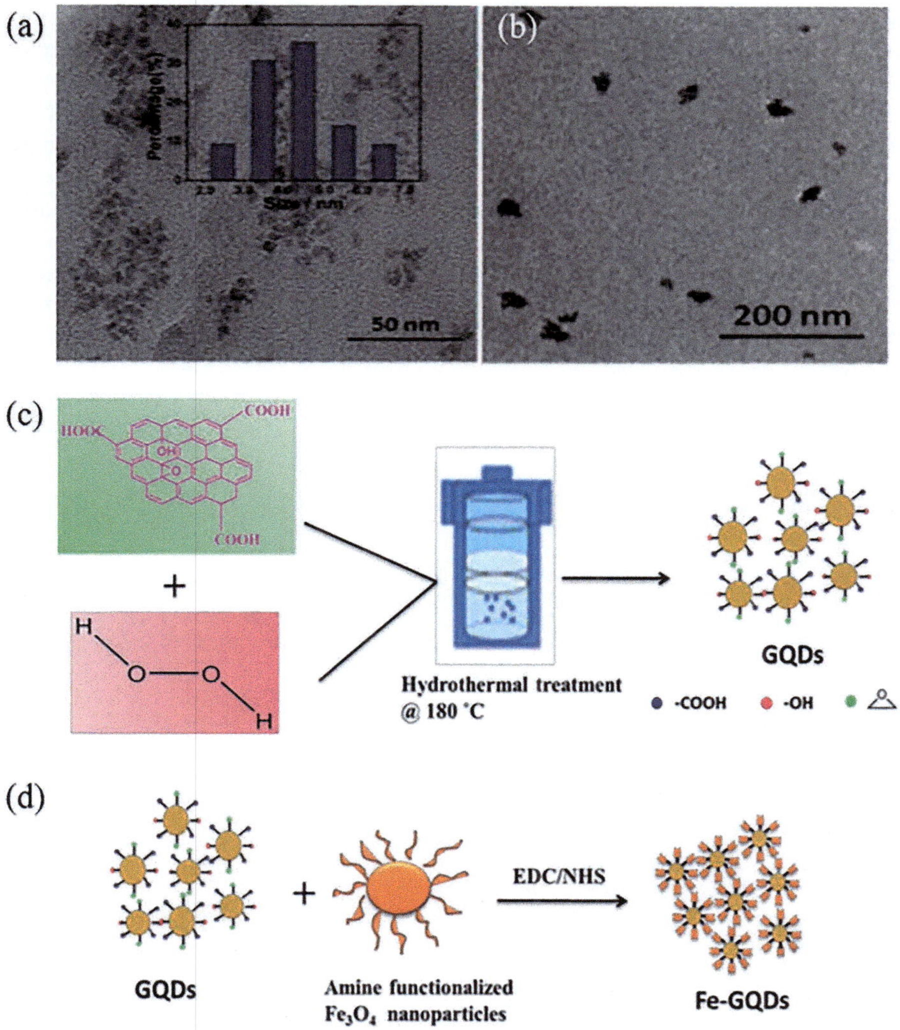

Figure 4.4 Schematic diagram of the fluorescence performance of graphene quantum dots (GQDs). (a, b) The TEM images of s-GQDs before and after addition of Al^{3+}. Reproduced from ref. 39 with permission from the Royal Society of Chemistry. (c, d) Schematic for the preparation of Fe-GQDs. Reproduced from ref. 43 with permission from the Royal Society of Chemistry.

effect. These results suggest that methods for sensing metal ions based on AIE effects are useful for practical samples.[39]

Pathan and co-workers reported magnetic GQDs (Fe-GQDs) for the detection of As^{3+} ions.[43] It was found that the Fe-GQDs showed significant fluorescence enhancement upon addition of As^{3+} ions. Notably, the Fe-GQDs displayed a linear response to the As^{3+}, with a limit of detection of 5.1 ppb. The Fe-GQDs were also found to have excellent selectivity and anti-interference for As^{3+} ions over other, coexisting ions (see Figure 4.4c and d). Moreover, the Fe-GQDs-As^{3+} showed a longer fluorescent lifetime than the Fe-GQDs. These results may be related to the unusual "aggregation-induced increase" effect. The zeta potential, DLS, X-ray photoelectron spectroscopy and Raman data further demonstrated that the fluorescence enhancement may be ascribed to restriction of intramolecular rotation caused by association of the Fe-GQDs-As^{3+}.

In addition, various carbon dots (CDs) have received extensive research interest due to their good water solubility and biocompatibility.[44] Recently, a novel CD was developed through the hydrothermal method using *o*-phenylendiamine (OPD) as the carbon source. The prepared OPD-CDs showed a fluorescence emission at 568 nm and an obvious yellow color under UV light. Upon addition of Cu^{2+}, the fluorescence emission of the OPD-CDs was significantly enhanced, revealing a typical AIE feature (see Figure 4.5a). This AIE phenomenon may be related to the coordination between Cu^{2+} and amino groups of the OPD-CDs, which can trigger aggregation of OPD-CDs and then restrict the intramolecular motion, together with reduction of the non-

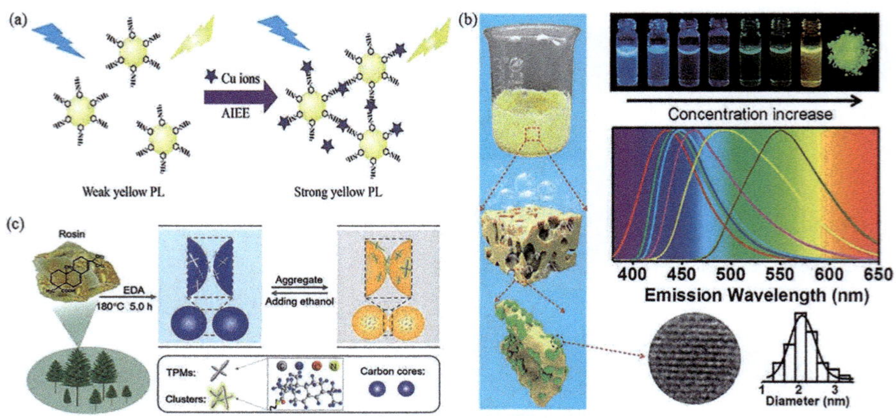

Figure 4.5 The fluorescence performance of carbon dots (CDs). (a) The fluorescence response of OPD-CDs to Cu^{2+}. Reproduced from ref. 45 with permission from Elsevier, Copyright 2019. (b) The fluorescence behavior of CDs. Reproduced from ref. 46 with permission from American Chemical Society, Copyright 2019. (c) Schematic diagram of the preparation of carbon dot-based AIEgens (CD-AIEgens) and their fluorescence behavior. Reproduced from ref. 47 with permission from Elsevier, Copyright 2021.

radiative rate. This mechanism was further confirmed by the quantum yield and lifetime results for the OPD-CDs. As a nanosensor, OPD-CDs were found to display great sensitivity to Cu^{2+}, with a limit of detection of 0.28 μmol L^{-1}. In addition, it was found that only Cu^{2+} can enhance the luminescence, thus showing high selectivity. The unique CD-AIE-based sensing characteristics enabled the OPD-CDs to be exploited for sensing Cu^{2+} in complex biological and environmental samples.[45]

Zhang and co-workers prepared yellow emissive CDs *via* a one-pot microwave heating approach.[46] As expected, these CDs showed an obvious red shift from 435 to 550 nm with increase in concentration. As shown in Figure 4.5b, the color of the CDs also changed from blue to bright yellow under UV light, showing a strong AIE behavior. In addition, the CDs were found to have an excellent quantum yield of 65% in the solid state. The absorption spectra and fluorescence lifetime of the CDs demonstrated that the observed AIE properties are associated with the occurrence of Förster resonant energy transfer (FRET) from dispersed CDs to agglomerations. Using this AIE mechanism, the CDs were used as a color-converting layer for constructing white LEDs, which provides a promising potential for a wide range of applications.

More recently, Ge and co-workers reported carbon dot-based AIEgens (CD-AIEgens) from sustainable rosin (see Figure 4.5c).[47] The prepared CD-AIEgens showed an obvious fluorescence enhancement upon addition of water up to 70%. Moreover, the CD-AIEgens were found to have an increased fluorescence lifetime and fluorescence quantum yield. Such a result may be related to the occurrence of clusters on the surface of the CD-AIEgens. The effect of temperature on the CD-AIEgens was also examined to investigate the fluorescence mechanism. It was found that the CD-AIEgens displayed reversible fluorescence under heating and cooling. At lower temperatures, the clusters of non-planar tricyclic phenanthrene moieties (TPMs) are promoted, which can increase radiative decay and lead to higher fluorescence intensity. Besides, with increasing the ratio of glycerol to ethanol solution, the CD-AIEgens generated an increased fluorescence intensity owing to the formation of clusters in the more viscous solutions. According to the clustering-induced emission, the CD-AIEgens displayed concentration-dependent fluorescence enhancement and a distinct blue emission under ethanol solutions and in the solid state, respectively. Furthermore, the switchable fluorescence emission of CD-AIEgens suggest that they can be used for robust two-switch-mode luminescence inks.

In addition to nanodots, a series of nano-based metal nanoparticles were developed that showed aggregation-induced enhancement. These AIE-active luminogens usually contain Au,[48,49] Cu nanoparticles[50,51] and Ag.[52] A highly luminescent Au-SR complex was reported by exploiting glutathione (GSH)-coordinated Au complexes and nanoclusters (NCs) as a model through a carbon monoxide reduction method.[53] The related molecular-level purity and cluster formula of the as-purified Au NCs were confirmed by using electrospray ionization mass spectrometry (ESI-MS) (see Figure 4.6a). The $[Au_{10}SR_{10}]$, $[Au_{15}SR_{13}]$, $[Au_{18}SR_{14}]$ and $[Au_{25}SR_{18}]$ were found to display fluorescence

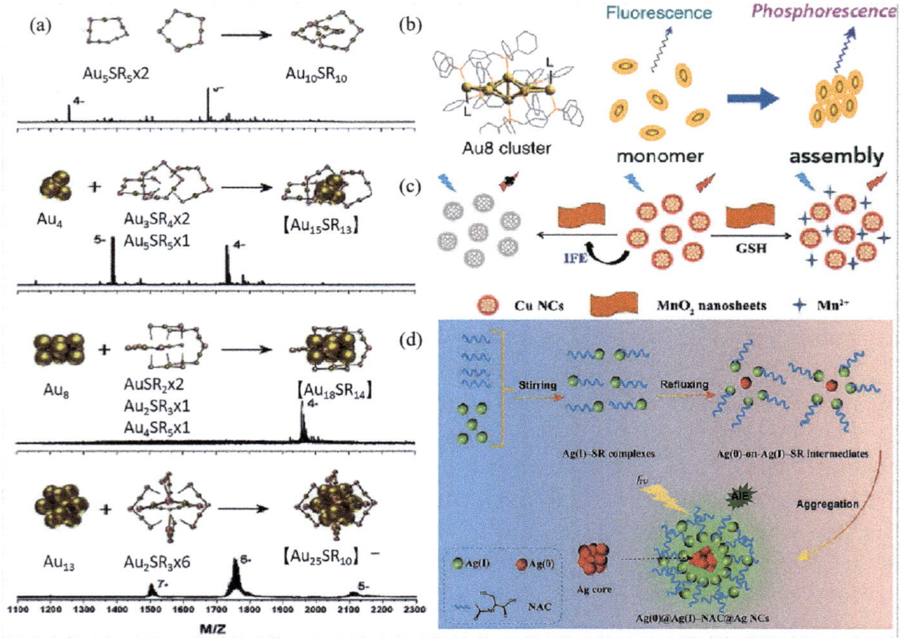

Figure 4.6 Schematic diagram of the structures and AIE performance of metal-based AIEgens. (a) Schematic illustration of the electrospray ionization mass spectrometry spectra for $[Au_{10}SR_{10}]$, $[Au_{15}SR_{13}]$, $[Au_{18}SR_{14}]$, and $[Au_{25}SR_{18}]$. Reproduced from ref. 53 with permission from John Wiley & Sons, Copyright © 2020 Wiley-VCH Verlag GmbH & Co. KGaA, Weinheim. (b) Fluorescence and phosphorescence-type performance of the Au_8 clusters (1–3) in solution. Reproduced from ref. 49 with permission from American Chemical Society, Copyright 2017. (c) The fluorescence behavior of Cu NCs for sensing of GSH. Reproduced from ref. 54 with permission from the Royal Society of Chemistry. (d) The preparation of Ag NCs. Reproduced from ref. 55 with permission from Elsevier, Copyright 2021.

emissions at 658, 660, 670 and 987 nm upon excitation at 450 nm. Moreover, with the addition of ethanol, $[Au_{10}SR_{10}]$, $[Au_{15}SR_{13}]$ and $[Au_{18}SR_{14}]$ showed an obvious fluorescence enhancement. $[Au_{25}SR_{18}]$ displayed no apparent dependence of emission intensity on the mixed solvent. These results indicated that Au^I-SR motifs were key for the AIE behavior of Au NCs. It was found that Au^I-SR motifs can not only reduce the AIE intensity of Au-SR complexes owing to the shortened Au^I-SR motifs, but also enhance the luminescence by inhibiting the non-radiative decay pathway. On the other hand, the aggregation of complexes can facilitate aurophilic $Au^I \cdots Au^I$ interactions, thus showing enhanced fluorescence.

The consequences of cluster aggregation have been well investigated for aggregation-induced emission. Sugiuchi and co-workers[49] reported Au_8 clusters ($[Au_8(dppp)_4L_2]X_2$, 1–3) using a bitetrahedral Au_6 core with two additional gold atoms ("[core + exo]-type"). The Au_8 clusters exist as monomers in solvents, exhibiting an obvious fluorescence intensity at ~600 nm. Upon aggregation, a red shift and a new fluorescence band at ~700 nm are observed, leading to variation of the photoluminescence emission mode from fluorescence to phosphorescence (see Figure 4.6b). The related fluorescence and phosphorescence characteristics were confirmed by the lifetime measurements. Moreover, the Au_8 clusters exhibited a higher quantum efficiency in the solid state than in solution. These results indicated that the effect of the phosphorescence does not simply rely on the monomer. This observation is probably a result of the formation of closely packed cluster assemblies, thus showing an enhanced phosphorescence emission.

Du and co-workers developed AIE-active red-emitting Cu NCs for the detection of glutathione (GSH) (see Figure 4.6c).[54] The Cu NCs were shown to have a large Stokes shift of 235 nm. Moreover, the Cu NCs were found to show an enhanced fluorescence emission upon addition of ethanol solution, a characteristic AIE feature. The TEM images demonstrated that the AIE behavior of Cu NCs may be related to Cu NC aggregation. According to the remarkable fluorescence properties, a rapid switch-on sensor for GSH was prepared by introducing MnO_2 nanosheets to the Cu NCs. The sensor showed great sensitivity and selectivity for sensing of GSH. The observed fluorescence enhancement may be ascribed to the redox reaction of MnO_2 nanosheets to Mn^{2+}, thus restoring the quenched fluorescence of the Cu NCs. The above observations provide a novel strategy for further applications of AIE-active Cu NCs in biological systems.

More recently, Zhang and co-workers reported silver nanoclusters (Ag NCs) with core–shell structure *via* a reflux method (see Figure 4.6d).[55] The synthesized Ag NCs showed a fluorescence emission at 494 nm and a distinct green emission under UV light. It was also found that the Ag NCs showed a longer lifetime of 11.07 μs, demonstrating a typical AIE performance. Such an AIE feature may be associated with aggregates of Ag^{I}-SR complexes in the Ag NCs, which was confirmed by the TEM images. In addition, the Ag NCs were exploited to recognize tetracycline (TC) based on the internal filtration effect and the static quenching effect. Notably, they exhibited excellent sensitivity and selectivity for TC, and were employed for the detection of TC in real milk samples.

Zhu and co-workers developed a chiral Ag–Au bimetallic nanocluster (Au_4Ag_5) through the co-reduction method.[56] The ESI-MS characterization of the Au_4Ag_5 nanocluster confirmed that its precise composition was $[Au_4Ag_5(Dppm)_2(SAdm)_6](BPh4)$, which was consistent with the crystal data. The prepared Au_4Ag_5 nanocluster was found to be non-fluorescent in solution, but demonstrated a strong fluorescence emission in the aggregated state and the solid state. In addition, the film state of the Au_4Ag_5 nanocluster was investigated and displayed a red photoluminescence under UV excitation. The AIE effects of the Au_4Ag_5 nanocluster may be associated with restriction of intramolecular motion. Namely, the tight binding of the left- and right-handed enantiomers can cause a strong restriction of intramolecular motion, thus resulting in aggregation-induced emission behavior (see Figure 4.7a and b).

Liu and co-workers reported Cu nanoclusters (S1), Ag(I)-doped Cu/Ag nanoclusters (S2) and D-penicillamine-coated Cu/Ag alloy nanoclusters (S3) *via* a self-assembly strategy (see Figure 4.7c).[57] The prepared metal nanoclusters (S1–S3) displayed a significant absorption band (Figure 4.7d). These metal nanoclusters showed a strong fluorescence emission with a hypsochromic shift from 615 to 570 nm upon excitation at 360 nm. The related luminescence quantum yields of S1–S3 were determined to be 11.4–14.2%. Moreover, they also displayed great emission, with a distinct color change from red to orange in the solid state. These results demonstrated that the introduction of Ag(I) ions contributed to the Cu/Ag NC structure becoming more compact from S1 to S3, which would further promote the radiative transition of ligand-to-metal charge transfer (LMCT) and/or ligand-to-metal–metal charge transfer (LMMCT), thus producing an AIE effect. Based on the selective

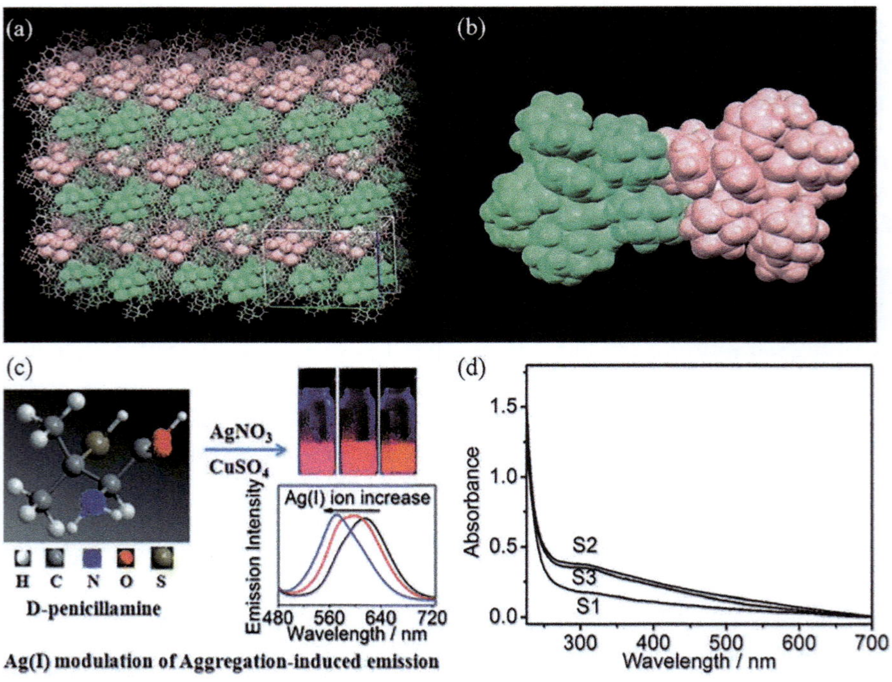

Figure 4.7 Schematic illustration of (a) ensemble packing of the Au_4Ag_5 nanocluster in the *C*12/*c*1 space group (left-handed enantiomers: pink; right-handed enantiomers: green), and (b) detailed packing information of single left- and right-handed enantiomers. Reproduced from ref. 56 with permission from John Wiley & Sons, Copyright © 2018 Wiley-VCH Verlag GmbH & Co. KGaA, Weinheim. (c) Schematic diagram for the preparation of Cu/Ag nanoclusters and the Ag(I) modulation effect of aggregation-induced emission. (d) Absorption spectra of metal nanoclusters. Reproduced from ref. 57 with permission from the Royal Society of Chemistry.

modulation property of Ag^+, the Cu/Ag NCs were employed to detect Ag^+ and halide anion in real water samples *via* a ratiometric luminescence method.

4.3.2 Natural Luminogens

Although important developments have been made, synthetic luminogens require complex preparation and expensive chemicals, which could compromise their potential applications. Motivated by this factor, various natural luminogens obtained from biomass have been reported for investigating fluorescence behavior. These luminogens can be mainly divided into conventional natural luminogens and non-conventional luminogens.

Lignin, as a natural biomass, has gained extensive attention in academic and industrial research owing to its abundant aromatic rings. To explore its fluorescence performance, Ma and co-workers reported cellulolytic enzyme lignin nanoparticles (CEL-NPs) using a one-step self-assembly method (see Figure 4.8a).[58] As shown in Figure 4.8b, a significant fluorescence

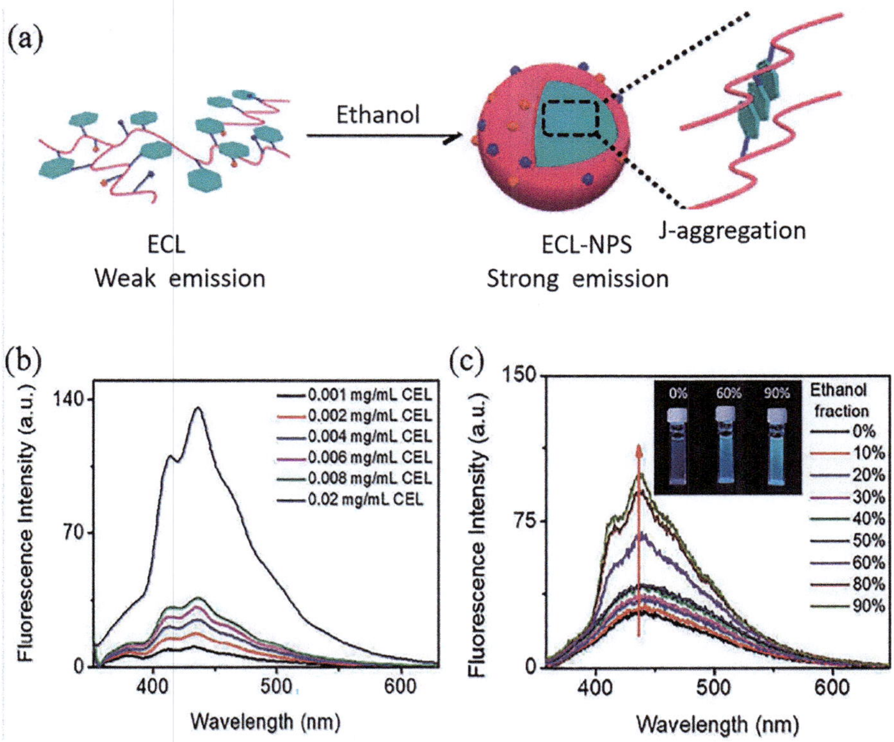

Figure 4.8 Schematic diagram of the fluorescence properties of cellulolytic enzyme lignin nanoparticles (CEL-NPs). (a) The preparation procedure for CEL-NPs. (b) Fluorescence intensity of different concentrations of CEL-NPs. (c) Fluorescence intensity of CEL-NPs in water and ethanol (inset: photographs of CEL-NPs in mixtures under UV light). Reproduced from ref. 58 with permission from American Chemical Society, Copyright 2018.

enhancement was observed upon addition of concentrations of CEL, which provides positive evidence for the J-aggregation and fluorescence behavior of the CEL-NPs. As expected, with increasing ethanol to water ratio, the CEL-NPs were found to have an increased fluorescence emission, showing a strong AIE-active fluorescence (see Figure 4.8c). This AIE feature may be associated with the formation of J-aggregates. Based on this great fluorescence behavior, CEL-NPs also displayed temperature dependence and good resistance to photobleaching compared to the commercially available 4′,6-diamidino-2-phenylindole (DAPI) dye. The observations for CEL-NPs offer insight for investigating the mechanism of natural luminogens.

The same group also developed a quercetin (QC) from *Sophora japonica*, showing a remarkable AIE and excited-state intramolecular proton transfer (ESIPT) (see Figure 4.9).[59] In the presence of water, the QC showed an obvious enhancement in the keto emission and reduction in the enol emission (see Figure 4.9b). This suggested that the keto emission is related to the AIE properties. With increasing concentration of QC, a similar effect was also found. More concentrated quercetin solutions displayed a higher AIE behavior (see Figure 4.9c) owing to the formation of aggregates. What is more, the QC showed a significant AIE fluorescence upon addition of water, which demonstrates that the fluorescence emission is related to the concentration of QC. The TEM images further confirmed that the AIE performance of QC

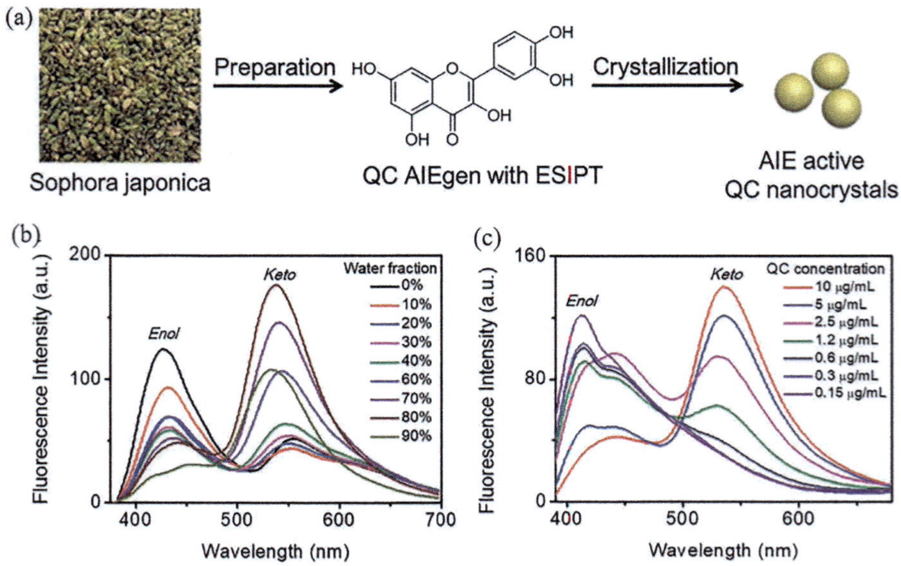

Figure 4.9 Schematic diagram of fluorescence performance for the quercetin (QC) nanocrystals. (a) Preparation of QC nanocrystals. (b) Fluorescence intensity of QC nanocrystals in different fractions of water. (c) Fluorescence intensity of different concentrations of QC. Reproduced from ref. 59 with permission from John Wiley & Sons, Copyright © 2018 WILEY-VCH Verlag GmbH & Co. KGaA, Weinheim.

was attributed to the crystallization-induced keto emission. Furthermore, the QC displayed strong resistance to photobleaching.

Lei and co-workers developed chelerythrine (Che) and sanguinarine (San) with AIE fluorescence.[60] Upon addition of a water fraction of up to 80%, San showed a significant fluorescence enhancement at 582 nm (see Figure 4.10a). Che also exhibited an increased fluorescence emission at 570 nm with an increase in the water fraction from 30 to 90% (see Figure 4.10b). In addition, both San and Che displayed a distinct color change form dark blue to yellow under UV light, showing a typical AIE feature. Such a result may be ascribed to the J-aggregation. Furthermore, San and Che showed a pH response owing to the presence of non-ionic forms and quaternary ammonium salt forms. These two natural products displayed great cycle performance for sensing pH (Figure 4.10c, d). According to the AIE and pH-sensitive properties, Che showed a great cellular permeability and good mitochondrial targeting.

Recently, Xu and co-workers reported a natural xanthan gum (Xag) from polysaccharide.[61] As expected, the Xag displayed an enhanced fluorescence emission and a bright blue emission with increase of the THF fraction. In Figure 4.11a, the Xag shows a similar fluorescence enhancement and a visible color change in the concentrated state. Moreover, the luminescence

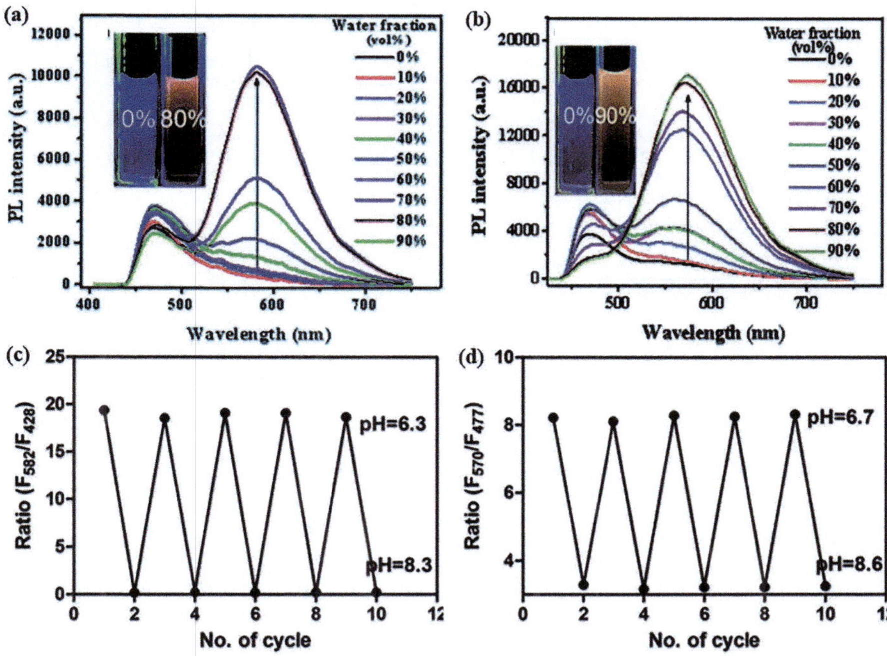

Figure 4.10 Schematic illustration of fluorescence performance of chelerythrine (Che) and sanguinarine (San). (a and b) Fluorescence intensity of San and Che at different water fractions. (c and d) Cycle performance of Che and San. Reproduced from ref. 60 with permission from the Royal Society of Chemistry.

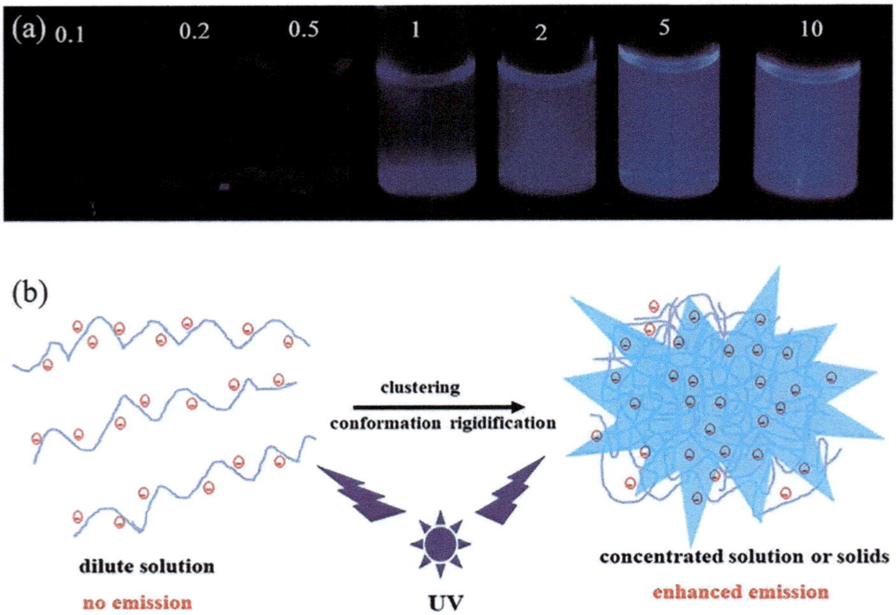

Figure 4.11 Schematic diagram of fluorescence property of xanthan gum. (a) Images of different concentrations of xanthan gum (mg L^{-1}) under UV light. (b) The fluorescence mechanism of xanthan gum. Reproduced from ref. 61 with permission from Elsevier, Copyright 2021.

lifetimes of Xag were calculated to be 3.27 ns and 17.23 ms in aqueous and solid state, respectively. Xag also showed a higher quantum yield in the solid and concentrated state. These results further demonstrate that Xag showed the unique AIE feature. To further verify the AIE mechanism, the SEM and rheological viscosity of Xag were investigated. In the presence of THF, a series of aggregates of about 100 nm can be found in the SEM images for Xag, displaying a clustering-triggered emission. The increased viscosity of Xag further indicated that the AIE performance may be related to electronic conjugation between the chains (see Figure 4.11b). Using this excellent fluorescence feature, Xag shows potential applications in intracellular imaging and metal ion sensing.

Cellulose-based AIEgens have also been developed for investigating the fluorescence behavior.[62] Recently, our group reported a non-conjugated natural luminogen, carboxymethylated nanocellulose (C-CNC).[63] The C-CNC showed a strong fluorescence emission in concentrated solution and in the solid state, displaying typical CTE features. In addition, the quantum yields of C-CNC in the solid state and aqueous solutions were calculated to be 7.8% and 12.3%, respectively. The DFT calculation results demonstrated that the $n \rightarrow \pi*$ interaction and through-space electronic interactions are responsible for the fluorescence emission (see Figure 4.12). To further explore the

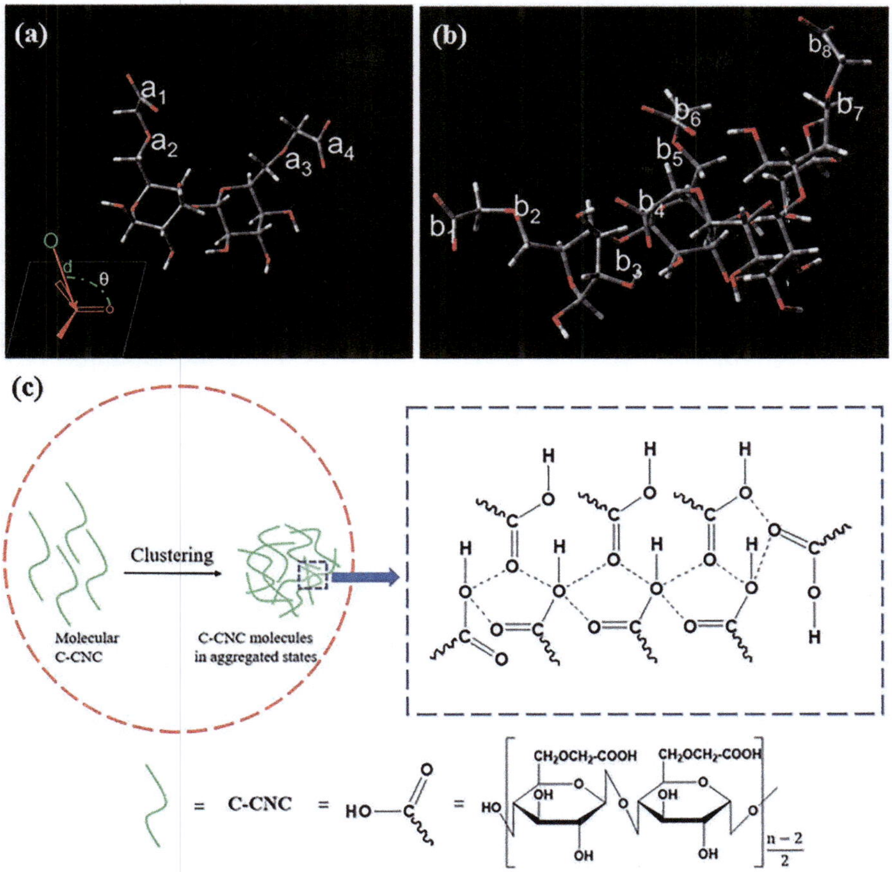

Figure 4.12 Schematic diagram of fluorescence mechanism of C-CNC. (a) Optimized conformations of C-CNC (a) (inset: proposed model of n $\rightarrow \pi^*$ interaction and C-CNC dimer) and in water (b) (red sticks, oxygen atoms; gray sticks, carbon atoms). (c) Images of the fluorescence mechanism of C–CNC. Reproduced from ref. 63, https://doi.org/10.3389/fchem.2019.00447, under the terms of the CC BY 4.0 license https://creativecommons.org/licenses/by/4.0/.

fluorescence mechanism, the C-CNC was "locked" using ethylenediamine. The result also verified that the CTE was related to the amide moieties and through-space conjugation between oxygen and carbonyl groups, which is consistent with the proposed mechanism. It is expected that C-CNC will provide a possible mechanism for preparing and understanding non-conjugated luminogens.

In addition, sodium alginate (SA) containing mannuronic (M) and guluronic acids (G) was developed for investigating emission behavior (see Figure 4.13a).[20] The SA was found to be non-emissive in dilute solution, but highly emissive when aggregated in the concentrated state, solid and film.

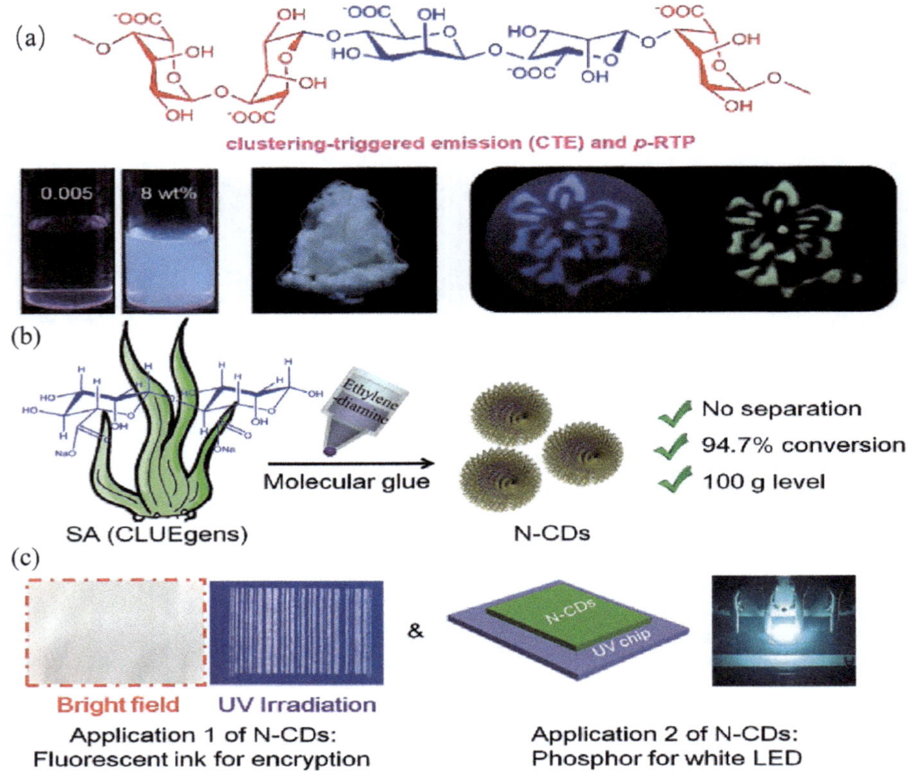

Figure 4.13 Schematic diagram of the fluorescence behavior of sodium alginate (SA) and N-CDs. (a) The clustering-triggered emission of SA. Reproduced from ref. 20 with permission from American Chemical Society, Copyright 2018. (b) Schematic diagram of preparation of N-CDs. Reproduced from ref. 64 with permission from American Chemical Society, Copyright 2019. (c) The applications of N-CDs. Reproduced from ref. 64 with permission from American Chemical Society, Copyright 2019.

In concentrated solutions, the GM_r and G_rM displayed a distinct grey-white emission under UV light with quantum yield values of 3.9%/3.1%. The good fluorescence properties may be ascribed to the CTE mechanism. Essentially, the clustering of oxygen atoms and carboxylate atoms could afford electronic interactions, facilitating the through-space electronic conjugation between the n and π electrons, together with conformation rigidification. In addition, the amorphous SA powders showed a strong blue and green phosphorescence after UV light irradiation and cessation of the UV light, displaying a room-temperature phosphorescence feature. Using this remarkable, excellent feature, SA exhibited potential applications for bioimaging.

Inspired by the clustering-induced emission of SA, Wang and co-workers reported carbon dots (N-CDs) using a molecular glue strategy (see Figure 4.13b).[64] The carbon dots were prepared by using sodium alginate and ethylenediamine *via* a hydrothermal method. It was found that a conversion yield of N-CDs of 94.7% can be achieved. The N-CDs displayed a fluorescence emission at 350–580 nm under excitation of 300–450 nm. Moreover, the N-CDs showed a high quantum yield of 39%. Such excellent properties may be ascribed to the through-space conjugation between oxygen atoms and carbonyl moieties, which was confirmed by DFT calculations. Furthermore, the N-CDs were employed for encryption and a white light LED (see Figure 4.13c).

4.4 Conclusions

AIE-based fluorescent luminogens have gained significant attention due to their structure tunability, biocompatibility and multiple functionalities. In this chapter, we have summarized recently recognized classes of fluorescent luminogens that exhibit AIE. Herein, the AIE fluorescence properties and the underlying mechanisms of these AIE luminogens containing organic molecules and clusteroluminogens are systematically discussed and investigated. It is hoped that this chapter can provide insight into such AIE-based fluorescent systems and contribute to a breakthrough in understanding of the luminescence mechanism. In addition, these studies offer promising directions for practical applications.

References

1. W. Z. Yuan and Y. Zhang, *J. Polym. Sci., Part A: Polym. Chem.*, 2017, **55**, 560–574.
2. H. Lu, Z. Hu and S. Feng, *Chem. – Asian J.*, 2017, **12**, 1213–1217.
3. R. Hu, N. L. C. Leung and B. Z. Tang, *Chem. Soc. Rev.*, 2014, **43**, 4494–4562.
4. M. Luo, S. Wang, M. Wang, S. Huang, C. Li, L. Chen and X. Ma, *Dyes Pigm.*, 2016, **132**, 48–57.
5. C. Zhan, X. You, G. Zhang and D. Zhang, *Chem. Rec.*, 2016, **16**, 2142–2160.
6. Y. Yuan, R. T. Kwok, G. Feng, J. Liang, J. Geng, B. Z. Tang and B. Liu, *Chem. Commun.*, 2014, **50**, 295–297.
7. M. Li, X. Li, H. N. Xiao and T. D. James, *Chemistryopen*, 2017, **6**, 685–696.
8. B. Fu, J. Huang, D. Bai, Y. Xie, Y. Wang, S. Wang and X. Zhou, *Chem. Commun.*, 2015, **51**, 16960–16963.
9. S. Li, X. Ling, Y. Lin, A. Qin, M. Gao and B. Z. Tang, *Chem. Sci.*, 2018, **9**, 5730–5735.
10. H. Wang and G. Liu, *J. Mater. Chem. B*, 2018, **6**, 4029–4042.
11. J. Mei, Y. Huang and H. Tian, *ACS Appl. Mater. Interfaces*, 2017, **10**, 12217–12261.
12. X. Y. Shen, W. Z. Yuan, Y. Liu, Q. Zhao, P. Lu, Y. Ma, I. D. Williams, A. Qin, J. Z. Sun and B. Z. Tang, *J. Phys. Chem. C*, 2012, **116**, 10541–10547.

13. J. H. Cheng, Y. X. Li, R. Sun, J. Y. Liu, F. Gou, X. G. Zhou, H. F. Xiang and J. Liu, *J. Mater. Chem. C*, 2015, **3**, 11099–11110.
14. X. Ma, R. Sun, J. Cheng, J. Liu, F. Gou, H. Xiang and X. Zhou, *J. Chem. Educ.*, 2016, **93**, 345–350.
15. J. Luo, Z. Xie, J. W. Y. Lam, L. Cheng, H. Chen, C. Qiu, H. S. Kwok, X. Zhan, Y. Liu, D. Zhu and B. Z. Tang, *Chem. Commun.*, 2001, **18**, 1740–1741.
16. Z. Zhao, S. Chen, C. Deng, J. W. Y. Lam, C. Y. K. Chan, P. Lu, Z. Wang, B. Hu, X. Chen, P. Lu, H. S. Kwok, Y. Ma, H. Qiu and B. Z. Tang, *J. Mater. Chem.*, 2011, **21**, 10949–10956.
17. B. He, H. Nie, L. Chen, X. Lou, R. Hu, A. Qin, Z. Zhao and B. Z. Tang, *Org. Lett.*, 2015, **17**, 6174–6177.
18. N. L. C. Leung, N. Xie, W. Z. Yuan, Y. Liu, Q. Wu, Q. Peng, Q. Miao, J. W. Y. Lam and B. Z. Tang, *Chemistry*, 2014, **20**, 15349–15353.
19. Z. Yang, W. Qin, N. L. C. Leung, M. Arseneault, J. W. Y. Lam, G. Liang, H. H. Y. Sung, I. D. Williams and B. Z. Tang, *J. Mater. Chem. C*, 2015, **4**, 99–107.
20. X. Dou, Q. Zhou, X. Chen, Y. Tan, X. He, P. Lu, K. Sui, B. Z. Tang, Y. Zhang and W. Z. Yuan, *Biomacromolecules*, 2018, **19**, 2014–2022.
21. R. Wang, W. Yuan and X. Zhu, *Chin. J. Polym. Sci.*, 2015, **33**, 680–687.
22. Y. Y. Gong, Y. Q. Tan, J. Mei, Y. R. Zhang, W. Z. Yuan, Y. M. Zhang, J. Z. Sun and B. Z. Tang, *Sci. China: Chem.*, 2013, **56**, 1178–1182.
23. H. Nie, B. Chen, J. Zeng, Y. Xiong, Z. Zhao and B. Z. Tang, *J. Mater. Chem. C*, 2018, **6**, 3690–3698.
24. H. Zhang, X. Zheng, N. Xie, Z. He, J. Liu, N. L. C. Leung, Y. Niu, X. Huang, K. S. Wong, R. T. K. Kwok, H. H. Y. Sung, I. D. Williams, A. Qin, J. W. Y. Lam and B. Z. Tang, *J. Am. Chem. Soc.*, 2017, **139**, 16264–16272.
25. L. Chen, Y. H. Wang, B. He, H. Nie, R. Hu, F. Huang, A. Qin, X. S. Zhou, Z. Zhao and B. Z. Tang, *Angew. Chem., Int. Ed.*, 2015, **54**, 4231–4235.
26. L. Chen, C. Zhang, G. Lin, H. Nie, W. Luo, Z. Zhuang, S. Ding, R. Hu, S. J. Su, F. Huang, A. Qin, Z. Zhao and B. Z. Tang, *J. Mater. Chem. C*, 2016, **4**, 2775–2783.
27. A. Shao, Y. Xie, S. Zhu, Z. Guo, S. Zhu, J. Guo, P. Shi, T. D. James, H. Tian and W. H. Zhu, *Angew. Chem.*, 2015, **127**, 7383–7388.
28. Z. Q. Xia, A. D. Shao, Q. Li, S. Q. Zhu and W. H. Zhu, *Acta Chim. Sin.*, 2016, **74**, 351–355.
29. S. Chen, W. Li, X. Li and W. H. Zhu, *J. Mater. Chem. C*, 2017, **5**, 2717–2722.
30. M. Wang, N. Yang, Z. Guo, K. Gu, A. Shao, W. H. Zhu, Y. Xu, J. Wang, R. K. Prud'homme and X. Guo, *Ind. Eng. Chem. Res.*, 2015, **54**, 4683–4688.
31. E. Zhao, J. W. Y. Lam, L. Meng, Y. Hong, H. Deng, G. Bai, X. Huang, J. Hao and B. Z. Tang, *Macromolecules*, 2015, **48**, 64–71.
32. X. Chen, X. Liu, J. Lei, L. Xu, Z. Zhao, F. Kausar, X. Xie, X. Zhu, Y. Zhang and W. Z. Yuan, *Mol. Syst. Des. Eng.*, 2018, **3**, 364–375.
33. Q. Zhang, Q. Mao, C. Shang, Y. N. Chen, X. Peng, H. Tan and H. Wang, *J. Mater. Chem. C*, 2017, **5**, 3699–3705.
34. Q. Zhou, B. Cao, C. Zhu, S. Xu, Y. Gong, W. Z. Yuan and Y. Zhang, *Small*, 2016, **12**, 6586–6592.

35. X. Chen, T. Yang, J. Lei, X. Liu, Z. Zhao, Z. Xue, W. Li, Y. Zhang and W. Z. Yuan, *J. Phys. Chem. B*, 2020, **124**, 8928–8936.
36. Z. Zhao, X. Chen, Q. Wang, T. Yang, Y. Zhang and W. Z. Yuan, *Polym. Chem.*, 2019, **10**, 3639–3646.
37. M. Mou, Y. Wu, Q. Niu, Y. Wang, Z. Yan and S. Liao, *Chem. Commun.*, 2017, **53**, 3357–3360.
38. L. Zhang, Z. Y. Zhang, R. P. Liang, Y. H. Li and J. D. Qiu, *Anal. Chem.*, 2014, **86**, 4423–4430.
39. B. B. Chen, R. S. Li, M. L. Liu, H. Z. Zhang and C. Z. Huang, *Chem. Commun.*, 2017, **53**, 4958–4961.
40. V. Gude, A. Das, T. Chatterjee and P. K. Mandal, *Phys. Chem. Chem. Phys.*, 2016, **18**, 28274–28280.
41. Z. X. Liu, Z. L. Wu, M. X. Gao, H. Liu and C. Z. Huang, *Chem. Commun.*, 2016, **52**, 2063–2066.
42. M. Mou, Y. Wu, H. Zou, J. Dong, S. Wu, Z. Yan and S. Liao, *Sens. Actuators, B*, 2019, **284**, 265–270.
43. S. Pathan, M. Jalal, S. Prasad and S. Bose, *J. Mater. Chem. A*, 2019, **7**, 8510–8520.
44. Y. Xu, H. Li, B. Wang, H. Liu, L. Zhao, T. Zhou, M. Liu, N. Huang, Y. Li, L. Ding and Y. Chen, *Mikrochim. Acta.*, 2018, **185**, 252.
45. W. Lv, M. Lin, R. Li, Q. Zhang, H. Liu, J. Wang and C. Huang, *Chin. Chem. Lett.*, 2019, **30**, 1410–1414.
46. Y. Zhang, P. Zhuo, H. Yin, Y. Fan, J. Zhang, X. Liu and Z. Chen, *ACS Appl. Mater. Interfaces*, 2019, **11**, 24395–24403.
47. M. Ge, Y. Han, J. Ni, Y. Li, S. Han, S. Li, H. Yu, C. Zhang, S. Liu, J. Li and Z. Chen, *Chem. Eng. J.*, 2021, **413**, 127457.
48. N. Goswami, F. Lin, Y. Liu, D. T. Leong and J. Xie, *Chem. Mater.*, 2016, **28**, 4009–4016.
49. M. Sugiuchi, J. Maeba, N. Okubo, M. Iwamura, K. Nozaki and K. Konishi, *J. Am. Chem. Soc.*, 2017, **139**, 17731–17734.
50. A. Dutta, U. Goswami and A. Chattopadhyay, *ACS Appl. Mater. Interfaces*, 2018, **10**, 19459–19472.
51. K. T. Prakash, N. Singh and V. Venkatesh, *Chem. Commun.*, 2019, **55**, 322–325.
52. X. Liu, C. Shao, T. Chen, Z. He and G. Du, *Sens. Actuators, B*, 2019, **278**, 181–189.
53. Z. Wu, Q. Yao, O. J. H. Chai, N. Ding, W. Xu, S. Zang and J. Xie, *Angew. Chem.*, 2020, **59**, 9934–9939.
54. Q. Du, X. Hu, X. Zhang, H. Cao and Y. Huang, *Anal. Methods*, 2019, **11**, 3446–3451.
55. Y. Zhang, M. Lv, P. Gao, G. Zhang, L. Shi, M. Yuan and S. Shuang, *Sens. Actuators, B*, 2021, **326**, 129009.
56. S. Jin, W. Liu, D. Hu, X. Zou, X. Kang, W. Du, S. Chen, S. Wei, S. Wang and M. Zhu, *Chem. -Eur. J.*, 2018, **24**, 3712–3715.
57. L. Kong, X. Chu, C. Wang, H. Zhou, Y. Wu and W. Liu, *Nanoscale*, 2018, **10**, 1631–1640.

58. Z. Ma, C. Liu, N. Niu, Z. Chen, S. Li, S. Liu and J. Li, *ACS Sustainable Chem. Eng.*, 2018, **6**, 3169–3175.
59. T. He, N. Niu, Z. Chen, S. Li, S. Liu and J. Li, *Adv. Funct. Mater.*, 2018, **28**, 1706196.
60. Y. Lei, L. Liu, X. Tang, D. Yang, X. Yang and F. He, *RSC Adv.*, 2018, **8**, 3919–3927.
61. L. Xu, X. Liang, S. Zhong, Y. Gao and X. Cui, *Colloids Surf., B*, 2021, **206**, 111961.
62. L.-L. Du, B.-L. Jiang, X.-H. Chen, Y.-Z. Wang, L.-M. Zou, Y.-L. Liu, Y.-Y. Gong, C. Wei and W.-Z. Yuan, *Chin. J. Polym. Sci.*, 2019, **37**, 409–415.
63. M. Li, X. Li, X. An, Z. Chen and H. Xiao, *Front. Chem.*, 2019, **7**, 447.
64. P. Wang, C. Liu, W. Tang, S. Ren, Z. Chen, Y. Guo, R. Rostamian, S. Zhao, J. Li, S. Liu and S. Li, *ACS Appl. Mater. Interfaces*, 2019, **11**, 19301–19307.

CHAPTER 5

Diversity-oriented Fluorescence Library Approach: Accelerating Probe Development for Biological and Environmental Applications

ANIMESH SAMANTA*[a], SUBRATA MUNAN[a], ANAL JANA[a] AND YOUNG TAE CHANG*[b,c]

[a]Molecular Sensors and Therapeutics Research Laboratory, Department of Chemistry, School of Natural Sciences, Shiv Nadar University Delhi-NCR, NH91, Tehsil Dadri, Gautam Buddha Nagar, Uttar Pradesh, 201314, India; [b]Center for Self-assembly and Complexity, Institute for Basic Science (IBS), Pohang 37673, Republic of Korea; [c]Department of Chemistry, Pohang University of Science and Technology (POSTECH), Pohang 37673, Republic of Korea
*E-mail: animesh.samanta@snu.edu.in, ytchang@postech.ac.kr

5.1 Introduction

5.1.1 Why Are Diversity-oriented Synthesis and DOFLA Important?

In the 20th century, after discovering the "central dogma of molecular biology", the pivotal roles of small molecules in living systems started to be explored.[1] Since then, small organic molecules have become indispensable

Monographs in Supramolecular Chemistry No. 34
Fluorescent Chemosensors
Edited by Luling Wu, Adam C. Sedgwick, Xiao-Peng He and Tony D. James

Published by the Royal Society of Chemistry, www.rsc.org

chemical tools to investigate molecular interactions in cellular biology. The small molecules perturb the protein function in biology, which enriches knowledge of the biological activity. Thus, the availability of potential small molecules is invaluable in drug discovery and advanced molecular biology.[2] An in-depth understanding of specific biomolecules essentially depends on the visualization of phenotypic responses to small molecules. However, mapping particular macromolecules in living systems is equal to mapping the stars in the galaxy. Thus, systematic strategies are required to generate the chemical tools to extract the information from 'biological activity space'. Optical imaging is indispensable for diagnostics, theranostics as well as restoring health.[3] In the 20th century, synthetic chemists developed an ample, potent chemical space to understand life's processes.

The first approach is to target the recognition moiety of the molecule, known as target-oriented synthesis (TOS).[4] However, it is limited to empirical knowledge of target moieties. It is therefore necessary to create libraries of small molecules with structural diversity capable of perturbing molecular functions to explore beyond the ability of TOS. Thus, diversity-oriented synthesis (DOS) has become an alternative approach for rendering these libraries.[5] These libraries of molecules, with huge unexplored chemical spaces, may contain exciting biological properties. If the molecules are optically active, fluorescence, in particular, can provide some cellular responses. It has been believed that flexible and hydrophobic fluorophores may bind to the hydrophobic pocket of proteins in the cells.[6] However, this has not been systematically examined because of the lack of diverse fluorophores. Moreover, other biomolecules, apart from proteins, may modulate the fluorescence properties. Herein we summarize the applicability of fluorescent small-molecule libraries for sensing and imaging various biomolecules, cell states, and *ex vivo* and *in vivo* imaging.

5.1.2 What Are Chemical Probes and Chemical Sensors?

Although "seeing is believing" is the key approach in science, living cells usually do not provide enough visual information to understand biological events, including protein expression, cellular signaling, and so on.[7] Thus, substantial efforts have been devoted to the development of fluorophores for both analytical sensing and imaging. The discovery of the first fluorophore, quinine sulfate, by Sir John Herschel in the mid-19th century,[8] and the fluorescence imaging of cells at the beginning of the 20th century by von Prowazek,[9] followed by tissue imaging with antibody-conjugated fluorescein with the help of fluorescence microscopy by Coons and co-workers in 1942, resulted in the emergence of a new era of fluorescence imaging in chemical biology.[10] The intensity and color can be altered reversibly or irreversibly depending on interactions with target analytes and fluorophores. An alteration of the optical signal is reversible and called the "sensor", while signal modulation is irreversible upon interaction with analytes and is known as the "probe".[11] However, the research community often noticed the interchangeable wording used for sensors and probes. The diversity-oriented fluorescent library approach (DOFLA) coined by Professor Chang has evolved, along with many significant breakthroughs in identifying chemical sensors/probes.[12] We will concentrate on probes from cell-based lead molecules due to irreversible signals interacting with cells.

5.1.3 Why Does Fluorescence Change?

Both sensors and probes alter their optical properties with various mechanisms. Several experimental and theoretical approaches have been well studied to investigate the mechanisms, and applied for probe design. Photoinduced electron transfer (PeT),[13] intramolecular charge transfer (ICT),[14] and chemoselective alteration of fluorophore scaffolds (CAFS)[15] are established mechanisms for TOS-based fluorescent probes (Figure 5.1). In the DOFLA, which offers paramount molecular diversity, few additional mechanisms are involved in altering the fluorescence intensity upon interaction with a range of target analytes.[5] The modulation of the optical properties is explained with conformational restriction,[16] motion-induced change in emission (MICE),[17] aggregation-induced emission (AIE),[18] and disaggregation from template aggregates (DATA).[19] A few of the mechanisms will be discussed in the individual case studies in the later sections of this chapter. In principle, the screening of such diversity-oriented fluorescence libraries

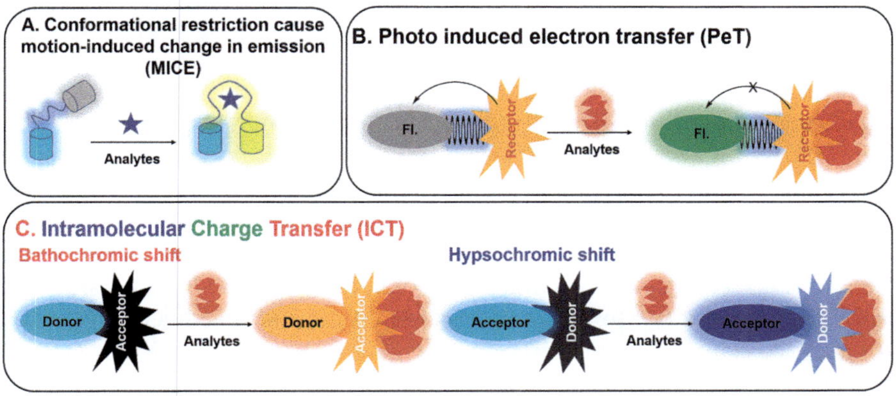

Figure 5.1 Schematic MICE, PeT and ICT mechanisms in designing fluorescent probes. (A) Flexible receptors bind with an analyte and lead to conformational restrictions that cause motion-induced change in emission (MICE). (B) Fluorescent probes composed of fluorophore–spacer–receptor, in which a short aliphatic spacer separates the fluorophore and receptor that cause the PeT. (C) The fluorophore consists of an electron-donating group directly connected to an electron-withdrawing group without a spacer. The electron density in the fluorophore is affected during the recognition of analytes, resulting in emission color changes of the fluorophores. ICT causes significant shifts in the absorption and fluorescence emission bands.

(DOFLs) accelerates the identification of chemical probes/sensors, including those more challenging in TOS, and with decreased time and cost.

5.1.4 Approaches for the Development of Fluorescent Probes

There are two major approaches to develop fluorescent probes/sensors. The first, TOS, is through tagging fluorophores with known recognition moieties (Figure 5.1). This activity-based design approach depends solely on empirical knowledge of the target analytes. The emission can be altered during specific interactions. However, this conventional approach is limited in its scope for sensor discovery due to a lack of traditional chemistry knowledge. Moreover, the lack of theoretical understanding in complex biology is equally challenging for probe development. Hence, predicting the structural requirements and designing new probes for unknown targets is limited. The DOS approach has been successfully utilized in drug discovery since the introduction of this concept by Stuart Schreiber in 1990.[20] More than a million small molecules are screened to identify lead molecules to further validate them as drug candidates. Recently, a similar approach was used to develop a library of small fluorescent molecules, to develop optically active chemical probes to elucidate complex scientific problems beyond the scope of the TOS approach. Inspired by this, Chang and co-workers, who pioneered DOFLA, have extensively explored probe discovery (Figure 5.2).

5.1.5 Generation of Fluorescent Libraries

The high sensitivity of the fluorescence technique is the key attraction in optical sensing and imaging. Despite the potential imaging of intracellular biomarkers, the fluorescent "core scaffolds" are still limited to fluorescein,[21] rhodamine,[22] cyanine,[23] coumarin,[24] chalcone,[25] BODIPY,[26] prodan,[27] naphthamide,[28] *etc.* The effort to synthesize a large set of libraries with sufficient structural and chemical diversity is exceptionally challenging. The development of chemical sensors or probes has not been accelerated in the same way as for drug discovery. However, a few research groups have been devoted to utilizing these important opportunities to generate fluorescent libraries. The structural and chemical diversity is explored either on the binding ligands, named receptors, or the core structure of the fluorophores. Finney and co-workers initiated the fluorescence library approach to reverse the discovery paradigm from conventional TOS at the beginning of the 21st century.[29] In this combinatorial approach, the diversity of ligands binds to the active target sites of analytes. The conformational restriction of the bond rotation induces a fluorescent emission that guides the probe development. Similarly, Sames and co-workers synthesized a library of triamine ionophores without disturbing the core structure of the dansyl fluorophore.[30] When the electron-donating receptors bind to the active target sites of analytes, transfer of electrons from the receptor to the fluorophore is diminished, which

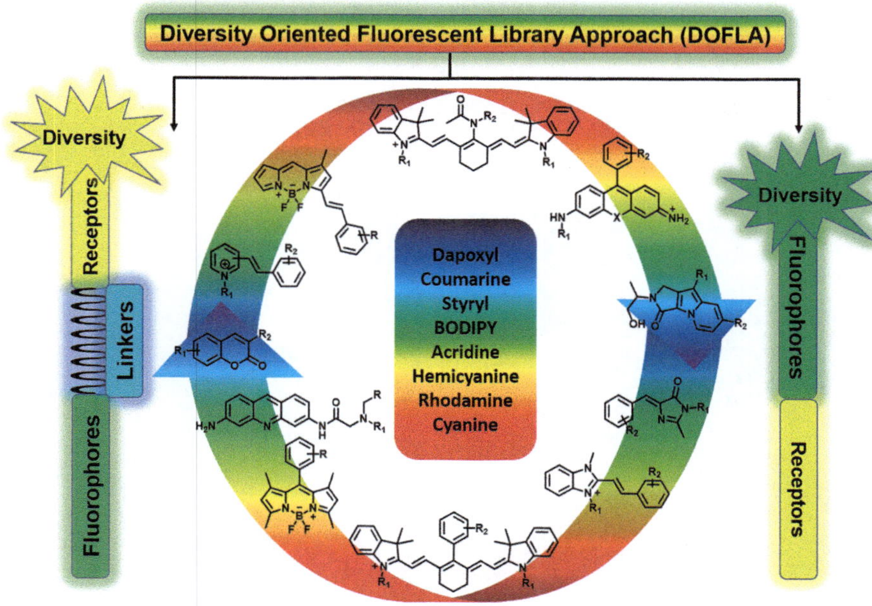

Figure 5.2 A general strategy for structure diversity of fluorescent scaffolds for diversity-oriented libraries. Both receptor sites and fluorophores can be diversified by combinatorial chemistry. Different fluorescent scaffolds with full visible color emission wavelengths are potential chemical toolboxes.

induces the regaining of the emission, resulting in turn-on fluorescence. This phenomenon is called photoinduced electron transfer (PeT)-based probe development.[31]

Each fluorescent core-scaffold cannot be easily modified to render a library of compounds. Two major limitations for DOFLA generation are the large-scale synthesis of the fluorophores and the ease of the accessibility of DOS with commercially available building blocks. Pyridinium, benzimidazolium, hemicyanine, BODIPY, xanthene, chalcone, and cyanine fulfill these requirements and Chang and co-workers synthesized a large set of DOFL compounds[32] (Figure 5.3). The authors accelerated probe development by generating >20 DOFL libraries composed of >10 000 compounds by modifying the core structure or receptors. Library synthesis was mostly performed by combinatorial chemistry in solution and solid phase using different reactions, such as aldol condensation, Knoevenagel condensation, Grignard reaction, nucleophilic substitution reaction ($S_N1/2/i$), amide coupling, click reaction, and so on[4] (Figure 5.3). Chang and co-workers generated the first styryl library composed of 574 styryl compounds by simple aldol reaction of 14 pyridinium salts and 41 commercial aldehydes.[33] Similarly, thousands of DOFL compounds were synthesized using different building blocks, such as aldehydes, amines, acids, acid chlorides, alkynes, and Grignard reagents.

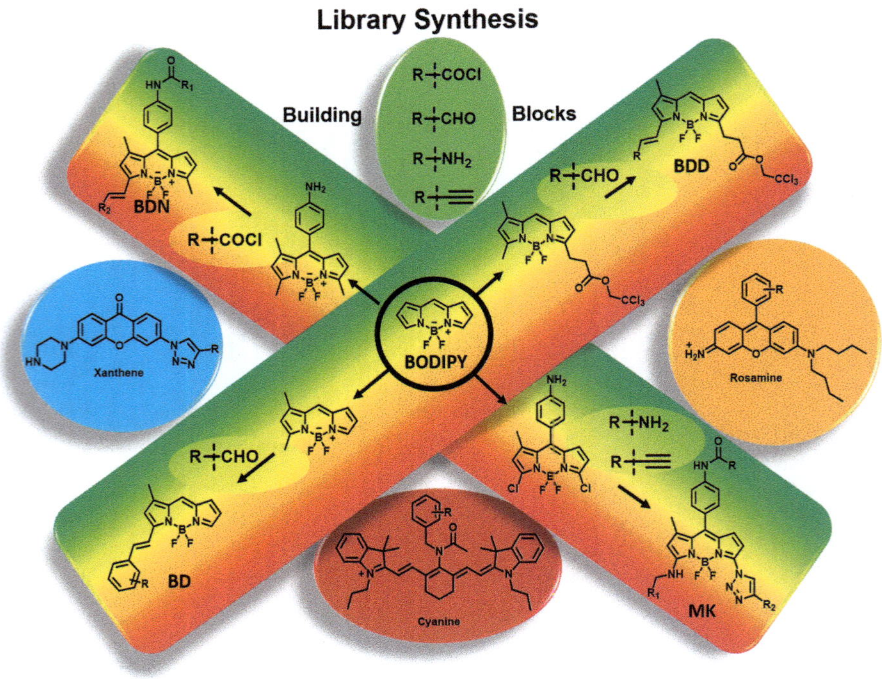

Figure 5.3 Construction of library using combinatorial chemistry in solid and solution phase. Different building blocks are used to derivatize known fluorescent scaffolds with an easy synthetic strategy.

The products can be easily purified by filtration and washing in solid-phase synthesis, in contrast to solution-phase synthesis. Thus, solid-phase synthesis is always attractive for library generation, and hence ~50% of the libraries were developed using solid-phase chemistry.

5.2 Selection of Screening Strategies

The selectivity of target-specific analytes is usually achieved with high sensitivity during *in vitro* studies. However, rationally designed receptors do not guarantee specific recognition targets in complex systems, including environmental, food and beverages, body fluids (urine or plasma), and living cells. The expected mechanisms for either turn-on or turn-off may not occur as several environmental factors, such as viscosity, pH, and temperature, also play crucial roles in complex systems. If chemical probes translate into bioimaging probes, these are often largely unfit due to cell permeability, cytotoxicity, and lack of specific recognition. Therefore, a systematic screening platform that can offer an advanced strategy to develop new chemosensors or chemical probes should be planned. Depending on the application, four different screening strategies are available, as shown

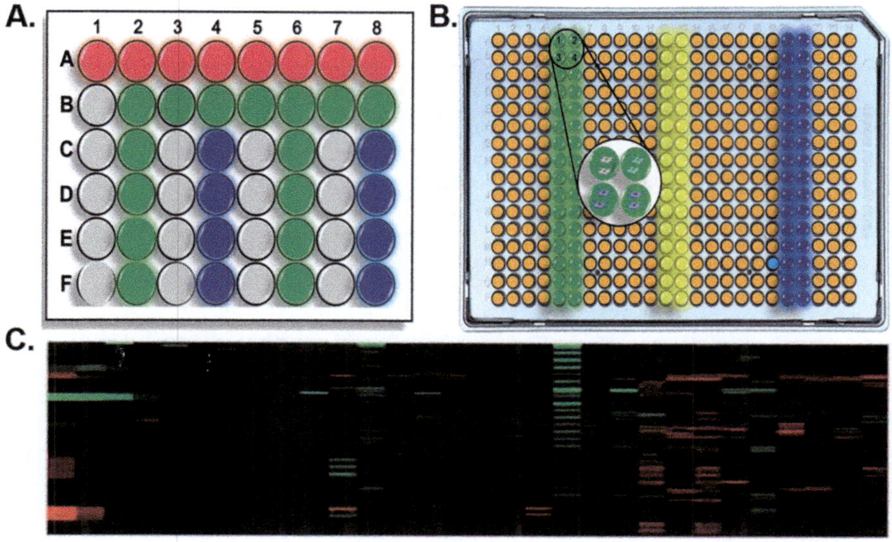

Figure 5.4 High-throughput screening strategy. (A, B) Both 96- and 384-well plates can be used for fast screening of DOFL molecules. (C) Analysis of the signal output compared to its control analytes and fluorophores. The green color indicates turn-on emission, and the red color suggests turn-off emission upon recognition of analyte. Reproduced from ref. 5 with permission from the Royal Society of Chemistry.

in Figure 5.4. (1) The simplest way of screening several analytes is *in vitro* spectra-based high-throughput screening. The identified chemical sensors or probes from spectral analysis cannot guarantee sensing of the same analytes in cells due to their complex environment. For example, 42 RNA selective probes from spectral analysis were active only by <5% in cells.[34] This revealed the challenges for the development of cell-imaging probes. (2) Thus, advanced cell-based unbiased screening is an alternative for identifying biocompatible cellular imaging probes.[12] Although this works elegantly for isolated cells, it also does not guarantee tissue imaging. (3) So, *ex vivo* tissue imaging and (4) *in vivo* whole-organism-based screening has led to the development of imaging probes for in-depth understanding in a living system.

5.2.1 Spectra-based High-throughput Screening *In Vitro*

A large set of fluorescent compounds, named briefly as DOFL, are an asset to probe discovery. High-throughput screening strategies are applied to uncover the activity of these optically active chemical tools. A spectra-based screening strategy is established in 96- or 384-wells plates in a high-throughput screening manner. A single compound can be screened against variable analytes or a single analyte with DOFL compounds. As chemists can

Table 5.1 Summary of spectra-based high-throughput screening for the discovery of chemosensors and chemical probes.

Bioanalytes	Fluorophore	Probe name	Reference
HSA	Rosamine	G13	39
GSH	Rosamine	H22	42
DNA	Styryl	A41B2	35
DNA	Styryl	C61	36
RNA	Styryl	E36, E144, and F22	34
Aβ amyloid	Styryl	STB-8	43
Chymotrypsin	Styryl	QN-33 and QN-49	40
GTP	Hemicyanine	GTP green	38
Heparin	Hemicyanine	Heparin orange and heparin blue	41
G-quadruplex	Cyanine	CyC-M716	37
Aβ plaques	Dapoxyl	2C40 and 2E10	43
Aβ oligomer	BODIPY	BD-Oligo	56
GHB and GBL	BODIPY	Green Date	48 and 49
Caffeine	BODIPY	Caffeine Orange (CO)	50
Fructose	BODIPY	Fructose Orange (FO)	51
Milk fat	BODIPY	Milk Orange (MO)	52
Bisphenol A	BODIPY	Bisphenol Orange (BPO)	55
HSA	BODIPY	BDC-9	45
IgG	BODIPY	BDM-69, Ig Orange	46
Cu/Zn superoxide dismutase	BODIPY	MK103-48, SODO	47
HSA	Dapoxyl	A41-S	44

generate a limitless number of compounds, the screening strategies should be against variable analytes.

Several successful candidates were found from the DOFL using the strategy mentioned above. Herein, we discuss a few case studies, including nucleic acids (DNA, RNA, GTP, ATP), proteins (bovine serum albumin (BSA), human serum albumin (HSA), immunoglobulin G (IgG), chymotrypsin), peptides (β-amyloid, glutathione (GSH)), carbohydrates (heparin, fructose), and food beverages (milk fat, caffeine, gamma-hydroxybutyrate (GHB), gamma-aminobutyric acid (GABA)) and environmental pollutants (water, metal ions, bisphenol A), which are summarized in Table 5.1. Here, we will first discuss cationic fluorophores, in which the donor–π-acceptor (D–π-A) system favors strong fluorescence properties, followed by neutral fluorophores.

5.2.1.1 *Cationic Fluorophores for Sensor/Probe Development (Styryl Dyes, Rosamine, Hemicyanine)*

5.2.1.1.1 Nucleic Acids

Styryl Dyes for DNA. The common fluorophores, including rhodamines, styryls, and cyanines, possess donor–π-acceptor (D–π-A) characteristics. Their cationic acceptors pull the electron from the donor, resulting in the enhancement of tunable bright emissive fluorescence properties. Cationic molecules tend to bind towards the negative components of cells and tissues, which include phosphate groups of nucleic acids (DNA and RNA) or sulfate groups of a few polysaccharides (*e.g.* glycosaminoglycan). Chang and co-workers identified a styryl-based double-strand DNA (dsDNA) probe, denoted as **A41B2**, after screening the synthesized styryl library[35] (Figure 5.5A). A structure–activity relationship (SAR) study revealed the significance of both 2,4,5-trimethoxy in benzaldehyde and adamantyl pyridinium analogs. Although **A41B2** was suitable for live-cell imaging, the fluorescence quantum yield ($\Phi^{\text{with DNA}} = 0.003$) is very low, even after binding to dsDNA. Thus, >500 μM of the probe was required for cellular imaging. Later, Feng *et al.* systematically screened thousands of styryl compounds against dsDNA and total RNA to discover a DNA-specific fluorescent probe named Green DNA (**C61**).[36] The selectivity of DNA over RNA was confirmed by the test for DNase and ribonuclease (RNase). The emission intensity of **C61** was drastically decreased after DNase treatment but unaffected with RNase treatment. Nocodazole was treated to arrest the G2 phase and quantified the DNA in normal and G2 phases. The fluorescence signal of **C61** in G2/M-arrested cells was doubled as twice the

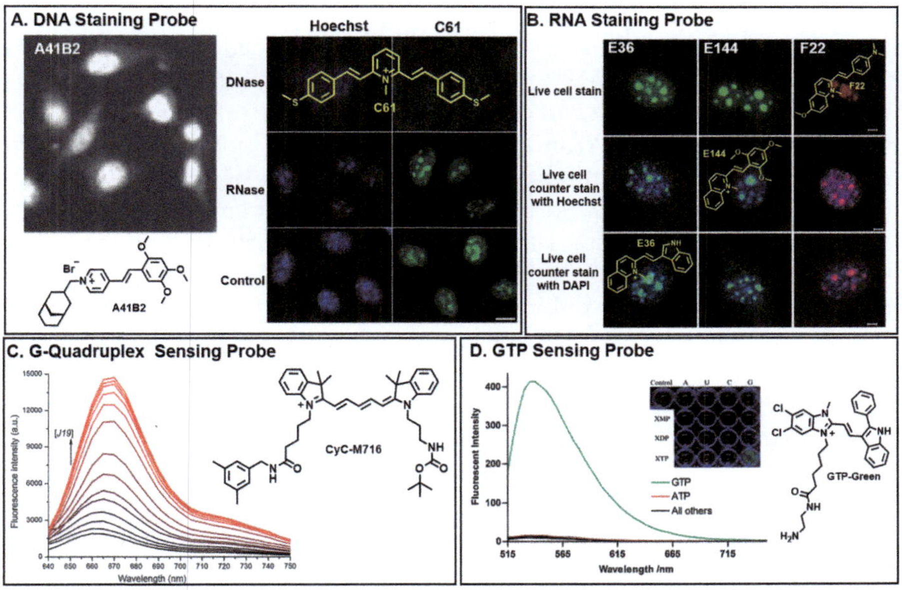

Figure 5.5 Discovery of fluorescent probes for, nucleic acids (A) dsDNA, (B) RNA, (C) G-quadruplex, and (D) GTP, from styryl and hemicyanine derivatives. Reproduced from ref. 35 with permission from the Royal Society of Chemistry; ref. 36 with permission from the Royal Society of Chemistry; ref. 34 with permission from Elsevier, Copyright 2006; ref. 37 with permission from the Royal Society of Chemistry; ref. 38 with permission from American Chemical Society, Copyright 2006.

amount of DNA is present compared with normal cells. Flow cytometry also revealed the distinct cell populations in normal and G2/M-arrested cells. **C61** works in both live and fixed cells and offers a distinct green color compared with the standard blue-emitting DAPI and Hoechst dyes.

Styryl for RNA. RNA is an essential nucleic acid in transcriptional and translation activity in protein synthesis. However, selective probes only for RNA in live cells, without intervention of DNA, is very rare. Only SYTORNASelect™, a commercial live-cell RNA probe, is available. To study the RNA distribution within the nucleus, Li *et al.* screened 1336 synthesized styryl compounds in living 3T3 cells, which led to identifying 88 primary hits. A secondary screening confirmed three RNA probes, **E36, E144,** and **F22,** composed of the same quinaldine motif.[34] Co-localization with these three probes was corroborated with SYTORNASelect™ and DNA staining with Hoechst dye (Figure 5.5B). In the enzymatic assay, the fluorescent intensity was significantly diminished after only RNase treatment, compared to DNase. DNA- and RNA-specific probes with different colors can be imaged simultaneously by selecting appropriate microscopic filter channels. Parallel high-throughput spectra and live-cell screening strategies are powerful for efficient probe discovery. Live-cell screening strategies are further discussed in Section 5.3.

Cyanine for G-quadruplexes (G4s). Non-canonical G-quadruplexes (G4s) are formed from nucleic acids rich in guanine. Four-stranded G4 structures have unique functions in human telomeres and gene promoters. The discovery of fluorescent probes that can detect and discriminate G4s from dsDNA and single strand DNA (ssDNA) and provide structural information is at the forefront of research. A few fluorescent probes are capable of differentiating G4s but are unable to provide structural information. Zhang *et al.* performed systematic screening of a 241-membered Cy5-based cyanine library to identify a novel fluorescent probe (**CyC-M716**) for parallel G4s[37] (Figure 5.5C). A remarkable fluorescence enhancement was observed for parallel-stranded J19, T95-2T, A95, and C95 G4s, but fluorescence quenching was observed in genomic dsDNA, ssDNA, and RNA. Molecular docking depicted plausible binding of the probe with the grooves of J19 and its two side chains embedding into the propeller loops of the quadruplex.

Hemicyanine for GTP. The fundamental function of nucleotides is to carry and store genetic information in cells. Guanosine 5′-triphosphate (GTP) is associated with several hydrolase enzyme activities. Malachite green has been used for analyzing GTPase activity by indirect measurement of the phosphate ion concentration. Still, specific detection of GTPase enzyme is always challenging due to interference from other nucleotides. Thus, Chang and co-workers screened a hemicyanine library containing a primary amine linker, which interacts with the negatively charged monomer of nucleic acids to identify a GTP-specific probe named **GTP-green.**[38] It showed a turn-on response to the only GTP, but remained silent for the 4 nucleic acids and 11 nucleotides. It can also be visualized using a CCD camera, which provides a novel platform for easy analysis of pathological samples.

5.2.1.1.2 Proteins and Peptides

Rosamine Dye for Human Serum Albumin. The broad structural diversity of DOFL compounds offers variable quantum yields (Φ). Compounds with low quantum yields are likely to display maximum turn-on response compared with those with moderate quantum yields, which will probably show both turn-on and turn-off responses after interacting with analytes. Thus, both types are suitable for probe development, whereas compounds with high quantum yields are best suited for fluorescent tagging and immunohistochemistry. Ahn *et al.* synthesized a flexible rosamine library with moderate quantum yields by removing 2′-carboxylic acids from high quantum-yield rhodamine. This rosamine library was expected to interact with biomolecules that may restrict the bond rotation, resulting in a turn-on fluorescent probe.[39] Thus, a high-throughput screening of 240 rosamine compounds against proteins, nucleic acids, and polysaccharides was tested to identify **G13** as an HSA-selective probe. A family of albumins from humans, bovine (BSA), rat (RSA), sheep (SSA), and porcine (PSA), were also tested, giving a maximum response of 36-fold to HSA, followed by 15-fold to RSA, and a very minimal response of 5–7-fold for the rest of the albumin species. **G13** binds to the binding site I of has, as both ibuprofen and warfarin displaced **G13** during the competition assay. **G13** enables measurement of HSA in urine samples and quantifies unknown HSA concentrations in patient samples. It may lead to diagnosis of malnutrition, microalbuminuria, and liver or kidney disease.

Chymotrypsin. Chymotrypsin is an enzyme that breaks down proteins into amino acids used in the small intestine. It specifically targets the aromatic amino acids phenylalanine, tyrosine, and tryptophan. Thus fluorescent probes for chymotrypsin may lead to monitoring for abscesses, ulcers, and critical illness (trauma). The interactions of quinaldinium derivatives with 12 proteins from different families (protease A, chymotrypsin, thermolysin, lipase AY, cellulase A, hemicellulase, lysozyme, deamizyme, ribonuclease A, cytochrome *c*, hemoglobin, and insulin) were evaluated *in vitro*. Wang *et al.* identified two compounds (**QN-33** and **QN-49**) that specifically respond to chymotrypsin compared with other enzymes.[40] A structure–activity relationship revealed that the binding affinity towards chymotrypsin was diminished for both quinoline and a neutral analog compared to its cationic quinaldinium derivatives.

Hemicyanine for Heparin. The power of DOFLA was further explored by utilizing the benzimidazolium library. Fluorescent probes for negatively charged heparin were investigated using these positive charged molecules. Heparin is a clinically approved anticoagulant to treat thrombosis. However, an overdose causes the adverse effect of hemorrhage and heparin-induced thrombocytopenia. Thus, monitoring heparin levels in a patient's blood is very important. Although a few standard assays are available, there is a lack of fast and simple methods, which are attractive for monitoring heparin. Chang and co-workers developed two-fluorescent probes for heparin using high-throughput screening of a benzimidazolium library in human plasma.[41] The probes are named **heparin orange** and **heparin blue**, depending on their turn-on emission color. Their potential application was validated in human plasma within the therapeutic concentration range.

Glutathione. The rosamine library was used to develop a HSA-selective probe. In addition, Ahn *et al.* examined the rosamine library against several small-molecule analytes, including metal ions and metabolites.[42] A promising GSH-selective probe, **H22**, exhibited fluorescence enhancement towards reduced GSH, but remained silent towards oxidized GSSG, along with β-mercaptoethanol, cysteine, and dithiothreitol (DTT).

Styryl for Aβ Amyloid. DOFL compounds were further evaluated to discover molecular probes that can detect peptides and proteins. Thirteen primary hits for Aβ-amyloid peptides were identified from high-throughput screening of 320 styryl compounds in insulin amyloid fibrils with similar amyloidogenic properties to Aβ peptides. These hits were further examined *in vitro* and mouse brain tissues against the plaque proteins of Alzheimer's disease having Aβ40 and Aβ42 peptides, to discover two specific fluorescent probes.[43] These probes target the core of the Aβ plaques compared to the conventional probe, thioflavin S (ThS), which images the peripheral area in mouse brain tissue.

5.2.1.2 Neutral Dyes (Dapoxyl Library, BODIPY)

The most well-known and popular neutral dyes are BODIPY (4,4-difluoro-4-bora-3*a*,4*a*-diaza-*s*-indacene), Dapoxyl, and so on. These hydrophobic neutral fluorophores aggregate in polar media and display fluorescence quenching. They reveal turn-on responses with analytes after interacting with lipophilic organelles and proteins. Several fluorescent probes have been developed, and are discussed below.

5.2.1.2.1 Human Serum Albumin. Dapoxyl (2-pyridyl-5-aryl-oxazole) is a class of neutral dye. The fluorescence intensity of dapoxyl is easily altered by pH change and solvent polarity. With the success of DOFLA in cationic structures, Min *et al.* explored neutral dapoxyl. An 80-member library with structural diversity was generated in both solid and solution chemistry. Systematic screening of the synthesized hydrophobic fluorescent library led to the identification of a HSA-selective probe, **A41-S**.[44] This probe showed a turn-on fluorescence response when it binds to the drug-binding site I of HSA protein, indicating that **A41-S** is a potential fluorescent probe for binding site I. The higher sensitivity of this **A41-S** in dansylamide displacement

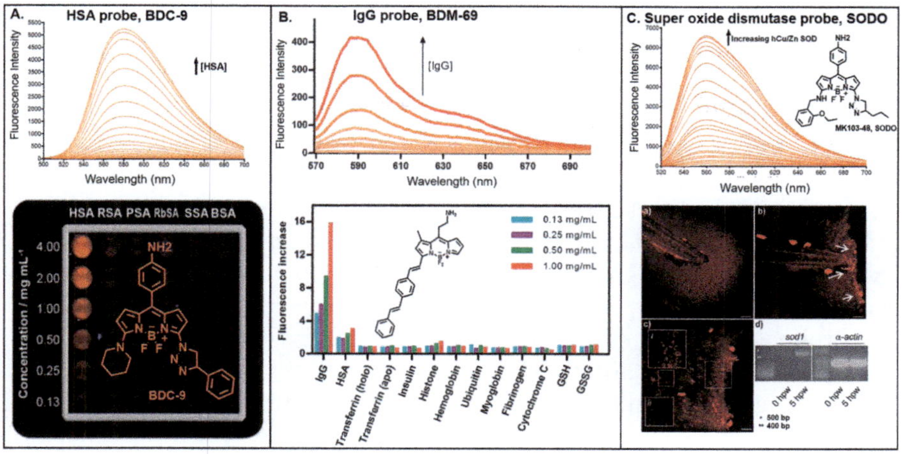

Figure 5.6 DOFLA for the probe discovery of proteins (A) HSA, (B) IgG, and (C) hSOD from neutral BODIPY derivatives. Reproduced from ref. 45 with permission from the Royal Society of Chemistry; ref. 46 with permission from the Royal Society of Chemistry; ref. 47 with permission from the Royal Society of Chemistry.

assays may offer a better probe. Next, a 180-member neutral BODIPY-triazole library was screened against different proteins, to identify **BDC-9** as an HSA probe[45] (Figure 5.6A). **BDC-9** displayed a remarkable turn-on emission 220-fold higher upon binding to the hydrophobic regions of HSA, and remained silent towards other species of albumins despite having 75% homology. The hemin, which is specific for the FA1 binding site, enables replacement by **BDC-9**. This indicates that **BDC-9** is an FA1 site-specific fluorescent probe. However, dansyl-L-norvaline (FA3/4 sites), ibuprofen (FA3/4 and FA6 sites), propofol (FA3/4 and FA5 sites), and warfarin (FA7 site) did not alter the fluorescence in the competition assays. The first fluorescent probe, **BDC-9** binds to the heme site in HSA. **BDC-9** quantified has concentrations within microalbuminuria limits in urine samples, which led to liver and renal disease diagnoses.

5.2.1.2.2 Immunoglobin. Solid-phase synthesis is relatively easy as vigorous purification pathways can be avoided. Vendrell *et al.* synthesized the first BODIPY library in solid-phase to discriminate among different macromolecules in a complex mixture of proteins, carbohydrates, and nucleic acids. Through a systematic screening of the BODIPY library, a selective turn-on fluorescent probe, **BDM-69**, was identified for immunoglobulin (Ig), and it was silent to non-peptide macromolecules[46] (Figure 5.6B). Further analysis suggested that the probe is active to commonly abundant immunoglobulins (IgA, IgG, and IgM). IgM exhibited a higher response than the other two. Based on emission color, **BDM-69** was named **Ig Orange**, and it is likely bound to the backbone structure of the Fab region, as opposed to the Fc fragments of the immunoglobulin. This small-molecule fluorescent probe can be an alternative to the ELISA kit as IgG is commonly used to quantify antibodies.

5.2.1.2.3 Cu/Zn Superoxide Dismutase. The BODIPY structure can be derivatized with maximum structural diversity in a different position with different building blocks. Chang and co-workers developed an **MK** library by introducing primary amines and PeT quenchers, namely, triazole groups. A preliminary screening of DOFL compounds against human Cu/Zn superoxide dismutase (SOD) led to identifying a SOD-selective probe, **MK103-48.** This probe was named superoxide dismutase orange (**SODO**) (Figure 5.6C), as it emits an orange color. It exhibited 150-fold fluorescence enhancement after binding to hCu/Zn SOD, but was silent over other reactive oxygen species (ROS)-related enzymes, such as Mn-SOD and Fe-SOD, and metabolites. **SODO** is the first small molecular fluorescent probe for hCu/Zn SOD and enabled imaging of an inflammatory response associated with SOD in an injured zebrafish model.[47]

5.2.1.2.4 Illicit Drugs GHB and GBL. Zhai *et al.* developed a BODIPY-based fluorescent probe, **Green Date** for gamma-butyrolactone (GBL), which is the pro-drug of gamma-hydroxybutyric acid (GHB) and is also used as an illicit drug in sexual assault.[48,49] Overdose of GBL usually leads to an adverse effect on the respiratory system, which causes unconsciousness, bradycardia, and even death. Due to the hydrophobic nature of **Green Date**, it is initially aggregated in an aqueous solution. When hydrophobic GBL molecules are inserted into the aggregated **Green Date**, the dissociate into monomers to show a turn-on response. A simple laser pointer can show an orange emission whenever GBL is present in common drinks. Hence, this fast and simple fluorescent probe can be used as a potential optical tool for forensic applications.

5.2.1.2.5 Caffeine. A neutral BODIPY compound was used as a traffic light detector to detect caffeine in beverages. Xu *et al.* screened a BODIPY library using high-throughput screening to discover a caffeine selective fluorescent probe **Caffeine Orange**[50] (Figure 5.7A). This probe exhibited a 250-fold fluorescence enhancement upon caffeine addition in aqueous media. **Caffeine Orange** showed aggregation-induced quenching initially due to π-stacking and intra/intermolecular hydrogen bonding. Then it showed a turn-on response upon interaction with caffeine in food beverages, through disaggregation from template aggregates (DATA). The authors also demonstrated a detection kit for naked-eye sensing using simple laser irradiation. A microfluidic device-based detection tool was used for ultrasensitive detection of caffeine in food beverages.

5.2.1.2.6 Fructose. The high-throughput screening of 160 derivatives of a trichloroethyl-activated ester BODIPY led to the selection of a disaccharide

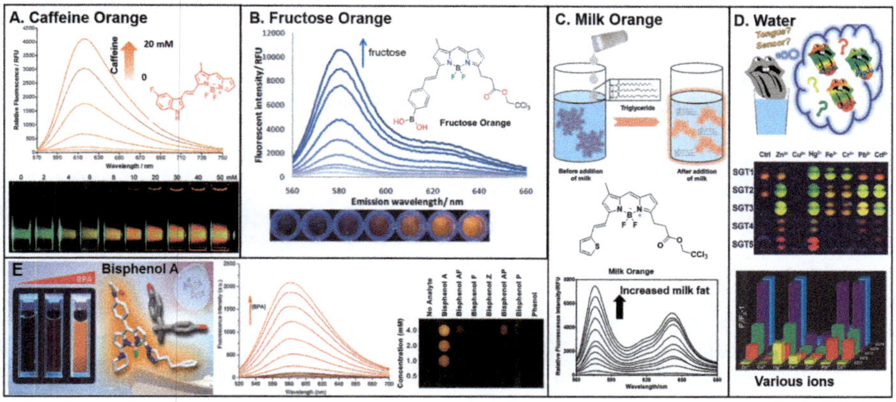

Figure 5.7 DOFLA for the probe discovery of small molecular analytes in the environment and food samples: (A) caffeine, (B) fructose, (C) milk, (D) water, and (E) bisphenol A from neutral BODIPY derivatives and arrays of dye molecules. Reproduced from ref. 50 with permission from Springer Nature, Copyright 2013; ref. 51 with permission from American Chemical Society, Copyright 2012; ref. 52 with permission from the Royal Society of Chemistry; ref. 54 with permission from American Chemical Society, Copyright 2014; ref. 55 with permission from Elsevier, Copyright 2014.

probe, **fructose orange**[51] (Figure 5.7B). The BODIPY probe with boronic acid detected fructose amongst 52 biologically relevant analytes. A 24-fold fluorescence enhancement occurred upon recognition of fructose without interfering with the other 23 mono- and disaccharides, resulting in the measurement of fructose in a matrix of soft drinks without interference. This simple fluorescence method can easily quantify 270 mM of fructose in regular Coca-Cola and was well-matched with expensive high-performance liquid chromatography analysis.

5.2.1.2.7 Milk Fat. The trichloroethyl-activated ester of BODIPY is not only limited to fructose sensing; it was extended to several applications. Another interesting probe from this library for milk fat analysis was named **milk orange** (**MO**)[52] (Figure 5.7C). A complete food, milk contains all essential nutrients, such as carbohydrates, proteins, fats, and vitamins except iron. Hence, it is has been regularly consumed by billions of people in the world for thousands of years. Among the nutrients, the amount of fat is a key indicator of milk quality and its market value. However, there was no easy method to measure fat with high precision. Xu *et al.* developed a fluorescence image-based high-throughput screening system using a black imaging box, and rapidly screened thousands of fluorescent molecules in a single day. Image analysis led to identifying an unprecedented milk fat sensor, **MO**. A mechanistic understanding derived from ^{19}F NMR and dynamic light scattering revealed that the sensing is based on DATA. Interestingly, highly abundant fat or proteins did not interfere. Later, the versatility of **MO** was established by combination with a simple fluorescence detector for on-site milk fat analysis at resource-limited locations.

5.2.1.2.8 Water. Drinking water with minerals is indispensable to life. The human body is made of 60% water, and minerals, which include cations and anions, play a vital role in nutrition. The ion concentrations in water determine the quality of the drinking water. In particular, hardness and pH are crucial for drinking water. However, high levels of hardness caused by heavy metal cations can impact health. Lee *et al.* developed an array of "New York Tongue 1" consisting of 47 dyes that are capable of discriminating 39 cations out of a total of 44 metal cations, with 5 alkali metal ions (Na^+, K^+, Li^+, Cs^+, and Rb^+) being almost silent.[53] Next, Chan *et al.* demonstrated another advanced chemosensor array named "Singapore Tongue" (SGT) consisting of 45 colorful dyes for water analysis[54] (Figure 5.7D). SGT was applied to analyze 63 tap waters from Singapore's mass rapid transit stations and 12 different nationwide brands of bottled water. Interestingly, SGT discriminated types of water using principal component analysis. While these dye-based arrays are difficult to interpret they can provide immediate results on whether samples are from similar or different sources.

5.2.1.2.9 Bisphenol A. A BODIPY library was used to discover several fluorescent probes. Zhang *et al.* performed a high-throughput screening of DOFL consisting of a BODIPY-triazoles library. A turn-on probe named bisphenol orange (**BPO**) was discovered for the water pollutant bisphenol A (BPA)[55] (Figure 5.7E). BPA is a monomer of a polycarbonate polymer plastic that is used globally in the packaging and paints industry. BPA is widely released from plastics and causes tremendous health problems through disruption of the endocrine system. To ensure water safety, detecting BPA with simple, sensitive, and convenient fluorescent techniques is important for monitoring water contamination. **BPO** was very selective over other phenol analogs and enabled the analysis of BPA in real water samples. Mechanistic studies for the selective turn-on response indicate a restriction of bond rotation for **BPO** while trapped within the BPA hydrophobic pocket.

5.3 Cell-based Unbiased Screening for Imaging Probes

As discussed, spectra-based high-throughput screening led to the discovery of many fluorescent probes/sensors. The selected fluorescent probes are <1% of the synthesized DOFL compounds. What happens to the remaining compounds in a library? Next, cell-based high-throughput screening was adopted for cell discrimination and the identified fluorescent probes are summarized in Table 5.2.

Table 5.2 Summary of cell-based high-throughput screening for the discovery of cell-specific imaging probes. A few of the small molecular probes are defined as 'compound of designation' with their emission color (CDcolor: CDy, CDg CDb, ..., and so on), which are alternatives for standard CD biomarkers for cell-specific identification and discrimination.

CD	Fluorophore	Target animal cells or bacterial cells	Targeted biomarkers	Reference
CDy1	Rosamine	Pluripotent stem cell probe or, more specifically, mESC probe	ABCB1 gene	57
CDy2	Rosamine	Muscle cells	Mitochondrial aldehyde dehydrogenase (ALDH2)	6
CDr3	BODIPY	Neural stem/progenitor cell (NSPC)	Fatty acid-binding protein 7 (FABP7)	63
CDg4	Chalcone	Mouse embryonic stem cell (mESC)	Glycogen	61
CDy5	Rosamine	NSPC	Acid ceramidase (AC)	64
CDy6	BODIPY	Mitosis and proliferating cells	Mitotic cells	83
CDnir7	Cyanine	Macrophage-targeted NIR probe	Alzheimer's Disease in mouse brain	78
CDb8	Xanthene	ESCs, also called PSCs	Embryonic stem cell	62
CDy9	BODIPY	mESC selective; ICM cells	Slc13a5 transporters	60
CDr10 a and b	BODIPY	Microglia	Microglia	67
CDy11	BODIPY	Bacterial biofilm probe	Functional amyloids in *Pseudomonas* (Fap)	87
CDb12	Xanthone	Cells in mitotic phase	DNA	84
CDg13	Fluorescein	NSPC	Abcg2 inhibitor	65
CDy14	BODIPY	Bacterial biofilm probe	Exopolysaccharide, Psl	88
CDr15	BODIPY	Bacterial biofilm probe, biofilm target, neutrophil extracellular trap	Exopolysaccharide, Psl	89
CDg16	Acridine	Active inflammation in atherosclerosis	SLC18B1 (description)	81
CDr20	BODIPY	Brain macrophage	Microglia	68
PiY	BODIPY	β-Cell probe	Islets of Langerhans	71
GY	BODIPY	Pancreatic alpha cells	Glucagon	69
PiF (PET)	Rosamine	Pancreatic beta cells	HOMX2	73
TiY	BODIPY	Tumor-initiating cell (TIC)	Vimentin	85
TPα	Acedan	Pancreatic alpha cells	Glucagon	70
TPβ	Coumarin	Pancreatic beta cells	Glucagon	72
KP-1	Rosamine	ihPSC		59
NeuO	BODIPY	Neuron		66
CDgB	BODIPY	B-cell-selective plasma membrane-staining probe (CDgB)	Plasma membrane	82
TiNIR	Cyanine	TIC	HOMX2	86

5.3.1 Organelle-staining Fluorescent Probes

Positively charged styryl compounds with high polarizability may interact strongly with the mitochondrial membrane, having a higher electronic potential than cytosol. However, 64 out of 276 developed styryl compounds explicitly localized to mitochondria and 55 molecules localized in the endoplasmic reticulum, nucleoli, and cytoplasm. These structure–localization relationships demonstrated the power of DOFL compounds to discover multicolor molecular probes for cell discrimination. Modulation of fluorescence of DOFL inside organelles suggested specific interaction with macromolecules, which offers an opportunity to elucidate the binding partners from empirical knowledge or high-throughput screening approaches. Cell-specific imaging probes are discussed case by case. However, when probes were localized in the nucleus, Chang and co-workers examined their specificity towards nucleic acids and nucleotides to identify their binding partners.

5.3.2 Aβ Amyloid

Histological staining of aggregated Aβ amyloid is crucial for diagnosing Alzheimer's disease, as 30 million people are affected worldwide and it is the leading cause of death in older populations. Despite enormous research on Alzheimer's disease (AD) over the last few years, florbetapir is the only US Food and Drug Administration-approved brain imaging agent. Chang and co-workers developed a cationic styryl-based Aβ plaque-specific probe. It was the complement to the standard thioflavin S. It also showed advantages for imaging of Aβ aggregates, whereas thioflavin T (ThT) and Congo Red (CR) bind to mature amyloid fibrils. However, both probes (**2C40** and **2E10**) were not suitable for *in vivo* application.[43] Second-generation neutral derivatives were generated without disturbing the core structure. The neutral lipophilic small molecules penetrate the blood–brain barrier *in vivo* and can image Aβ amyloid plaques in the brain. A recent study suggests that the Aβ oligomer is more responsible for AD disease progression than Aβ plaque deposition, due to high toxicity. Hence, the development of probes for Aβ oligomers is significant. Thus, we discuss probe development for Aβ oligomers in the next section.

5.3.2.1 Aβ Oligomer

Several methods have been used for Aβ oligomer detection, including peptide-flash systems, fluorescent protein, antibody, and ELISA. Despite the existence of these methods, their inability to cross the blood–brain barrier makes then unfit for *in vivo* application. Inspired by the success story of developing a neutral hydrophobic probe for *in vivo* β-amyloid plaques, Chang and co-workers focused on developing a probe for Aβ oligomers utilizing neutral DOFL compounds. Systematic cell-based screening was carried out on 3500 compounds to identify an Aβ oligomer-specific probe, named **BD-Oligo.**[56] This chemical probe displayed a turn-on response towards Aβ oligomers compared to either monomers or fibrils. This probe enables the monitoring of the dynamics of Aβ oligomers from monomers to oligomers, which ultimately form fibrils. The fluorescence intensity increases over time and decreases after fibril formation. The molecular docking study revealed a conformational change from planar to twisted geometry to maximize π–π interaction with the aromatic ring of **BD-Oligo** and F19/V36 residues in the Aβ oligomer. The neutral **BD-Oligo** facilitated crossing of the blood–brain barrier to detect oligomers in AD transgenic mouse brains without toxicity. **BD-Oligo** is believed to be a potential optical *in vivo* imaging probe for Aβ oligomers for disease diagnosis and treatment monitoring.

5.3.3 Cell-based High-throughput Screening

The discovery of cell-type-selective probes is more complex than the development of chemosensors/probes. The mainstream of cell-type-specific probes is entirely dependent on cell-surface biomarkers. After the nomenclature designation of these cell-surface molecules as "cluster of differentiation (CD)" at the *1st International Workshop and Conference on Human Leukocyte Differentiation Antigens (HLDA)* in 1982, CD markers were used to differentiate one cell type from another. Until now, more than 350 CD cell-surface markers, known as CD antigens, have been serially named with the prefix CD (CD1, CD2, ... to CD371). CD antibodies identify CD antigens through immunolabeling. However, CD antibodies are still limited for quite a few cell types. The intracellular markers are different from one to another. Still, they cannot differentiate the cell using antibody methods, as the massive size of the antibody (~150 kDa) cannot diffuse through the cell membrane, unless the fixing method is used. Small molecules can quickly penetrate cell membranes and discriminate cell type by the gating-oriented live-cell distinction approach.

A small molecule can interact with intracellular molecules, which are undetectable by the CD antibodies. Thus, Chang and co-workers screened >12 000 DOFL compounds in various cell types, such as immune cells, muscle cells, stem cells, endocrine cells, neuronal cells, cancer cells, and bacterial cells, to develop cell-type-specific probes. However, the critical challenge was target identification due to non-covalent interactions with target-binding ligands. With the loss of target-binding ligands during proteome or transcriptome in-gel analysis, primary chemical probes may need further modification for *in situ* bio-orthogonal labeling. The following sections summarize several cell-type-specific probes determined between 2000 and 2021 from the screening of synthetic libraries and their target identification. The mimicked small-molecule-based antibody generation strategy by combinatorial screening is an advanced technology. So far, Chang and co-workers have successfully achieved ~17 small-molecule-based CD markers.

5.3.3.1 Stem Cell Probes

Stem cell research is at the heart of biology and modern regenerative medicine due to the properties of stem cells. Pluripotent stem cells (PSCs) can differentiate into any cell type, such as induced pluripotent stem cells (iPSCs), embryonic stem cells (ESCs), and various other cell types. These can be isolated from a wide range of tissues in life science. The iPSC can be generated from reprogramming somatic cells. Similarly, ESCs are derived from the inner cell mass of blastocysts. PSCs have been used extensively to study model human diseases, developmental biology, and cell-based therapies. Thus, a cell-type-specific imaging probe for their identification, isolation, and characterization is of great interest. Stem-cell-selective CD antibodies are highly useful for detecting PSCs, but suffer from high costs and often require cell fixation and permeabilization. Thus, the small-molecule fluorescent probe is an unprecedented alternative tool for live stem cells research. Chang and co-workers have developed five fluorescent probes for PSC using DOFLA.

5.3.3.2 Pluripotent Stem Cell Probes

Im *et al.* first developed a small-molecule-based fluorescent probe for mouse embryonic stem cells (mESCs) by high-throughput screening of 280 DOFL compounds of a rosamine library. In screening of mESCs and mouse embryonic fibroblasts as a negative control in 384-well plates, the authors identified an mESC cell-type-specific imaging probe with yellow emission, defined as the compound of designation yellow 1 (**CDy1**)[57] (Figure 5.8). **CDy1** was stained to human ESCs and reprogrammed fibroblasts into iPSCs compared to the control fibroblasts. It can efficiently discriminate iPSC cells at early stages during fibroblast reprogramming, whereas Nanog and Oct4-GFP genetic reporters cannot discriminate. **CDy1** can identify reprogramming iPSCs through quick and simple screening without immunostaining or genetically modified fluorescent proteins in cells. As evidence, Vendrell *et al.* established a **CDy1**-based screening of 240 hydroxamic acids in OSK-transduced cells (transfected fibroblasts with Oct4, Sox2, and Klf4).[58] A positive inducer was discovered by tracking cells that undergo reprogramming. Hirata *et al.* also identified KP-1 (Kyoto probe 1) for induced human iPSC (hiPSC) after screening of 326 rosamine compounds.[59] Unlike **CDy1**, this **KP-1** probe was not efficient for early detection of iPSC during reprogramming. The **KP-1** probe bound reversibly to ALDH2 in mitochondria, depending on probe concentration and incubation time (Figure 5.8).

Later, Chandran *et al.* also discovered another efficient fluorescent probe for mESCs named compound of designation yellow 9 (**CDy9**)[60] (Figure 5.8). It was applied for highly selective detection and isolation of mESCs using fluorescence microscopy and flow cytometry. **CDy9** is positively stained to the mouse embryo's inner cell mass (ICM), suggesting high specificity for

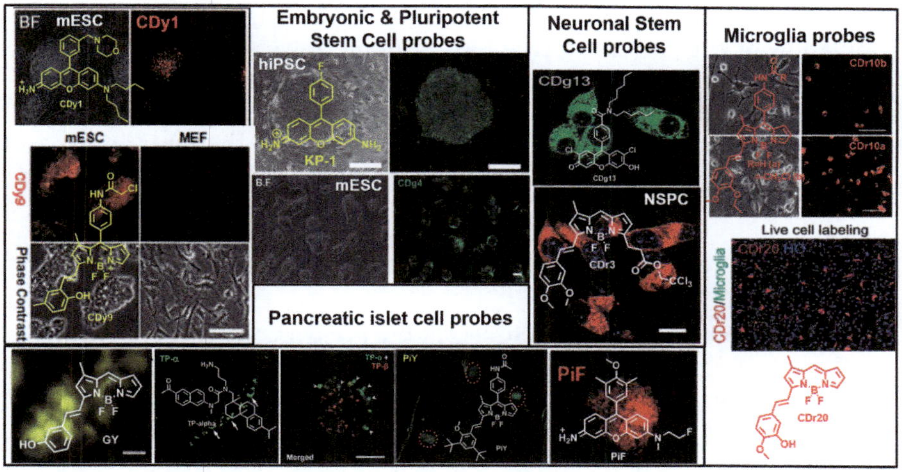

Figure 5.8 The discovery of small molecular fluorescent probes for stem cells, microglia, and pancreatic islets from rosamine, BODIPY, acedan, and fluorescein derivatives. Reproduced from ref. 57 with permission from John Wiley & Sons, Copyright © 2010 WILEY-VCH Verlag GmbH & Co. KGaA, Weinheim; ref. 60 with permission from Elsevier, Copyright 2015; ref. 59 with permission from Elsevier, Copyright 2014; ref. 61 with permission from the Royal Society of Chemistry; ref. 65 with permission from John Wiley & Sons, Copyright © 2016 Wiley-VCH Verlag GmbH & Co. KGaA, Weinheim; ref. 63 with permission from National Academy of Sciences; ref. 67 with permission from the Royal Society of Chemistry; ref. 68 with permission from John Wiley & Sons, Copyright © 2019 Wiley-VCH Verlag GmbH & Co. KGaA, Weinheim; ref. 69 with permission from American Chemical Society, Copyright 2009; ref. 70 with permission from American Chemical Society, Copyright 2015; ref. 72 with permission from American Chemical Society, Copyright 2017; ref. 71 with permission from John Wiley & Sons, Copyright © 2013 WILEY-VCH Verlag GmbH & Co. KGaA, Weinheim; ref. 73 with permission from American Chemical Society, Copyright 2020.

naïve mESCs. It can isolate naïve mESCs from a mixture of naïve and primed mESCs by fluorescent-activated cell sorting (FACS) analysis. **CDy9** is limited to naïve cells, but did not work in fixed cells. Gene ontology analysis of membrane transporters, *Slc13a5*, identified a target gene from 1402 overexpressed genes in naïve stem cells. This *Slc13a5* gene encodes a citrate influx transporter.

The emission color was the same for **CDy1**, **KP-1**, and **CDy9**. Multicolor imaging probes are frequently required, from fundamental research to clinics, considering their wide range of applications. Therefore, Lee *et al.* screened 160 chalcone derivatives to discover a green color fluorescent probe for mESCs, which was named compound of designation green 4 (**CDg4**).[61] It stained the plasma membrane and around the ESC colony, differently from **CDy1** and **CDy9**, which stained mitochondria and the ICM. Target-binding ligands of **CDg4** were found to be glycogen, and were further confirmed from modulation of the fluorescence intensity upon amylase treatment, which hydrolyses glycans. Ghosh *et al.* screened 160 blue-emitting xanthone compounds and developed an ESC-specific fluorescent probe named as a compound of designation blue 8 (**CDb8**).[62] Depending on the emission color needed, **CDy1**, **CDy9**, **CDg4**, and **CDb8** can effectively detect, isolate, and characterize PSCs, especially ESCs.

5.3.3.3 Neural Stem Cells

Research on neural stem cells (NSCs) is at the forefront of regenerative medicine for neurological disorders. NSCs derived from the ectoderm germ layer can differentiate into neurons and four types of neuroglia to generate the central nervous system (CNS). Discrimination between NSCs and PSCs is essential for cell replacement therapy and basic research in human neurons. Yun *et al.* successfully developed the first small molecular probe for both human and mouse NSCs, named as compound of designation red 3 (**CDr3**)[63] (Figure 5.8). Systematic high-throughput screening was employed for 3160 DOFL compounds in mouse NSCs (**NS5**) along with negative control ESCs (**E14**), astrocytes, and fibroblasts. The specificity towards NSCs was confirmed in a mixed cell culture of primary mouse brain cells and human cells differentiated from ESCs. Fatty acid-binding protein 7 (Fabp7) was identified as a target protein and successfully isolated by two-dimensional (2D) gel electrophoresis and analyzed by matrix-assisted laser desorption/ionization (MALDI) mass spectrometry (MS). It was further confirmed by staining with **CDr3** to the recombinant FABP7-expressing HEK293 cells. NSCs proliferated after being stained with **CDr3**. However, **CDr3** was limited in its selectivity as it cannot discriminate NSCs from astrocytes, which also overexpressed Fabp7 target-binding protein. Hence, **CDr3**-stained isolated NSCs were distinct phenotypes from CD antibodies (CD133, CD15) and Aldefluor. Thus, developing a neurogenic NSC fluorescent probe that can target different NSC biomarkers with different colors would be helpful for isolation and characterization. Kim *et al.* systematically screened in-house DOFL compounds to discover neural stem/progenitor cells (NSPCs) and identified two efficient fluorescent probes, named **CDy5**[64] and **CDg13**.[65] The developed **CDy5** probe stains NSPCs by binding to acid ceramidase in mouse neurospheres. Another new biomarker of mouse and human NSPCs was determined to be the low-expressed Abcg2. Hence, the probe was used to isolate cells with $Abcg2^{low}$ from heterogeneous cells. **CDg13**-stained NSPCs are more prone to differentiate into neurons (Figure 5.8).

5.3.3.4 Neuron Cells

After developing the NSC probe, Teoh *et al.* developed a chemical probe for live neuron labeling after in-house screening of 5040 DOFL compounds derived from BODIPY, xanthone, rosamine, and chalcone scaffolds.

Twenty-one derivatives from the hit compounds were further evaluated carefully in primary neurons, astrocytes, and isolated microglia from P1–P3 neonatal mouse brains. The image-based analysis identified a neuron-specific probe with orange emission, named neuron orange (**NeuO**).[66] This **NeuO** was selective only towards live cells, and not to dead neurons. After 1 hour post-injection of **NeuO**, *in vivo* brain tissues showed strong fluorescence signals from the neurons in the cerebellum, hippocampus, and spinal cord. This **NeuO** is a potential optical tool for valuable neurodegenerative studies and neuronal networking.

5.3.3.5 Microglia Probe

The CNS comprises two major cell types: neurons and glial cells. We discussed the importance of neurons and the corresponding chemical probe (**NeuO**) above. Here, we discuss glial cells, including astrocytes, oligodendrocytes, and microglia. About 5–12% of total glial cells, microglia are the prime immune cells in the CNS and play a pivotal role in maintaining brain functions. However, over-activation may lead to inflammation in the brain, resulting in infectious, ischemic, traumatic, and degenerative disease progression. Thus, microglia are also now a prime target for imaging and therapy. Fluorescent-labeled CD antibodies, lectin, and fluorescence proteins with genetic manipulation are the common imaging agents limited to disruption of native cellular function, targeting only surface markers.

A small molecular probe that specifically stains live microglia in heterogeneous populations of brain cells is key to the investigations to address these issues. Leong *et al.* systematically screened microglia, astrocytes, and primary neurons in-house with 5000 DOFL compounds to identify a microglia-specific red-emissive fluorescent probe from the BODIPY library, which was named compound of designation red binding (**CDr10b**)[67] (Figure 5.8). A minimal structural change of this **CDr10b** from a thiol-reactive chloroacetyl to a non-reactive acetyl group, **CDr10a**,[67] was also specific to microglia. **CDr10a** was washed out much faster than its parent **CDr10b**, which can covalently link with target-binding ligands, but non-specificity was the critical concern. The specificity of **CDr10a** and **CDr10b** was further confirmed by overlapping images of primary neural cells derived from CX3CR1-GFP transgenic mice and BV2 microglia. Both fluorescent probes (**CDr10a** and **CDr10b**) were potential chemical tools for live imaging microglia but failed to stain microglia in brain tissues. Thus, Kim *et al.* further screened 880 asymmetric mono-styryl BODIPY compounds against microglia, astrocytes, and oligodendrocytes with neurons. A new BODIPY derivative was revealed as a red-emissive microglia probe and designated as **CDr20**[68] (Figure 5.8). The structure–activity relationship of 36 BODIPY derivatives showed that at least one phenol group is essential for labeling microglia. *In vitro* spectral analysis demonstrated that **CDr20** did not emit in aqueous solution due to excited-state proton transfer from phenol, suggesting no background during fluorescent imaging. The UDP-glucuronosyltransferase family protein,

Ugt1a7c, was induced to form CDr20-Glu, which led to a turn-on response in microglia. **CDr20** was efficient for microglia isolation from heterogeneous cell populations compared with **CDr10** and the commercial probe, fluorescence-dye-conjugated isolectin B4 (IB4) using FACS. IB4 and **CDr10** were revealed to be non-specific to the endothelial cells and other glial cells. Transcriptome profiles of CDr20-based cell-sorted microglia nicely matched with microglia-specific antibody CD11b^{+}/CD45^{low}-gated cells. Moreover, **CDr20** showed promise as a label of microglia *in vivo* upon intravenous injection in the aged Alzheimer's disease mouse model. Hence, **CDr20** is a powerful chemical tool for identifying microglia in neurodegenerative disorders.

5.3.3.6 *Pancreatic Islet Cell Probe*

The pancreas is a unique gland that excretes hormones and enzymes into the blood through ducts. The pancreatic islets of Langerhans are composed of three major cell types, α-cells, β-cells, and δ-cells. Insulin-secreting β-cells and glucagon-secreting α-cells play a crucial role in regulating glucose levels in the blood by functioning in opposition to each other. Dysfunction of β-cells leads to hyperglycemia, which is known to cause diabetes, of which there are two types. Type 1 (does not produce insulin) and type 2 (does not produce enough insulin) affect 8.8% of the adult population worldwide. Biochemical assays easily quantify the levels of these hormones in the plasma. However, the functions of the live pancreatic islets, especially α-cells and β-cells are still challenging as the conventional antibody method interrupts native cellular function. Thus, Lee *et al.* screened 160 BODIPY-based DOFL in alphaTC1, betaTC6, and other cell types as negative control, and identified the first α-cell-selective turn-on fluorescent probe, named Glucagon Yellow (**GY**) based on emission color[69] (Figure 5.8). Immunohistochemistry (IHC) staining of pancreatic tissue nicely overlapped with anti-glucagon but not with anti-insulin. **GY** exhibited a remarkable selectivity toward glucagon peptide hormone compared to 18 biologically relevant analytes, including serum albumins, and visualized glucagon in live cells and partially to the pancreatic tissue. However, it is limited to deep-tissue imaging of islets due to the negligible two-photon (TP) property. Agrawal *et al.* synthesized an 80-member acedan-based TP library through solid-phase synthesis and screened it with TC1, TC6, and exocrine acinar 266-6 cells. The image-based analysis revealed the first TP probe for live α-cell imaging in intact mice. The name of this probe was **TP-α**, and it is suitable for TP pancreatic islet imaging.[70]

GY was nicely stained in isolated α-cells, and **TP-α** was effective for pancreatic tissue and may be effective for monitoring the survival of α-cells during islet transplantation. However, targeting small-population α-cells (~15–20%) is more challenging than large-population β-cells (75–80%) in pancreatic islets. Thus, Kang *et al.* screened 1200 DOFL compounds in pancreatic islet cells (betaTC-6, alphaTC-1, and control acinar cells) to identify β-cell-selective two-fluorescent probes. The *N*-acetyl BODIPY derivative, pancreatic islet yellow (PiY), was specific to β-cells, but the corresponding chloroacetyl derivative was non-specific.[71] The molecular mechanism of the non-specificity remains to be elucidated. PiY was a potential probe for viewing damaged islets induced by streptozotocin (STZ) in a diabetic mouse model. The PiY was still limited to damaged islets but unable to perform deep-tissue imaging. Thus, Agrawalla *et al.* synthesized TP probes attached with a variable linker to the 2-glucasamine, a derivative of STZ that targets β-cells. The six-carbon (C6) linker of the TP probe, named **TP-β**, exhibited maximum uptake into primary β-cells compared with other probes in the mouse model (Figure 5.8).[72] Interestingly, **TP-β** displayed an orange emission, which was accompanied by the green-emissive **TP-α** for multiplexing. The authors demonstrated dual-color imaging of alpha and beta cells in live islets using both **TP-α** and **TP-β** probes. **TP-α** stained the periphery, whereas **TP-β** stained at the core of the pancreatic islet. **TP-β** did not interfere with insulin secretion. Although these developed pancreatic probes (**GY**, **PiY**, **TP-α**, **TP-β**) are useful for *in vitro* and *in vivo* imaging of pancreatic islets in mouse models, there are still limits to their use in human islets due to the lack of penetration depth of visible light. The authors further screened the DOFL library to identify the β-cell-selective rosamine derivative, designated as **PiF**, which can visualize and isolate β-cells from pancreatic islets without any immunostaining[73] (Figure 5.8). **PiF** can be easily modified to a positron emission tomography (PET) imaging probe for deep-tissue imaging by labeling with ^{18}F radioisotope. **PiF** can efficiently image pancreatic tissue without fixation and removal of tissue freezing media. The ^{19}F isotope of **PiF** was replaced with the radioisotope ^{18}F to obtain [^{18}F]**PiF** and successfully demonstrated deep-tissue imaging in mouse pancreatic islets. The discovery of fluorescent probes from DOFLA and their subsequent modification for MRI, PET, and SPECT may provide advanced clinical diagnosis.

5.3.3.7 Myocyte Probe in Muscle Cell

Muscle cells, known as myocytes, can be classified into three types in the human body: cardiac, skeletal, and smooth. Skeletal muscle is formed by multinucleated myocytes, which are long and thread-like and known as myotubes. In the case of myogenesis, the myoblasts fuse with myotubes and develop skeletal muscular tissue. Myocytes dedifferentiate into myoblasts (embryonic progenitor cells) in amphibians and regenerate muscle cells in lost tails and limbs. Thus, insight into the mechanism for determining the cell state may offer therapeutic advances in muscle regeneration. Myotube-specific proteins are capable of distinguishing myoblasts from differentiated myotubes during myogenesis. However, a small molecular probe would provide advances in high-throughput screening. Wagner *et al.* screened 1606 DOFL compounds in differentiated C2C12 myoblasts and myotube cells to identify six myotube-selective compounds that stain myotubes more brightly than myoblasts.[74] Of the six compounds, **E26** was the best probe and was applied for further screening of 84 kinase inhibitors. A total of 17 new inhibitors were found along with rapamycin, a known myogenesis inhibitor.

However, **E26** was not photostable and so lacks real-time imaging. Later, Kim *et al.* incubated C2C12 myoblasts with DOFL compounds for five days under differentiation conditions. Mononucleated myoblasts differentiated into myotubes in almost every well, but **B25**-treated cells were undifferentiated.[75]

Due to the lack of target binding with **B25**, the authors explored further, for advanced probe discovery. Interestingly, similarly to the **B25** structure, **A25**, with a primary amine derivative, formed cylindrical multinucleated myotubes. Kim *et al.* rescreened the DOFL compounds under extended light exposure to reveal two more myotube-selective probes, **I25** and **I31**, from the rosamine library[76] (Figure 5.9). Interestingly, the **I25** structure is almost the

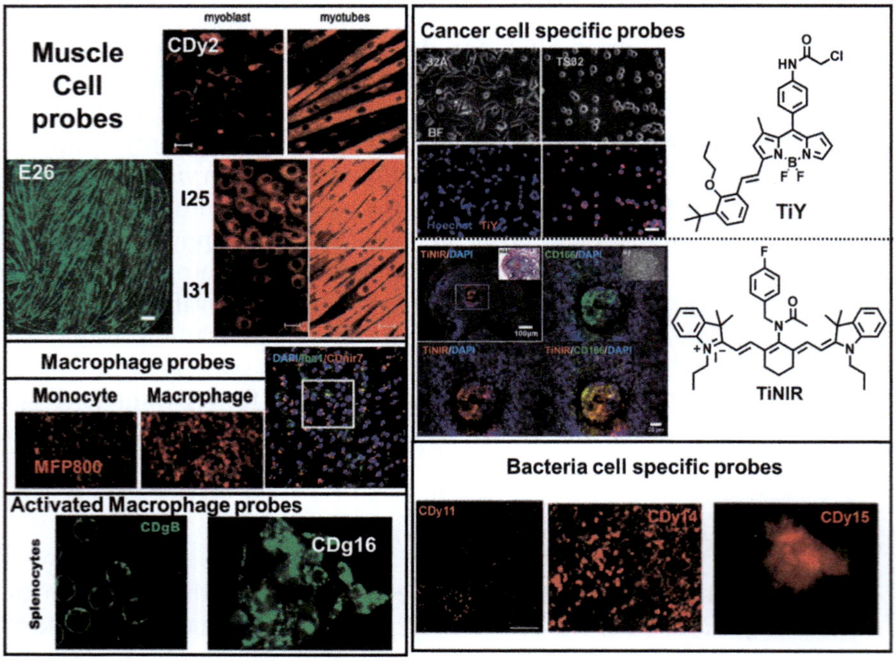

Figure 5.9 The discovery of small molecular fluorescent probes for muscle cell, macrophage, activated macrophage, cancer stem cells, and bacteria cells from rosamine, cyanine, chalcone, and BODIPY derivatives. Reproduced from ref. 76 with permission from John Wiley & Sons, Copyright © 2011 WILEY-VCH Verlag GmbH & Co. KGaA, Weinheim; ref. 74 with permission from American Chemical Society, Copyright 2008; ref. 78 with permission from Springer Nature, Copyright 2019; ref. 85 with permission from John Wiley & Sons, Copyright © 2018 Wiley-VCH Verlag GmbH & Co. KGaA, Weinheim; ref. 86 with permission from American Chemical Society, Copyright 2019; ref. 82 with permission from American Chemical Society, Copyright 2021; ref. 81 with permission from Springer Nature, Copyright 2019; ref. 87 with permission from American Chemical Society, Copyright 2016; ref. 88 with permission from the Royal Society of Chemistry; ref. 89 with permission from the Royal Society of Chemistry.

same as **B25** except for the long alkyl chain on the donor site. **I31** differs from the **I25** in the phenyl ring. Both compounds were localized in mitochondria, but the target-binding protein was tubulin, a cytoskeletal protein. This controversial target binding led to the synthesis of the thiol-reactive derivative of **I25**, named as compound of designation yellow 2 (**CDy2**).[76] A new binding protein, aldehyde dehydrogenase 2 (ALDH2), was identified and confirmed by short interfering (si) RNA experiments. Depending on the experimental conditions, **CDy2** binds ALDH2 and tubulin. Target protein identification is the most challenging aspect in chemical biology as it differs between experiments.

5.3.3.8 Inflammation and Macrophage-specific Probes

When the body is infected with pathogens (bacteria, viruses, and parasites) an alarm is raised in the immune system for sending out two armies of cells: innate and adaptive cells. The body's first line of defense are the innate cells, which include monocytes, macrophages, and neutrophils that are deployed to fight invading pathogens. Once an invading pathogen crosses this first line of defense, B and T lymphocytes provide adaptive immunity to fight against the specific pathogen. These different immune cell types are distinctive in their surface biomarkers. Each cell type can be discriminated by targeting specific biomarkers with the corresponding antibodies. As an alternative to cell-surface-targeting anti-CD antibodies, small molecules can also differentiate cell types by targeting intracellular biomarkers. Thus, Chang and co-workers explored DOFL compounds to discover new imaging probes that can discriminate immune cell types using high-throughput screening. These imaging probes can study the immune response of tissues triggered by injury, heat shock, and allergic response. During inflammation, immune cells, especially monocytes and lymphocytes (B cells, T cells, and natural killer (NK) cells) migrate from the bone marrow to the site of the damaged tissue. Once monocytes are recruited to the tissues, they differentiate into macrophages (Mφ or MΦ). Macrophages remove invading pathogens, apoptotic cells, and cellular debris ($\sim 2 \times 10^{11}$ erythrocytes per day) or recycle their content (~3 kg of iron per year).

Hence, macrophages are crucial for metabolism. Macrophage imaging provides advanced knowledge of the immune system and plays a pivotal role in elucidating disease progression and effective therapies. The microglia-specific probe, **CDr10b**, was a potential chemical probe to study neuroinflammatory diseases.[67] Later, Gu *et al.* evaluated the effect of **CDr10b** on the immune system and identified suppression of two of the most important genes, cycloxygenase-2 (COX-2) and induced nitric oxide synthase (iNOS), which regulate certain inflammatory diseases.[77] **CDr10b** also suppresses lipopolysaccharide (LPS)-induced NF-κB activation in murine macrophages and regulates toll-like receptor (TLR4) signaling pathways. It also modulates NF-κB and IRF3 signaling pathways that induce inflammatory interferons and cytokines to defend against invading pathogens. In particular, COX-2, iNOS, IFNβ, and IP-10 were downregulated upon treatment with **CDr10b**, resulting in reduced chronic inflammation. **CDr10b** offers a potential therapeutic target for developing anti-inflammatory drugs.

However, **CDr10b** is still limited to *in vivo* imaging of inflamed tissue due to low tissue penetration of the visible light-absorbing probe. Thus, the high-throughput screening was restricted to near-infrared light-absorbing DOFL compounds. Kang *et al.* systematically screened the first reported near-infrared (NIR) libraries composed of 320 compounds to identify a final hit molecule, named as compound of designation near-infrared 7 (**CDnir7**) with excellent uptake in macrophages over other immune cells[78] (Figure 5.9). Fluorescence molecular tomography imaging in an animal model of inflammation demonstrated macrophage influx into an inflamed paw. Furthermore, photoacoustic imaging corroborated the potential application of **CDnir7** for imaging tumor-associated macrophages *in vivo* in mice bearing 4T1 breast tumors. **CDnir7** revealed efficient *in vivo* molecular probes with fast response and rapid plasma clearance ($t_{1/2}$ = 21 min) compared to the available nanoparticle-based contrast agents.

When cancer spreads from the primary site to other parts of the body through the bloodstream or the lymph system, known as metastasis, there is limited imaging modality for the metastatic and inflamed lymph nodes. As DOFLA revealed powerful chemical tools, Yoo *et al.* screened 1153 DOFL compounds against immune cells, MOLT-4 for lymphocytes, HL-60, U-937 for monocytes, and U-937-DM for monocyte-derived macrophages, to discriminate between closely related cell states.[79] A monocyte- and macrophage-specific fluorescent probe, designated **MFP**, was discovered to visualize sentinel lymph nodes during surgery. The emission from the probe displayed abnormalities in the tumor and inflamed tissue. *In vivo* macrophage imaging in lymph nodes offers real-time monitoring of lymph node metastases, inflammation-associated adenopathy without lymph node biopsy, and histology. However, **MFP** emits light in visible wavelengths, resulting in a high background signal. To discover an efficient NIR-based macrophage-specific imaging probe, the authors extended the screening towards NIR cyanine and tetra-arylazadipyrromethene libraries of 520 compounds. From a rigorous cell-based screening with the established protocol and flow cytometry assay, a promising macrophage-specific probe (**MF800**) was discovered[80] (Figure 5.9). Specific staining of U-937 and U-937-DM with MF800 revealed visualization of inflamed tissue in the hind limb ischemia model. **MF800** co-localized in the femoral artery with genetically labeled macrophages (csf1r-EGFP) in the mouse model. Moreover, good correlation in the immunofluorescence of macrophage surface markers (CD68 and CD169) with **MF800** in the inflamed paw and the ischemic hind limb confirmed the high selectivity of the probe to macrophages.

5.3.3.9 Activated Macrophage

Macrophages (Mφ) that either stimulate or mitigate inflammation are known to be M1 or M2 macrophages, respectively. Mφ can be induced to M1 by treating with LPS or interferon-γ (IFN-γ) and induced to M2 by interleukin-4 (IL-4) treatment. Both M1 and M2 macrophages are known as activated macrophages (Mφ*). Selective detection of M1 activated macrophage (Mφ*) leads to the advanced diagnosis of inflammatory diseases. However, discrimination between M1 activated macrophages (Mφ*) from non-activated macrophages (Mφ) is challenging. Park *et al.* performed unbiased cell-based high-throughput imaging of over 8000 DOFL compounds using Mφ* and Raw 264.7 (Mφ) as a negative control. A Mφ*-specific fluorescent probe was discovered from the acridine-based library. The name of the Mφ* probe was compound of designation green (**CDg16**), which was accumulated in atherosclerotic plaques in an animal model, resulting in imaging of active inflammation[81] (Figure 5.9). Its staining mechanism was elucidated by next-generation sequencing analysis, to uncover the SIC18B1 transporter. Simple intravenous injection of **CDg16** can visualize the active inflammation. It provides a unique diagnostic tool for several inflammatory diseases, including allergy, asthma, hepatitis, and inflammatory bowel disease.

5.3.3.10 *Live-cell Distinction of B and T Lymphocytes*

The body's second line of defense are the adaptive immune cells, B and T lymphocytes. These cells fight against foreign invaders, *i.e.*, pathogens, by producing antibodies and cytokines. As both B and T cells originate from the same progenitor cells, they are similar in size and shape. Hence, there is a considerable challenge in distinguishing them. Kwon *et al.* systematically screened over 10 000 DOFL compounds in B and T cells. These cells were isolated from the spleen by negative MACS (magnetically activated cell sorting) with biotin-labeled antibodies and a chemical probe for B cells was identified, named **CDgB** (compound of designation green for B cell).[82] Isolated **CDgB**-stained cells in flow cytometry were nicely matched with anti-$CD45R^+$ cells, which distinguish B cells from T cells. **CDgB** localized on the plasma membrane was analyzed by LCMS-MRM (liquid chromatography-mass spectrometry-multiple reaction monitoring) (Figure 5.9). **CDgB** preferentially binds the short carbon chain 34 of phosphatidyl choline (C34-PC) in a flexible B-cell membrane rather than the more rigid T-cell membrane. The B-cell model giant unilamellar vesicle (GUV) suggests that nano-aggregates of **CDgB** are stained at the membrane by a fusion-induced staining mechanism. CyTOF (cytometry with time-of-flight) mass spectrometry analysis with and without **CDgB,** comparing with 28 antibodies, confirmed that $CDgB^+$ cells are B cells and $CDgB^-$ cells are T cells. The mechanism of discrimination lies in lipid-oriented live-cell distinction.

5.3.3.11 *Cancer Cell, Mitotic Cell and Cancer Stem Cell Probes*

Both genomic and proteomic studies of cancer demonstrated cancer heterogeneity. Cancer is significantly associated with aberrations of structural chromosomal genes. Lack of DNA repair in solid tumors leads to genetic instability, resulting in the alteration of millions of proteins. Lee *et al.* examined the heterogeneity of NCI-60 human cancer cells by treating, in-house, 557 DOFL compounds. The fluorescence intensity change was analyzed by linear discriminant analysis to distinguish the phenotypes of each cancer cell type. From careful research, a rosamine-based fluorescent probe was found to be specific to KM12 colon cancer and three rosamine compounds were specific for ACHN renal cancer cells. The heterogenicity of cancer cells can be phenotypically classified by employing DOFL compounds. Hence, DOFL offers new chemical probes for identifying cancer cells and may provide advanced diagnostic tools.

To identify the mitosis process, Jeong *et al.* developed a long-term photostable mitotic or proliferating marker, CDy6, which has the ability to label a mouse xenograft model in real time.[83] Subsequently, a xanthone-based low-toxicity nucleus-staining fluorescent probe, **CDb12**, was also developed by the DOFL approach for mitosis imaging in live cells.[84]

Cancer stem cells, also known as tumor-initiating cells (TICs), produce various differentiated progeny to balance the rapidly dividing tumor. TICs activate DNA repair proteins that oppose DNA damage and result in faster growth of cancer cells. TICs are also associated with metastasis and resistance to conventional anti-cancer therapy. A few cancer cell surface markers and corresponding antibodies have been developed, but these antibodies have fundamental limitations to their intracellular markers. Additionally, a combination of multiple antibodies is required to target TIC selectively. The discovery of a cell-selective small molecular probe is an alternative. Lee *et al.* methodically screened thousands of DOFL compounds in tumor sphere 32 (TS32 cells) with CD166$^+$ as positive and CD166$^-$ as negative adherent cells (32A) in the primary screening.

Furthermore, image-based phenotypic screening and flow cytometry assays in patient-derived lung TS (TS10) and normal lung epithelial cells (LuEpi) were performed to identify the first probe for TIC, named tumor-initiating cell yellow (**TiY**).[85] It was stained at cytosol of a TS32- xenograft (TDX 32) tumor cell and exhibited greater propensity for *in vitro* sphere formation. It also showed more specificity towards tumorigenic cells, irrespective of any cancers, than the CD166$^+$ antibody. A new cytosolic marker, vimentin, was identified from mass spectrometry and confirmed from reverse transcriptase–quantitative polymerase chain rection (RT–qPCR) and siRNA knockdown. The **TiY** is an alternate and advanced imaging probe for visualization and isolation of TICs, and can be applied for drug discovery targeting vimentin biomarkers (Figure 5.9). However, **TiY** is also limited for deep-tissue imaging *in vivo*. Next, Kim *et al.* performed a cell-based unbiased screening of DOFL compounds with CD166$^+$ TIC cells (TS32, TS10) and CD166$^-$ (32A, LuEpi) to discover an NIR-emitting TIC-specific probe named as **TiNIR.**[86] Patient-derived xenograft tumor cells were transplanted to NOD scid gamma (NSG™) mice and two distinct populations of the stained cells using both **TiNIR** and anti-CD166 were isolated by flow cytometry. Sphere formation was 1.5 times better in the **TiNIR**$^{\mathbf{high}}$ fraction compared to the CD166^{high} fraction. *In vivo* TIC imaging was further confirmed by multispectral optoacoustic tomography imaging. Next, **TiNIR**-treated cell extracts were analyzed by LC–MS/MS analysis to identify the novel biomarker, HMOX2. At a high concentration of **TiNIR**, **TiNIR**$^{\mathbf{high}}$, accumulation of ROS was induced in the TIC cells by blocking the function of HMOX2. Thus, **TiNIR** suppresses the growth of TIC cells at high concentrations, resulting in tumor treatment. Therefore, **TiNIR** is a potential theranostic NIR-fluorescent agent for TIC imaging and therapy.

5.3.3.12 Bacteria Cell Targeting Biofilm Probe

Despite the successful application of the first antibiotic, penicillin, used for controlling bacterial infections for soldiers in 1940 during the Second World War, bacterial infections remain one of the major threats to human society. One of the reasons for antibiotic resistance is biofilm formation. Biofilms are self-generated extracellular polymeric substances (EPS) that protect bacterial cells from oxidative stress, nutrient starvation, and extreme pH. Biofilms develop on the surface of medical instruments, including implant devices, urinary catheters, *etc.* Thus, biofilm detection is essential. Fluorescent tag lectin proteins and *in situ* hybridization (FISH) are often used for biofilm detection. A small molecular fluorescent probe alternative to the traditional approach for biofilm detection was discovered, and named compound of designation yellow 11 (**CDy11**), from primary screening of thousands of DOFL compounds against two different strains of *Pseudomonas aeruginosa*, PAO1 wild type and PAO1Δfap.[87] Kim *et al.* first elucidated that **CDy11** binds with amyloid proteins of *P. aeruginosa* by surface plasmon resonance and antibody staining. Thus, amyloid protein was identified to be one of the key components of EPS of *P. aeruginosa* during biofilm formation. *In vivo* imaging of biofilms in corneal infection and implant animal models revealed the importance of **CDy11** for visualization of bacterial infections. Furthermore, unbiased screening was extended by Kwon *et al.* to the EPS DNA, having polysaccharides in addition to proteins, which were limited initially. Two strains of *P. aeruginosa*, Δ*wspF* mutant and p*YhjH* mutant, were chosen as a positive and negative control to set up the screening of thousands of in-house DOFL compounds. A potential probe was identified, named **CDy14** (compound of designation yellow 14). **CDy14** binds exopolysaccharides in *P. aeruginosa* as targeting biomarkers.[88] The *in vivo* corneal infection model demonstrated the efficiency of biofilm detection in chronic infections. Both **CDy11** and **CDy14** provide new biomarkers in bacteria EPS for early detection of biofilms (Figure 5.9), which may lead to advice on appropriate antibiotic use for the treatment of infectious diseases.

Biofilm detection is a counter strategy to visualize the camouflaged bacteria. Kwon *et al.* performed image-based screening of 2000 red-absorbing (λ_{Em} >700 nm) DOFL compounds against highly elevated cyclic-diGMP, *P. aeruginosa* Δ*wspF* and p*YhjH* strains. A selected molecule, compound of designation red 15 (**CDr15**) exhibited high fluorescence contrast towards Δ*wspF* biofilm over p*YhjH*.[89] The turn-on fluorescence intensity of **CDr15** was significantly increased upon binding with only extracellular DNA (eDNA). A confocal image of nuclei in the live mammalian cells was not stained. Therefore, **CDr15** enabled the detection of bacteria in infected host cells. **CDr15** demonstrated *in vivo* imaging of biofilm in the infected cornea in an animal model, indicating an exciting guiding tool for efficient diagnosis and treatment of biofilm-infected patients. Thus, biofilm types can be revealed by combining these developed probes (**CDy11**, **CDy14**, and **CDr15**).

5.3.3.13 Gram-positive Bacteria

The structural diversity of DOFL compounds exhibits promising chemical tools for targeting different macromolecules in mammalian to bacterial cells. BODIPY derivatives have been applied for the discovery of many imaging and sensing probes. Boronic acids are likely to bind with the glycoprotein of Gram-positive bacteria. BODIPY derivatives of boronic acids in DOFL compounds were examined against Gram-positive bacteria and two BODIPY probes were identified (**B2** and **B3**). However, boronic acid-containing rhodamine and styryl libraries were unable to discriminate Gram-positive and -negative bacteria. The authors further synthesized a series of boronic acid derivatives of BODIPY. They tested against nine Gram-positive and seven Gram-negative bacteria to identify orange-emissive **BacGO** (bacteria Gram-positive orange).[90] This has several advantages over the standard detecting tool, WGA (wheat-germ agglutinin). **BacGO** detects single bacteria, bacteria colonies, and activated bacterial sludge by cytometry and microscopic images (Figure 5.9). *In vivo* imaging in infected eyes exhibited the potential for clinical diagnosis of Gram-positive bacteria.

5.4 Tissue Elastin Imaging

Near-infrared (NIR) imaging is the most powerful tool for deep-tissue imaging due to its high tissue penetration with the least autofluorescence and reduced scattering. NIR-absorbing fluorophores with electrically neutral surface charge (zwitterionic) lead to non-specific interactions with biomolecules and easy renal clearance. The zwitterionic compounds are attractive for *ex vivo* and *in vivo* studies in various imaging modalities, such as fluorescence and optoacoustic. Su *et al.* first synthesized a zwitterionic NIR cyanine library (CyZW) in solid-phase synthesis to perform image-based tissue screening of a library against frozen mouse aorta tissues. It revealed an elastin-specific imaging probe named **ElaNIR** (elastin NIR).[91] **ElaNIR** bound selectively to elastin in the aorta, lungs, and skin (Figure 5.10). It can also monitor changes of elastin in tissue and quantify elastin levels in different ages of mice. **ElaNIR** is a unique chemical tool for discovering the changes of elastin in live models, by both fluorescence and photoacoustic imaging (PAI).

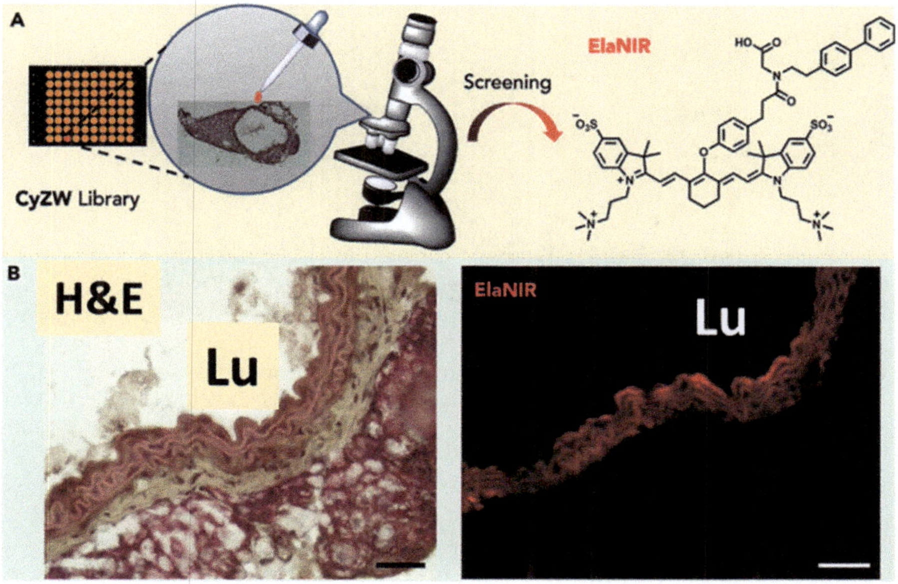

Figure 5.10 (A) Image-based tissue screening to identify the elastin imaging probe CyZW from zwitterionic cyanine fluorophores. (B) Hematoxylin and eosin (H&E) staining of the aorta (left) and ElaNIR staining of elastin (right) of Lu (lumen). Reproduced from ref. 91 with permission from Elsevier, Copyright 2018.

5.5 Animal Imaging

Non-invasive *in vivo* fluorescence imaging plays a crucial role in the precise monitoring of animal phenotypes. Animal imaging offers an ample variety of diagnostic and therapeutic applications. Interest in fluorescent techniques in translational research has been growing substantially due to its biological safety, high sensitivity, and wide acceptance by investigators of its relatively easy use. The major applications are quantitative screening, non-invasive disease diagnosis, and post-treatment monitoring of patients using exogenous contrast agents. Compared to analyte-based screening, cell-based screening always provides greater ability for mimicking the physiological environment. Although whole animal-based imaging is challenging due to poor probe delivery to the specific target, it is always better than analyte- or cell-based screening. To demonstrate this, a water-borne organism, zebrafish larvae, are usually selected for screening due to their transparency. Thus, Li *et al.* screened a rosamine library against zebrafish larvae, which can be easily maintained in large stocks for high-throughput screening. Two neuron-specific *in vivo* imaging probes were identified. Depending on the emissive color of the neural tracers, these were named **ZeN-Green** (zebrafish neurotracer-green) and **ZeN-Red** (zebrafish neurotracer-red).[92] Both neurotracers specifically stained the neurons of

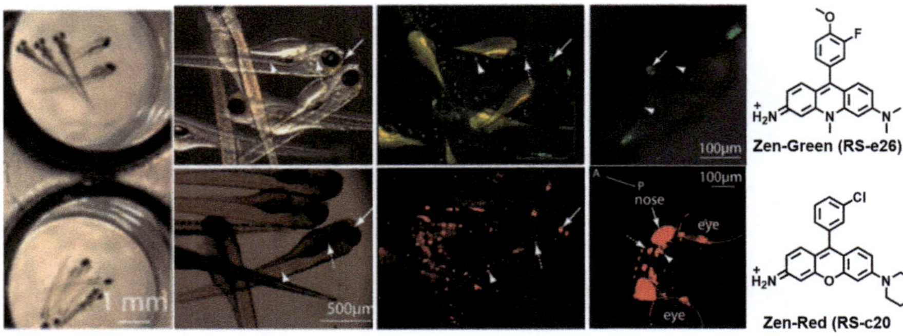

Figure 5.11 Larvae in a 96-well plate for high-throughput screening. High-resolution reconstruction of confocal stacks showing that ZeN-Green and ZeN-Red labeled neuron in the nose, as well as their projections (arrowheads). Reproduced from ref. 92 with permission from the Royal Society of Chemistry.

the internal ear, nose, and neuromasts located at the body surface (Figure 5.11). Good *in vivo* biocompatibility also revealed the superiority of both. A neurotoxin, gentamicin-treated neural damage, led to turn-off fluorescence intensity. Later Teoh *et al.* performed *in vivo* neuronal imaging by simple immersion of zebrafish embryos in neuron orange (**NeuO**) solution and injection of **NeuO** into zebrafish brain ventricles.[66] **NeuO** was discovered as a neuron-imaging probe in the brain, as discussed in Section 5.3.3.4. **NeuO** exhibited potential as an *in vivo* organism imaging probe for studying neuronal development and degeneration.

Similarly, another BODIPY-based probe, **CDr10b**, a potential probe for microglia, was applied in zebrafish larvae. Their external development allows examining DOFL compounds by simply adding them to the water in the embryo of the larvae. **CDr10b** stained explicitly to the mid-intestinal segment and intestinal lumen in larval development. In contrast, standard probes, Neutral Red and Acridine Orange were stained to the macrophages and necrotic/apoptotic cells, respectively. The non-toxic **CDr10b** allows fast detection of intestinal patterning defects. It has become an additional advanced chemical probe to study intestinal biology and injury and repair in the *in vivo* zebrafish model. Small organism-based imaging by small molecular probes could play a substantial role in biomedical research.

5.6 Conclusions and Future Perspective

The DOFLA methodology has accelerated the development of novel chemosensors and chemical probes. The critical limitations in target-oriented synthesis can easily be addressed by DOFLA, as little or even no information about the molecular recognition process is required. Based on the different core fluorescent scaffolds, a diverse range of small molecules are synthesized, which provide a potential chemical toolbox for hunting for molecular interactions in the unknown world of complex biology (summarized in Tables 5.1 and 5.2). The application of DOFLA has shown the way to explore and utilize chemical diversity for detection in diverse areas, including stem cell research, cell state discrimination of immune cells, pancreatic cells, bacteria cells, *etc.* DOFLA is not confined only to complex biology. It has also been used to explore the possibilities for detection of environmental entities and food items, such as bisphenol A, milk fat, water, and other relevant analytes. The discovery of chemosensors and chemical probes capable of sensing and distinguishing novel analytes or biomarkers will undoubtedly offer the scientific community unique and valuable chemical tools. DOFLA has mainly concentrated on structures with known fluorescent core scaffolds. There are plenty of possibilities to diversify using new core structures, for further development. These new fluorescent architectures and new screening strategies will undoubtedly boost the discovery of more sophisticated chemosensors and chemical probes.

Acknowledgements

A.S. is grateful to the DST-SERB, India, for the Ramanujan Fellowship (SB/S2/RJN148/2015), DST-Core Research Grant (CRG/2021/002313) and Shiv Nadar Foundation for funding. PhD students S. M. and A. J. are grateful to Shiv Nadar University and Shiv Nadar Foundation for their research fellowship. YTC was supported by the Institute for Basic Science (IBS) (IBS-R007-A1).

References

1. M. Cobb, *PLoS Biol.*, 2017, **15**, 1–8.
2. J. P. Hughes, S. Rees, S. B. Kalindjian and K. L. Philpott, *Br. J. Pharmacol.*, 2011, **162**, 1239–1249.
3. A. Arranz and J. Ripoll, *Front. Pharmacol.*, 2015, **6**, 189.
4. M. Vendrell, D. Zhai, J. C. Er and Y. T. Chang, *Chem. Rev.*, 2012, **112**, 4391–4420.
5. N. Y. Kang, H. H. Ha, S. W. Yun, Y. H. Yu and Y. T. Chang, *Chem. Soc. Rev.*, 2011, **40**, 3613–3626.
6. D. L. Jasinski, H. Yin, Z. Li and P. Guo, *Hum. Gene Ther.*, 2017, **29**, 77–86.
7. F. Zanella, J. B. Lorens and W. Link, *Trends Biotechnol.*, 2010, **28**, 237–245.
8. *Principles of Fluorescence Spectroscopy*, ed. J. R. Lakowicz, Springer US, Boston, MA, 2006, pp. 1–26.
9. C. Denz, *Ann. Phys.*, 2015, **527**, A77–A80.
10. J. Duraiyan, R. Govindarajan, K. Kaliyappan and M. Palanisamy, *J. Pharm. BioAllied Sci.*, 2012, **4**, 307–309.
11. E. A. Lemke and C. Schultz, *Nat. Chem. Biol.*, 2011, **7**, 480–483.

12. S. W. Yun, N. Y. Kang, S. J. Park, H. H. Ha, Y. K. Kim, J. S. Lee and Y. T. Chang, *Acc. Chem. Res.*, 2014, **47**, 1277–1286.
13. R. A. Bissell, A. Prasanna de Silva, H. Q. Nimal Gunaratne, P. L. Mark Lynch, G. E. M. Maguire, C. P. McCoy and K. R. A. Samankumara Sandanayake, *Photoinduced Electron Transfer V*, ed. J. Mattay, Springer Berlin Heidelberg, Berlin, Heidelberg, 1993, pp. 223–264.
14. R. Misra and S. P. Bhattacharyya, *Intramolecular Charge Transfer*, 2018, pp. 29–69.
15. X. Zhou, L. Lesiak, R. Lai, J. R. Beck, J. Zhao, C. G. Elowsky, H. Li and C. I. Stains, *Angew. Chem., Int. Ed.*, 2017, **56**, 4197–4200.
16. J. V. Mello and N. S. Finney, *J. Am. Chem. Soc.*, 2005, **127**, 10124–10125.
17. D. Su, C. L. Teoh, L. Wang, X. Liu and Y.-T. Chang, *Chem. Soc. Rev.*, 2017, **46**, 4833–4844.
18. J. Luo, Z. Xie, J. W. Y. Lam, L. Cheng, H. Chen, C. Qiu, H. S. Kwok, X. Zhan, Y. Liu, D. Zhu and B. Z. Tang, *Chem. Commun.*, 2001, 1740–1741.
19. D. Zhai, W. Xu, L. Zhang and Y.-T. Chang, *Chem. Soc. Rev.*, 2014, **43**, 2402–2411.
20. S. L. Schreiber, M. Ciba and S. L. Schreiber, 2016, **287**, 1964–1969.
21. F. Yan, K. Fan, Z. Bai, R. Zhang, F. Zu, J. Xu and X. Li, *TrAC, Trends Anal. Chem.*, 2017, **97**, 15–35.
22. M. Beija, C. A. M. Afonso and J. M. G. Martinho, *Chem. Soc. Rev.*, 2009, **38**, 2410–2433.
23. H. A. Shindy, *Dyes Pigm.*, 2017, **145**, 505–513.
24. X. Liu, J. M. Cole, P. G. Waddell, T.-C. Lin, J. Radia and A. Zeidler, *J. Phys. Chem. A*, 2012, **116**, 727–737.
25. A. Gupta, S. Garg and H. Singh, *Anal. Methods*, 2020, **12**, 5022–5045.
26. A. Loudet and K. Burgess, *Chem. Rev.*, 2007, **107**, 4891–4932.
27. E. K. Krasnowska, E. Gratton and T. Parasassi, *Biophys. J.*, 1998, **74**, 1984–1993.
28. H. Shaki, K. Gharanjig, S. Rouhani and A. Khosravi, *J. Photochem. Photobiol., A*, 2010, **216**, 44–50.
29. N. S. Finney, *Curr. Opin. Chem. Biol.*, 2006, **10**, 238–245.
30. A. Singh, Q. Yao, L. Tong, W. Clark Still and D. Sames, *Tetrahedron Lett.*, 2000, **41**, 9601–9605.
31. A. P. De Silva, T. S. Moody and G. D. Wright, *Analyst*, 2009, **134**, 2385–2393.
32. Y. K. Choi, J. J. Kim and Y. T. Chang, *Acc. Chem. Res.*, 2019, **52**, 3097–3107.
33. G. R. Rosania, J. W. Lee, L. Ding, H.-S. Yoon and Y.-T. Chang, *J. Am. Chem. Soc.*, 2003, **125**, 1130–1131.
34. Q. Li, Y. Kim, J. Namm, A. Kulkarni, G. R. Rosania, Y.-H. Ahn and Y.-T. Chang, *Chem. Biol.*, 2006, **13**, 615–623.
35. J. W. Lee, M. Jung, G. R. Rosania and Y. T. Chang, *Chem. Commun.*, 2003, **3**, 1852–1853.
36. S. Feng, Y. Kyung Kim, S. Yang and Y. T. Chang, *Chem. Commun.*, 2010, **46**, 436–438.
37. L. Zhang, J. C. Er, X. Li, J. J. Heng, A. Samanta, Y. T. Chang and C. L. K. Lee, *Chem. Commun.*, 2015, **51**, 7386–7389.
38. S. Wang and Y. T. Chang, *J. Am. Chem. Soc.*, 2006, **128**, 10380–10381.

39. Y. H. Ahn, J. S. Lee and Y. T. Chang, *J. Comb. Chem.*, 2008, **10**, 376–380.
40. S. Wang, Y. K. Kim and Y.-T. Chang, *J. Comb. Chem.*, 2008, **10**, 460–465.
41. S. Wang and Y. T. Chang, *Chem. Commun.*, 2008, 1173–1175.
42. Y. H. Ahn, J. S. Lee and Y. T. Chang, *J. Am. Chem. Soc.*, 2007, **129**, 4510–4511.
43. Q. Li, J. S. Lee, C. Ha, B. P. Chan, G. Yang, B. G. Wen and Y. T. Chang, *Angew. Chem., Int. Ed.*, 2004, **43**, 6331–6335.
44. J. Min, J. W. Lee, Y. H. Ahn and Y. T. Chang, *J. Comb. Chem.*, 2007, **9**, 1079–1083.
45. J. C. Er, M. K. Tang, C. G. Chia, H. Liew, M. Vendrell and Y. T. Chang, *Chem. Sci.*, 2013, **4**, 2168–2176.
46. M. Vendrell, G. G. Krishna, K. K. Ghosh, D. Zhai, J. S. Lee, Q. Zhu, Y. H. Yau, S. G. Shochat, H. Kim, J. Chung and Y. T. Chang, *Chem. Commun.*, 2011, **47**, 8424–8426.
47. L. Zhang, J. C. Er, H. Jiang, X. Li, Z. Luo, T. Ramezani, Y. Feng, M. K. Tang, Y. T. Chang and M. Vendrell, *Chem. Commun.*, 2016, **52**, 9093–9096.
48. D. Zhai, Y. Q. Elton Tan, W. Xu and Y. T. Chang, *Chem. Commun.*, 2014, **50**, 2904–2906.
49. D. Zhai, B. K. Agrawalla, P. S. Fronia Eng, S. C. Lee, W. Xu and Y. T. Chang, *Chem. Commun.*, 2013, **49**, 6170–6172.
50. W. Xu, T. H. Kim, D. Zhai, J. C. Er, L. Zhang, A. A. Kale, B. K. Agrawalla, Y. K. Cho and Y. T. Chang, *Sci. Rep.*, 2013, **3**, 1–7.
51. D. Zhai, S. C. Lee, M. Vendrell, L. P. Leong and Y. T. Chang, *ACS Comb. Sci.*, 2012, **14**, 81–84.
52. W. Xu, J. Bai, J. Peng, S. Animesh, D. Divyanshu and Y. T. Chang, *Chem. Commun.*, 2014, **50**, 10398–10401.
53. J. W. Lee, J. S. Lee, M. Kang, A. I. Su and Y. T. Chang, *Chem. – Eur. J.*, 2006, **12**, 5691–5696.
54. W. Xu, C. Ren, C. L. Teoh, J. Peng, S. H. Gadre, H. W. Rhee, C. L. K. Lee and Y. T. Chang, *Anal. Chem.*, 2014, **86**, 8763–8769.
55. L. Zhang, J. C. Er, W. Xu, X. Qin, A. Samanta, S. Jana, C. L. K. Lee and Y. T. Chang, *Anal. Chim. Acta*, 2014, **815**, 51–56.
56. C. L. Teoh, D. Su, S. Sahu, S. W. Yun, E. Drummond, F. Prelli, S. Lim, S. Cho, S. Ham, T. Wisniewski and Y. T. Chang, *J. Am. Chem. Soc.*, 2015, **137**, 13503–13509.
57. C. N. Im, N. Y. Kang, H. H. Ha, X. Bi, J. J. Lee, S. J. Park, S. Y. Lee, M. Vendrell, Y. K. Kim, J. S. Lee, J. Li, Y. H. Ahn, B. Feng, H. H. Ng, S. W. Yun and Y. T. Chang, *Angew. Chem., Int. Ed.*, 2010, **49**, 7497–7500.
58. M. Vendrell, S. J. Park, Y. Chandran, C. L. K. Lee, H. H. Ha, N. Y. Kang, S. W. Yun and Y. T. Chang, *Stem Cell Res.*, 2012, **9**, 185–191.
59. N. Hirata, M. Nakagawa, Y. Fujibayashi, K. Yamauchi, A. Murata, I. Minami, M. Tomioka, T. Kondo, T. F. Kuo, H. Endo, H. Inoue, S. I. Sato, S. Ando, Y. Kawazoe, K. Aiba, K. Nagata, E. Kawase, Y. T. Chang, H. Suemori, K. Eto, H. Nakauchi, S. Yamanaka, N. Nakatsuji, K. Ueda and M. Uesugi, *Cell Rep.*, 2014, **6**, 1165–1174.
60. Y. Chandran, N. Y. Kang, S. J. Park, S. Husen Alamudi, J. Y. Kim, S. Sahu, D. Su, J. Lee, M. Vendrell and Y. T. Chang, *Bioorg. Med. Chem. Lett.*, 2015, **25**, 4862–4865.

61. S. C. Lee, N. Y. Kang, S. J. Park, S. W. Yun, Y. Chandran and Y. T. Chang, *Chem. Commun.*, 2012, **48**, 6681–6683.
62. K. K. Ghosh, H. H. Ha, N. Y. Kang, Y. Chandran and Y. T. Chang, *Chem. Commun.*, 2011, **47**, 7488–7490.
63. S. W. Yun, C. Leong, D. Zhai, Y. L. Tan, L. Lim, X. Bi, J. J. Lee, H. J. Kim, N. Y. Kang, S. H. Ng, L. W. Stanton and Y. T. Chang, *Proc. Natl. Acad. Sci. U. S. A.*, 2012, **109**, 10214–10217.
64. S. W. Yun, C. Leong, X. Bi, H. H. Ha, Y. H. Yu, Y. L. Tan, G. Narayanan, S. Sankaran, J. Y. Kim, S. Hariharan, S. Ahmed and Y. T. Chang, *Chem. Commun.*, 2014, **50**, 7492–7494.
65. B. Kim, S. Feng, S. W. Yun, C. Leong, R. Satapathy, S. Y. D. Wan and Y. T. Chang, *ChemBioChem*, 2016, **17**, 2118–2122.
66. C. L. Teoh, J. C. Er, P. Mukherjee and Y.-T. Chang, *Tomography*, 2015, **1**, 30–36.
67. C. Leong, S. C. Lee, J. Ock, X. Li, P. See, S. Jin Park, F. Ginhoux, S. W. Yun and Y. T. Chang, *Chem. Commun.*, 2014, **50**, 1089–1091.
68. B. Kim, M. Fukuda, J. Y. Lee, D. Su, S. Sanu, A. Silvin, A. T. T. Khoo, T. Kwon, X. Liu, W. Chi, X. Liu, S. Choi, D. S. Y. Wan, S. J. Park, J. S. Kim, F. Ginhoux, H. S. Je and Y. T. Chang, *Angew. Chem., Int. Ed.*, 2019, **58**, 7972–7976.
69. J.-S. Lee, N. Kang, Y. K. Kim, A. Samanta, S. Feng, H. K. Kim, M. Vendrell, J. H. Park and Y.-T. Chang, *J. Am. Chem. Soc.*, 2009, **131**, 10077–10082.
70. B. K. Agrawalla, Y. Chandran, W. H. Phue, S. C. Lee, Y. M. Jeong, S. Y. D. Wan, N. Y. Kang and Y. T. Chang, *J. Am. Chem. Soc.*, 2015, **137**, 5355–5362.
71. N. Y. Kang, S. C. Lee, S. J. Park, H. H. Ha, S. W. Yun, E. Kostromina, N. Gustavsson, Y. Ali, Y. Chandran, H. S. Chun, M. Bae, J. H. Ahn, W. Han, G. K. Radda and Y. T. Chang, *Angew. Chem., Int. Ed.*, 2013, **52**, 8557–8560.
72. B. K. Agrawalla, H. W. Lee, W. H. Phue, A. Raju, J. J. Kim, H. M. Kim, N. Y. Kang and Y. T. Chang, *J. Am. Chem. Soc.*, 2017, **139**, 3480–3487.
73. N. Y. Kang, J. Y. Lee, S. H. Lee, I. H. Song, Y. H. Hwang, M. J. Kim, W. H. Phue, B. K. Agrawalla, S. Y. D. Wan, J. Lalic, S. J. Park, J. J. Kim, H. Y. Kwon, S. H. Im, M. A. Bae, J. H. Ahn, C. S. Lim, A. K. K. Teo, S. Park, S. E. Kim, B. C. Lee, D. Y. Lee and Y. T. Chang, *J. Am. Chem. Soc.*, 2020, **142**, 3430–3439.
74. B. K. Wagner, H. A. Carrinski, Y.-H. Ahn, Y. K. Kim, T. J. Gilbert, D. A. Fomina, S. L. Schreiber, Y.-T. Chang and P. A. Clemons, *J. Am. Chem. Soc.*, 2008, **130**, 4208–4209.
75. Y. K. Kim, H. H. Ha, J. S. Lee, X. Bi, Y. H. Ahn, S. Hajar, J. J. Lee and Y. T. Chang, *J. Am. Chem. Soc.*, 2010, **132**, 576–579.
76. Y. K. Kim, J. S. Lee, X. Bi, H. H. Ha, S. H. Ng, Y. H. Ahn, J. J. Lee, B. K. Wagner, P. A. Clemons and Y. T. Chang, *Angew. Chem., Int. Ed.*, 2011, **50**, 2761–2763.
77. G. J. Gu, S. J. Lim, S. Il Ahn, S. C. Lee, Y. T. Chang, T. H. Choi, B. S. Kim, Y. Bin Eom, N. K. Lee and H. S. Youn, *Eur. J. Pharmacol.*, 2014, **742**, 42–46.
78. S. J. Park, C. J. H. Ho, S. Arai, A. Samanta, M. Olivo and Y. T. Chang, *Sci. Rep.*, 2019, **9**, 5–12.

79. J. S. Yoo, S. C. Lee, Z. Y. Jow, P. Y. X. Koh and Y. T. Chang, *Cancer Res.*, 2014, **74**, 44–55.
80. J. S. Yoo, R. K. Das, Z. Y. Jow and Y. T. Chang, *PLoS One*, 2014, **9**(7), e103721.
81. S. J. Park, B. Kim, S. Choi, S. Balasubramaniam, S. C. Lee, J. Y. Lee, H. S. Kim, J. Y. Kim, J. J. Kim, Y. A. Lee, N. Y. Kang, J. S. Kim and Y. T. Chang, *Nat. Commun.*, 2019, **10**, 1–7.
82. H. Y. Kwon, R. Kumar Das, G. T. Jung, H. G. Lee, S. H. Lee, S. N. Berry, J. K. S. Tan, S. Park, J. S. Yang, S. Park, K. Baek, K. M. Park, J. W. Lee, Y. K. Choi, K. H. Kim, S. Kim, K. P. Kim, N. Y. Kang, K. Kim and Y. T. Chang, *J. Am. Chem. Soc.*, 2021, **143**, 5836–5844.
83. Y. M. Jeong, Z. Duanting, H. Hennig, A. Samanta, B. K. Agrawalla, M. A. Bray, A. E. Carpenter and Y. T. Chang, *Chem. Biol.*, 2015, **22**, 299–307.
84. K. K. Ghosh, Y. M. Jeong, N. Y. Kang, J. Y. Lee, W. Si Yan Diana, J. Y. Kim, J. Yoo, D. Kim, Y. K. Kim and Y. T. Chang, *Chem. Commun.*, 2015, **51**, 9336–9338.
85. Y. A. Lee, J. J. Kim, J. Lee, J. H. J. Lee, S. Sahu, H. Y. Kwon, S. J. Park, S. Y. Jang, J. S. Lee, Z. Wang, W. L. Tam, B. Lim, N. Y. Kang and Y. T. Chang, *Angew. Chem., Int. Ed.*, 2018, **57**, 2851–2854.
86. J. J. Kim, Y. A. Lee, D. Su, J. Lee, S. J. Park, B. Kim, J. H. Jane Lee, X. Liu, S. S. Kim, M. A. Bae, J. S. Lee, S. C. Hong, L. Wang, A. Samanta, H. Y. Kwon, S. Y. Choi, J. Y. Kim, Y. H. Yu, H. H. Ha, Z. Wang, W. L. Tam, B. Lim, N. Y. Kang and Y. T. Chang, *J. Am. Chem. Soc.*, 2019, **141**, 14673–14686.
87. J. Y. Kim, S. Sahu, Y. H. Yau, X. Wang, S. G. Shochat, P. H. Nielsen, M. S. Dueholm, D. E. Otzen, J. Lee, M. M. S. Delos Santos, J. K. H. Yam, N. Y. Kang, S. J. Park, H. Kwon, T. Seviour, L. Yang, M. Givskov and Y. T. Chang, *J. Am. Chem. Soc.*, 2016, **138**, 402–407.
88. H. Y. Kwon, J. Y. Kim, J. Y. Lee, J. K. H. Yam, L. D. Hultqvist, W. Xu, M. Rybtke, T. Tolker-Nielsen, J. J. Kim, N. Y. Kang, L. Yang, S. J. Park, M. Givskov and Y. T. Chang, *Chem. Commun.*, 2018, **54**, 11865–11868.
89. H. Y. Kwon, J. Y. Kim, X. Liu, J. Y. Lee, J. K. H. Yam, L. Dahl Hultqvist, W. Xu, M. Rybtke, T. Tolker-Nielsen, W. Heo, J. J. Kim, N. Y. Kang, T. Joo, L. Yang, S. J. Park, M. Givskov and Y. T. Chang, *Biomater. Sci.*, 2019, **7**, 3594–3598.
90. H. Kwon, X. Liu, E. G. Choi, J. Y. Lee, S. Choi, J. Kim, L. Wang, S. Park, B. Kim, Y. Lee, J. Kim, N. Y. Kang and Y. Chang, *Angew. Chem.*, 2019, **41061**, 8514–8519.
91. D. Su, C. L. Teoh, S.-J. Park, J.-J. Kim, A. Samanta, R. Bi, U. S. Dinish, M. Olivo, M. Piantino, F. Louis, M. Matsusaki, S. S. Kim, M. A. Bae and Y.-T. Chang, *Chem*, 2018, **4**, 1128–1138.
92. J. Li, H. H. Ha, L. Guo, D. Coomber and Y. T. Chang, *Chem. Commun.*, 2010, **46**, 2932–2934.

CHAPTER 6

Two-photon Fluorescent Probes

VINAYAK JUVEKAR[a] AND HWAN MYUNG KIM*[a]

[a]Ajou University, Department of Chemistry and Department of Energy Systems Research, Suwon 443–749, South Korea
*E-mail: kimhm@ajou.ac.kr

6.1 Introduction

The development of fluorescence probes and the modernization of microscopy have played key roles in the evolution of biology.[1,2] Over the years, multiphoton imaging based on probe excitation using near-infrared (NIR) excitation sources has proven to be one of the most effective tools in biomedical applications.[3–13] The general fluorescence imaging process in one-photon microscopy (OPM) uses ultraviolet or visible (UV/Vis) light as an excitation source. In contrast, two-photon (TP) microscopy (TPM) excites the fluorophore with two photons simultaneously, thus using only half the photon energy compared to OPM (see Figure 6.1a). TPM, which uses two light quanta to excite a fluorophore, was envisioned by Nobel laureate Maria Göppert-Mayer.[14] Denk *et al.* utilized this technique to obtain cellular images by simultaneously treating fluorophores with TPs of NIR wavelength (700–1100 nm) using a femtosecond (fs) pulsed laser.[15] Therefore, the use of a low-energy NIR window as an excitation source results in minimized autofluorescence, tissue absorption, and scattering by native fluorophores, such as NADH (nicotinamide adenine dinucleotide) and

Monographs in Supramolecular Chemistry No. 34
Fluorescent Chemosensors
Edited by Luling Wu, Adam C. Sedgwick, Xiao-Peng He and Tony D. James

Published by the Royal Society of Chemistry, www.rsc.org

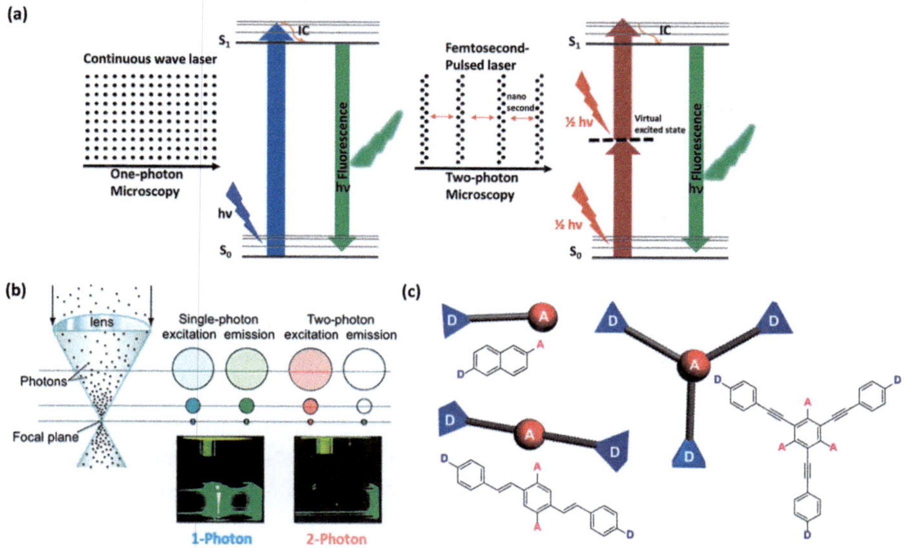

Figure 6.1 (a) Energy diagrams showing one-photon absorption (OPA), two-photon absorption (TPA), and fluorescence for a molecule. (b) Comparing the emission profiles of OPA and TPA excitation of a cuvette solution of a fluorophore. (c) Representative structural motifs of an electron donor–acceptor (D–A) substituted dipole, quadrupole (D–A–D), and two-dimensional octupole. Reproduced from ref. 5 with permission from American Chemical Society, Copyright 2011; Reproduced from ref. 3 with permission from American Chemical Society, Copyright 2015.

flavin adenine dinucleotide, yielding tissue images at enhanced depths of 1000 μm and improved stability under physiological conditions.[16,17] Additionally, the use of NIR light at a localized region further reduces photobleaching and photodamage to the specimen, compared to one-photon excitation.[3] Most importantly, the probability of simultaneous excitation of fluorophores outside the focal plane is so low that the focused excitation results in submicrometric *z*-sectional images along the *z*-axis, reducing the out-of-focus optical output. Therefore, the combined *z*-sectional images by TP excitation (TPE) will reveal the three-dimensional architecture of the scattering tissue samples under investigation.[3,18] The low background optical output is a unique feature of using a femtosecond pulsed laser (TP) compared to a continuous laser (one-photon). In addition, deep-tissue imaging through TPM is receiving increased attention due to continuous innovation in the field of microscopy and developments in the optical output algorithms. Hence, these applications through TPM have instigated the continuous development of an array of innovative TP probes with enhanced TP features for biomedical imaging applications.[3–13]

6.1.1 Two-photon Absorption

The two-photon absorption (TPA) property of the TP probe is based on its ability to reach the excited state through simultaneous absorption of TPs at fs intervals. The probability of absorption of TP light by the TP probe (S) is directly proportional to the square of the incident light intensity (I), that is, $S \propto \delta I^n$ ($n = 2$), where δ corresponds to the TP absorption cross-sections expressed in GM units (1 GM = 1×10^{-50} cm^4 s per photon per molecule). In contrast, the efficiency of the TPE significantly decreases with the distance from the incident focal plane (r), that is, efficiency $\propto 1/r^2$. Therefore, the probability of TPs arriving simultaneously on the same molecule outside the focal point is essentially zero, indicating that out-of-focus fluorescence is almost minimized, and undesirable background fluorescence is diminished (see Figure 6.1b). Thus, the fluorophores are efficiently excited only in the focal region of the incident light, which is beneficial for obtaining fluorescent images with reduced signal to background noise ratio, with improved spatial resolution, for localized photodynamic therapy, and for three-dimensional (3D) optical images.

TPM is limited by the need for a high-energy pulsed laser with an incident laser power, $I = 10$ GW cm^{-2}, to excite the TP fluorophore. This high-power incident light might cause photodamage to the specimen under investigation. This drawback can be addressed by improving the probability of TP light absorption by the TP reporter (TP fluorophore with high δ values) with increased donor–acceptor strength using proper π-bridges representative of dipolar, quadrupolar, and octupolar molecules (see Figure 6.1c). In addition, even large δ values (greater than 10 000 GM) cannot be used in TP bioimaging because of the relatively low brightness. Therefore, for use in bioimaging applications, TP fluorophores with a high fluorescence quantum yield (Φ) and excellent δ values are required. The measure of TP brightness is expressed in the TP action cross-section ($\Phi\delta$), which indicates the photons absorbed through the TP process and the efficiency of the fluorescence emission. Hence, it is critical to focus on developing TP fluorophores with excellent $\Phi\delta$ values for use in biological applications, which will help to maintain the TP laser power within an acceptable range. This chapter discusses the various chemical strategies developed to design TP fluorescent sensors for bioimaging applications.

6.2 Design of Two-photon Probes for Bioimaging Applications

The basic design of TP probes is based on similar strategies to those used for OP fluorescent molecules. A typical TP probe consists of a TP fluorophore connected to a recognition unit with or without a linkage. The optical output varies through up/downregulated fluorescence intensity and/or emission wavelength changes based on the interaction between a target analyte and the recognition receptor. The signal should have an appropriate signal-to-noise ratio to enhance the analyte sensitivity. Additional parameters to be considered during TP probe design include a TP reporter with high $\Phi\delta$ values, which is critical to obtain TPM images at low laser power and probe concentrations. In addition, the developed TP probe should possess high biocompatibility, such as minimum cytotoxicity, enhanced cell permeability, and good solubility in aqueous conditions, impressive sensitivity and selectivity toward target analytes under investigation, very low photobleaching rates for lifetime TPM imaging, and the ability to visualize target analytes in the subcellular region of the specimen.

The appropriate $\Phi\delta$ values could be obtained by optimizing the donor-to-acceptor strength and choosing an appropriate conjugation length. $\Phi\delta$ values greater than 10 GM are required to obtain bright TP images at low laser power. TP photostability can be achieved by introducing rigidity into the designed fluorophore, or by converting π-bridges to proper heterocycles. The solubility of the TP probe under aqueous conditions could be improved by reducing the probe size, conjugating the probe to a water-solubilizing agent, or encapsulating the probe into micelles.

6.2.1 Two-photon Probes for Organelles

Most eukaryotic cells contain subcellular regions, such as the nucleus, mitochondria, lysosomes, lipid droplets, endoplasmic reticulum, and the Golgi apparatus, which play key roles in numerous physiological functions, including enzyme activation, protein degradation, and energy provision. Therefore, TP probe development is imperative for visualizing specific cell regions, such as proteins, antibodies, and organelles, to study their biological roles. The usual TP probe design involves conjugating TP fluorophores to specific targeting units. By observing the TP excited fluorescence (TPEF), one could easily detect specific regions with high sensitivity and selectivity. Therefore, to analyze these organelles within cells and tissues, various TP probes have been developed.[3,10,12]

The widely used strategy to visualize lysosomes involves conjugating TP fluorophores with an amino group, by which protonation results in both aggregations around lysosomes and significantly enhances TP properties. Probe **1** is an early example of a TP lysotracker consisting of acedan as the TP fluorophore and a tertiary amine with $pK_a \approx 10$ as the protonation site (see Scheme 6.1).[19] The probe showed impressive TP properties ($\Phi\delta_{max}$ = 93 GM, which is nine times higher than commercial lysoTracker Red) with an emission maximum at 496 nm in the universal buffer. Probe **1**-labeled fresh rat hippocampal tissue images with TPM showed abundant acidic vesicles in the CA3 and dentate gyrus regions compared to the CA1 region. For mitochondrial visualization, triphenylphosphonium salt (TPP), a mitochondrial-targeting moiety, could be used. A red-emissive TP sensor, **2**, consisting of a TPP unit and a 2*H*-benzo[*g*]chromene-2-one as the TP fluorophore showed TP emission at 626 nm (Φ = 0.07) and a $\Phi\delta_{max}$ value of 20 GM in EtOH.[20] TPM images of probe **2**-labeled rat hippocampal tissue at depths of 90–200 μm showed evenly distributed mitochondria in the

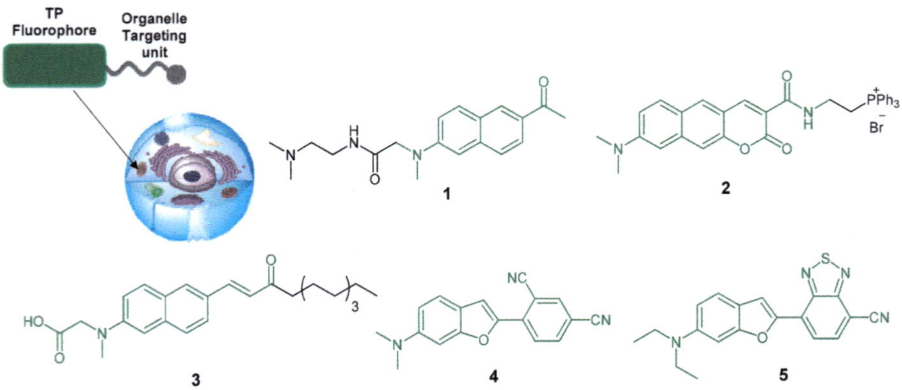

Scheme 6.1 Schematic diagram illustrating the sensing mechanism of representative organelle-targeted TP probes **1–5**.

CA1–CA3 region, and mitochondria were observed individually in the pyramidal neuronal layer of the CA1 region. Lipid rafts are plasma membranes with rigid compartments that play crucial roles in key cellular events, such as pathogen invasion, signal transduction, neurodegenerative diseases, cholesterol homeostasis, and angiogenesis. To visualize such a long-chain domain installed on probe **3**, 6-[(*E*)-3-oxo-1-dodecenyl]-2-[*N*-methyl-*N*-(carboxymethyl)amino]naphthalene was used to accommodate it on the lipid raft (see Scheme 6.1).[21] In a hydrophobic environment, the probe photoluminescence properties showed a bathochromic shift in solvent polarity and a much higher $\Phi\delta_{max}$ value. The probe was efficiently utilized in lipid raft visualization in live cells and the pyramidal neuronal layer of the CA1 region in rat hippocampal tissue at depths of 100–250 μm. Lipid droplets (LD) are crucial organelles that are involved in maintaining the proper storage of neutral lipids. For localization in LDs, probes should possess the following criteria: low molecular weight, conjugated bond number less than 24, high lipophilicity ($\log P > 5.0$), and low amphiphilicity. Based on this, dipole molecules (donor–acceptor) **4** and **5** were developed with exceptional $\Phi\delta_{max}$ values of 166 and 43 GM, respectively.[22] The narrow emission in the range 460–520 and 620–680 nm by probes **4** and **5** is used in multicolor imaging in LD-associated signaling. Additionally, a novel diagnostic tool was developed using probes **4** and **5** with LysoTrackers to estimate drug-induced steatosis and/or phospholipidosis in LD-associated liver injuries or diseases (see Figure 6.2).

6.2.2 Two-photon Turn-on Probes

The sensor contains receptors either with or without a spacer to detect protons, small molecules, or metal ions through turn-on or turn-off processes. In the turn-on/turn-off process, TP probes usually undergo signal enhancement/detraction due to the photoinduced electron transfer (PeT) process in the presence of target analytes.

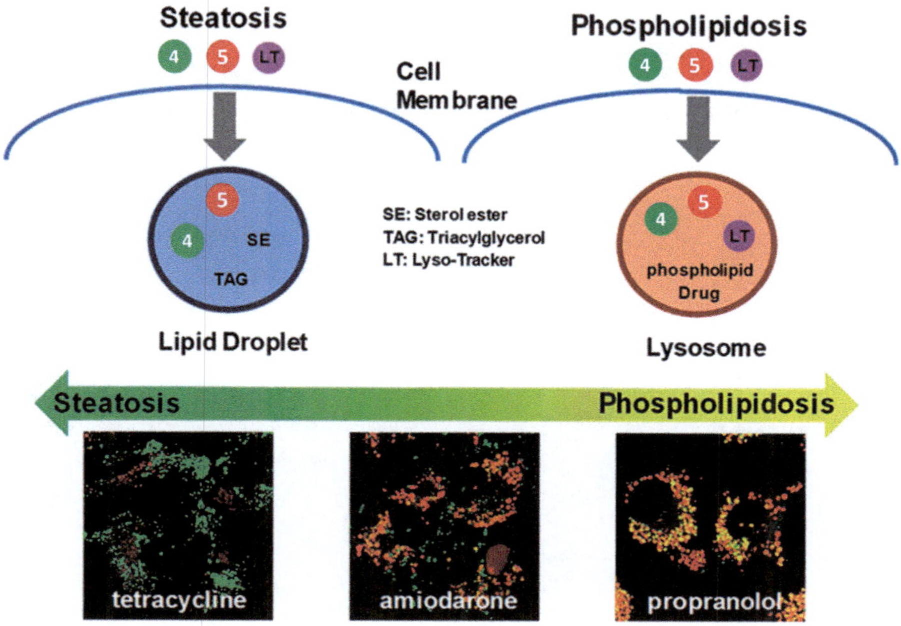

Figure 6.2 Two-photon microscopy (TPM) images of probe **4** and **5** in Huh-7 cells treated with indicated inducers of steatosis and/or phospholipidosis. Reproduced from ref. 22 with permission from American Chemical Society, Copyright 2020.

A novel TP probe, **6**, was devised to detect zinc ions in mitochondria. Probe **6**, consisting of *N*,*N*-di-(2-picolyl)ethylenediamine (DPEN) ligand as the Zn^{2+} receptor, inhibits fluorescence emission through the PeT process from the receptor to the TP reporter (see Scheme 6.2).[23] At the Zn bound state, probe **6** exhibited a seven-fold increase in TPEF intensity and a massive enhancement in $\Phi\delta_{max}$ values (75 GM). The probe's affinity toward Zn^{2+} was measured using the dissociation constant (KdTP), which was estimated to be 3.1 nM. Probe **6**-stained rat hippocampal slice showed Zn^{2+} ions at a depth of 100–200 μm without interference from other biologically relevant metal ions.

Viscosity is a critical microenvironment state in live cells. The common strategy for TP viscosity sensors involves restricting the rotatable group in molecules under increased viscous media, and the induced rigidity will enhance $\Phi\delta_{max}$ values. The TP turn-on probe **7** consists of a carbazole unit as the donor and a hemocyanin core as the molecular rotor. The mitochondrial-targeting group could detect mitochondrial viscosity within the living system using TPM images (see Scheme 6.2).[24] As predicted, probe **7** with low viscosity (1.4 cP) showed weak fluorescence emission due to free rotation between the indolenium and carbazole derivatives. However, upon increased polarity (956 cP), probe **7** showed a 32-fold increase in fluorescence emission and δ_{max} values of 121 GM at 740 nm. Furthermore, probe **7** in

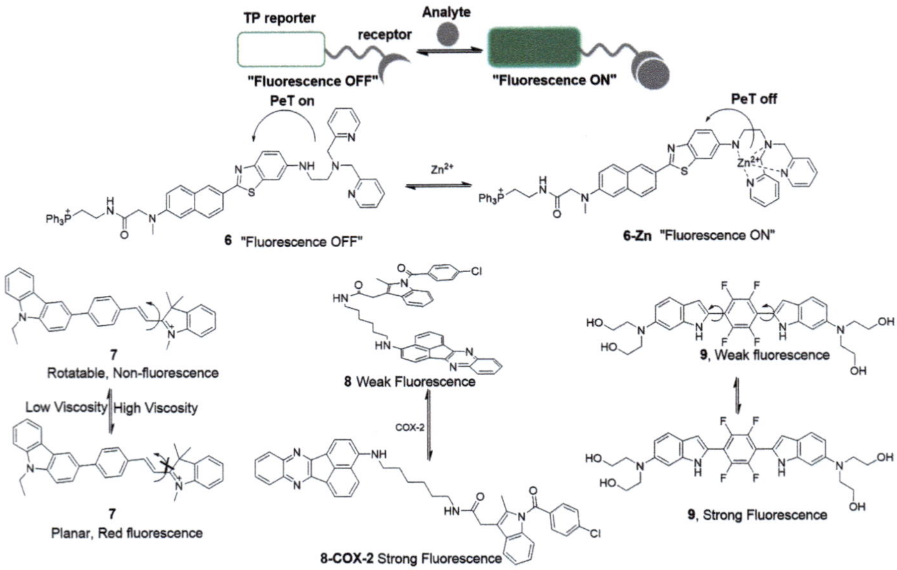

Scheme 6.2 Structures and schematic diagram illustrating the sensing mechanism of representative turn-on TP probes **6–9**.

monensin or nystatin pretreated HeLa cells and zebrafish showed increased fluorescence emission due to the increased viscosity of the biological system.

Various turn-on TP probes have been developed to measure enzyme location and expression levels in physiological systems for various bioimaging applications.[25] For example, TP fluorescent probe **8** was developed to analyze COX-2 expression in the Golgi apparatus for early cancer diagnosis and to measure dynamic changes in the Golgi apparatus during cancer cell apoptosis.[26] The fluorescence off–on sensor design is based on the efficient PeT transfer between acenaphtho[1,2-*b*]quinoxaline (ANQ, TP fluorophore) and indomethacin (IMC, COX-2 receptor). In the solution containing COX-2, the probe undergoes an unfolded state, which restricts PeT transfer between the fluorophore and receptor, resulting in a 25-fold increase in TPEF emission and a $\Phi\delta_{max}$ value of 118 GM at 800 nm (see Table 6.1). Using bright TPM images, probe **8** could visualize COX-2 enzyme overexpression in the Golgi apparatus of live HeLa cells and sarcoma 180 tissue slices at 50–550 μm depth.

The TP turn-on probe with a centrosymmetric molecule having a quadrupole moment (D–A–D) was devised to visualize misfolded protein aggregates such as beta-amyloid (Aβ) plaques in a neurodegenerative disorder. Probe **9** showed a $\Phi\delta_{max}$ value greater than 420 GM, selective binding affinity towards beta amyloids, high photostability, and good aqueous solubility.[27] These impressive properties of probe **9** allowed visualization of Aβ plaques and cerebral amyloid angiopathy in living mouse brains through a good signal-to-noise ratio for a prolonged time.

Table 6.1 Photophysical properties of TP probes.

Probe	Solvent	$\lambda_{max}^{(1)\ a}$	$\lambda_{max}^{fl\ b}$	Φ^c	Sensitivity^d	$\lambda_{max}^{(2)\ e}$	δ^f	$\Phi\delta^g$	Ref.
1	Buffer	364	496	0.76	—	780	122	93	19
2	Buffer	470	626	0.07	—	920^h	n.a.	20	20
3	EtOH	401	580	0.09	—	800	250	20	21
4	Toluene	444	513	0.98	—	940	167	166	22
5	Toluene	487	618	0.22	—	970	195	43	22
6	Buffer	388	500	0.15	—	n.a.	n.a.	n.a.	23
6-Zn		375	493	0.92	3.1 nM^h	760	82	75	
7	MeOH	n.a.	n.a.	0.002	—	n.a.	n.a.	n.a.	24
	Glycerol	490	610	0.29		740	120	35	
8	Tris-HCl	457	547	0.013	n.a.	n.a.	n.a.	n.a.	26
8-COX-2		457	547	0.39		800	303	118	
9	EtOH	426	508	0.73	—	750	575	420	27
10	H_2O	378	458	0.25		750	184	46	28
10-Ca		378	495	0.49		750	284	139	
11	H_2O/EtOH	379/465	585	0.22	21	750	145	32	29
11-Cu		n.a.	470	n.a.		n.a.	n.a.	n.a.	
12^i	pH 3.5	368	490	0.72	5.82^j	750	389	280	30
	pH 7.2	341	448	1.00		750	95	95	
	pH 10	341	446	1.00		750	80	80	
13	HEPES	403	454	n.a.	n.a.	n.a.	n.a.	n.a.	31
13-ATP	DMSO	530	557	n.a.		n.a.	n.a.	n.a.	
14	HEPES	343	464	0.24	—	740	71	17	32
14-H_2S		383	545	0.12		750	458	55	
15	PBS	540/630	578/651	n.a.	65.3^k	n.a.	n.a.	n.a.	33
15-BACE1		540/630	578/651	n.a.		820	n.a.	18.8	
16	MeOH	356	n.a.	0.001	16.6^l	750	n.d.	n.d.	34
16-HOCl		356	500^l	0.40		750	205	82	
16-Mito	MeOH	356	n.a.	0.001	17.2^m	750	n.d.	n.d.	34
16-Mito-HOCl		356	500^l	0.41		750	202	83	
16-Lyso	MeOH	356	n.a.	0.001	19.6^n	750	n.d.	n.d.	34
16-Lyso-HOCl		356	500^l	0.35		750	211	74	
17	PBS/ACN^o	375	485	0.08	66^k	800	30	2	35
17-H_2O_2		405	585	0.26		800	96	25	
18	PBS	430	555	0.011	11.96^p	840	109	1.2	36
18-MAO		530	618	0.097		840	367	35.6	

$^a\lambda_{max}$ of one-photon absorption spectra in nm.
$^b\lambda_{max}$ of one-photon emission spectra in nm.
^{c}Fluorescence quantum yield.
^{d}Probe efficiency measurements.
$^e\lambda_{max}$ of TPE spectra in nm.
^{f}The peak TP cross-section at 10^{-50} cm^4 s per photon (GM units).
^{g}Two-photon action cross-section.
^{h}Dissociation constants for Zn^{2+} (nM).
^{i}The measurements were performed using EtOH.
^{j}pK_a value measured in the one-photon mode.
^{k}Detection limit measured in the corresponding solvent (nM).
^{l}Detection limit measured in PBS buffer/EtOH (1 : 1) in nM.
^{m}Detection limit measured in PBS (pH 7.4).
^{n}Detection limit measured in PBS buffer (pH 5.0) in nM.
^{o}PBS buffer/ACN = 9 : 1.
^{p}Michaelis–Menten kinetics (K_m) in μm. n.a., not applicable; n.d., not determined.

6.2.3 Two-photon Turn-on Probes with Internal Reference

The internal reference approach, in which a dual fluorophore containing a turn-on/turn-off probe is connected to an always-on probe through a linker in which Förster resonance energy transfer (FRET) between fluorophores is negligible during target analytes, induced TPEF variation. Usually, during the binding event between target analytes with a turn-on/turn-off probe, the analyte can be quantitatively estimated by measuring the TPEF ratio of both fluorophores.

The initial Ca^{2+} sensing systems based on the turn-on sensing strategy have limitations during quantitative Ca^{2+} estimation. Therefore, for quantitative Ca^{2+} visualization, TP probe **10** consists of BAPTA as the Ca^{2+} chelator and two luminating agents with similar TP absorption wavelengths but emitting at different wavelengths.[28] Both fluorophores are connected through a glycinamide linker, resulting in insignificant FRET interference (see Scheme 6.3). In the presence of Ca^{2+}, the fluorescence intensity at 500–600 nm gradually increased without interfering with the fluorescence emission in the range 400–450 nm at 750 nm TPE. The remarkable TP properties of sensor **10** were used to measure Ca^{2+} in real time in the live neurons and rat spinal cord injury (SCI) model. The fluorescence images of the SCI model showed a relationship between such injuries and the different Ca^{2+} levels with time after injury. The western blot studies showed similar results, with varying fluorescence intensities at different time intervals (see Figure 6.3).

A similar strategy was used to design TP probe **11** for copper ions (Cu^{2+}), which consists of a two-TP reporter coumarin derivative as an internal reference and chromine with a Cu^{2+} chelator 2-picolylmethylamine entity, which are connected through a piperazine linker (see Scheme 6.3).[29] Upon addition

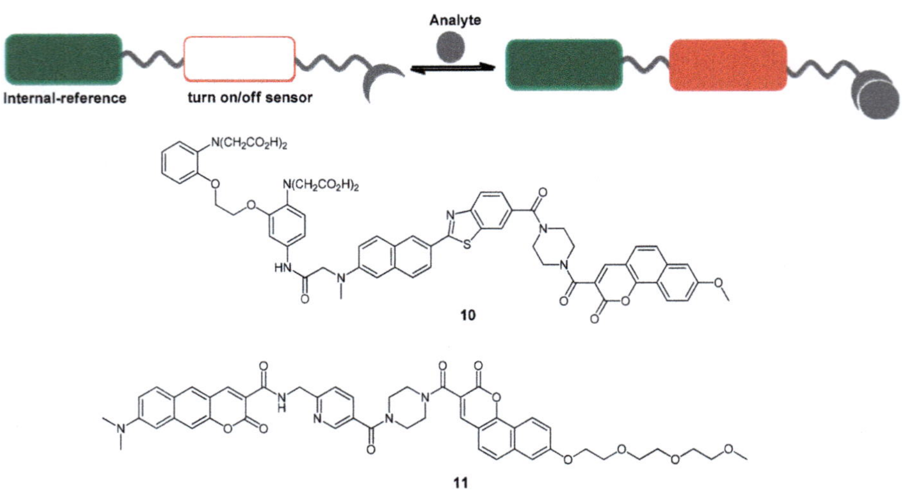

Scheme 6.3 Structures and schematic diagram illustrating of representative TP probes **10** and **11** with an internal reference.

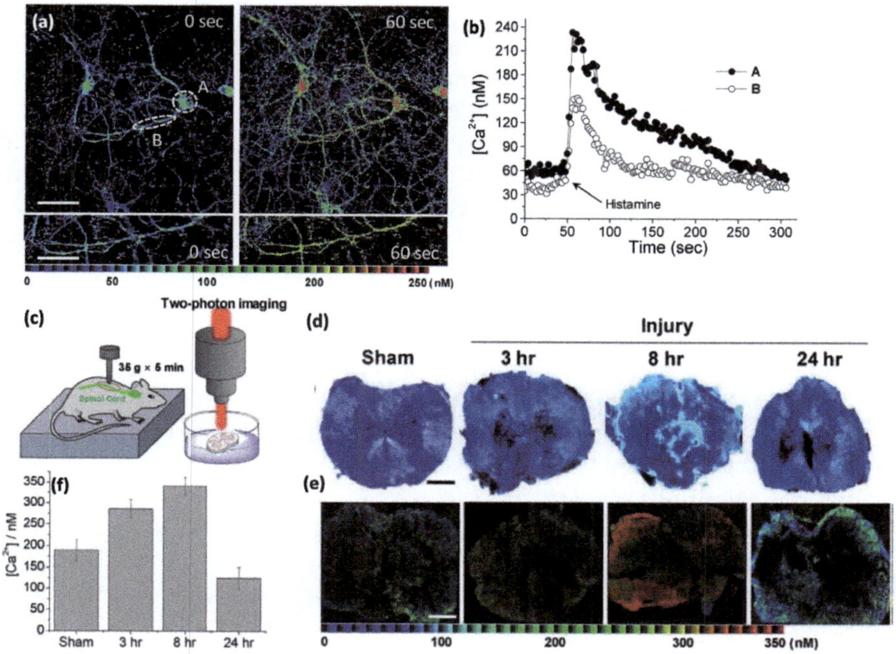

Figure 6.3 (a) TPM ratiometric images of neuronal cells stained with probe **10** before and after the addition of histamine. (b) Relative concentration of Ca^{2+} with time. (c) Graphic illustration of spinal cord injury (SCI) induction. (d) Confocal images of luxol fast blue (LFB) stained at SCI rat tissue epicenter. (e) Ratiometric TPM images of rat spinal cord slice tissue stained with **10** for 40 min. (f) Relative Ca^{2+} concentration in SCI rat tissue. Reproduced from ref. 28 with permission from Elsevier, Copyright 2017.

of Cu^{2+}, the emission intensity of the chromine derivative decreased gradually without any appreciable change in the coumarin emission. Furthermore, the probe was used to measure the Cu^{2+} concentration quantitatively in human colon tissues for colon cancer diagnosis at depths between 90 and 160 μm by dual-color TPM imaging.

6.2.4 Two-photon Ratiometric Probes

In the ratiometric approach, the emission wavelength of a TP probe changes in the presence of target analytes owing to the variation in ICT or the FRET process. This classical approach involves the complexation of protons or metal ions with TP probes in real time, giving a shorter or longer wavelength hypsochromic or bathochromic shift in emission that will be helpful to detect dynamic concentration variations of analytes.

Dysregulation of pH can lead to cellular abnormalities and cause various disorders, such as neurodegenerative diseases and cancer. The ratiometric TP probe **12** for acidic pH was developed to measure pH variation within subcellular regions of the cells, such as lysosomes (pH 4.5–5.5).[30] The design consists of a benzimidazole-based acceptor as the proton receptor, an amino-naphthalene core as the donor, and a tertiary amine substitute to accumulate the probe in the acidic region of the cells (see Scheme 6.4). The probe showed a blue-to-green fluorescence emission shift towards an altered pH due to reversible protonation at the benzimidazole nitrogen, leading to strong ICT from the donor to the acceptor. At pH 7.2, the maximum emission was observed at 448 nm, whereas at pH 3.5, the maximum emission red shifted by 490 nm (see Table 6.1). At 750 nm, the $\Phi\delta_{max}$ values at pH 3.5 and 7.2 were 280 and 95 GM, respectively. The three-fold increase in GM values at pH 3.5 were attributed to increased ICT from donor to acceptor. The ratiometric TPM images of live cells and tissues stained with probe **12** revealed a real-time analysis of the pH values inside the lysosomal compartments.

A FRET-based ratiometric TP probe **13** was developed to help evaluate biological mechanisms and signaling pathways mediated by ATP to detect lysosomal ATP quantitatively. The probe consists of rhodamine 6G and BODIPY dyes separated by a tetraamine chain (see Scheme 6.4).[31] In the absence of ATP, probe **13** emitted at 454 nm; however, in its presence, probe **13** emitted strong fluorescence at 557 nm. Most importantly, probe **13** exhibited strong fluorescence at 557 nm in lysosomal pH (4.5 to 5.5), whereas the ratio values were low at the cytosolic pH. The increase in fluorescence of probe **13** in ATP is probably due to the π–π stacking interaction between rhodamine and the adenosine base. Moreover, the ionic interaction between the tetraamine

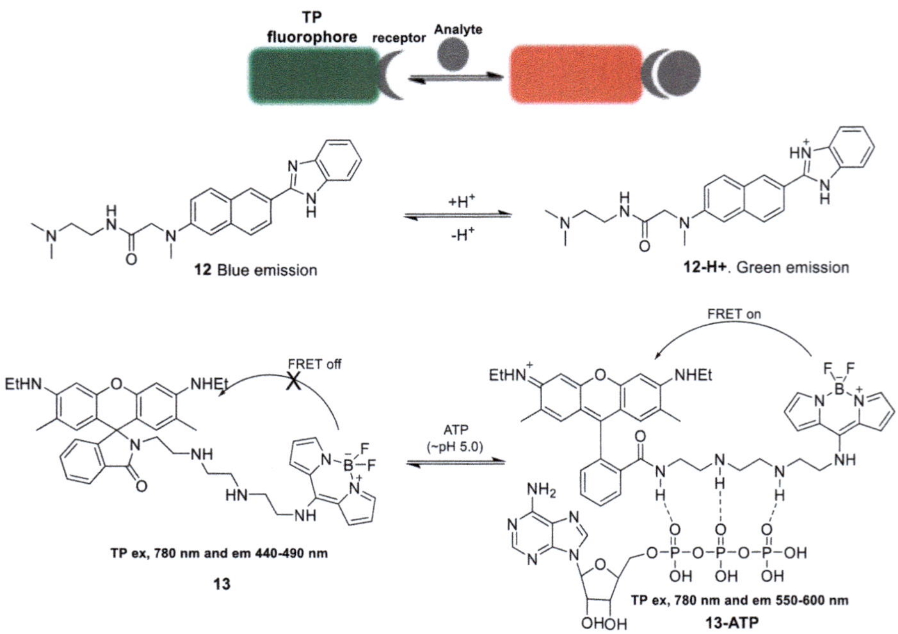

Scheme 6.4 Structures and sensing mechanism of representative ratiometric TP probes **12** and **13**.

linker and triphosphate restricts the FRET transfer. Probe **13** used to obtain TPM images of lysosomal ATP in mouse brain tissue indicated that ATP was primarily distributed in the corpus callosum (CC) area (see Figure 6.4). Additionally, probe **13** showed two different lysosomal changes in ATP levels during the "kiss-and-run" and the "full-collapse membrane fusion" processes.

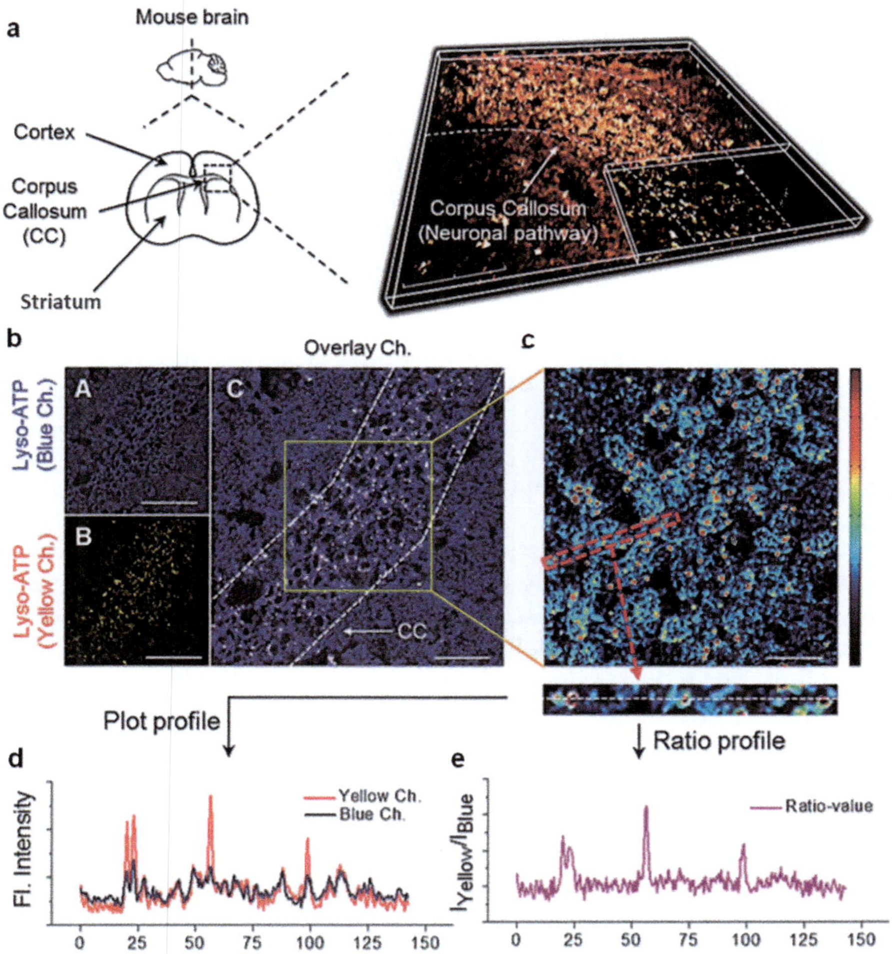

Figure 6.4 (a) 3D tissue two-photon image of brain slice in CC constructed based on the yellow channel of **13** (30 μm). (b) Cross-section images of the blue channel (A), yellow channel (B), and overlay (D). (c) Ratiometric image representing the ratio of the yellow to the blue channel from (b). (d) Plot profile of blue and yellow channels obtained from (c). (e) Ratio value profile of (d). Scale bars = 25 μm (a and b), 80 μm (e), 40 μm (f), and 20 μm (g). Reproduced from ref. 31 with permission from John Wiley & Sons, Copyright © 2018 Wiley-VCH Verlag GmbH & Co. KGaA, Weinheim.

6.2.5 Two-photon Reaction-based Probes

In an irreversible reaction-based turn-on/ratiometric approach, analytes of small molecules or enzymes react with the receptor, leading to the cleavage of the receptor from the TP probe, which results in optical output variation. This approach is highly selective toward analytes and useful for qualitative (in the case of turn-on) and quantitative (in the case of ratiometric) measurements.

The representative reaction-based ratiometric TP probe **14** was developed to study neurodegenerative diseases caused by defects in H_2S production. The probe design consists of 6-(benzo[*d*]thiazol-2′-yl)-2-(*N*,*N*-dimethylamino)naphthalene (BTDAN) as the TP reporter, 4-azidobenzyl carbamate moiety as the H_2S receptor, and TPP as the mitochondrial-targeting site (see Scheme 6.5).[32] In aqueous buffer conditions, probe **14** showed fluorescence emission at 464 nm ($\Phi\delta_{max}$ = 17 GM, at 740 nm), whereas in the H_2S receptor, probe emission showed a maximum bathochromic shift at 545 nm and $\Phi\delta_{max}$ values increased to 55 GM at 750 nm. The greater red shift in **14-H$_2$S** is due to the presence of a stronger electron-donating amine group, resulting in improved ICT within the fluorophore. Probe **14**-stained wild type and DJ-1-KO rat brain slices ratiometric TPM images showed different expression levels of endogenous H_2S and cystathionine β-synthase (an enzyme that catalyzes H_2S production) at depths of 90–180 μm without interference. This result indicates that genetically mutated Parkinson's disease genes can affect H_2S production in the brains of patients with Parkinson's disease.

A FRET-based TP ratiometric probe **15** for the β-secretase (BACE1) enzyme was devised to analyze pathological processes, including oxidative stress, in the brain tissue of patients with Alzheimer's disease. Sensor **15** consists of Alexa Fluor 633 (AF633) as a FRET acceptor and TP fluorophore merocyanine derivative (mCyd) as a FRET donor, which is connected through a BACE1-sensitive peptide substrate (EVNLDAEFRHDSGYK) (see Scheme 6.5).[33] The probe showed marked sensitivity and selectivity towards BACE1 with an increase in emission at 578 nm, whereas fluorescence at 651 nm decreased owing to specific cleavage of the peptide substrate between the donor and acceptor. Probe **15** was successfully utilized in sensing BACE1 in Alzheimer's disease mice at depths greater than 300 μm (see Figure 6.5).

A reaction-based TP turn-on TP sensor **16** was designed to measure hypochlorous acid (HOCl, a key oxidant in several physiological and pathological processes) in living systems. The presence of oxathiolane on the amino-naphthalene core causes loss of the "push–pull" structure of acedan (see Scheme 6.5).[34] In excess HOCl, oxathiolane will be oxidized to form sulfoxide, which is subsequently oxidized into sulfone by HOCl. The unstable sulfone, upon hydrolysis, restores the push–pull character, thereby restoring TPEF properties ($\Phi\delta_{max}$ = 82 GM at 750 nm) in the acedan (see Table 6.1). Additionally, targeting units, triphenylphosphine and

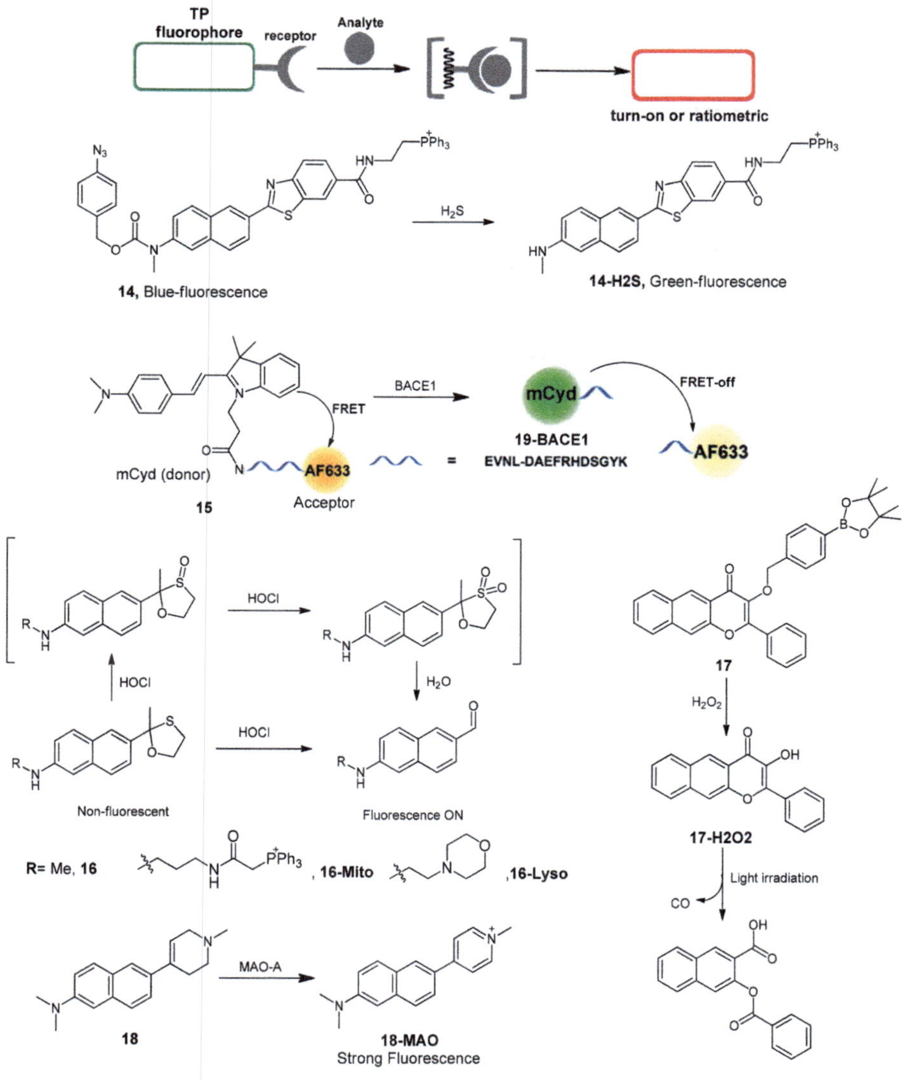

Scheme 6.5 Structures and sensing mechanism of representative reaction-based TP probes **14–18**.

morpholine, were introduced onto the TP fluorophore, acedan, and HOCl reactive site, oxathiolane, to generate mitochondrial and lysosomal probes **16-Mito** and **16-Lyso**, respectively. The TP probes **16-Mito** and **16-Lyso** were used to measure HOCl overproduction in the lipopolysaccharide-induced inflammation model, which showed strong fluorescence in inflamed tissues, whereas fluorescence signals were primarily absent in normal tissues.

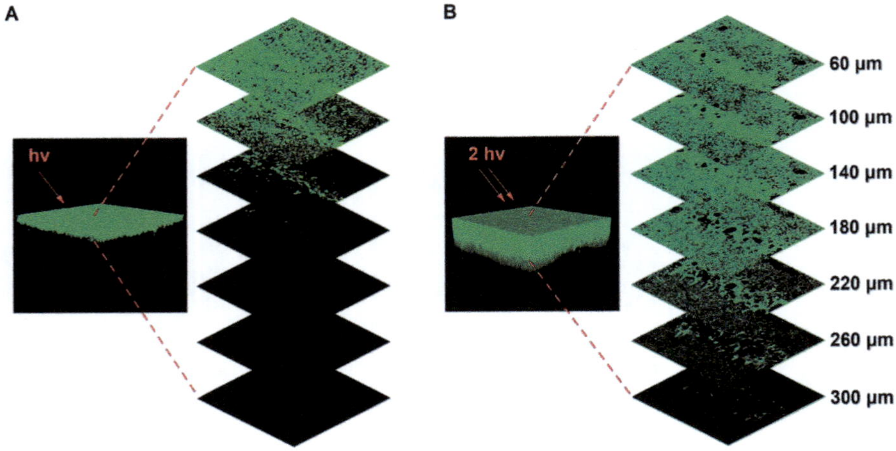

Figure 6.5 Three-dimensional (A) one-photon or (B) two-photon microscope images of the hippocampus region in Alzheimer's disease mouse brain labeled with probe **15**, excited at 552 and 820 nm, respectively. Reproduced from ref. 33 with permission from the Royal Society of Chemistry.

The reaction-based ratiometric TP probe **17** was designed as a dual platform for simultaneous photoinduced carbon monoxide (CO) release and H_2O_2 sensing system (see Scheme 6.5).[35] Probe **17** consists of a flavonol-borate system as an H_2O_2 TP sensor, which triggers H_2O_2-mediated recovery of the excited-state intramolecular proton transfer (ESIPT) effect through borate deprotection. The resulting **17-H_2O_2** showed a 19-fold ratiometric response and bathochromic shift in both absorption (from 375 to 405 nm) and emission (from 480 to 585 nm) spectra, and $\Phi\delta_{max}$ values of 25 GM at 800 nm (see Table 6.1). Subsequent light irradiation of **17-H_2O_2** showed a gradual decrease in emission intensity at 585 nm, indicating CO release. The CO-releasing system was evaluated in live zebrafish pretreated with angiotensin II. Zebrafish loaded-probe **17** and angiotensin II showed an increased fluorescence ratio ($I_{585\,nm}/I_{485\,nm}$), suggesting that angiotensin II-mediated oxidative stress and subsequent TP irradiation released CO, which caused significant vasodilatation, showing its hemangiectasis effect.

The reaction-based TP probe **18** was developed to study the abnormal enzyme activities of monoamine oxidases (MAOs) in neurodegenerative diseases and cancer. The sensor consists of tetrahydropyridine as the new warhead to detect MAO-A (see Scheme 6.5).[36] The probe **18** undergoes MAO-A-dependent oxidation, subsequently converting to a push–pull system **18-MAO** to give bright TPEF emission at 618 nm ($\Phi\delta_{max}$ = 35 GM, at 840 nm). The deep-tissue TPM studies using probe **18** in fresh human glioma and human precancerous tissues showed increased endogenous MAO-A activity in human glioma tissues through high-resolution images at a depth of 220 µm (see Figure 6.6).

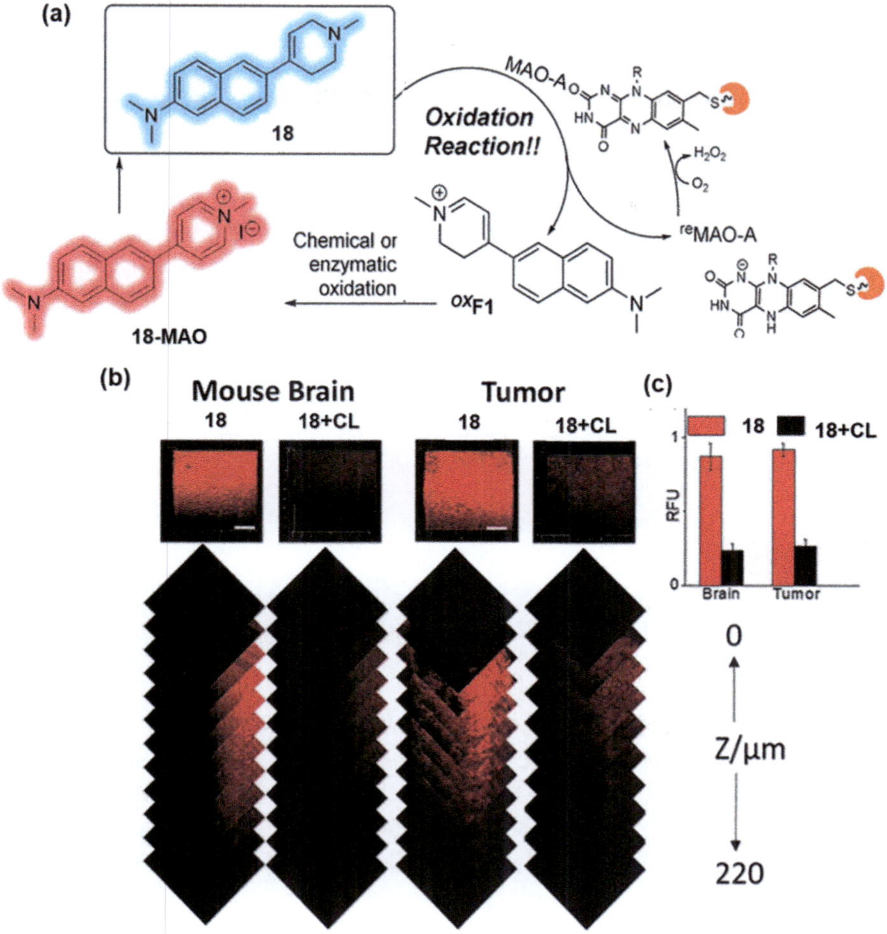

Figure 6.6 (a) Schematic mechanism of MAO reactivity against probe **18**. (b) TPM imaging of endogenous MAO-A activity with probe **18**-treated fresh mouse brain and SH-SY5Y tumor slices. CL pretreatment (MAOs inhibitor) was performed where indicated. (c) Average fluorescence intensity profiles of reconstructed tissues image in (b). Reproduced from ref. 36 with permission from John Wiley & Sons, Copyright © 2020 Wiley-VCH Verlag GmbH & Co. KGaA, Weinheim.

6.3 Conclusions and Outlook

In this chapter, the basic principles of two-photon fluorescent probes and their biomedical applications were briefly introduced. The existing TP probes were designed by combining a dye with high two-photon activity and a recognition moiety. Today, many TP probes have been developed for imaging targets in live tissues and animals. Organelle-targeted probes and turn-on probes can be analyzed qualitatively, while ratiometric probes and internal reference-based probes allow quantitative approaches. Since the existing TP dyes have a two-photon action value of 10–100 GM, there is a need for various TP scaffolds with large TPA (>100 GM) that can complement and improve on these existing probes. These will permit the dynamic investigation of appropriate TP probes for precise, multicolor detection to better understand the complex nature of biology in a living system. Ultimately, the primary thrust of future research in TP probes should focus on their possible biomedical applications, particularly for disease diagnosis, drug screening, and treatment, such as photodynamic therapy. In all these efforts, it will be important for chemists and clinical scientists to collaborate closely.

The development of TP probes is continually expanding. The increasing demand from the biomedical field and the growing number of research groups participating in the TP probe community indicate that this research area has a bright future.

References

1. J. Chan, S. C. Dodani and C. J. Chang, *Nat. Chem.*, 2012, **4**, 973.
2. J. Mei, N. L. C. Leung, R. T. K. Kwok, J. W. Y. Lam and B. Z. Tang, *Chem. Rev.*, 2015, **115**, 11718.
3. H. M. Kim and B. R. Cho, *Chem. Rev.*, 2015, **115**, 5014.
4. H. M. Kim and B. R. Cho, *Acc. Chem. Res.*, 2009, **42**, 863.
5. G. C. R. Ellis-Davies, *ACS Chem. Neurosci.*, 2011, **2**, 185.
6. S. Sumalekshmy and C. J. Fahrni, *Chem. Mater.*, 2011, **23**, 483.
7. D. Kim, H. G. Ryu and K. H. Ahn, *Org. Biomol. Chem.*, 2014, **12**, 4550.
8. A. R. Sarkar, D. E. Kang, H. M. Kim and B. R. Cho, *Inorg. Chem.*, 2014, **53**, 1794.
9. L. Qian, L. Li and S. Q. Shao, *Acc. Chem. Res.*, 2016, **49**, 626.
10. X. Lou, Z. Zhao and B. Z. Tang, *Small*, 2016, **12**, 6430.
11. V. Juvekar, S. J. Park, J. Yoon and H. M. Kim, *Coord. Chem. Rev.*, 2021, **427**, 213574.
12. L. Wu, J. Liu, P. Li, B. Tang and T. D. James, *Chem. Soc. Rev.*, 2021, **50**, 702.
13. H. Jin, M. Yang, Z. Sun and R. Gui, *Coord. Chem. Rev.*, 2021, **446**, 214114.
14. M. Göppert-Mayer, *Ann. Phys.*, 1931, **401**, 273.
15. W. Denk, J. H. Strickler and W. W. Webb, *Science*, 1990, **248**, 73.
16. P. There and W. Denk, *J. Opt. Soc. Am. A*, 2006, **23**, 3139.
17. D. Kobat, N. G. Horton and C. Xu, *J. Biomed. Opt.*, 2011, **16**, 106014.
18. P. S. Tsai, B. Friedman, A. I. Ifarraguerri, B. D. Thompson, V. Lev-Ram, C. B. Schaffer, Q. Xiong, R. Y. Tsien, J. A. Squier and D. Kleinfeld, *Neuron*, 2003, **39**, 27.
19. H. M. Kim, M. J. An, J. H. Hong, B. H. Jeong, O. Kwon, J. Y. Hyon, S. C. Hong, K. J. Lee and B. R. Cho, *Angew. Chem., Int. Ed.*, 2008, **47**, 2231.
20. A. R. Sarkar, C. H. Heo, E. Kim, H. W. Lee, H. Singh, J. J. Kim, H. Kang, C. Kang and H. M. Kim, *Anal. Chem.*, 2014, **86**, 5638.

21. H. M. Kim, B. H. Jeong, J. Y. Hyon, M. J. An, M. S. Seo, J. H. Hong, K. J. Lee, C. H. Kim, T. Joo, S. C. Hong and B. R. Cho, *J. Am. Chem. Soc.*, 2008, **130**, 4246.
22. M. K. Cho, M. J. Seo, V. Juvekar, J. H. Jo, W. Kim, K. S. Choi and H. M. Kim, *Anal. Chem.*, 2020, **92**, 11223.
23. G. Masanta, C. S. Lim, H. J. Kim, J. H. Han, H. M. Kim and B. R. Cho, *J. Am. Chem. Soc.*, 2011, **133**, 5698.
24. J. Yin, M. Peng and W. Lin, *Anal. Chem.*, 2019, **91**, 8415.
25. V. Juvekar, H. W. Lee and H. M. Kim, *ACS Appl. Bio Mater.*, 2021, **4**, 2957.
26. H. Zhang, J. Fan, J. Wang, S. Zhang, B. Dou and X. Peng, *J. Am. Chem. Soc.*, 2013, **135**, 11663.
27. C. H. Heo, A. R. Sarkar, S. H. Baik, T. S. Jung, J. J. Kim, H. Kang, I. Mook-J. and H. M. Kim, *Chem. Sci.*, 2016, **7**, 4600.
28. H. J. Kim, C. S. Lim, H. W. Lee, H. S. Lee, Y. J. Um, H. Kumar, I. Han and H. M. Kim, *Biomaterials*, 2017, **141**, 251.
29. D. E. Kang, C. S. Lim, J. Y. Kim, E. S. Kim, H. J. Chun and B. R. Cho, *Anal. Chem.*, 2014, **86**, 5353.
30. H. J. Kim, C. H. Heo and H. M. Kim, *J. Am. Chem. Soc.*, 2013, **135**, 17969.
31. Y. W. Jun, T. Wang, S. Hwang, D. Kim, D. Ma, K. H. Kim, S. Kim, J. Jung and K. H. Ahn, *Angew. Chem., Int. Ed.*, 2018, **57**, 10142.
32. S. K. Bae, C. H. Heo, D. J. Choi, D. Sen, E. H. Cho, B. R. Cho and H. M. Kim, *J. Am. Chem. Soc.*, 2013, **135**, 9915.
33. L. Ge, Z. Liu and Y. Tian, *Chem. Sci.*, 2020, **11**, 2215.
34. L. Yuan, L. Wang, B. K. Agrawalla, S.-J. Park, H. Zhu, B. Sivaraman, J. Peng, Q.-H. Xu and Y.-T. Chang, *J. Am. Chem. Soc.*, 2015, **137**, 5930.
35. Y. Li, Y. Shu, M. Liang, X. Xie, X. Jiao, X. Wang and B. Tang, *Angew. Chem., Int. Ed.*, 2018, **57**, 12415.
36. H. Fang, H. Zhang, L. Li, Y. Ni, R. Shi, Z. Li, X. Yang, B. Ma, C. Zhang, Q. Wu, C. Yu, N. Yang, S. Yoa and W. Huang, *Angew. Chem.*, 2020, **132**, 7606.

CHAPTER 7

Ratiometric Fluorescent Chemosensors: Photophysical/Chemical Mechanism Principles and Design Strategies

JINWOO SHIN[a], JUSUNG AN[a], JUNGRYUN KIM[a], YUVIN NOH[a], PARAMESH JANGILI*[a] AND JONG SEUNG KIM*[a]

[a]Department of Chemistry, Korea University, Seoul 02841, Korea
*E-mail: paramesh.iict@gmail.com, jongskim@korea.ac.kr

7.1 Introduction

A fluorescent chemosensor that uses the measurement of the ratio between two or more analytes is called a ratiometric fluorescent chemosensor. It has significant advantages in that quantitative analysis of target analytes is possible due to the stoichiometry of the matter and photophysical variation according to the amount of analyte. Furthermore, with the need for photophysical quantitative analysis of biologically/environmentally important analytes, the importance of ratiometric fluorescent chemosensors has exploded due to the inspirational and pioneering discoveries of a number of researchers, along with rapid advancements in microscopic imaging technologies.[1–3]

Monographs in Supramolecular Chemistry No. 34
Fluorescent Chemosensors
Edited by Luling Wu, Adam C. Sedgwick, Xiao-Peng He and Tony D. James

Published by the Royal Society of Chemistry, www.rsc.org

Attempts to perform quantitative analysis using a fluorophore with a single emission feature have been frustrated by various analyte independencies, such as the effect of the microenvironment around the probe, variations in the local probe concentration, light scattering caused by the sample matrix, and photobleaching, which interfere with precise analysis. Unlike single-emissive fluorescent probes, ratiometric fluorescent chemosensors harness analyte-induced intensity changes with respect to two or more emission wavelengths. Due to the ratio obtained from comparing two or more emission bands, an effective internal standard system is produced, thereby enabling more enhanced selectivity and reliability, as well as improved quantification. Consequently, ratiometric approaches have become the most effective fluorescence method for measurements for quantitative analysis, compensating for several analyte-independent parameters and eliminating most ambiguities that might affect the fluorescence signal.[3–6]

This chapter explains the basic principles of ratiometric fluorescent chemosensors, including the photophysical/chemical sensing mechanism and design strategies for constructing ratiometric fluorescent chemosensors.

7.2 Principles of Ratiometric Fluorescent Chemosensors

To achieve a ratiometric fluorescent chemosensor, the molecular design should include the following: (1) combination of a reference probe, which should have no signal change in the presence of analytes, and a dynamic probe, which shows intensity shift in signal upon presence of the analyte; (2) two dynamic fluorophore systems that modify their fluorescence intensities in opposite directions (Figure 7.1) in response to the analyte. Similar effects to the internal standard method applied in several other analytical methods are possible through the ratio obtained from different signals, which are normalized signals that came from the ratiometric measurement. Therefore, the measurement is less affected by external factors that can affect the analysis results, enabling a more accurate quantitative assay.

To date, with the dedication of numerous researchers, various ratiometric fluorescent chemosensors have been developed that enable quantitative analysis of various analytes, such as biologically and environmentally essential cations, anions, small molecules, and biomolecules. This development has become feasible through harnessing one or more photophysical/chemical mechanisms for the detection of analytes *via* ratiometric fluorescent chemosensors, and the number of strategies is still being developed and expanded. In the next section, we discuss photophysical/chemical mechanisms used to build the ratiometric fluorescent chemosensors.

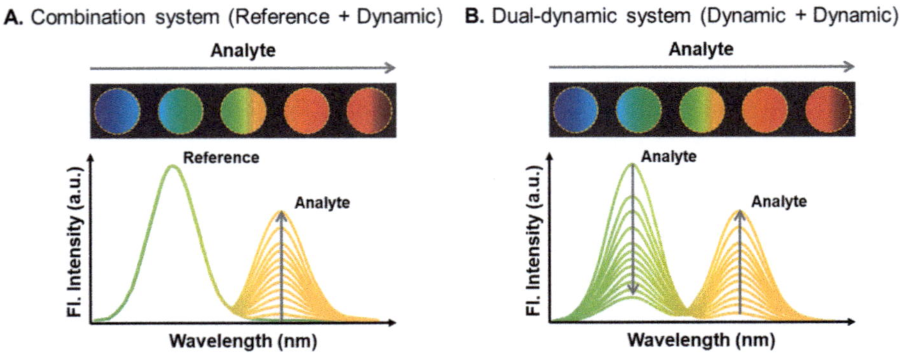

Figure 7.1 Ratiometric signal acquisitions from different ratiometric system designs. (A) Combination system. (B) Dual-dynamic system.

7.2.1 Photophysical/Chemical Processes for Ratiometric Sensing Mechanisms

The molecular design of ratiometric fluorescent chemosensors takes advantage of various photophysical processes, namely: (1) intramolecular charge transfer (ICT);[7,8] (2) excited-state intramolecular proton transfer (ESIPT);[9] (3) Förster resonance energy transfer (FRET);[10] (4) bioluminescence resonance energy transfer (BRET);[11,12] (5) thermally activated delayed fluorescence (TADF);[13,14] and (6) excimers.[15,16]

7.2.1.1 Intramolecular Charge Transfer

The phenomenon known as ICT occurs when the donor and acceptor are from the same molecule and this process takes place when a molecule becomes photoexcited as a response to the absorption of light of the appropriate wavelength.[7,17] ICT probes have an electron-donating unit conjugated to an electron-accepting unit in one molecule, resulting in a "push–pull" π-electron system in the excited state, which has been widely employed for cation sensing.[18,19] Nowadays, a large number of these ICT probes are being developed for ratiometric fluorescence sensing by altering the electron-donating, electron-withdrawing, or π-conjugation degree of the fluorophore to interact with the target analyte (Figure 7.2).[20]

For example, a Hg^{2+} ratiometric fluorescence probe, **ATC-Hg**, was developed by utilizing the Hg^{2+}-promoted 1,3-dithiolane cleavage. Because of this cleavage, an aldehyde group is produced, which condenses with the amino group of the *ortho*-2-aminophenyl moiety.[21] This produces a heterocyclic aromatic group in conjunction with an ICT-induced fluorescence alteration of 7-aminocoumarin (Figure 7.3A). Consequently, it shows a ratiometric response in fluorescence intensity with the emission ratios (I_{572}/I_{492}) and Hg^{2+} concentration (in a dose-dependent manner). Another ICT-based

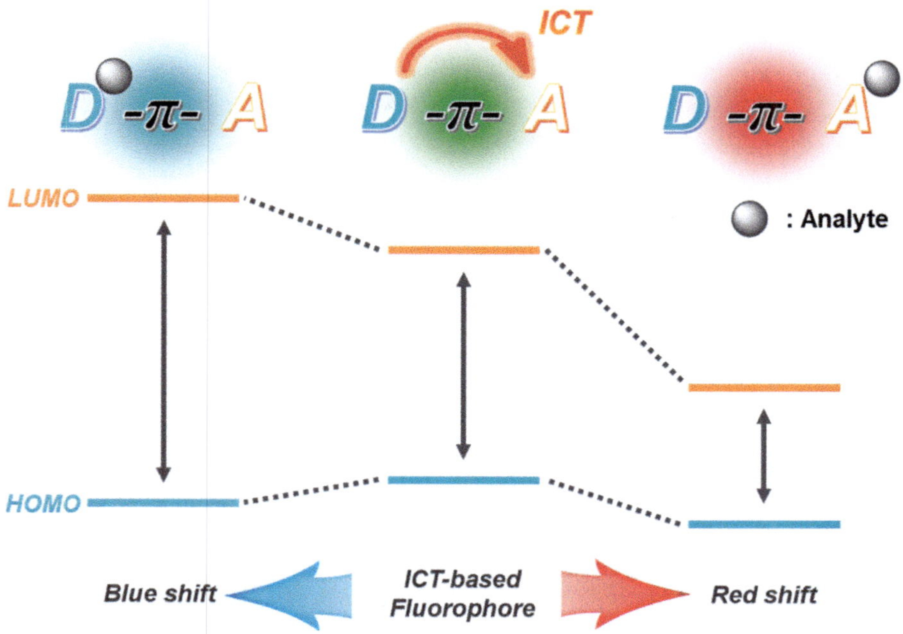

Figure 7.2 Schematic illustration of intramolecular charge transfer (ICT) mechanism in ratiometric fluorescent chemosensor design.

probe, **NAP-PS,** is used for the ratiometric sensing of Hg^{2+} ions based on the ICT process between the strong electron-donating oxyanion and the strong electron-withdrawing naphthalimide group.[22] The introduction of a thiophosphinate ester P–O linking moiety into a known 4-hydroxy-1,8-naphthalimide fluorophore, results in emission at a longer wavelength (Figure 7.3B). In addition to ratiometric detection of mercury, several ICT-based ratiometric fluorescent probes are also used for the detection of Zn^{2+} ions. For example, for the ICT-based fluorescent probe **L1**, Zn^{2+} binding is attributed to the amplification of ICT from the amide nitrogen close to the conjugated quinoline moiety and phenylethynyl groups. This results in a 1 : 1 complexation between the quinoline backbone with tetrahydrofuran-carboxamide and Zn^{2+} ions (Figure 7.3C).[23] The addition of Zn^{2+} exhibits a gradual decrease in the emission intensity of **L1** at 430 nm, along with a new emission band appearing at 525 nm in a ratiometric manner. Another ICT-based ratiometric two-photon fluorescent probe **P-Zn** shows a red shift of the absorption peak after 2,2′-dipyridyl conjugation with Zn^{2+} due to increase in the ICT effect (Figure 7.3D).[24] Subsequently, the ratiometric response is achieved by the fluorescence appearing, followed by increase in the new emission peak at 550 nm while attenuating the intensity at 465 nm towards Zn^{2+} ions.

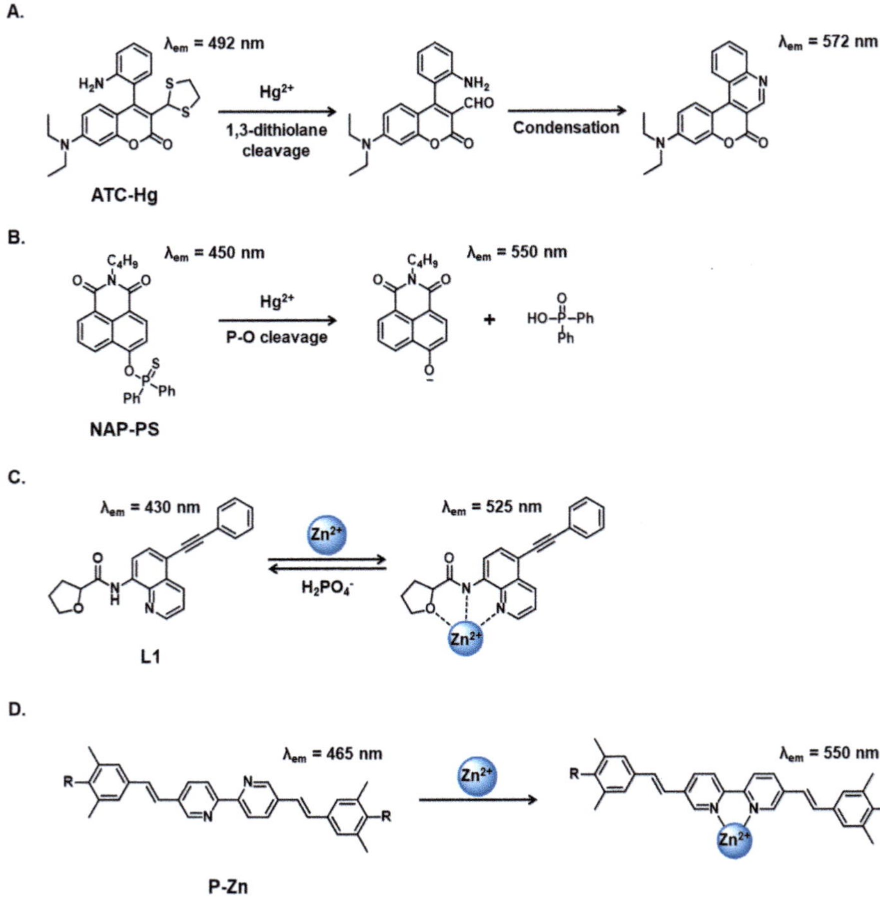

Figure 7.3 Structure examples of ICT-based ratiometric fluorescent chemosensors. Chemical structure and sensing mechanism of (A) **ATC-Hg**, (B) **NAP-PS**, (C) **L1**, and (D) **P-Zn**.

7.2.1.2 Excited-state Intramolecular Proton Transfer

ESIPT fluorophores have attracted much attention as ratiometric fluorescent sensors due to their fast four-level photochemical processes and efficient photophysical properties of (1) intense emission, (2) light stability, and (3) high Stokes shift (~200 nm). Because ESIPT fluorophore dual-emission spectra are derived from excited-state enol (E*) and keto (K*) emissions, the overall technique for developing the design of ESIPT-based fluorescent probes is based on enabling the blockage of the ESIPT fluorophore's hydrogen bond donor with a reaction-specific unit that hinders the ESIPT process (Figure 7.4).[2,25,26] Several ESIPT-based fluorescent probes show the ratiometric fluorescence response with different analytes, including ions, small molecules, and biomacromolecules.

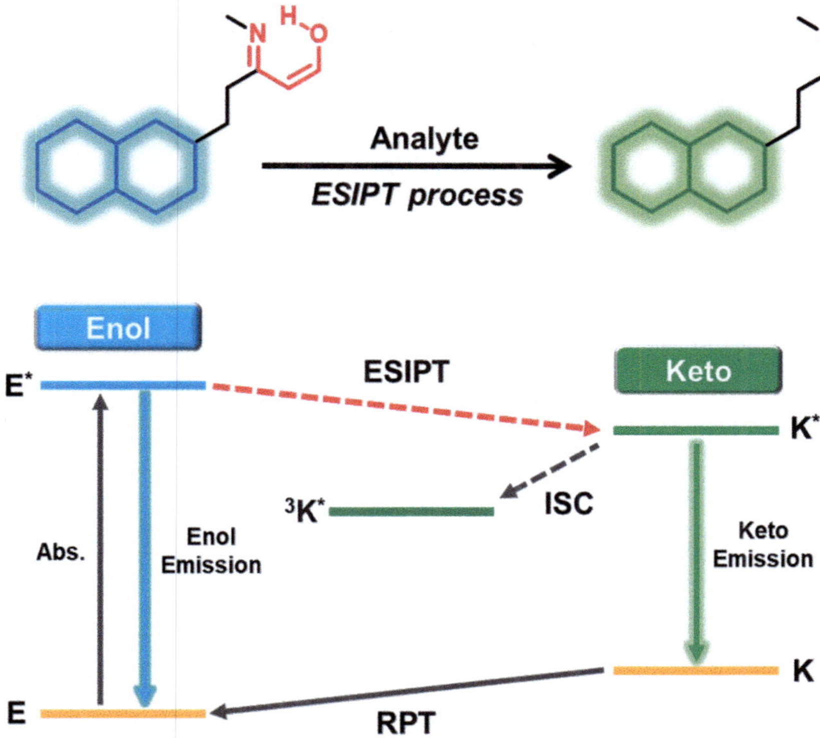

Figure 7.4 Schematic illustration of excited-state intramolecular proton transfer (ESIPT) mechanism in a ratiometric fluorescent chemosensor.

7.2.1.2.1 Ions. Several ESIPT-based fluorescent probes are used for the detection of diverse ionic analytes. For example, an ESIPT-based fluorescent probe **Pvi** shows a ratiometric response toward Hg^{2+} ions by cleaving the enol ether moiety to liberate a free hydroxyl group, which causes improved ESIPT (Figure 7.5A).[27] Another ESIPT fluorescent probe (probe **1** in Figure 7.5A) can be used to detect palladium (Pd) ions by the Pd-triggered cleavage process. This shows a large Stokes shift due to the ESIPT process of 3-hydroxylflavone during rapid photoinduced proton transfer, which results in tautomerization and dual emission (Figure 7.5B).[28] The flavone-based ratiometric fluorescent probe, probe **25**, demonstrates ESIPT characteristics to detect F^- ions relying on Si–O bond cleavage by F^- ions to produce hydroxyflavone, which causes the strong ESIPT process (Figure 7.5C).[29] Probe **25** exhibits a ratiometric response toward F^- ions with diminishing fluorescence at 400 nm, along with enhancement at 530 nm.

7.2.1.2.2 Small Molecules. ESIPT-based fluorescent probes can detect small molecules quantitatively in a selective and sensitive ratiometric manner. For example, nitric oxide (NO), a vital physiological signal gasotransmitter, can be successfully detected by an ESIPT ratiometric fluorescent probe

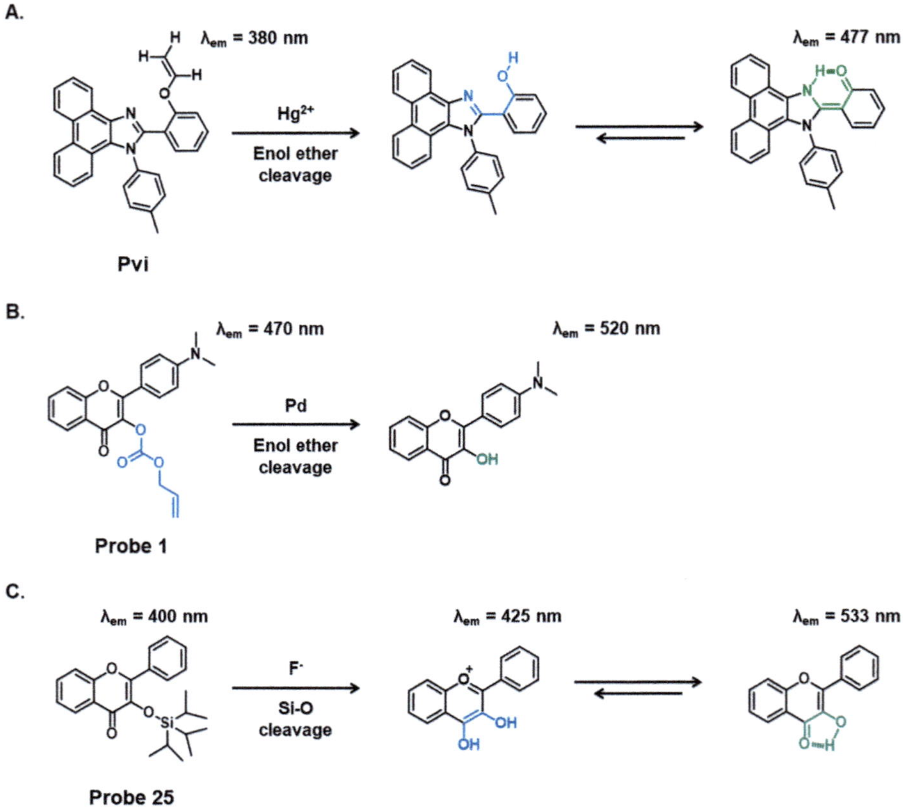

Figure 7.5 Structure examples of ESIPT-based ratiometric fluorophores for sensing ions. Chemical structure and sensing mechanism of (A) **Pvi**, (B) probe **1**, and (C) probe **25**.

(probe **1** in Figure 7.6A). Probe **1** is endowed with several beneficial features, including avoidance of self-absorption (*i.e.*, inner filter effect), which can be attributed to its significant Stokes shift (Figure 7.6A).[30] Another ratiometric fluorescent probe **TPE-TE** is used to ratiometrically monitor sulfur dioxide (SO_2) with combination of the ESIPT mechanism and the aggregate fluorescence method (Figure 7.6B).[31] In addition, the ratiometric fluorescent probe **HMBT** is used to detect superoxide ($O_2^{\bullet-}$) ions. The trifluoromethanesulfonate moiety that blocks the ESIPT process can be deprotected in the presence of $O_2^{\bullet-}$ and the resulting product attains the ESIPT process (Figure 7.6C).[32]

7.2.1.2.3 Biomacromolecules. The ESIPT process is also involved in the photophysics of ratiometric fluorophores for monitoring various biomacromolecules in biological systems. For example, a ratiometric fluorescent probe **HCAP** is developed based on the phosphorylated chalcone derivative and is used for alkaline phosphatase (ALP) activity assay and visualization in living cells. The chalcone product, formed by cleavage of a phosphorylated

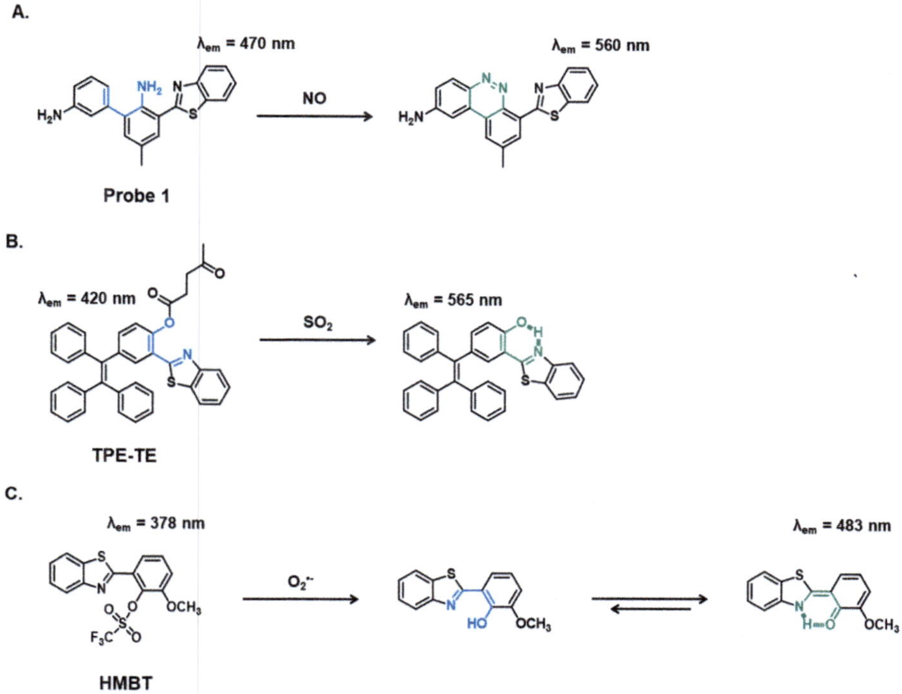

Figure 7.6 Structure examples of ESIPT-based ratiometric fluorophores for sensing small molecules. Chemical structure and sensing mechanism of (A) probe **1**, (B) **TPE-TE**, and (C) probe **HMBT**.

group with ALP, undergoes the ESIPT process through keto–enol tautomerization, resulting in a well-conjugated structure with alternating bonds (Figure 7.7A).[33] In addition, based on the ESIPT mechanism, an ester ratiometric fluorescent probe (probe **1**) derived from 3-hydroxyflavone provides selective and sensitive detection of carboxylesterase 2 (CE2) (Figure 7.7B).[34]

7.2.1.3 Förster Resonance Energy Transfer

For the design of ratiometric fluorescent probes, FRET is one of the most extensively used sensing mechanisms. FRET is a non-radiative energy transfer process involving long-range dipole–dipole interactions between the donor (FRET donor) and acceptor (FRET acceptor). The electronic excitation energy of the donor in its excited state can be transferred to the acceptor in its ground state after photoexcitation. Several standard approaches have been used to construct ratiometric FRET probes (Figure 7.8), and here we provide a few ideas with appropriate examples from the available literature.[10]

To construct ratiometric FRET probes, one option is to modify the chemical structure of an acceptor from weakly fluorescent to strongly fluorescent (Figure 7.8A). The absorption band of the non-fluorescent acceptor and the

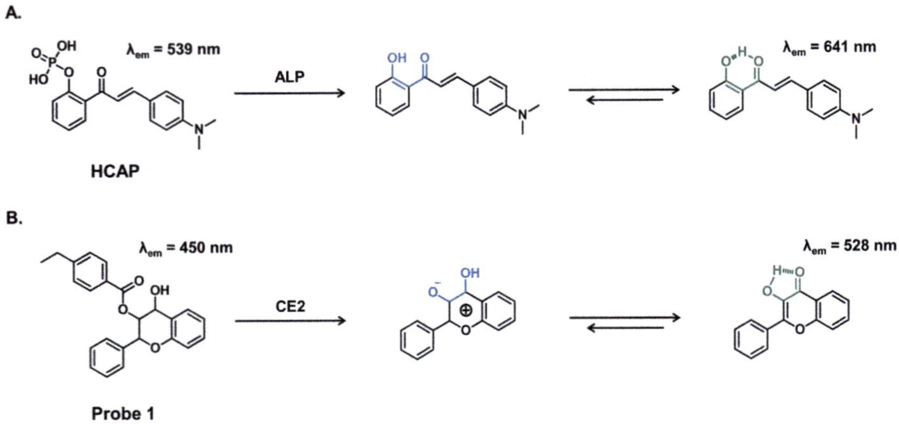

Figure 7.7 Structure examples of ESIPT-based ratiometric fluorophores for sensing biomacromolecules. Chemical structure and sensing mechanism of (A) **HCAP** and (B) probe **1**.

emission band of the donor overlap relatively little at first. Following contact between the non-fluorescent form of the acceptor and the target analyte, the absorption band of the acceptor is shifted, increasing the overlap with the donor emission band and promoting FRET. For example, a ratiometric fluorescent probe, **IRP**, can be used to detect hypochlorite (OCl^-) based on the FRET process. An imidazo[1,5-*a*]pyridine moiety (FRET donor) is combined with a rhodamine moiety (FRET acceptor) to generate the probe **IRP**. The fluorescence intensity ratio (I_{575}/I_{467}) shows considerable improvement when the OCl^- concentration is increased from 0 to 5 M. Notably, the experimental result reveals that I_{575}/I_{467} is linearly related to the OCl^- concentration from 0.5 to 3.5 M (R^2 = 0.998). The structural creation of a ring-opening form of the rhodamine moiety followed by FRET, which can be generated by the reaction of **IRP** with OCl^-, could explain the spectral variations (Figure 7.9A).[35] A two-photon fluorescent probe, **P-Np-Rhod**, shows a quick and selective ratiometric response to nitroxyl (HNO) following the FRET mechanism (Figure 7.9B).[36] The free probe **P-Np-Rhod** displayed a cyan color in the absence of HNO, with an emission peak at 448 nm. A new fluorescence emission peak at 541 nm was exhibited and steadily increased after treatment with increasing concentrations of HNO, while the characteristic donor emission peak (λ_{em} = 448 nm) showed a decreasing tendency. With the concentration of HNO varying from 0 to 100 M, the fluorescence intensity ratio between the two wavelengths (I_{541}/I_{448}) changes from 0.13 to 4.67. Furthermore, in 2018, by changing the core of the known adenosine triphosphate (ATP) detection system based on rhodamine B into rhodamine 6G, and introducing a BODIPY moiety at the end of the tetramine chain, a FRET probe for lysosomal ATP (**Lyso-ATP**) was constructed (Figure 7.9C).[37] At an excitation wavelength (λ_{ex} = 403 nm), **Lyso-ATP** generates blue fluorescence (λ_{em} = 454 nm). However, when ATP is added to the **Lyso-ATP** solution in acidic HEPES buffer,

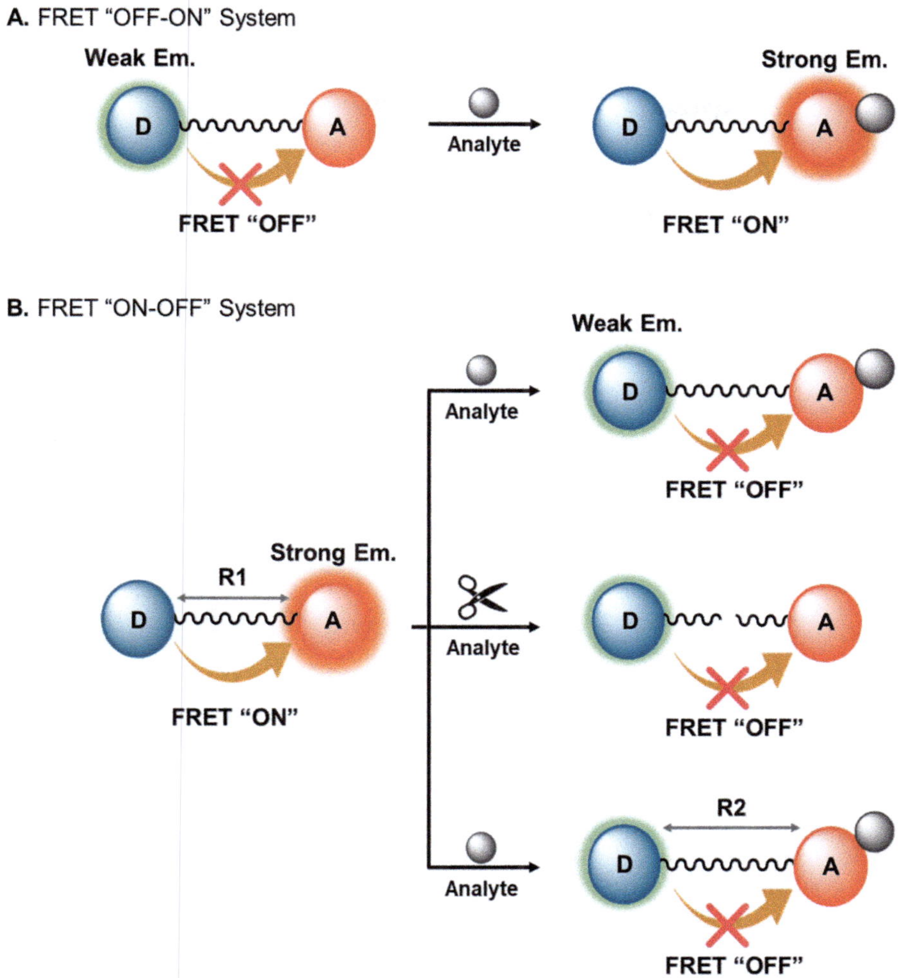

Figure 7.8 Schematic illustration of Förster resonance energy transfer (FRET) mechanism in ratiometric fluorescent chemosensor design. (A) FRET "OFF–ON" system. (B) FRET "ON–OFF" system.

the resultant solution exhibits a bright yellow fluorescence (λ_{em} = 557 nm) from rhodamine 6G due to FRET from the amino-BODIPY moiety. Notably, I_{454}/I_{557} is proportional to the ATP content up to the millimolar level. The first ratiometric probe (**RDC1**) for cisplatin detection was developed to investigate Pt^{IV} anticancer complexes in biological systems. The design of this ratiometric probe is based on the coumarin-466 (FRET donor) and rhodamine 6G (FRET acceptor) (Figure 7.9D).[38] **RDC1** has a maximum emission at 470 nm in the absence of Pt^{2+}. When **RDC1** is titrated with Pt^{2+} (0–10 equiv.), the fluorescence at 470 nm declines in intensity while the fluorescence at 565 nm increases, indicating FRET after spiro ring opening. In the absence of Pt^{2+}, the ratio of emission intensities I_{565}/I_{470} increases from 0.07 to 4.8 (68-fold) with an increase in energy transfer efficiency of 81%.

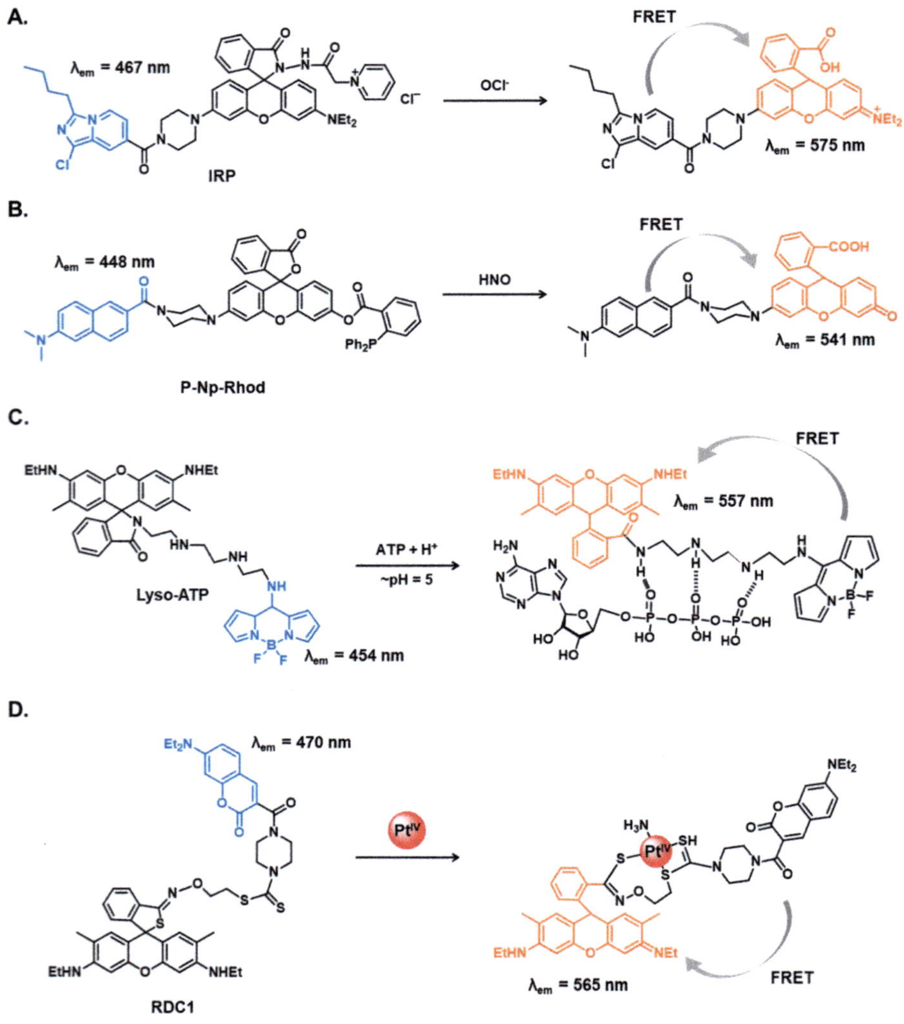

Figure 7.9 Structure examples of FRET "OFF–ON" system ratiometric fluorophores. Chemical structure and sensing mechanism of (A) **IRP**, (B) **P-Np-Rhod**, (C) **Lyso-ATP**, and (D) **RDC1**.

Another method to construct ratiometric FRET probes is by changing the chemical structure of an acceptor from fluorescent to non-fluorescent (Figure 7.8B). To give an example, the fluorescent probe that could detect matrix metalloproteinase (MMP) in a dual-color mode was constructed. The FRET pairs of fluorescent probes (**FixD and LA**) contain cationic groups to target mitochondria, while the FRET donor (**FixD**) consists of an additional phenyl chloride group to localize onto mitochondria. In mitochondria with high MMP, the FRET pair shows a deep-red emission (λ_{em} = 675 nm), whereas in mitochondria with low MMP it shows green fluorescence (λ_{em} = 565 nm) (Figure 7.10A).[39] Furthermore, in 2021, to detect sulfur dioxide (SO_2)

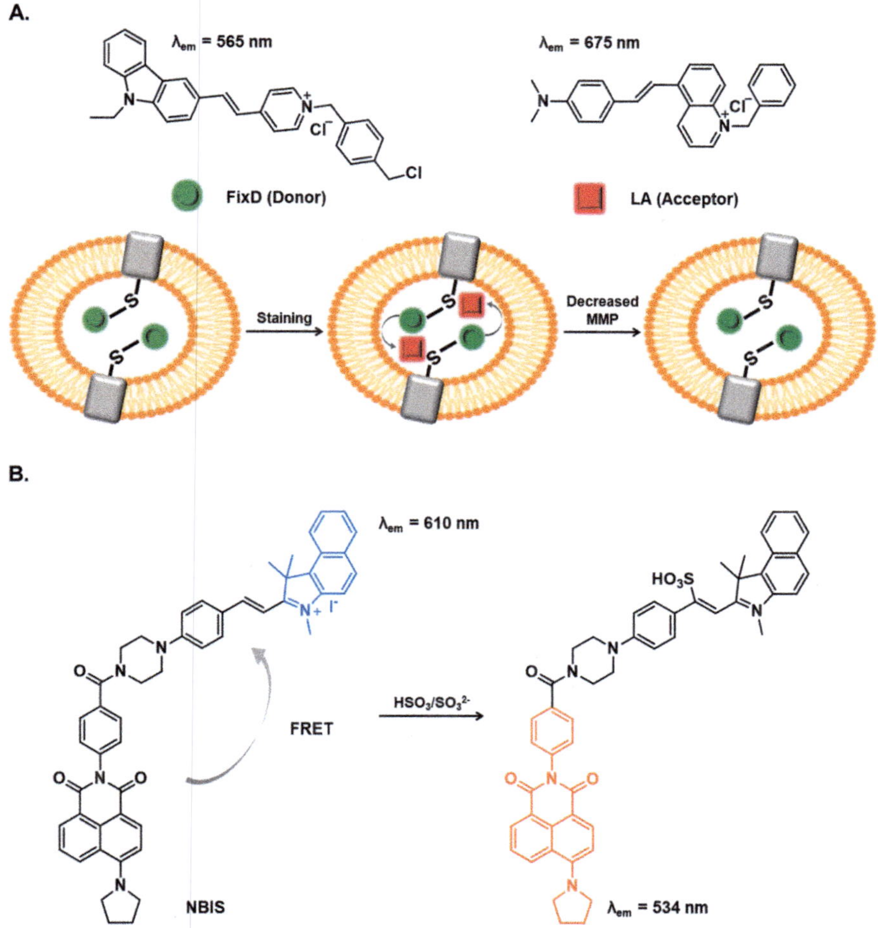

Figure 7.10 Structure examples of FRET "ON–OFF" system ratiometric fluorophores. Chemical structure and sensing mechanism of (A) **FixD/LA** and (B) **NBIS**.

derivatives, a new FRET-based fluorescent probe, **NBIS**, was developed using benzoindole-based hemicyanine as an acceptor and naphthalimide derivatives as a donor (Figure 7.10B).[40] As the ratio of HSO_3^-/SO_3^{2-} increases, the fluorescence intensity at 610 nm gradually decreases, and the fluorescence intensity at 534 nm increases. The fluorescence intensity ratios (I_{534}/I_{610}) and the concentrations of HSO_3^-/SO_3^{2-} from 0 to 0.9 equiv. show a linear relationship.

In addition, there are other turn-off-based FRET systems. One way to disturb the FRET process is by cleaving the covalent linker between the donor and the acceptor. For example, the first-generation ratiometric FRET copper probe, **FCP-1**, is used for activity-based sensing of labile Cu(I) pools in live cells. **FCP-1** shows an increase in fluorescein emission at 526 nm, with a corresponding decline in rhodamine fluorescence at 576 nm when Cu(I) is

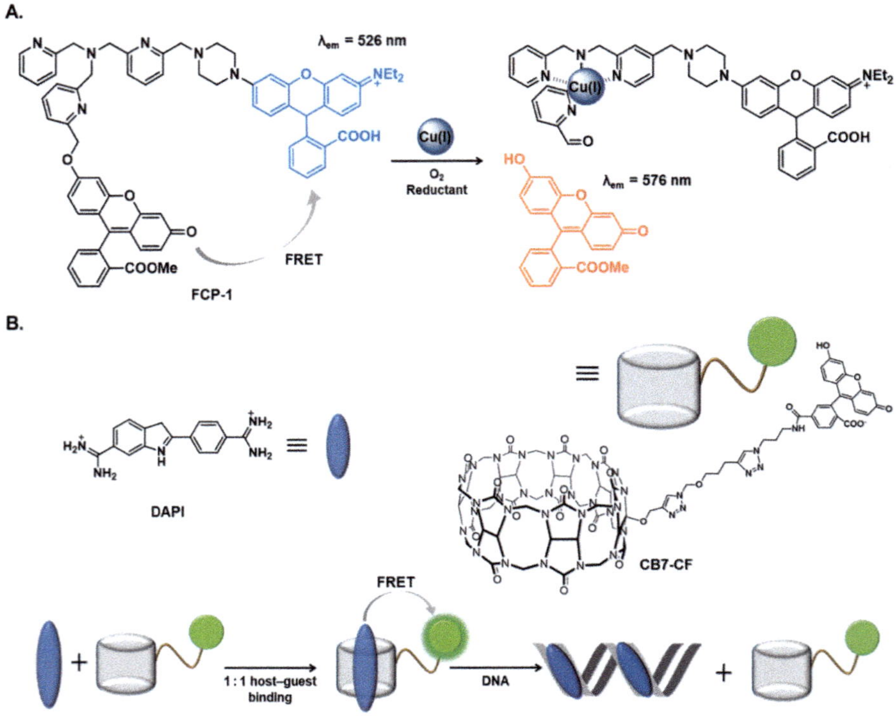

Figure 7.11 Structure examples of FRET "ON–OFF" system ratiometric fluorophores. Chemical structure and sensing mechanism of (A) **FCP-1** and (B) **CB7-CF/DAPI**.

added. Cu(I) promotes the oxidative cleavage of the benzyl ether C–O bond on the tris[(2-pyridyl)methyl]amine (TPA) linker of **FCP-1**, resulting in separation of the fluorescein donor from the rhodamine acceptor and a decrease in intramolecular FRET efficiency with a ratiometric emission spectrum change (Figure 7.11A).[41] The other way is by dissociation of the FRET donor and acceptor. For example, a supramolecular host–guest FRET pair based on carboxyfluorescein-labeled cucurbit[7]uril (CB7-CF, FRET acceptor) and the fluorescent dye 4′,6-diamidino-2-phenylindole (DAPI, FRET donor) is used for DNA sensing (Figure 7.11B).[42] With increasing DNA concentrations, DAPI is transferred from CB7-CF, where it no longer acts as a FRET donor and reduces CF fluorescence at 520 nm, to DNA, where its own fluorescence at 450 nm is enhanced. In the picomolar range (up to 20 μg mL^{-1}), the ratio of fluorescence intensity values (I_{450}/I_{520}) increases linearly.

7.2.1.4 Bioluminescence Resonance Energy Transfer

BRET is very similar to FRET, but the difference is in the energy derivation. In FRET, the donor is a fluorescent protein that needs to be excited, whereas in BRET the energy derived from a luciferase reaction can be utilized to excite a fluorescent protein if it is close to the luciferase enzyme. Luminescence is

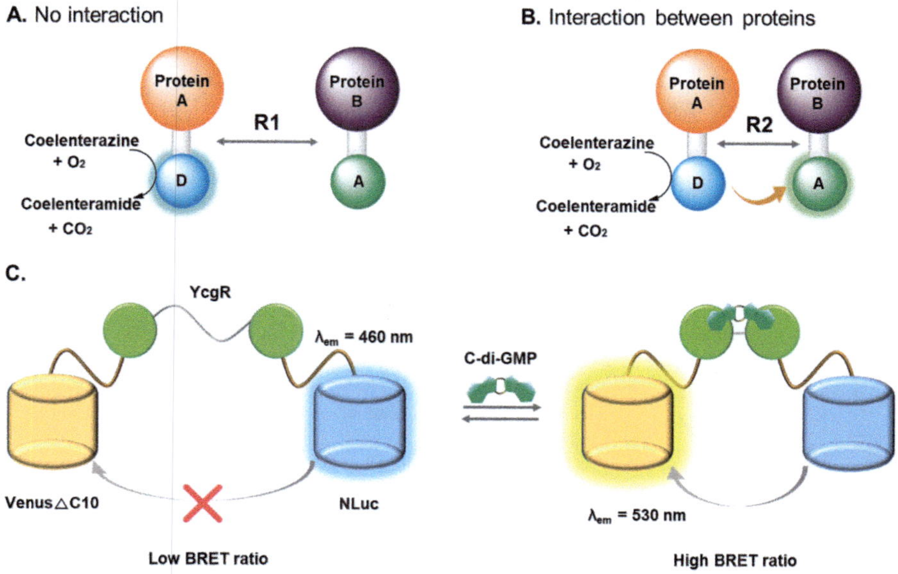

Figure 7.12 Schematic illustration of bioluminescence resonance energy transfer (BRET) mechanism in ratiometric fluorescent chemosensor design. (A) No interaction between proteins. (B) Interaction between proteins. (C) Sensing mechanism of **VYN**.

an alternative optical technology that avoids the problems associated with autofluorescence (Figure 7.12A and B). For example, the Venus-YcgR-NLuc (**VYN**) biosensors, with substantial variations in the BRET signal, can image bis-(3′-5′)-cyclic dimeric guanosine monophosphate (c di-GMP) in tissues (Figure 7.12C).[12] YcgR proteins typically undergo substantial conformational changes upon c-di-GMP binding, which has been harnessed to generate genetically encoded sensors for c-di-GMP. The binding of c-di-GMP to the **VYN** sensor produces a change in energy transfer efficiency between Venus and NLuc, changing the BRET ratio.

7.2.1.5 Thermally Activated Delayed Fluorescence

The TADF transition is a unique radiative transition of excitons from the S1 state, which exhibits intersystem crossing (ISC), and a delayed emission with a longer lifetime than rapid fluorescence. Reverse intersystem crossing (RISC) from T1 to S1 yields the final emissive singlet excitons involved in the TADF process (Figure 7.13A). For example, a series of red-emissive TADF acrylic monomers based on naphthalimide (**NAI**) was developed (Figure 7.13B).[14] The NAI core (λ_{em} = 390 nm) and the charge transfer (CT) process (λ_{em} = 660 nm) emit roughly identical amounts of light at T < LCST (lower critical solution temperature). The peak emission ratio at 390 to 660 nm (I_{390}/I_{660}) grows linearly with temperature when T > LCST.

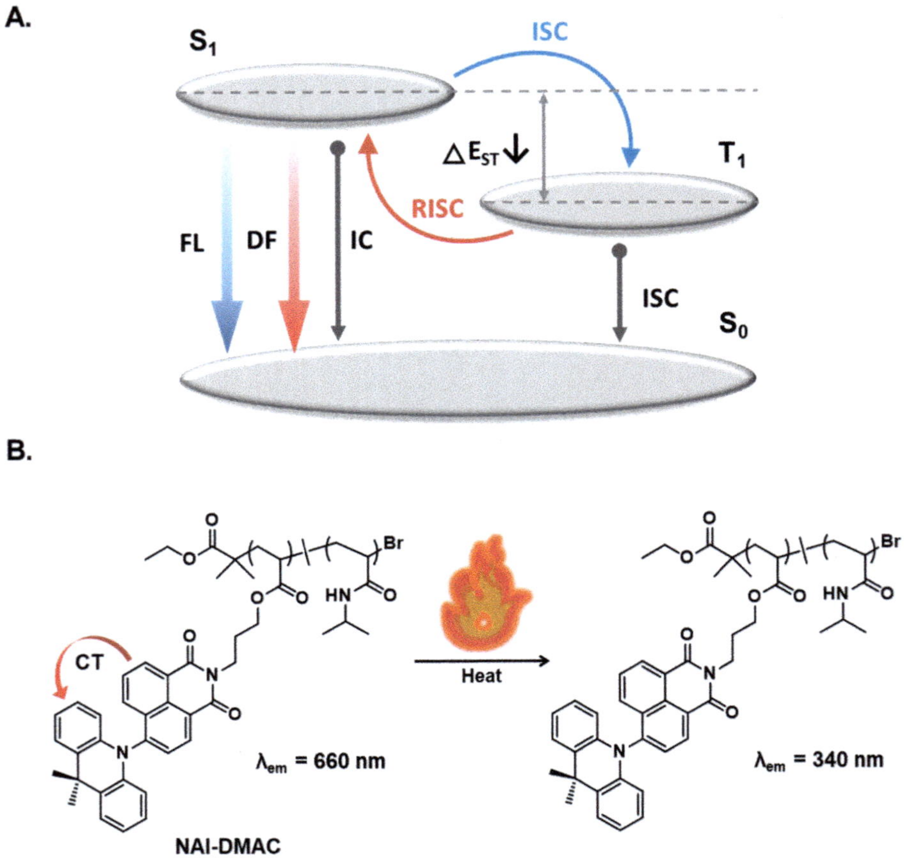

Figure 7.13 (A) Schematic illustration of thermally activated delayed fluorescence (TADF) mechanism in ratiometric fluorescent chemosensor design. The sensing mechanism of (B) **NAI–DMAC**.

7.2.1.6 Excimer

When an excited molecule might come close to another during the lifespan of the excited state, several fluorophores, such as anthracene and pyrene, can form an excimer (excited dimer) (Figure 7.14A).[15] For example, a perylene probe monomer–excimer transition has been used to produce a new sensing strategy. Monomers of compound **1** are produced when single-stranded DNA is introduced to the mixture of compound **1** and cationic polymer, and an excimer–monomer emission transition is observed. The effects of four distinct cationic polymers on the aggregation of compound **1** were investigated. The results reveal that compound **1** excimer emission can be visible in all situations, regardless of the cationic polymer utilized. With increasing polycation concentration, the emission intensity ratio (IE/IM value) of the excimer (IE at 680 nm) to the monomer (IM at 548 nm) rapidly rises. Also, the polymer structure has a considerable impact on the strength of excimer

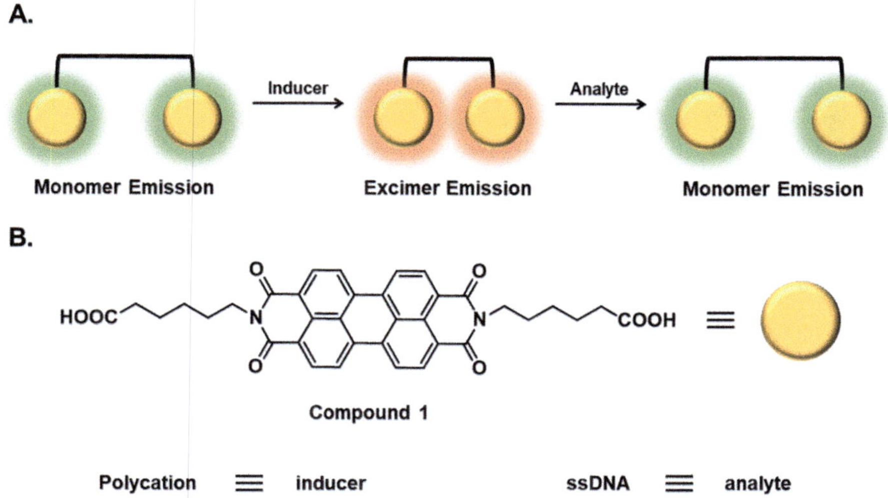

Figure 7.14 (A) Schematic illustration of excimer mechanism in ratiometric fluorescent chemosensor design. (B) The sensing mechanism of compound **1**.

emission. The creation of excimers appears to be aided by hydrogen-bonding interactions between the polymer and compound **1** (Figure 7.14B).[16]

7.3 Design Strategies for Ratiometric Fluorescent Chemosensors

7.3.1 Small-molecule-based Ratiometric Fluorophores

Molecular fluorophores have been used for analytical detection and optical imaging due to their rapid response time, significant sensitivity, and technical simplicity.[5,43–47] Among them, small-molecule-based ratiometric fluorophores (SMRFs) can provide intuitive visual information through direct optical change, which simultaneously relies on analyte-caused fluorescent emission changes in both wavelength and intensity.[1,2,48] The performance of SMRFs is significantly controlled by the chemical architectures of the optical-changing trigger and the fluorophore backbone used. To design appropriate SMRFs, the important spectroscopic characteristics of fluorophores that need to be considered are: (1) wavelength; (2) molar absorption coefficient (ε); (3) the width and shape of the absorption and emission spectrum; (4) the Stokes shift; (5) fluorescence lifetime; (6) fluorescence quantum yield; and (7) emission anisotropy. These features significantly affect the reliability, sensitivity, reproducibility, and dynamic range of fluorescence detection using SMRFs.

Depending on the trigger that induces change and which fluorophore is employed, the mechanism by which SMRFs interact with an analyte and

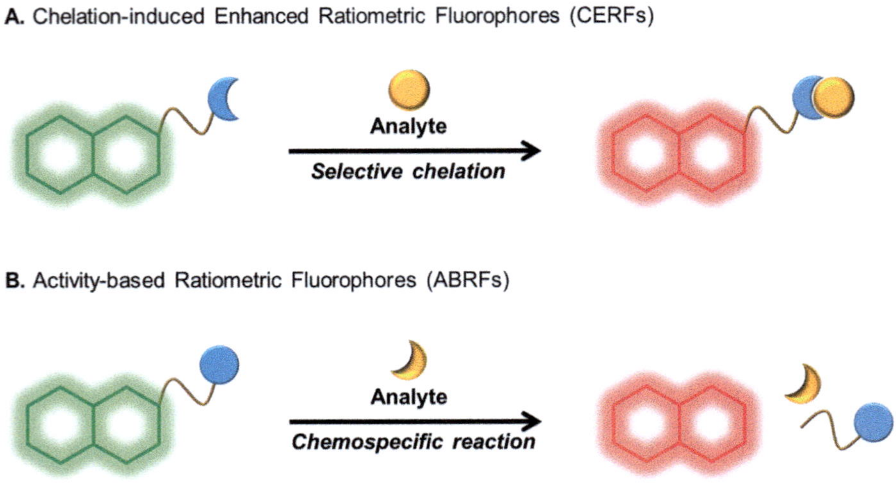

Figure 7.15 General design principles for ratiometric optical activation mechanisms. (A) Chelation-induced enhanced ratiometric fluorophores (CERFs). (B) Activity-based ratiometric fluorophores (ABRFs).

demonstrate performance is comprehensively classified into two categories: chelation-induced enhanced ratiometric fluorophores (CERFs) or activity-based ratiometric fluorophores (ABRFs) (Figure 7.15). With these mechanisms, SMRFs can be widely applied to environmental and biological systems to analyze specific ions, small molecules, and biological processes. This section broadly covers several examples of SMRFs with fluorescence-change triggering mechanisms with different analytes.

7.3.1.1 Chelation-induced Enhanced Ratiometric Fluorophores

CERFs for an analyte, especially ions, are designed by linking a coordinating ligand to a luminescent responsive unit: in other words, the coordination complexation between "the hosts and the guests". The chemical architecture of CERFs is dichotomously constructed by both essential moieties: (1) the ligands that can bind ion analytes; and (2) the small molecular fluorophores, which sensitively change the fluorescence response upon metal ion chelation or anionic molecular interaction.[2,48–53] Fortunately, the design of the ligand moiety of CERFs with excellent selectivity and tuned binding for target ions has been brilliantly developed on account of the knowledge accumulated by decades of organic molecular chemistry and coordination chemistry so far.[54] Typical examples are listed below.

Metal cations with abundant components that play an essential role in nature can be ideal analytes for CERFs. Mainly these are sodium (Na^+), potassium (K^+), zinc (Zn^{2+}), calcium (Ca^{2+}), and magnesium (Mg^{2+}) ions, and a few kinds of heavy metal ions, including mercury (Hg^{2+}), lead (Pb^{2+}), rubidium (Rb^+), cadmium (Cd^{2+}), and cesium (Cs^+), which show significant metal–ligand chelating interactions. Therefore, the CERF-based metal ion sensing mechanism can be achieved by the chelating interaction between small molecular ligands and cationic metal analytes.

Several types of metal ion chelators have been developed with the CERF system. Common examples include crown ether, aza-crown ether, or amine-based chelating ligands, such as dipicolylamine (DPA) (Figure 7.16A). Ordinarily, the complexation of K^+, Rb^+, Mg^{2+}, Ca^{2+}, or Cs^+ ion with fluorescence sensor conjugated with metal ion chelating ligands can induce rigidification and conformational effects to block the non-radiative relaxation of the potential energy of the excited state. Furthermore, for Rb^+ and Cs^+, the observed fluorescence changes can be ascribed to an ion complexation-induced change and/or the heavy atom effect in triplet energy corresponding to the ground and excited singlet state energies. Consequent to these mechanisms, CERFs

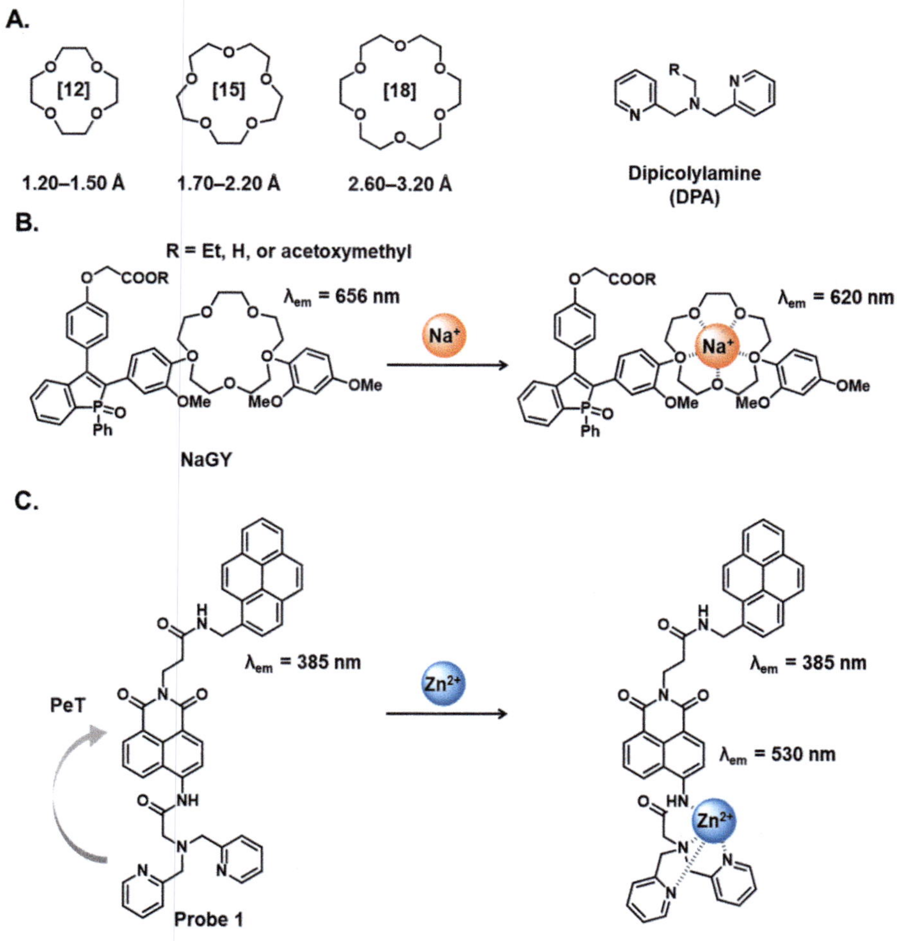

Figure 7.16 Structure examples of CERFs. (A) Metal chelating ligands, including crown ethers or dipicolylamine (DPA). Chemical structure and sensing mechanism of (B) **NaGY** and (C) **probe 1**.

finally demonstrate a significant change in the photophysical properties for ratiometric fluorescence response.[5]

For example, a benzophosphole *P*-oxide fluorophore conjugated with an aza-crown ether scaffold (**NaGY**) shows a ratiometric detection ability for Na^+ ion (Figure 7.16B).[49] Probe **NaGY** demonstrates a hypochromic shift of the emission spectrum upon complexation with Na^+ ion under violet light excitation (λ_{ex} = 405 nm). Additionally, the ratiometric analysis exhibited a Na^+ dissociation constant (K_d = 16.0 mM), which is valuable for monitoring the variation of intracellular Na^+ ion concentrations. As another example, the amine-based ligand employing pyrene-appended naphthalimide-DPA (**probe 1**) is used to detect Zn^{2+} ions (Figure 7.16C).[51] The naphthalimide-DPA moiety forms a 1 : 1 stoichiometric complex with Zn^{2+} ions, which causes a ratiometric fluorescence change through modulation of the PeT process. This probe provides a ratiometric fluorescence change at 385 and 530 nm; hence, an overall output fluorescent colorimetric change can be detected from blue to green in the coordination of Zn^{2+} ions.

7.3.1.2 Activity-based Ratiometric Fluorophores

ABRFs provide an outstanding suite of analytes that react selectively with the functionally active form to measure its activity state within a specific environment.[16] Also, for monitoring biological activity, ABRFs have noticeable advantages in evaluating and identifying the potential binding affinity of analytes, such as an enzyme. In the past decade, various ABRFs have been developed to detect ions, small molecules, or enzymes essential to understanding the physiological/environmental processes.[55–58] Furthermore, ABRFs for multi-modal imaging applications have extended their potential in biology; however, more studies are needed to increase the effectiveness of ABRFs in these applications. Fundamentally, ABRFs show a response by means of chemospecific reaction with analytes, and this follows complicated mechanisms based on ICT, PeT, or FRET to show the change in fluorescence emission. This section focuses on the development and application of ABRFs and provides a basic idea of the structural features and their use in ratiometric fluorescence imaging and multi-modal applications.

7.3.1.2.1 Ions. Using chemospecific reactions to detect transition metal ions has broadened the universe of ABRFs in combination with the FRET-based ratiometric system. To trigger ratiometric fluorescence change with the FRET system, the energy acceptor should usually be "turn-on" under chemospecific reaction. The spirolactam ring-opening processes of rhodamine derivatives, used as a fluorescent turn-on/off system, are a representative strategy for ratiometric ABRFs for metal ions. Typically, naphthalimide dyes (λ_{em} = 450–650 nm) have gained substantial attention due to their outstanding features of introducing a ratiometric system with rhodamine derivative pairing (λ_{ex} = 450–600 nm). Convenient overlap between the absorption range of rhodamine derivatives and the emission range of naphthalimide derivatives means the naphthalimide moiety can work as a suitable energy donor to rhodamine in the FRET system. A chemospecific reaction with *d*-block metal ions demonstrates the ring-opening response of the inert rhodamine, which eventually achieves the emission change from green (naphthalimide) to red (rhodamine).

As shown in Figure 7.17A, the rhodamine-derived fluorescent dye conjugated with naphthalimide (**L**) can undergo a selective reduction reaction through the catalytic activity of Cu^{2+}. Consequently, the FRET process occurs between two fluorophores, which leads to a high fluorescence yield from rhodamine dye in an active form.[21] This reaction mechanism shows significant selectivity and sensitivity to Cu^{2+} over other metal ions (*ca.* 3.2 ppb in an aqueous medium). In addition, in a similar mechanism, mercury (Hg^{2+}) can show a desulfurization reaction, with its unique thiophilic character, to give the spirolactam ring-opening system with rhodamine fluorophore.[22] The desulfurization reaction by Hg^{2+} makes rhodamine activate irreversibly and provides a FRET system, which can be applied to ratiometric detection of Hg^{2+}. Also, in the naphthalimide/rhodamine-based FRET system, various energy donors (*e.g.*, coumarin, quinolone-benzothiazole, BODIPY, and naphthalene derivatives) introduced to different energy acceptors, such as rhodamine or other fluorophores, can be used to detect various metal ions in a ratiometric manner.[2,48] All these sensing mechanisms are based on the ring-opening reaction of rhodamine with the catalytic activity of metal ions.

7.3.1.2.2 **Small Molecules.** Various small molecules in physiological and environmental nature can show specific reactivity with ratiometric fluorophores. Similar to ions, functional groups with selective reactivity to small molecules can be applied to the ratiometric sensor. Thus, ABRFs for sensing small molecules have been widely used to detect subtle alteration of super-reactive chemicals in mild conditions, such as reactive oxygen-nitrogen species (RONS), reactive carbonyl species (RCS), and thiols.[55,56]

Recently, various recognition triggers for reliable and selective detection of RONS have been manipulated. A significant part of the recognition trigger is found through the chemospecific reaction with RONS and substrates under physiological conditions by exploiting the unique chemical properties of RONS, including nucleophilicity, vigorous oxidation, and nitration. The recognition triggers dominate the chemospecific reaction (sensitivity and reaction kinetics) of the probe toward the analyte, and effectively adjust the fluorescent output measures (fluorescence quantum yields, emission wavelengths, and lifetimes) through assorted fluorescence regulation fundamentals, as described in Section 7.2. As for metal ions, chemospecific reaction can directly affect the fluorescence emission wavelength or activate the FRET system to induce ratiometric fluorescence emission. For example, under reaction with peroxynitrite ($ONOO^-$), one of the BODIPY-derived ratiometric chemosensors ($\mathbf{F_{482}}$) shows a C–C bond cleavage reaction to shorten the π-conjugation, which induces a blue-shifted fluorescence emission (Figure 7.17B).[59] The probe $\mathbf{F_{482}}$ demonstrates a 50-fold increase in ratiometric

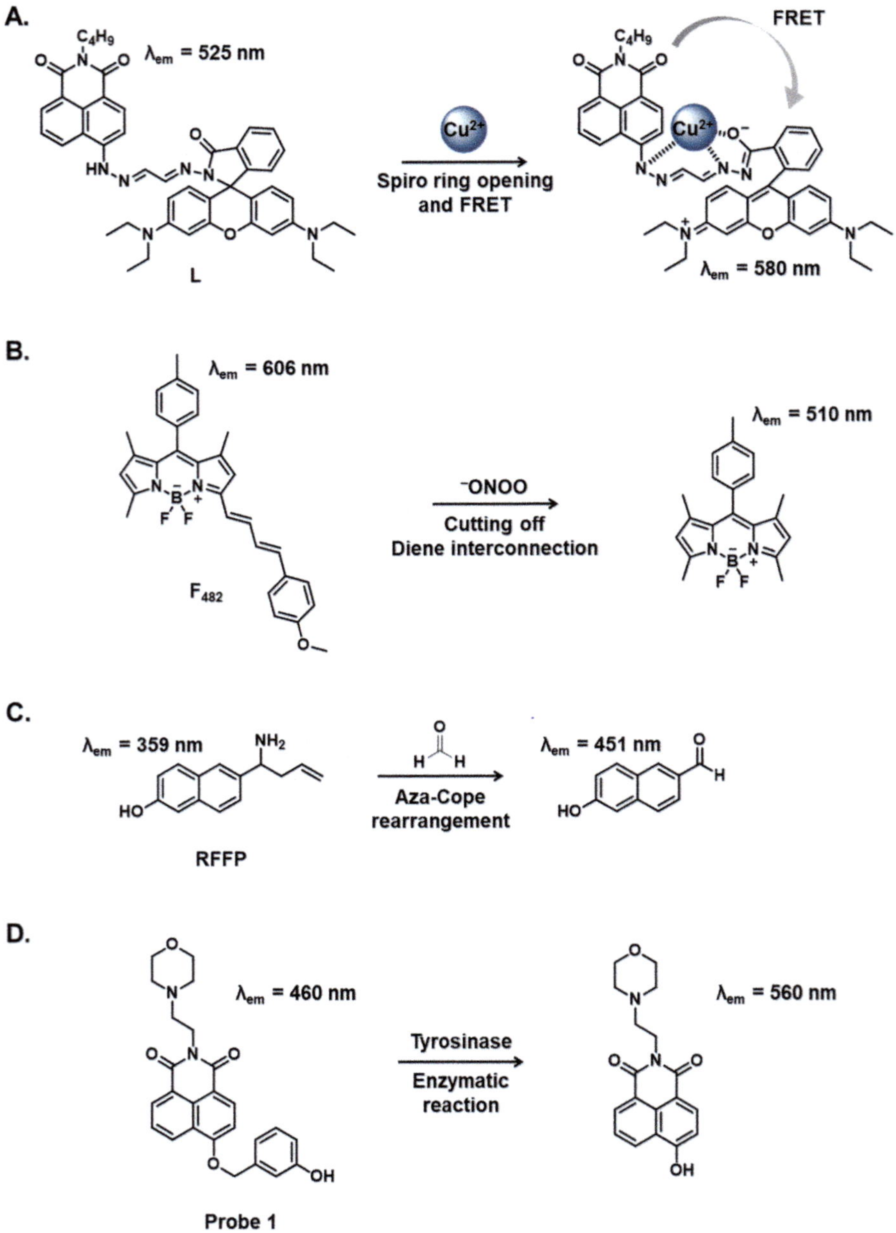

Figure 7.17 Structure examples of ABRFs. (A) Chemical structure and sensing mechanisms of (A) **L**, (B) $\mathbf{F_{482}}$, (C) **RFFP**, and (D) **probe 1**.

fluorescence signal to $ONOO^-$ in 1 min, and the ratio is kept constant for 30 min.

Generally, RCS show exquisite chemospecific reactivity with activity-based probes to induce ratiometric fluorescence emission. For example, one of the simplest aldehyde molecules, formaldehyde, offers the typical aza-Cope rearrangement reaction mechanism that shifts single and double bonds between two allylic components *via* a [3,3]-σ-rearrangement. As an example, the ratiometric fluorescence probe (**RFFP**), consisting of 6-hydroxynaphthalene, exhibits a fluorescent emission in the near-UV range (λ_{em} = 350 nm) due to the lack of an ICT system (Figure 7.17C).[60] The product formed by a chemospecific reaction with formaldehyde can undergo the aza-Cope rearrangement reaction. Thus, the homoallylamino group on the probe can be altered into an aldehyde, contributing to a robust electron-withdrawing effect. Hence, the electron push–pull intensity of the fluorophore structure is enhanced, inducing a ratiometric fluorescence change with a bathochromic shift (approximately 90 nm) of the emission wavelength from 359 to 451 nm.

7.3.1.2.3 Enzymes. Several enzymes that induce chemical reactions under physiological conditions as a catalyst can provide chemospecific reactions with ABRFs for optical imaging. Most small molecular enzymatic fluorophores are designed based on selective interaction with the active sites of enzymes: catalytic sites (reduce chemical activation energy) or binding sites (bind and orient substrates). Hence, the enzymes can distinguish very small differences between diverse substrates to bind and catalyze the specific reaction. ABRFs can work with enzymes to exhibit sensing performance under biological conditions, such as physiological temperature (37 °C), pH (7.4), and aqueous solution, with sufficient reaction yields. Furthermore, biologically applied ABRFs have tremendous superiority regarding specificity, biocompatibility, and design flexibility.[58]

Enzymes possess an inherent ability for catalytic conversion of their specific substrates into products. Thus, it is possible to produce substrate-derived fluorophores by conjugating an optimum substrate-derived moiety to a probe, whose physicochemical properties change upon alteration by an enzymatic reaction. Substrate-derived ABRFs minimize the background signal by generating a ratiometric fluorescence response. Typical substrate-derived ratiometric fluorescent dyes include two characteristic moieties: a fluorophore backbone and its corresponding enzyme-recognition unit. Occasionally, some substrate-derived ratiometric probes are employed with self-immolative linkers to connect between enzyme-recognition units and fluorophores.

As shown in Figure 7.17D, a morpholine–naphthalimide employed with the 3-hydroxy benzyl moiety (**probe 1**) exhibits significant performance in detecting tyrosinase activity in melanosomes.[61] The 3-hydroxy benzyl unit of **probe 1** is cleaved in the presence of tyrosinase to produce a ratiometrically active form of the fluorophore by the self-immolative reaction. This enzymatic reaction eventually provides a ratiometric fluorescence change from blue emission (λ_{em} = 460 nm) to yellow emission (λ_{em} = 560 nm). Therefore, the ABRFs allow quantitative and accurate monitoring of enzymatic activity, and this system will be useful for studying and understanding physiological/pathological processes.

7.3.1.3 Ratiometric Two-photon Fluorophores

The two-photon microscopy (TPM) imaging technique has been applied to the development of ratiometric fluorescence chemosensors due to its inherent advantages of: (1) deeper penetration depth due to near-infrared wavelength excitation; (2) less-responsive to light scattering and internal reflection, which can be prevented by non-linear excitation; (3) reduced photobleaching and phototoxicity by harnessing relatively small amounts of energy (Figure 7.18A).[62,63] Numerous studies have enhanced the research on ratiometric two-photon fluorescence (RTPF) chemosensors, which present up-and-coming imaging techniques, especially in analyte sensing, bioimaging, and disease diagnosis.[64]

To facilitate RTPF chemosensors, classic fluorophores (*e.g.*, naphthalene, coumarin, quinoline, fluorene, carbazole, naphthalimide, *etc.*) with extensively conjugated structures, such as donor–acceptor (D–A) dipoles and D–π-A/D–A–D quadrupoles, can be introduced to achieve an appreciable two-photon action cross-section (δ_Φ), which is considered as the most common factor to evaluate two-photon absorption of TPF (Figure 7.18B).[65]

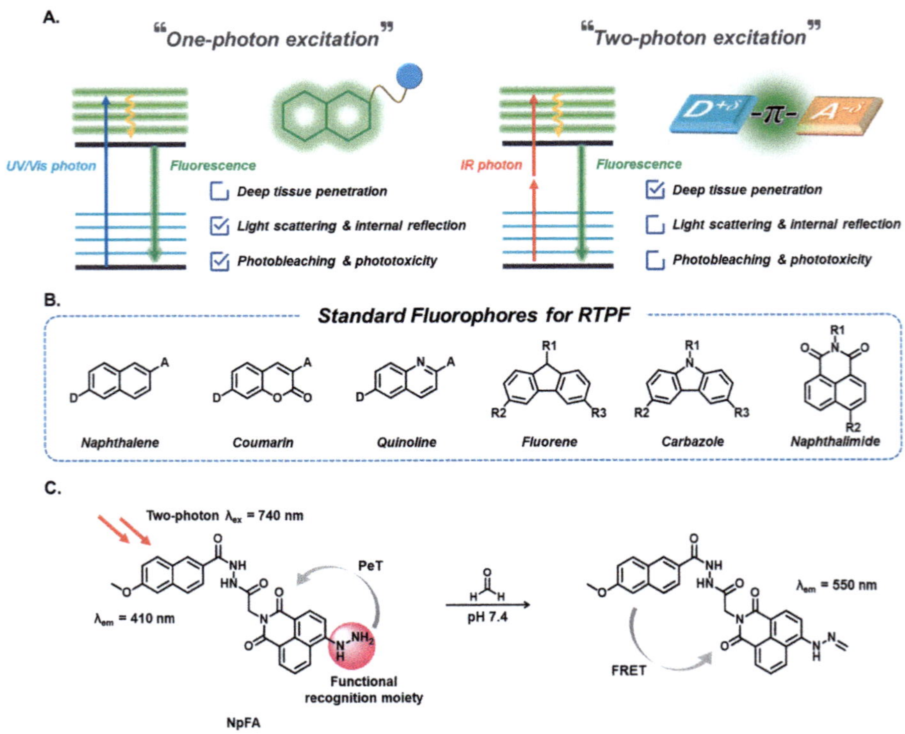

Figure 7.18 (A) Comparison of Jabłoński diagram based on one-photon and two-photon excitation. (B) Standard fluorophores for constructing ratiometric two-photon fluorescence chemosensor (RTPF). (C) Chemical structure and sensing mechanisms of **NpFA**.

In addition, a recent trend for designing effective RTPF probes is to incorporate functional recognition moieties that can selectively and sensitively target various biomarkers in the biosystem. Since design strategies for RTPF chemosensors closely resemble a classic type of two-photon probe, chelation-induced enhanced ratiometric and activity-based ratiometric characteristics are mainly utilized. One of the standard RTPF chemosensors, **NpFA**, is designed for ratiometric detection of formaldehyde in zebrafish by harnessing FRET and PeT (Figure 7.18C).[66] **NpFA** contains a hydrazide-responsive moiety for formaldehyde detection, which shows rapid response, high selectivity, and sensitivity. In the absence of formaldehyde, blue fluorescence (λ_{em} = 410 nm) from naphthalene derivatives with dual activation of FRET and PeT processes is increased significantly under two-photon excitation at 760 nm. On the other hand, the reaction between hydrazine and formaldehyde enables an increase in yellow fluorescence (λ_{em} = 550 nm) from the naphthalimide fluorophore, which allows **NpFA** to be suitable for RTPF imaging of formaldehyde.

7.3.2 Ratiometric Fluorescent Nanoprobes

Owing to the rapid development of nanotechnology and materials science, the design of fluorescent probes based on nanomaterials now plays a vital role in diagnostic applications.[67] The inherent limitations in using small-molecule fluorescent probes, such as targeting ability, specificity, and sensitivity, can be effectively improved by the use of nanoprobes because of their intrinsic optical and physicochemical properties.[68] In addition, multiplexing and multi-modal sensing is possible with nanoprobes. The smaller dimensions (usually <100 nm) and prolonged blood circulation times of nanoprobes permit efficient tissue accumulation, as well as allowing them to freely traverse the whole body through systemic circulation.[69] Also, the optical properties of nanoprobes are distinct according to their surface functionalization, composition, shape, and size, which can be controlled and modified for diverse applications.[70] Additionally, their large specific surface area can be used to conjugate the required targeting molecules, and the high payload capacity may result in high signal intensity.[71]

Although fluorescent nanoprobes can be used as significant alterations to small-molecule fluorescent probes in sensing applications, certain limitations remain. Conventional nanoprobes usually depend on "always-on" or "always-off" signal transduction modes for sensing and imaging analyte concentrations and molecular events.[72] But physiological or experimental factors may influence the absolute intensity-dependent signal acquisition for a single-targeted nanoprobe, which can cause misleading and strong non-specific background signals in molecular sensing and imaging applications.[3] The simple alternative to minimize these non-specific effects is a ratiometric measurement strategy. Ratiometric measurement is a self-calibrated method for recording two or more analyte-induced signals, in which one signal is a reference for normalization of other signals.[73] It is independent of

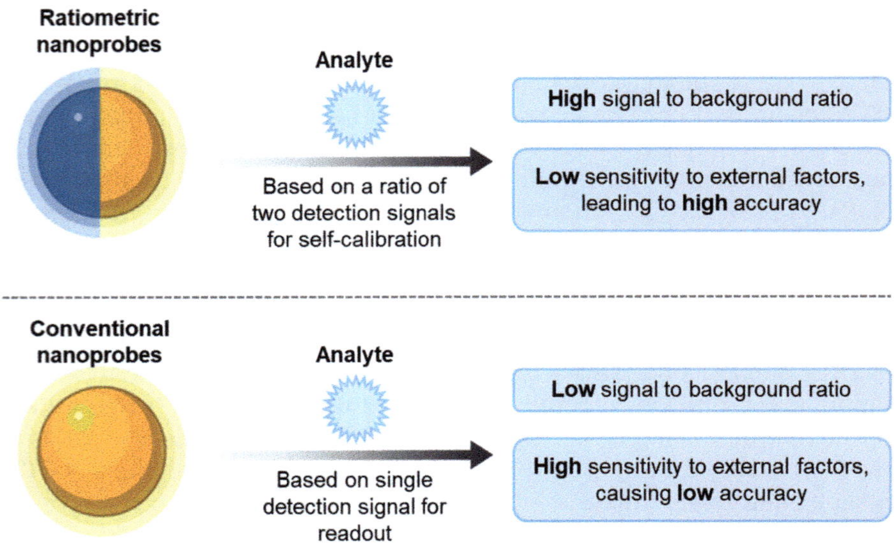

Figure 7.19 Comparison of conventional nanoprobes and ratiometric nanoprobes.

nanoprobe concentration and facilitates more reliable and accurate quantitation. Figure 7.19 provides the comparison between conventional nanoprobes and ratiometric nanoprobes.

Based on the advantages of more accurate imaging contrast and higher detection sensitivity of ratiometric fluorescence, researchers have utilized different nanoparticles to develop ratiometric fluorescent nanoprobes and evaluated their molecular sensing and imaging activities. In this section, we discuss different ratiometric nanoprobes, which are categorized based on the type of nanomaterial used for their construction.

7.3.2.1 Carbon/Quantum Dots-based Ratiometric Nanoprobes

Owing to their fascinating properties, such as high photostability, water dispersibility, and biocompatibility, carbon/quantum dots (CDs/QDs) have attracted significant attention in biomedical, biosensing, and bioimaging applications.[74] Different kinds of CDs/QDs can be used as ratiometric nanoprobes for sensing and imaging applications. For example, label-free a CD-based dual-emissive fluorescent nanoprobe is simultaneously used for ratiometric detection of Pb^{2+} and pyrophosphate (PPi).[75] The single-wavelength excitation (λ_{ex} = 391 nm) of these CDs shows dual-emission bands at 477 nm and 651 nm. They exhibit highly reliable and accurate simultaneous sensing toward Pb^{2+} and PPi ions. The emission band of CDs at 477 nm increases slightly, and the band at 651 nm gradually decreases for Pb^{2+} ions (CDs-Pb^{2+}). The limit of detection (LOD) is as low as 0.055 μM for Pb^{2+} and 0.089 μM for PPi without interference from other relevant analytes. An amine-rich carbon quantum dot (CQD)-based nanoprobes

(**AS1411–Ce6–CQDs**) with a nucleolin-targeting aptamer, AS1411, and a cathepsin B cleavable peptide, Gly-Arg-Arg-Gly-Lys-Gly-Gly-Cys–COOH, which is tethered with a chlorin e6 (Ce6, a near-infrared (NIR) fluorophore)–peptide is used for the ratiometric sensitive and specific detection of cathepsin B.[76] Cathepsin B is one of the most significant biomarkers in various malignant tumors. The FRET process from the CQDs to Ce6 in AS1411–Ce6–CQD nanoaggregates causes remarkable fluorescence quenching at ~450 nm, while the Ce6 NIR fluorescence at ~650 nm is enhanced. Nucleolin-mediated endocytosis provides the **AS1411–Ce6–CQDs** nanoprobe with specific entry into the cancer cells, in which overexpressed cathepsin B can cleave Ce6–peptide and subsequently trigger the Ce6 fluorophore disassociation from **AS1411–Ce6–CQDs**. Consequently, FRET termination occurs, with a large emission shift of ~200 nm from the NIR to the blue region, and thus a cathepsin B ratiometric fluorescence response can be achieved. This shows an ultralow LOD of 0.096 ng mL^{-1}. This nanoprobe can be used to visualize the cathepsin B in living HeLa cells.

A dual-emitting Mn^{2+}-doped ZnS QD (orange emitting) containing a metal-methylsalicylaldimine complex (blue emitting) on its surface (**QDC**) can be used for ratiometric fluorescence sensing of Hg^{2+} and Cu^{2+} ions.[77] The dual-emitting nanoprobe **QDC** initially shows two emission bands at 600 nm and 430 nm, whereas, in the presence of Hg^{2+} ions (concentration range 0.67–16.67 μM) and Cu^{2+} ions (concentration range 0.67–13.33 mM), the emission intensity at 600 nm decreases along with an increasing intensity at 430 nm. At the same time, the visual color changes from purple to blue for both ions. **QDC** shows a LOD of 85.5 nM for Hg^{2+} ions, with a linear calibration range of 0.67 to 3.33 μM (intensity ratio (I_{430}/I_{600}) *vs.* concentration). The corresponding values are 34.9 nM and 0.67 to 5.33 μM for Cu^{2+} ions.[78] The antimonene quantum dot (**AMQD**)-based nanoprobe, which is prepared by the probe sonication-assisted liquid-phase exfoliation method, is used for pH-mediated dual-channel ratiometric tetracycline detection. **AMQDs** originally show a strong blue fluorescence at ~423 nm (λ_{ex} = 365 nm) with a quantum yield of 3.64%. The blue fluorescence of **AMQDs** changes to red (λ_{em} = 550 nm) with tetracyclines under acidic conditions (pH 2–7) because of the inherent fluorescence of tetracyclines. In contrast, the blue fluorescence converts to yellow-green under alkaline conditions due to the decomposition of tertacyclines. The LOD is about 27×10^{-9} M, and the linear calibration concentration range is 0.1×10^{-6}–150×10^{-6} M under acidic conditions. Under basic conditions, the corresponding values are LOD of 74×10^{-9} M and the range is 0.1×10^{-6}–10×10^{-6} and 20×10^{-6}–80×10^{-6} M.

7.3.2.2 Polymer-based Ratiometric Nanoprobes

Polymer-based nanomaterials are promisingly employed for versatile applications in the biosensing and bioimaging fields because of their remarkable advantages, including high photostability, flexible structure design and modification, good biocompatibility, good water solubility, and signal amplification ability.[79] For example, a semiconducting polymer-based fluorescent far-red to NIR Pdot nanoprobe $PFOBT36SeTBT_x$ is used for ratiometric sensing and imaging of hypochlorous acid (HOCl) *in vitro* and *in vivo*.[80] Pdot is comprised of a benzoxadiazole (OBT) as the energy donor (ClO^- inert) and a fluorophore 4,7-bis(2-thienyl)-2,1,3-benzoselenadiazole (SeTBT) as the energy acceptor (ClO^- sensitive), in which the FRET process occurs from the OBT to SeTBT and shows an emission peak at 680 nm. ClO^- oxidizes SeTBT which results in disturbing the FRET process of Pdots; consequently, the OBT emission is enhanced in the green channel along with decreasing emission of SeTBT in the far-red to NIR region. Thus, the fluorescence emission peak of Pdot at 540 nm gradually increases with attenuating the peak intensity at 680 nm for ClO^- ions. The intensity ratio (I_{540}/I_{680}) shows a linear relationship to a wide range of ClO^- ion concentrations, *i.e.*, 0–50 μM and 50–250 μM for $PFOBT36SeTBT_5$ and $PFOBT36SeTBT_7$, respectively, and the LOD is 0.5 μM for $PFOBT36SeTBT_5$. The **$PFOBT36SeTBT_5$ Pdot** probe could visualize the HOCl fluctuations in macrophage cells and lipopolysaccharide (LPS)-induced HOCl production in the peritonitis of living mice.

Another semiconducting polymeric nanoprobe (**SPNP**), consisting of a dye-labeling (IR800) and granzyme B cleavable peptide in its side chain, is used for the ratiometric fluorescence and photoacoustic imaging of granzyme B associated with cytotoxic T cells (CTLs).[81] **SPNP** shows two fluorescence peaks at 740 nm and 820 nm, corresponding to the SP backbone moiety and IR800. Granzyme B-mediated cleavage of dye-labeled peptide in the presence of granzyme B results in the release of the SP core and attenuates the intensity at 820 nm (IR800 window) (eight-fold) while the intensity at 740 nm (SP window) remains the same. The ratiometric fluorescence signal ($NIRF_{740}/NIRF_{820}$) shows four-fold enhancement with the granzyme B. The self-assembling property of amphiphilic SP into the **SPNP** nanoprobe provides passive targeting to the tumor. **SPNP** could visualize granzyme B in the granzyme B-positive cytotoxic ($CD8^+$) T cells (*in vitro*) and 4T1 tumor-bearing BALB/c mice (*in vivo*) after immunotherapy treatment.

7.3.2.3 Upconversion-based Ratiometric Nanoprobes

The remarkable properties of upconversion-based nanoprobes (UCNPs), such as biocompatibility, convenient surface functionalization, adjustable multiple emissions, and high stability, make them suitable nanoprobes for ratiometric sensing and imaging applications.[82] For example, a two-photon excitation (TPE)-based ratiometric Tm^{3+}-doped UCNP nanoprobe (**UCNP@DMI-mSiO2@βCD**) is used to visualize intracellular pH changes in living cells and tissues.[83] It consists of the following components: (1) a pH-insensitive UCNP core with a fluorescence emission at 810 nm; (2) a mesoporous SiO_2 ($mSiO_2$) shell coated on the UCNP core, and a pH-sensitive DMI encapsulated on $mSiO_2$ pores; and (3) further decoration of the $mSiO_2$ shell with beta-cyclodextrin (*β*CD). The TPE fluorescence spectrum of this nanoprobe exhibits upconverted emissions at 610 nm (TPE dye) and 810 nm (Tm^{3+}-doped UCNPs) with excitation at 920 nm. As the pH increases, the emission intensity of DMI (610 nm) dramatically increases, while the intensity of the UCNPs (810 nm) remains the same (Figure 7.20A). The ratiometric signal intensity ratio I_{610}/I_{810} shows a linear relationship with a pH range of 4.0–6.5. This nanoprobe can be used to visualize the chemically stimulated intracellular pH changes in living MCF-7 cells and rat liver tissue slices.

Another $mSiO_2$ shell and *β*CD-decorated nanoprobe, which is embedded with Pd-RBDs (palladium-bound rhodamine B derivatives), is used to sense CO in live cells and tissue slices.[84] This nanoprobe comprises a NaYF4 : Yb, Er/Tm UCNP core, which shows two luminescent upconversion peaks at 540 and 800 nm. This UCNP nanoprobe detects CO through the FRET process between Pd-RBDs and the UCNP core. Pd is released from the Pd-RBDs by CO reduction, followed by decrease of the Pd-RBDs absorption (range: 500–600 nm), which overlaps with the UCNPs luminescence spectrum. Consequently, the UCNPs luminescence peak at 540 nm increases. Therefore, the I_{540}/I_{800} ratio of the nanoprobe (which increases with CO) allows the quantitative determination of CO. The linear calibration range is about 0–1 μM (between I_{540}/I_{800} ratio and CO concentration), and the LOD is 0.029 μM. This nanoprobe is used to image CO in living HepG2 and L02 cells and mouse liver tissue sections.

7.3.2.4 Aggregation-induced Emission-based Ratiometric Nanoprobes

Unlike traditional fluorescent probes, aggregation-induced emission (AIE)-active molecules can become highly fluorescent in the aggregated state and possess salient features, including higher photostability and brightness, good water solubility, excellent biocompatibility, and improved targeting ability.[85] Thus, AIE-based nanoprobes have been used for ratiometric sensing and imaging applications. For example, glutathione (GSH)-capped Au nanoclusters (NCs) having negative carboxylic groups interact with positively charged hyaluronic acid (HA) and fluorescein isothiocyanate (FITC)-functionalized polymer (**FITC-EDA-HA**) to induce self-assembly, followed by generation of the AIE effect. Self-assembled **FHAuNCs** show two emission peaks at 520 nm and 570 nm corresponding to FITC and AuNCs, respectively (excitation at 430 nm), and can be used for ratiometric sensing and imaging of $^{\bullet}OH$ in live cells.[86] The $^{\bullet}OH$ can cleave the HA polymer chain resulting in the disassembly of AuNCs, and consequently reducing the AuNC AIE effect and decreasing their intensity at 570 nm. At the same time, the FITC emission at 520 nm changes little (internal reference). The linear relationship of the ratiometric signal intensity (I_{570}/I_{520}) with $^{\bullet}OH$ concentration range is between 0 and 20 μM, and the LOD is about 0.13 μM. **FHAuNCs** can visualize the $^{\bullet}OH$ in living RAW264.7 cells.

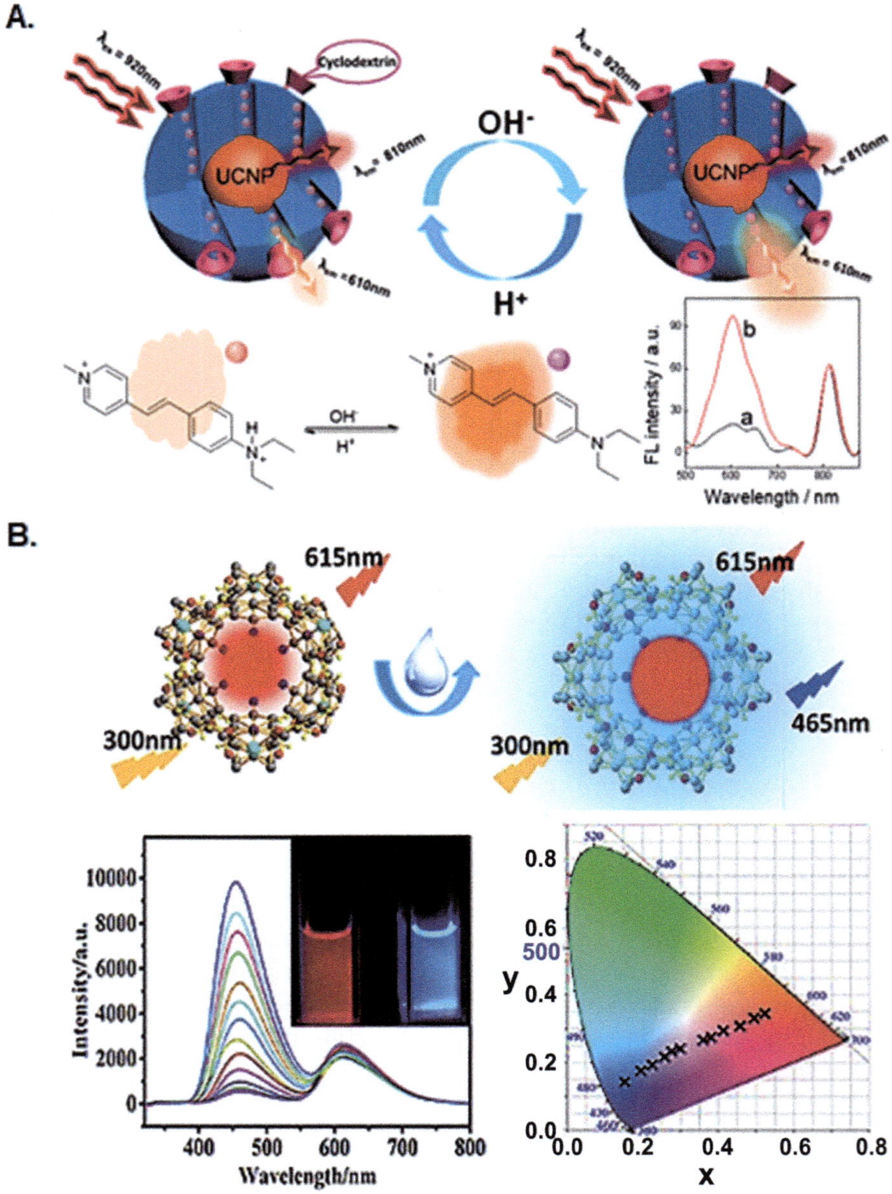

Figure 7.20 (A) Schematic illustration of sensing principle of the **UCNP@DMI-mSiO2@βCD** ratiometric fluorescent nanoprobe for intracellular pH measurement. Reproduced from ref. 83 with permission from American Chemical Society, Copyright 2019. (B) Schematic illustration for ratiometric response of **Ru@MIL–NH$_2$** toward water. Emission spectra and corresponding CIE chromaticity coordinates of **Ru@MIL–NH$_2$** in ethanol with different water contents (excitation at 300 nm). Reproduced from ref. 89 with permission from American Chemical Society, Copyright 2017.

7.3.2.5 Transition Metal-based Ratiometric Nanoprobes

Due to the excellent complexing nature of transition metal ions with ligands, numerous transition metal-based nanoprobes have been developed for ratiometric fluorescent sensing and imaging applications.[87] For example, a dual-ratiometric fluorescent nanoprobe, which is composed of a ratiometric pH-sensitive ANNA dye, a near-infrared fluorescent dye Cy5.5, and Fe_3O_4 nanoparticles, is used for the simultaneous detection of MMP-9 activity and the pH of tumors.[23] The pH-sensitive ANNA dye was connected to Fe_3O_4 nanoparticles through a peptide substrate, and it is in the quenched state due to the FRET process from ANNA to Fe_3O_4 nanoparticles. Cy5.5 is always in an "on" state and serves as an internal reference. Cleavage of the peptide linker by MMP-9 results in an increase of the ANNA fluorescence emission at ~510 nm (excitation: 450 nm), while the Cy5.5 intensity at ~700 nm remains constant (excitation: 650 nm). The LOD is about 0.74 nM. This nanoprobe could simultaneously map the MMP-9 protease activity and microenvironmental pH in LS180 cells (*in vitro*) and LS180 tumor xenografts (*in vivo*).

7.3.2.6 Metal–Organic Framework-based Ratiometric Nanoprobes

Metal–organic frameworks (MOFs) are hybrid materials that can be built through the chemical coordination of metal ions and organic bridge ligands. Owing to their well-defined structure with uniform pores and multiple functionalities, they have been considered ideal candidates for biomedical applications.[88] By modulating the selection of metal ions and organic ligands, structural and functional tunability of MOFs is possible. A ratiometric fluorescent MOF nanoprobe, **Ru@MIL–NH$_2$**, prepared from $Ru(bpy)_3^{2+}$, $AlCl_3$, and 2-aminoterephthalic acid, is used for the ratiometric sensing of water in organic solvents.[89] It shows two emission peaks at 465 nm (MIL–NH_2) and 615 nm ($Ru(bpy)_3^{2+}$), with excitation at 300 nm. A ratiometric response of **Ru@MIL–NH$_2$** is achieved with water due to protonation of the "N" atom of MIL–NH_2 and the stable fluorescence of ($Ru(bpy)_3^{2+}$). Consequently, the emission intensity at 465 nm is gradually enhanced, along with the stable intensity at 615 nm (Figure 7.20B). The intensity ratio (I_{465}/I_{615}) shows a linear relationship with water content, and the LOD is about 0.02% v/v.

7.3.2.7 Covalent–Organic Framework-based Ratiometric Nanoprobes

Covalent–organic frameworks (COFs) are crystalline porous polymers with strong covalent bonds, and they possess high stability, low density, and a large surface area.[90] Their characteristics of tunable functionality and regular pore structure make many COFs useful for ratiometric sensing and imaging applications. For example, a high-crystallinity COF, **TzDa**, is prepared *via* imine-linking of 4,4′,4″-(1,3,5-triazine-2,4,6-triyl)trianiline (Tz) and 2,5-dihydroxyterephthalaldehyde (Da) to trace the water in organic solvents.[91] The docking site of Tz and the hydrogen-bonding interactions between imines and Da causes ICT and ESIPT effects, respectively, in the **TzDa** structure. As a result, **TzDa** shows dual fluorescent emission peaks at 500 nm and 590 nm. As the water content increases, it affects the ICT and ESIPT processes, which consequently leads to enhancement in the fluorescence emission at 590 nm, while the emission at 500 nm decreases. This ratiometric response of the **TzDa** nanoprobe could be used to sense water in various organic solvents. The LOD in IPA, ethanol, THF, acetone, and EtOAc are 0.085, 0.034, 0.026, 0.022, 0.006, respectively.

7.4 General Conclusions

As discussed in this chapter, ratiometric fluorescent chemosensors have attracted considerable attention in analyte sensing, bioimaging, and disease diagnosis, and are of particular interest because they permit significant precision under quantitative detection. Taking a more structural approach when designing a ratiometric chemosensor allows us to design a customized quantitative diagnostic agent that exhibits proper characteristics tailored to its application and purpose. Moreover, understanding the photophysical/chemical properties used in ratiometric fluorophores is also crucial for effective ratiometric fluorophore design. Therefore, in this chapter, various photophysical/chemical properties (*e.g.*, ICT, ESIPT, FRET, BRET, TADF, and excimer), including general examples used to design the ratiometric fluorescent chemosensor, have been presented. In addition, several design strategies, as well as sensing mechanisms for constructing SMRFs, were proposed. Because of the specific intrinsic optical and physicochemical properties of nanomaterials, nanoprobes are considered to be suitable ratiometric chemosensors for sensing and imaging applications. Thus, we categorized different ratiometric nanoprobes based on the type of nanomaterial used for constructing the nanoprobes and discussed their mechanism and optical changes in ratiometric sensing, with standard examples.

Understanding how photophysical/chemical properties affect the ratiometric fluorescence emission, and determining probe designs that meet the purpose, may help to further improve tactics for quantitative detection of biologically/environmentally important analytes and widen their range of applications.

List of Abbreviations

$AlCl_3$	Aluminium chloride
CO	Carbon monoxide
ANNA	*N*-carboxyhexyl derivative of 3-amino-1,2,4-triazole fused 1,8-naphthalimide
Ce6	Chlorin e6

°C	Celsius
DF	Delayed fluorescence
K_d	Dissociation constant
DMI	*trans*-4-[*p*-(*N*,*N*-diethylamino) styryl]-*N*-methylpyridinium iodide
EtOAc	Ethyl acetate
FL	Fluorescence
AuNCs	Gold nanoclusters
˙OH	Hydroxy radical
IC	Internal conversion
ISC	Intersystem crossing
Fe_3O_4	Iron oxide
IPA	Isopropyl alcohol
MMP-9	Matrix metalloprotease-9
μM	Micromolar
NIR	Near-infrared
nm	Nanometer
nM	Nanomolar
ng mL^{-1}	Nanograms per millilitre
BODIPY	Dipyrrometheneboron difluoride
pH	Potential of hydrogen
SiO_2	Silica or silicon dioxide
THF	Tetrahydrofuran
T	Temperature
ZnS	Zinc sulfide

Acknowledgements

This work was supported by the CRI project (2018R1A3B1052702, J.S.K.) and the Korea University Grant (P.J.).

References

1. M. H. Lee, J. S. Kim and J. L. Sessler, *Chem. Soc. Rev.*, 2015, **44**, 4185–4191.
2. S.-H. Park, N. Kwon, J.-H. Lee, J. Yoon and I. Shin, *Chem. Soc. Rev.*, 2020, **49**, 143–179.
3. A. Bigdeli, F. Ghasemi, S. Abbasi-Moayed, M. Shahrajabian, N. Fahimi-Kashani, S. Jafarinejad, M. A. Farahmand Nejad and M. R. Hormozi-Nezhad, *Anal. Chim. Acta*, 2019, **1079**, 30–58.
4. J. Chan, S. C. Dodani and C. J. Chang, *Nat. Chem.*, 2012, **4**, 973–984.
5. D. Wu, A. C. Sedgwick, T. Gunnlaugsson, E. U. Akkaya, J. Yoon and T. D. James, *Chem. Soc. Rev.*, 2017, **46**, 7105–7123.
6. H. Zhu, J. Fan, B. Wang and X. Peng, *Chem. Soc. Rev.*, 2015, **44**, 4337–4366.
7. A. P. de Silva, H. Q. N. Gunaratne, T. Gunnlaugsson, A. J. M. Huxley, C. P. McCoy, J. T. Rademacher and T. E. Rice, *Chem. Rev.*, 1997, **97**, 1515–1566.
8. J. S. Kim and D. T. Quang, *Chem. Rev.*, 2007, **107**, 3780–3799.

9. A. C. Sedgwick, L. Wu, H.-H. Han, S. D. Bull, X.-P. He, T. D. James, J. L. Sessler, B. Z. Tang, H. Tian and J. Yoon, *Chem. Soc. Rev.*, 2018, **47**, 8842–8880.
10. L. Wu, C. Huang, B. P. Emery, A. C. Sedgwick, S. D. Bull, X.-P. He, H. Tian, J. Yoon, J. L. Sessler and T. D. James, *Chem. Soc. Rev.*, 2020, **49**, 5110–5139.
11. J. Yang, D. Cumberbatch, S. Centanni, S.-q. Shi, D. Winder, D. Webb and C. H. Johnson, *Nat. Commun.*, 2016, **7**, 13268.
12. A. B. Dippel, W. A. Anderson, J. H. Park, F. H. Yildiz and M. C. Hammond, *ACS Chem. Biol.*, 2020, **15**, 904–914.
13. X. Li, G. Baryshnikov, C. Deng, X. Bao, B. Wu, Y. Zhou, H. Ågren and L. Zhu, *Nat. Commun.*, 2019, **10**, 731.
14. C. J. Christopherson, D. M. Mayder, J. Poisson, N. R. Paisley, C. M. Tonge and Z. M. Hudson, *ACS Appl. Mater. Interfaces*, 2020, **12**, 20000–20011.
15. J. Wu, Y. Zou, C. Li, W. Sicking, I. Piantanida, T. Yi and C. Schmuck, *J. Am. Chem. Soc.*, 2012, **134**, 1958–1961.
16. Y. Wang, J. Chen, Y. Chen, W. Li and C. Yu, *Anal. Chem.*, 2014, **86**, 4371–4378.
17. B. Valeur and I. Leray, *Coord. Chem. Rev.*, 2000, **205**, 3–40.
18. H. Fang, R. Shi, D. Chen, Y. Qu, Q. Wu, X. Yang, X. Lu, C.-W. Zhang, L. Li and K.-L. Lim, *Chem. Commun.*, 2021, **57**, 11260–11263.
19. M. Liu, X. Yu, M. Li, N. Liao, A. Bi, Y. Jiang, S. Liu, Z. Gong and W. Zeng, *RSC Adv.*, 2018, **8**, 12573–12587.
20. F. Yu, P. Li, P. Song, B. Wang, J. Zhao and K. Han, *Chem. Commun.*, 2012, **48**, 2852–2854.
21. S.-L. Pan, K. Li, L.-L. Li, M.-Y. Li, L. Shi, Y.-H. Liu and X.-Q. Yu, *Chem. Commun.*, 2018, **54**, 4955–4958.
22. L. Chen, S. J. Park, D. Wu, H. M. Kim and J. Yoon, *Chem. Commun.*, 2019, **55**, 1766–1769.
23. K. Du, S. Niu, L. Qiao, Y. Dou, Q. Zhu, X. Chen and P. Zhang, *RSC Adv.*, 2017, **7**, 40615–40620.
24. W. Li, B. Fang, M. Jin and Y. Tian, *Anal. Chem.*, 2017, **89**, 2553–2560.
25. J. Zhao, S. Ji, Y. Chen, H. Guo and P. Yang, *Phys. Chem. Chem. Phys.*, 2012, **14**, 8803–8817.
26. A. Y. Chumak, V. O. Mudrak, V. M. Kotlyar and A. O. Doroshenko, *J. Photochem. Photobiol., A*, 2021, **406**, 112978.
27. B. Gu, L. Huang, N. Mi, P. Yin, Y. Zhang, X. Tu, X. Luo, S. Luo and S. Yao, *Analyst*, 2015, **140**, 2778–2784.
28. W. Luo, J. Li and W. Liu, *Org. Biomol. Chem.*, 2017, **15**, 5846–5850.
29. Y. Zhang, Y. Deng, N. Ji, J. Zhang, C. Fan, T. Ding, Z. Cao, Y. Li and Y. Fang, *Dyes Pigm.*, 2019, **166**, 473–479.
30. L. Chen, D. Wu and J. Yoon, *Sens. Actuators, B*, 2018, **259**, 347–353.
31. Y. Liu, J. Nie, J. Niu, W. Wang and W. Lin, *J. Mater. Chem. B*, 2018, **6**, 1973–1983.
32. L. Wu, L. Liu, H.-H. Han, X. Tian, M. L. Odyniec, L. Feng, A. C. Sedgwick, X.-P. He, S. D. Bull and T. D. James, *New J. Chem.*, 2019, **43**, 2875–2877.
33. Z. Song, R. T. K. Kwok, E. Zhao, Z. He, Y. Hong, J. W. Y. Lam, B. Liu and B. Z. Tang, *ACS Appl. Mater. Interfaces*, 2014, **6**, 17245–17254.

34. L. Feng, Z.-M. Liu, J. Hou, X. Lv, J. Ning, G.-B. Ge, J.-N. Cui and L. Yang, *Biosens. Bioelectron.*, 2015, **65**, 9–15.
35. S.-L. Shen, X.-F. Zhang, Y.-Q. Ge, Y. Zhu and X.-Q. Cao, *RSC Adv.*, 2017, **7**, 55296–55300.
36. X. Zhu, M. Xiong, H.-w. Liu, G.-j. Mao, L. Zhou, J. Zhang, X. Hu, X.-B. Zhang and W. Tan, *Chem. Commun.*, 2016, **52**, 733–736.
37. Y. W. Jun, T. Wang, S. Hwang, D. Kim, D. Ma, K. H. Kim, S. Kim, J. Jung and K. H. Ahn, *Angew. Chem., Int. Ed.*, 2018, **57**, 10142–10147.
38. J. X. Ong, C. S. Q. Lim, H. V. Le and W. H. Ang, *Angew. Chem., Int. Ed.*, 2019, **131**, 170–173.
39. R. Feng, L. Guo, J. Fang, Y. Jia, X. Wang, Q. Wei and X. Yu, *Anal. Chem.*, 2019, **91**, 3704–3709.
40. Z.-Y. Li, X.-L. Cui, Y.-H. Yan, Q.-L. Che, J.-Y. Miao, B.-X. Zhao and Z.-M. Lin, *Dyes Pigm.*, 2021, **188**, 109180.
41. C. Y. Chung, J. M. Posimo, S. Lee, T. Tsang, J. M. Davis, D. C. Brady and C. J. Chang, *Proc. Natl. Acad. Sci. U. S. A.*, 2019, **116**, 18285–18294.
42. S. Zhang, K. I. Assaf, C. Huang, A. Hennig and W. M. Nau, *Chem. Commun.*, 2019, **55**, 671–674.
43. T. Ueno and T. Nagano, *Nat. Methods*, 2011, **8**, 642–645.
44. A. Zelmer and T. H. Ward, *J. Microsc.*, 2013, **252**, 8–15.
45. M. Won, M. Li, H. S. Kim, P. Liu, S. Koo, S. Son, J. H. Seo and J. S. Kim, *Coord. Chem. Rev.*, 2021, **426**, 213608.
46. Y. Fu and N. S. Finney, *RSC Adv.*, 2018, **8**, 29051–29061.
47. L. M. Wysocki and L. D. Lavis, *Curr. Opin. Chem. Biol.*, 2011, **15**, 752–759.
48. R. Zhang, F. Yan, Y. Huang, D. Kong, Q. Ye, J. Xu and L. Chen, *RSC Adv.*, 2016, **6**, 50732–50760.
49. M. Taki, H. Ogasawara, H. Osaki, A. Fukazawa, Y. Sato, K. Ogasawara, T. Higashiyama and S. Yamaguchi, *Chem. Commun.*, 2015, **51**, 11880–11883.
50. P. A. Panchenko, Y. V. Fedorov, O. A. Fedorova and G. Jonusauskas, *Phys. Chem. Chem. Phys.*, 2015, **17**, 22749–22757.
51. S. A. Yoon, J. Lee and M. H. Lee, *Sens. Actuators, B*, 2018, **258**, 50–55.
52. W. Qin, W. Dou, V. Leen, W. Dehaen, M. Van der Auweraer and N. Boens, *RSC Adv.*, 2016, **6**, 7806–7816.
53. A. Kanegae, Y. Takata, I. Takashima, S. Uchinomiya, R. Kawagoe, K. Usui, A. Yamashita, J. Wongkongkatep, M. Sugimoto and A. Ojida, *Commun. Chem.*, 2021, **4**, 104.
54. Z. Liu, W. He and Z. Guo, *Chem. Soc. Rev.*, 2013, **42**, 1568–1600.
55. Z. Mao, J. Xiong, P. Wang, J. An, F. Zhang, Z. Liu and J. Seung Kim, *Coord. Chem. Rev.*, 2022, **454**, 214356.
56. H. W. Liu, L. Chen, C. Xu, Z. Li, H. Zhang, X. B. Zhang and W. Tan, *Chem. Soc. Rev.*, 2018, **47**, 7140–7180.
57. Y. Tang, Y. Ma, J. Yin and W. Lin, *Chem. Soc. Rev.*, 2019, **48**, 4036–4048.
58. J. Zhang, X. Chai, X. P. He, H. J. Kim, J. Yoon and H. Tian, *Chem. Soc. Rev.*, 2019, **48**, 683–722.
59. Z. Li, S. H. Yan, C. Chen, Z. R. Geng, J. Y. Chang, C. X. Chen, B. H. Huang and Z. L. Wang, *Biosens. Bioelectron.*, 2017, **90**, 75–82.

60. L. He, X. Yang, Y. Liu, X. Kong and W. Lin, *Chem. Commun.*, 2016, **52**, 4029–4032.
61. S. Y. Park, M. Won, J. S. Kim and M. H. Lee, *Sens. Actuators, B*, 2020, **319**, 128306.
62. W. Denk, J. Strickler and W. Webb, *Science*, 1990, **248**, 73–76.
63. F. Helmchen and W. Denk, *Nat. Methods*, 2005, **2**, 932–940.
64. H. Jin, M. Yang, Z. Sun and R. Gui, *Coord. Chem. Rev.*, 2021, **446**, 214114.
65. L. Wu, J. Liu, P. Li, B. Tang and T. D. James, *Chem. Soc. Rev.*, 2021, **50**, 702–734.
66. G. Yuan, H. Ding, L. Peng, L. Zhou and Q. Lin, *Food Chem.*, 2020, **331**, 127221.
67. N. Liao, L. Su, Y. Zheng, B. Zhao, M. Wu, D. Zhang, H. Yang, X. Liu and J. Song, *Angew. Chem., Int. Ed.*, 2021, **60**, 20888–20896.
68. I. Brigger, C. Dubernet and P. Couvreur, *Adv. Drug Delivery Rev.*, 2012, **64**, 24–36.
69. O. S. Wolfbeis, *Chem. Soc. Rev.*, 2015, **44**, 4743–4768.
70. H. Koo, M. S. Huh, J. H. Ryu, D.-E. Lee, I.-C. Sun, K. Choi, K. Kim and I. C. Kwon, *Nano Today*, 2011, **6**, 204–220.
71. S. T. Selvan, T. T. Y. Tan, D. K. Yi and N. R. Jana, *Langmuir*, 2010, **26**, 11631–11641.
72. B. Rathore, K. Sunwoo, P. Jangili, J. Kim, J. H. Kim, M. Huang, J. Xiong, A. Sharma, Z. Yang, J. Qu and J. S. Kim, *Biomaterials*, 2019, **211**, 25–47.
73. X. Huang, J. Song, B. C. Yung, X. Huang, Y. Xiong and X. Chen, *Chem. Soc. Rev.*, 2018, **47**, 2873–2920.
74. P. Wu, X. Hou, J. J. Xu and H. Y. Chen, *Nanoscale*, 2016, **8**, 8427–8442.
75. Y. Gao, Y. Jiao, H. Zhang, W. Lu, Y. Liu, H. Han, X. Gong, L. Li, S. Shuang and C. Dong, *J. Mater. Chem. B*, 2019, **7**, 5502–5509.
76. Y. Shen, T. Wu, Y. Wang, S. L. Zhang, X. Zhao, H. Y. Chen and J. J. Xu, *Anal. Chem.*, 2021, **93**, 4042–4050.
77. M. Manna, S. Roy, S. Bhandari and A. Chattopadhyay, *J. Mater. Chem. C*, 2020, **8**, 6972–6976.
78. Y. Liu, Y. Xiao, M. Yu, Y. Cao, Y. Zhang, T. Zhe, H. Zhang and L. Wang, *Small*, 2020, **16**, 2003429.
79. P. Wang, C. Zhang, H. W. Liu, M. Xiong, S. Y. Yin, Y. Yang, X. X. Hu, X. Yin, X. B. Zhang and W. Tan, *Chem. Sci.*, 2017, **8**, 8214–8220.
80. L. Wu, I. C. Wu, C. C. DuFort, M. A. Carlson, X. Wu, L. Chen, C. T. Kuo, Y. Qin, J. Yu, S. R. Hingorani and D. T. Chiu, *J. Am. Chem. Soc.*, 2017, **139**, 6911–6918.
81. Y. Zhang, S. He, W. Chen, Y. Liu, X. Zhang, Q. Miao and K. Pu, *Angew. Chem., Int. Ed.*, 2021, **60**, 5921–5927.
82. X. Ai, Z. Wang, H. Cheong, Y. Wang, R. Zhang, J. Lin, Y. Zheng, M. Gao and B. Xing, *Nat. Commun.*, 2019, **10**, 1087.
83. N. Wang, X. Yu, T. Deng, K. Zhang, R. Yang and J. Li, *Anal. Chem.*, 2020, **92**, 583–587.

84. N. Wang, Z. Li, W. Liu, T. Deng, J. Yang, R. Yang and J. Li, *ACS Appl. Mater. Interfaces*, 2019, **11**, 26684–26689.
85. M. Xue, H. Wang, J. Chen, J. Ren, S. Chen, H. Yang, R. Zeng, Y. Long and P. Zhang, *Sens. Actuators, B*, 2019, **282**, 1–8.
86. X. Ran, Z. Wang, F. Pu, Z. Liu, J. Ren and X. Qu, *Chem. Commun.*, 2019, **55**, 15097–15100.
87. X. Deng, X. Liu, S. Wu, S. Zang, X. Lin, Y. Zhao and C. Duan, *ACS Appl. Mater. Interfaces*, 2021, **13**, 45214–45223.
88. X. J. Kong, X. Ji, T. He, L. H. Xie, Y. Z. Zhang, H. Lv, C. Ding and J. R. Li, *ACS Appl. Mater. Interfaces*, 2020, **12**, 35375–35384.
89. H. Q. Yin, J. C. Yang and X. B. Yin, *Anal. Chem.*, 2017, **89**, 13434–13440.
90. A. Mal, R. K. Mishra, V. K. Praveen, M. A. Khayum, R. Banerjee and A. Ajayaghosh, *Angew. Chem., Int. Ed.*, 2018, **57**, 8443–8447.
91. H. L. Qian, C. Dai, C. X. Yang and X. P. Yan, *ACS Appl. Mater. Interfaces*, 2017, **9**, 24999–25005.

CHAPTER 8

Chirality Sensing with UV-visible Absorbance, Fluorescence, and Circular Dichroism Spectroscopy

JAMES R. HOWARD[a], JONGDOO LIM[a], SARAH R. MOOR[a] AND ERIC V. ANSLYN*[a]

[a]Department of Chemistry, University of Texas at Austin, Austin, Texas, USA
*E-mail: Anslyn@austin.utexas.edu

8.1 Introduction

Chirality is a key molecular feature considered when designing and developing modern pharmaceuticals.[1] With the advent of combinatorial chemistry and parallel synthesis, there is a growing interest in the development of high-throughput screening (HTS) platforms to monitor and evaluate the asymmetric synthesis of chiral drugs.[2] In this chapter, we will discuss both historical and modern examples of chirality sensing *via* UV-visible, fluorescence, and circular dichroism spectroscopy. Each of these techniques possess unique advantages that are leveraged to tailor sensing schemes to a particular application. These optical signaling methods are well suited for HTS platforms as they can be performed in parallel with multiwell plates.

Monographs in Supramolecular Chemistry No. 34
Fluorescent Chemosensors
Edited by Luling Wu, Adam C. Sedgwick, Xiao-Peng He and Tony D. James

Published by the Royal Society of Chemistry, www.rsc.org

This allows for rapid detection at low cost, which is advantageous compared to intrinsically serial conventional chromatographic methods.

8.2 UV-visible Absorbance

UV-visible (UV-vis) spectroscopy is a widely used optical technology for the quantitative analysis of conjugated organic compounds, transition metal complexes, proteins, and nucleic acids. Because fluorescence has low detection limits and circular dichroism (CD) has a unique sensitivity to chirality (see Section 8.3 and 8.4), UV-vis spectroscopy has become less appealing as an optical technique for chirality analysis. However, it is still of use by virtue of its ease of operation, potential for naked-eye detection, and the availability of simple and affordable UV-vis spectrophotometers.[3]

UV-vis spectroscopy works when a chromophore is present in the sensor. Thus, suitable chromophoric receptors are required for the analysis of non-UV-absorbing analytes. Such chromophoric receptors can be complexed with analytes by either covalent bonding or non-covalent interactions.[4] Figure 8.1 displays two general strategies using a chromophoric receptor or a receptor–indicator scaffold to generate a UV-vis absorption change. The first approach utilizes receptors that can enantioselectively interact with enantiomeric analytes, resulting in quantifiable UV-vis absorbance changes (Figure 8.1A). The latter employs a reversible receptor–indicator combination. When the indicator is displaced with the analyte, favoring one enantiomer over the

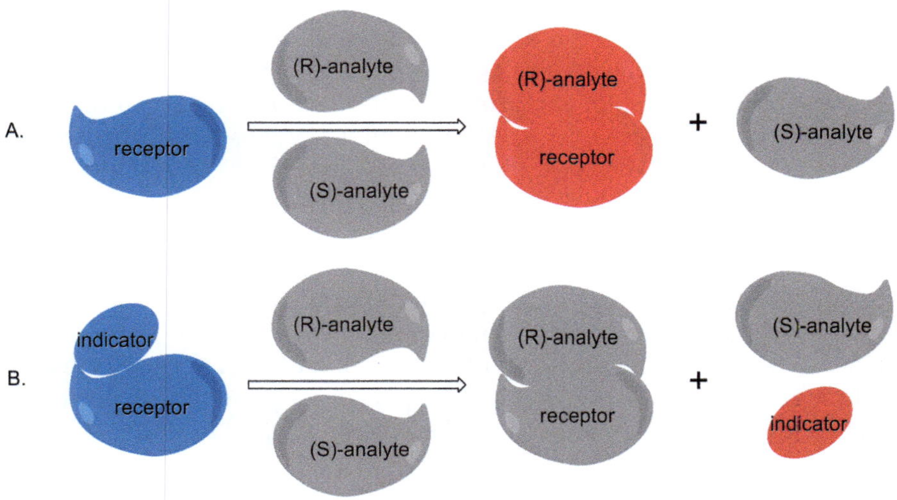

Figure 8.1 The rapid determination of the enantiomeric excess of chiral analytes *via* UV-vis spectroscopy. Enantioselective recognition utilizing (A) different binding affinities between the chromophoric receptor and the enantiomeric analyte and (B) indicator displacement by the enantiomeric analyte.

other, a distinct optical signal is produced (Figure 8.1B). This strategy is generally referred to as an indicator displacement assay (IDA).[2]

The host–guest systems for the enantioselective recognition of chiral analytes use a wide variety of molecular recognition driving forces, including hydrogen bonding, electrostatic/hydrophobic interactions, π electronic effects, van der Waals forces, dynamic covalent bonding, and coordinate covalent bonding metal chelation.[5–9] Steric effects, structural rigidity, and shape also play an important role in chiral recognition and determination of enantiomeric excess (ee). Thus, all these factors are considered in the design and development of HTS optical systems. This section presents the exemplary applications of chiral recognition systems based on UV-vis absorbance detection.

8.2.1 UV-vis Optical Sensing Using Different Interactions and Bindings Between Chromophoric Receptors and Chiral Analytes

As the assay relies on a calibration curve using absorbance, it is essential to design and synthesize a chromophoric receptor that can produce a change in absorbance when bound to the non-absorbing chiral analyte.[10–14]

Early on, Kubo and co-workers developed dual optical sensor systems based on calixarenes.[15] The phenol-derived macrocyclic oligomer displays favorable characteristics in colorimetric sensing, which include hydrophobic cavities for host–guest systems, facile derivatization with chromophores, and well-established synthetic procedures. As shown in Figure 8.2A, the receptors comprise a binaphthyl unit and a bis(indophenol)-derived calix[4] crown unit differing in the length of the ether spacer (**1** and **2**). Only **1** appeared to be enantioselective as the binaphthyl unit of **1** was closer to the indophenols than that of **2**, leading to better complexation with the amine in the cavity. The receptor **1** was red in ethanol, showing an absorbance maximum at 515.5 nm. When (*R*)-phenylglycinol as a chiral analyte was added, the solution immediately turned blue-violet, displaying two new absorption peaks at 538 nm (short-wavelength band) and at 652.5 nm (long-wavelength band). The color change was clearly observed by the naked eye. On the other hand, the addition of (*S*)-phenylglycinol did not create any discernible color change. The proposed mechanism of chiral recognition was as follows: the ether oxygens and the phenolate of the indophenol unit of the receptor stabilize the host–guest complex *via* hydrogen bonding and ion–dipole interactions, respectively. Simultaneously, the binaphthyl unit causes a significant steric strain for the (*S*) isomer. While other (*R*)-isomers of phenylalaninol and phenylglycine produced a visual color change, no enantioselectivity for phenylethylamine was observed, indicating the OH group also participated in the chiral recognition.

Crown ethers have been extensively studied in host–guest chemistry because of their binding capacity and open cavity. The Fuji group developed

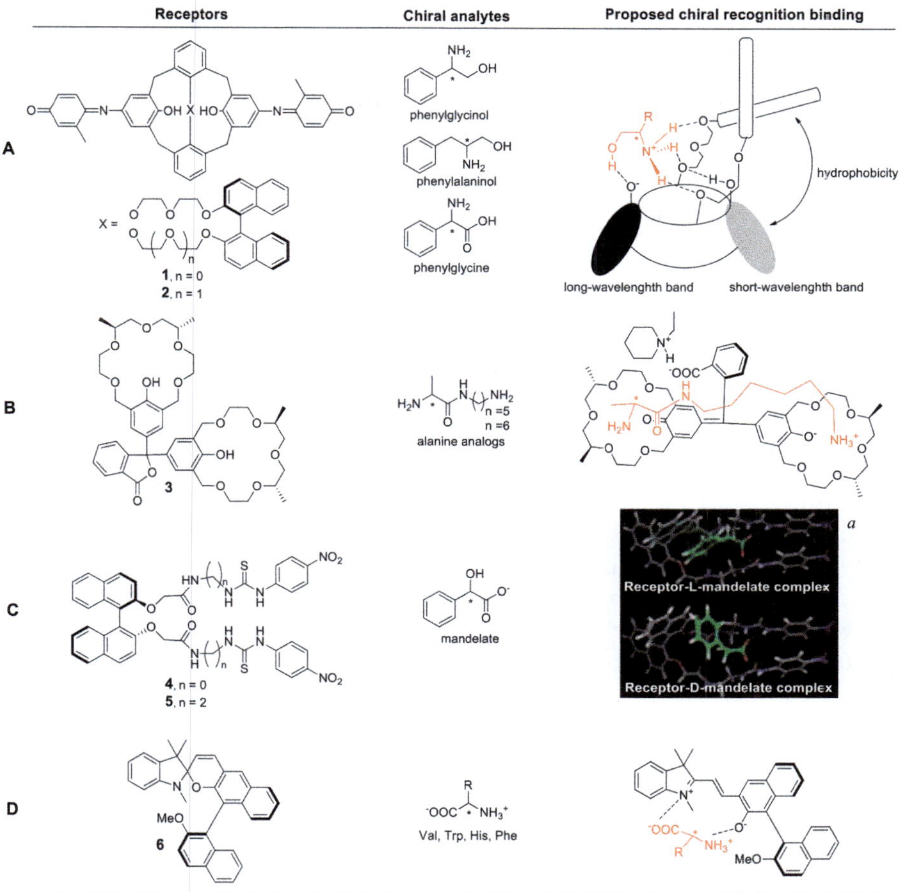

Figure 8.2 Various chromophoric receptors, chiral analytes, and their proposed chiral recognition bindings. Reproduced from ref. 17 with permission from Elsevier, Copyright 2009.

optically active crown ether hosts derived from a phenolphthalein, and investigated their enantioselective recognition of alanine analogs (Figure 8.2B).[16] The solution containing the crown ether derivative **3** was colorless in the presence of *N*-ethylpiperidine in methanol. When the (*R*)- or (*S*)-enantiomer of alanine analogs (n = 5 or 6) were added to the receptor, the solution with the (*R*)-isomer turned purple, whereas the solution with the (*S*)-isomer remained colorless. As shown in Figure 8.2B, the two terminal amine groups of the guest bridge the two phenolic crown ether rings by hydrogen bonding. Thus, the length of the analyte should be appropriate for this bridging interaction. The carboxylate generated from the lactone was stabilized by the *N*-ethylpiperidinium cation. The diastereomeric complex with the (*S*)-isomer was less stable due to the steric repulsion between the methyl group at

C-7 of the crown ring and the methyl group of the alanine analog, leading to chiral recognition of the analyte.

Chiral 1,1′-bi-2-naphthol (BINOL)-based derivatives have been broadly studied for enantioselective recognition and asymmetric synthesis due to their unique geometric and electronic properties. He and co-workers designed and synthesized an (*S*)-BINOL-based colorimetric sensor for mandelate anions (Figure 8.2C).[17] The two chiral receptors comprised an (*S*)-BINOL unit and a thiourea unit, differing in the length of the diamino linker (**4** and **5**). Although both chiral receptors were enantioselective for the mandelate anions, compound **5**, with the longer linker ($n = 2$), displayed the best chiral differentiation, which was presumably derived from the rigid structure and steric effects. When L-mandelate was added to the solution of **5** in dimethyl sulfoxide (DMSO), the color changed from light yellow to red, displaying a new absorption peak at 480 nm. The ratio of association constants for L/D-mandelates was K_a (L)/K_a (D) = 13.6. To understand such a high enantioselectivity, molecular modeling studies were conducted. As shown in Figure 8.2C, the energy-minimized complex structure indicated that the oxygen atoms of L-mandelate interacted with the amine groups of thiourea, forming three hydrogen bonds. At the same time, π–π stacking interactions between the binaphthyl group and the aromatic ring of L-mandelate were predicted. Instead, D-mandelate formed only two hydrogen bonds with **5** and lacked the π–π stacking interaction due to the spatial constraint.

The Fuji group developed a chiral spiropyran derivative for the enantiomeric recognition of amino acid (Figure 8.2D).[18] The chiral spiropyran was derivatized with the binaphthyl group to afford **6**, which could transform to a colored merocyanine form by UV irradiation. When mixed with an amino acid, the zwitterionic merocyanine form can be stabilized by electrostatic interactions. The decolorization half-lives of the merocyanine were measured in the presence of various L/D-amino acids, including alanine (Ala), valine (Val), tryptophan (Trp), histidine (His), and phenylalanine (Phe). The D-amino acids, except for alanine, interacted more strongly with the zwitterionic form of **6** compared to the L-amino acids, thus exhibiting slower decolorization. This work indicated that the steric repulsion arising from the bulky side chain of amino acids was a key element for the chiral discrimination.

8.2.2 Enantioselective Indicator Displacement Assays

Enantioselective indicator displacement assays (eIDAs) have become an attractive optical signaling technology for HTS in parallel, asymmetric synthesis, as they may be conducted using 96-well plate readers, potentially without purification steps.[19] Different from conventional host–guest chemistry, eIDAs have several unique advantages. For instance, a chromophoric unit does not need to be covalently attached to the receptor. Instead, an indicator is reversibly bound to the receptor for the enantioselective displacement by the chiral analyte. When the indicator is displaced, the occurrence can be quantified by a UV-vis or fluorescence spectrophotometer and observed by the naked eye.[2,20,21] In addition, eIDAs allow relatively facile optimization in the chiral sensing system by either changing or modifying the receptor and indicator. Lastly, the receptors and indictors are often commercially available, or derivatized using simple synthetic routes.

In 1959, Lorand and Edwards discovered the formation of cyclic boronic esters when saccharides were added to boronic acids in aqueous media.[22] While the boronic ester bonds form rapidly and are stable, they are reversible under certain conditions. Such dynamic covalent bonds derived from boronic acids have been utilized broadly for the creation of receptors to recognize anions, diols, and sugars/polyols.[23] With this precedent, the Anslyn group developed a variety of chiral receptors based on boronic acids for eIDAs (Figure 8.3A).[14,24] The boronic acid receptors were prepared either from 2-formylphenylboronic acid or 2-acetylphenylboronic acid in two steps. Catechol-containing dyes, such as alizarin complexone (**AC**), pyrocatechol violet (**PV**), pyrogallol red, bromopyrogallol red, 4-methylesculetin, and alizarin red, were installed on the receptor *via* reversible boronic ester formation and investigated for chiral recognition. Enantiomeric hydrobenzoins produced by Sharpless asymmetric dihydroxylation were used as the analyte for creation of the HTS protocol. The protocol used to determine the ee values and reaction yield is as follows.

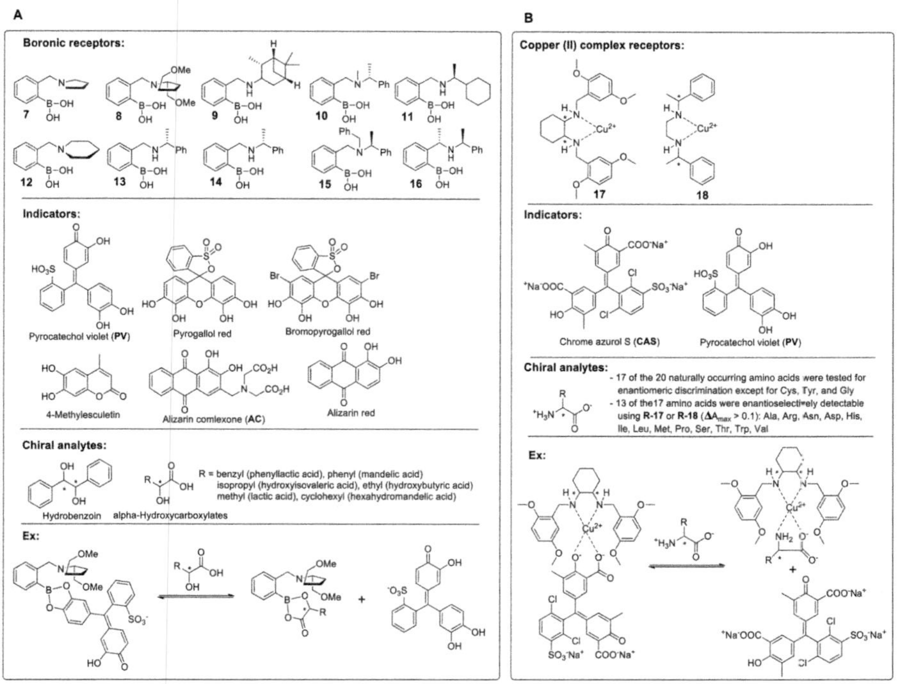

Figure 8.3 The eIDAs based on boronic acid receptors (A) and Cu^{2+}-binding diamine ligand receptors (B).

1. The proper concentrations of the receptor–indicator combinations are determined using UV-vis spectroscopy and binding isotherms.
2. The best receptor–indicator combination for the ee measurement of a target is determined by using a 96-well plate screening.
3. The optical data of the analytes are used to train an artificial neural network (ANN).
4. Using a 96-well plate automatically loaded with unknowns, the optical data are obtained.
5. The data are analyzed by the trained ANN, disclosing the ee values and reaction yield (total guest concentration $[G]_t$).

The unknowns produced in Sharpless asymmetric dihydroxylation reactions were analyzed with the receptors (**7**, **8**) and the indicator (bromopyrogallol red). The error was ~3.4% for ee and ± 0.17 mM for $[G]_t$. Thus, the protocol employed turned out to be practical for use in HTS assays.

The eIDAs based on boronic acid receptors were also used to determine the ee and concentration of α-hydroxycarboxylates (Figure 8.3A).[25] These indicator displacement assays for α-hydroxycarboxylates employed achiral receptor **7**, chiral receptor **8**, and the indicators (**PV** and **AC**). As achiral receptor **7** displays the same affinity to both enantiomers, the total concentration of an unknown sample can be determined using **7** in an IDA. Simultaneously, a chiral complex (**8** and **PV**/**AC**) was used to determine the ee of the analyte in the absence of an empirical ee calibration curve. The unknown samples of phenyl-lactic acid were evaluated using this two-step analysis. The error was within ±20% for the ee and ±10% for the yield. Overall, this protocol turned out to be practical, simple, and inexpensive, as it utilized simple UV-vis spectroscopy, omitting the need for time-consuming derivatization/synthesis and an empirical calibration curve.

Another eIDA developed by the Anslyn group was based on chiral receptors (**17**, **18**) and indicators (**PV** and chrome azurol S (**CAS**)) with coordinate-covalent bonds (Figure 8.3B).[25] The chiral coordination center generated from a diamino ligand and Cu^{2+} complex allows rapid exchange between the indicator and the amino acid *via* coordinate-covalent interactions. The resulting diastereomeric equilibria were utilized to determine the ee values of 17 naturally occurring α-amino acids, except of cysteine (solubility issue), tyrosine (solubility issue), and glycine (no stereocenter). The two diastereomeric complexes have different stabilities, presumably due to steric hindrance between the benzene ring of the chiral receptor and the side chain of the amino acid, resulting in different equilibrium constants ($K_R \neq K_S$). Accordingly, the absorbance and color of the solution are influenced by the stereochemistry of the analyte. When the chiral receptor–indicator ensemble in the solution was titrated with the same concentrations of D- and L-amino acid, the largest difference in absorbance observed was defined as ΔA_{max}. If the system with $\Delta A_{max} > 0.1$ was arbitrarily considered to be enantioselective, 13 of the 17 α-amino acids were recognized enantioselectively using $[\mathbf{17}(\mathbf{CAS})]^{2-}$ and $[\mathbf{18}(\mathbf{CAS})]^{2-}$ (Figure 8.3B). The average errors of the ee values measured were at an acceptably low level (±10–14%). This eIDA was implemented on a simple and affordable UV-vis spectrophotometer and could be combined with a microplate reader, making it suitable for rapid determination of ee values.

8.3 Fluorescence

Fluorescence has been used as a versatile analytical tool for over a hundred years with applications ranging from intramolecular reaction characterization, medical diagnostics, to even determining information about Earth's terrain. Over this time, the technology has advanced in leaps and bounds, but even in its fledgling days it was being used for incredible discoveries. In 1867, Goppelsröder published the first fluorescence-based assay used to determine Al(III) in solution using its morin chelate.[19,66] From this development came what is considered the foundation of analytical chemistry. A decade later, in 1877, ten liters of fluorescein were poured into the Danube river to determine if it was connected *via* underground currents to the Rhine.[26] After 50 hours, the fluorescein was observed in a lake that feeds immediately into the Rhine, thus proving the theory. After hundreds of years of work and discovery, fluorescence has become a widely used tool to detect minute quantities of analytes in solution, with recent developments continuing to expand this area of chemistry.

Fluorescence makes an excellent method for the quantification of trace analytes due to the large signals produced and the high sensitivity associated with the method. Fluorescence occurs when an excited molecule relaxes back to S_0 (ground state) from S_1 (excited state), with the simultaneous emission of a photon (Figure 8.4). Molecules that fluoresce, or have a portion of their population that fluoresces, are known as fluorophores. Due to the low background associated with this technique, it is particularly sensitive and widely used for analytical sensing.[19,27] For these reasons, monitoring fluorescence is ideal for measuring minute quantities of molecules very accurately. This method has been used widely for sensing a vast array of molecules using a combination of both indicator displacement assays and other binding motifs. The discrimination between the two enantiomers has been a rapidly growing area of investigation due to the importance of these molecules to everyday life.[28] New design strategies, particularly focusing on enantioselective fluorescent discrimination between the two configurations of amino acids, along with chiral amines and alcohols, will be covered in this section.

8.3.1 Binaphthyl Probes

Due to their extensive adaptability, indicator displacement assays make excellent assays for the detection of target analytes. The Pu group has demonstrated several innovative uses of IDAs, in one instance creating a BINOL-based bis(naphthylamine), **21**, with fluorescence responses at two emission wavelengths in response to several functional groups, including amines, diamines, amino alcohols, and amino acids (Figure 8.5).[29]

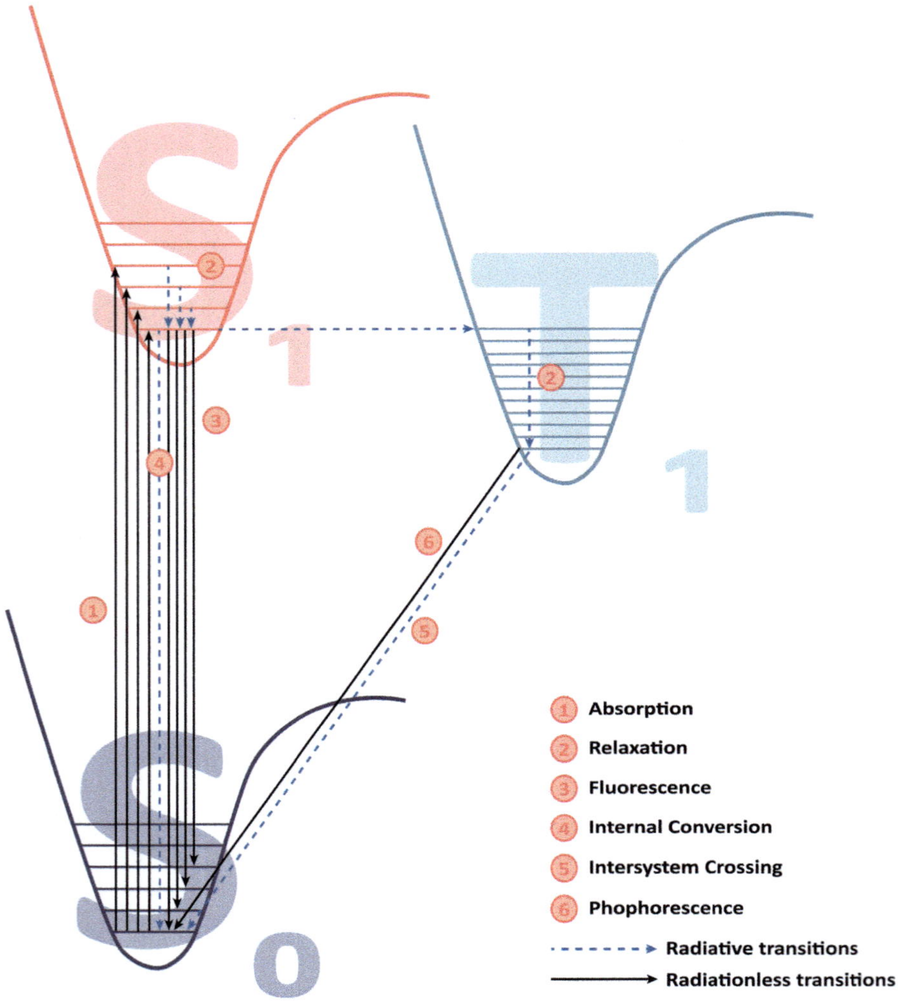

Figure 8.4 Jabłoński diagram.

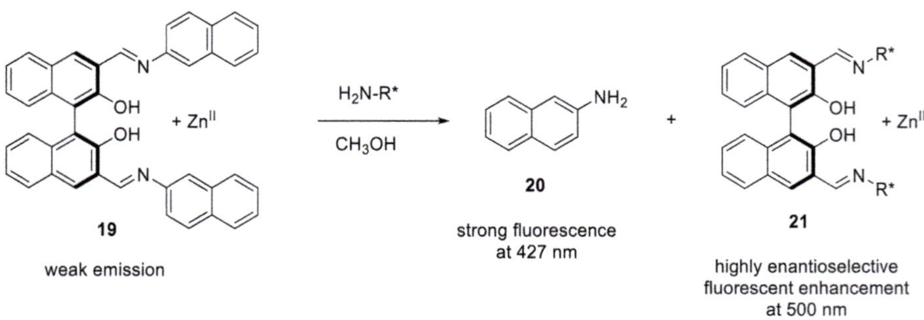

Figure 8.5 Binaphthyl-based sensor designed to give duel fluorescent readings to determine yield and ee.

When excited, one emission wavelength (λ_1) corresponds to the amount of targeted molecule in solution, while the next emission wavelength (λ_2) corresponds to its enantiopurity. The 2-naphthylamine **20**, when bound to the BINOL core, exhibits fluorescence quenching. After displacement by the target analyte, a significant fluorescence turn-on can be seen at λ_1 = 427 nm by the displaced 2-napthylamine. Due to the inherent chirality of the BINOL core, when the chiral analyte is added a highly enantioselective fluorescence enhancement is seen at λ_1 > 500 nm. Thus, with a single fluorescence measurement it is possible to determine the yield and enantiomeric excess (ee) of a reaction. This was the first fluorescent sensor able to determine ee and yield concomitantly, using only one fluorescence measurement. Anslyn and co-workers developed a method that similarly relied on the displacement of 2-napthylamine in solution to concomitantly determine the yield and ee in conjunction using CD measurements, along with fluorescence with a previously developed Fe-based octahedral complex.[30]

Anzenbacher and co-workers recently published a new method for the ee determination of chiral amino alcohols and amines for asymmetric reaction screening using BINOL- and VANOL-based probes, while diol moieties were screened using the tryptophanol-based probe (Figure 8.6).[31] The chiral amines, 2-formylphenylboronic acid, and chiral diols undergo dynamic self-assembly in acetonitrile to form a series of diastereomeric complexes created by the interaction of the analyte enantiomer and the enantiopure fluorescent ligand. These complexes give off unique fluorescent responses, in both their wavelength and intensity, allowing for determination of both the ee and diastereomeric excess (de) with an average error of less than 1% ee in only 4 to 6 hours. These assays were optimized using linear discriminate analysis, which uses a training set of data to further predict untested values. Using automated liquid dispensing and fluorescent-well plate readers, this protocol has been optimized for use in high-throughput experimentation along with real-time determination of ee and de. Due to its speed and versatility, this platform presents an exciting technique to use for reaction optimization and drug discovery.

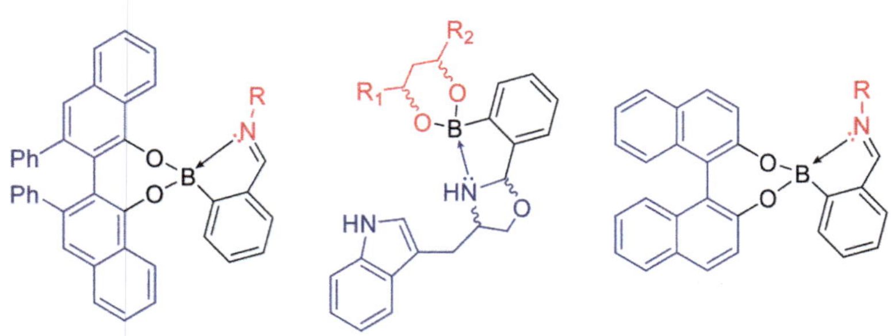

22 23 24

Figure 8.6 Components for enantiomeric excess determination of chiral amines, amine derivatives, diols, and sugars.

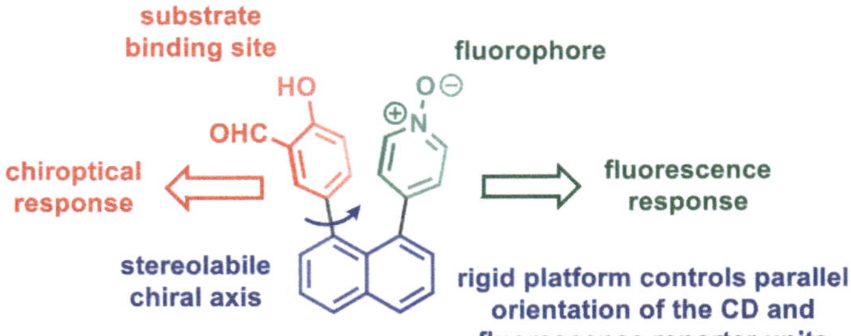

25

Figure 8.7 Salicylaldehyde-based probe with fluorescence and ee sensing units.

8.3.2 Salicyladehyde Probes

Similarly, Wolf and co-workers developed a fluorescent probe based on the use of salicylaldehyde with an adjacent pyridyl *N*-oxide fluorophore (Figure 8.7).[32] When the target substrate binds to the receptor, a corresponding change in CD effects and fluorescence signal can be used for determination of absolute configuration, enantiomeric composition, and total concentration of amino alcohols. One key design feature of this probe is the phenol moiety that is in close proximity to the formyl group. This has been shown, by Anslyn and co-workers, to significantly accelerate condensation by amines to form imines.[33] Another crucial feature is the proximate *N*-oxide group, which was shown, *via* the crystal structure, to hydrogen bond intramolecularly with the amino alcohol of interest. This, along with the ring's propensity to minimize steric interactions, creates the strong CD effect seen between 250 and 400 nm, as well as the fluorescent changes when the binding event occurs. From these effects, it is possible to determine the absolute configuration, ee, and total concentration of unknown reactions.

8.3.3 Cyclodextrin Probes

One particular system that has been extensively used for the detection of neutral molecules is with α-, β-, and γ-cyclodextrins, which are comprised of 6, 7, or 8 glucose units, respectively.[34] Crystal structures have shown that these units create a hydrophobic cavity and hydrophilic exterior (Figure 8.8).

Corradini and co-workers used a modified β-CD unit with a dansyl-based fluorophore attached to detect the presence of L- *versus* D-amino acids. Having the terminal dansyl moiety directed inside the hydrophobic cavity of the cyclodextrin ring gave rise to more intense fluorescent signals with significantly longer lifetimes. It was found in this study that when D-phenylalanine was used as a linker to the dansyl group *versus* L-phenylalanine, a significantly higher fluorescence signal was observed ($\Delta F_L/\Delta F_D$). To

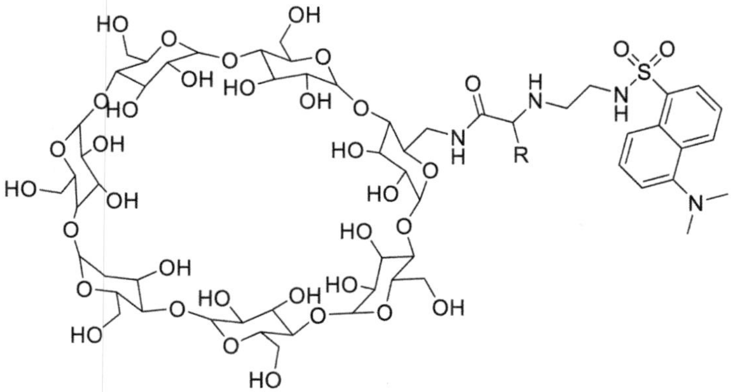

26

Figure 8.8 Cyclodextrin ring with dansyl moiety attached.

test the sensing capabilities of this complex further, a series of 11 amino acids were screened and a strong fluorescent response was seen in all cases to allow for chiral discrimination.[35] This response was found to occur due to ligand exchange between the dansyl group and the amino acid introduced, where the amino acid creates a β-CD + Cu^{II} + amino acid ternary complex. As a control, the dansyl-bearing side arm without the β-CD ring was synthesized and showed very little fluorescence response to the amino acids tested. This unique system demonstrated the first enantioselective fluorescent system using cyclodextrin to detect D-*versus* L-amino acids.

8.4 Circular Dichroism

One of the most extensive applications of circular dichroism spectroscopy is for the characterization of biomolecule secondary structure.[36–38] Biomolecules are built from optically active building blocks and adopt secondary structures, such as alpha helices and beta sheets, with unique circular dichroic spectral signals, allowing one to quantify these secondary structures. While the usefulness of CD is well established in biological contexts, small-molecule sensing *via* CD is a growing application in both the environmental and pharmaceutical fields. In these areas, quantitation of heavy metals,[39] amino acids,[40] ions,[41] and many other analytes have been achieved. The sensitivity and accuracy of CD spectroscopy has made it particularly useful for high-throughput experimentation (HTE). Modern pharmaceutical discovery and reaction optimization relies on HTE to screen thousands of reaction conditions that must be analyzed as quickly as they are screened to avoid bottlenecks during reaction discovery and optimization.[42] In this regard, the use of plate readers for CD analysis has recently come of age[43] in an instrument called the Ekko, manufactured by Hinds Inc.

CD is defined as the difference in absorbance of left-handed circularly polarized light (LCPL) and right-handed circularly polarized light (RCPL), and can be measured at any wavelength (λ).[44]

$$CD_\lambda = \Delta A = A_{LCPL} - A_{RCPL} \quad (8.1)$$

The tendency for a molecule to absorb LCPL or RCPL preferentially results from the same diastereomeric interactions that enable other solution-phase chiroptical phenomena, such as optical rotation (OR). In OR, the different circular polarizations of light propagate at different speeds proportional to the anisotropy of the solution, which is imparted by the configuration of the chiral solute. This phenomenon can be observed at any wavelength using a technique known as optical rotatory dispersion. CD, however, occurs only at wavelengths at which the chiral molecule absorbs light. This provides more structural information and is therefore suitable for sensing applications.[45] Many CD phenomena discussed in this section arise from the coupling of asymmetrically oriented chromophores, referred to as exciton-coupled circular dichroism (ECCD). The dipoles of the two or more chromophore excited states couple and, depending on the orientation of these dipoles, produce the signature bisignate curve associated with circular dichroism.

Figure 8.9 depicts a simplified view of the dipole orientation required for ECCD, which is invoked frequently for explaining CD spectral features. While most sensing applications utilize ECCD, it is one of several types of circular dichroism. Vibrational circular dichroism (VCD) and magnetic circular dichroism (MCD) are two notable phenomena that have extensive applications in biomolecule and metal complex characterization. Sensing schemes utilizing VCD and MCD are relatively scarce compared to UV and visible wavelength circular dichroism.[46,47]

8.4.1 Determination of Absolute Configuration

Although CD spectra may be interpreted for small chiral molecules, the interactions between different chromophores are complex and may lead to incorrect assignments of relevant stereocenters.[49] Additionally, small molecules with minimal conjugation exhibit weak CD signals and often overlap with the absorption bands of ligands and additives present in the applications. Early chirality sensing with ECCD relied on derivatization with chromophores to increase the magnitude and wavelength of the resulting CD signal. This approach is robust and continues to find use in the sensing of reactive functional groups.[4] Most modern sensors based on CD utilize host–guest chemistry to amplify analyte chirality and empirical trends for absolute stereochemistry assignment.

A variety of host–guest assemblies exist in which chiral analytes are bound through both non-covalent and covalent interactions. One of the first examples of these is the zinc porphyrin tweezers **27** (Figure 8.10A) developed by Nakanishi *et al.* for the determination of the absolute stereochemistry of

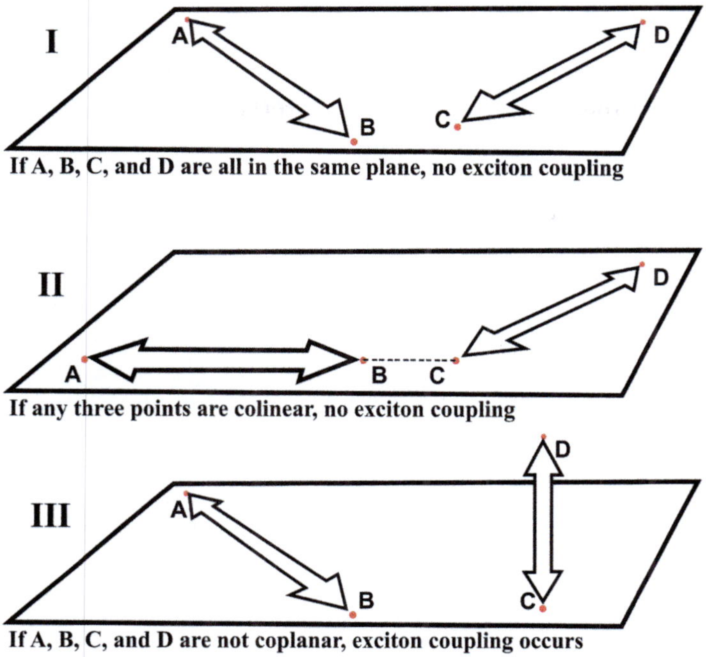

Figure 8.9 Resulting ECCD from four points constituting two transition dipoles.[48]

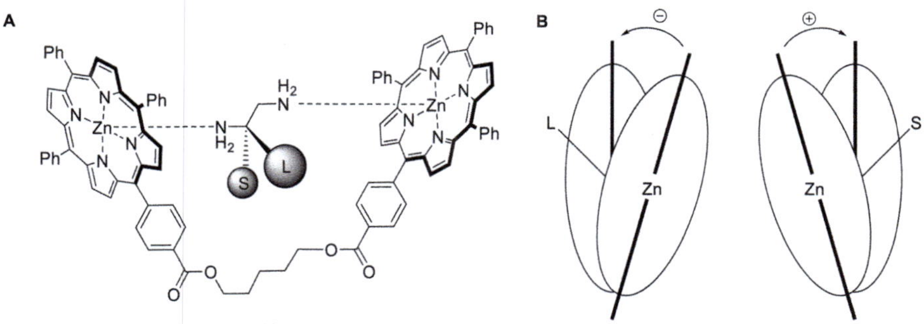

27

Figure 8.10 (A) Zinc porphyrin tweezer with flexible linker complexed with chiral diamine and (B) schematic demonstrating the preferred porphyrin-substituent sandwich.

chiral diamines.[50] By applying the rules above, the schematic drawing (Figure 8.10B) shows the two chromophores (in this case porphyrins), which are oriented to enable exciton coupling. The tweezers sandwich the chiral diamine between the two porphyrin moieties and adopt a helical conformation in which one of the substituents is "clamped" directly in the inter-porphyrin space. Because of steric interactions between the porphyrin moieties and the clamped substituent, there is an energetic preference for the smaller substituent (S) to be clamped over the larger substituent (L). This energetic preference creates a predominance of one helical tweezer conformation, thus transferring the chiral information of the point stereocenter of the diamine to the axial stereocenter within the host molecule. This point-to-axial chiral transfer is a prerequisite for real-world application of most sensors because it precludes derivatization of functional groups with chromophores and enables a broader analyte scope.[51]

Biaryl sensors comprise a large portion of host–guest sensors in both the covalent and non-covalent domains. Superchi *et al.* introduced a biaryl sensor in which an α-chiral primary amine is reacted with bis-bromomethylbiphenyl **28** to give 1,3-didehydroazepines (Figure 8.11A).[52] This covalent substrate-binding approach yields a substantial increase in both wavelength and the amplitude of the Cotton effects compared to those of the amine alone. The CD signal is primarily governed by the size of the substituents at the stereogenic center. The relative size of these substituents imparts varying steric pressure on the stereodynamic biaryl moiety, giving rise to a preferred torsion and thus a preferred sign of the bisignate curve. Much like the porphyrin tweezers, simple mnemonics are used to predict the sign of the resulting Cotton effect enabling absolute stereochemistry determination (Figure 8.11B).

It is important to note that the rules applied to such systems are empirical and thus do not always follow Cahn–Ingold–Prelog priority rules. Additionally, Superchi and co-workers observed an inversion of this mnemonic in amines containing an aryl substituent. This observation was due to

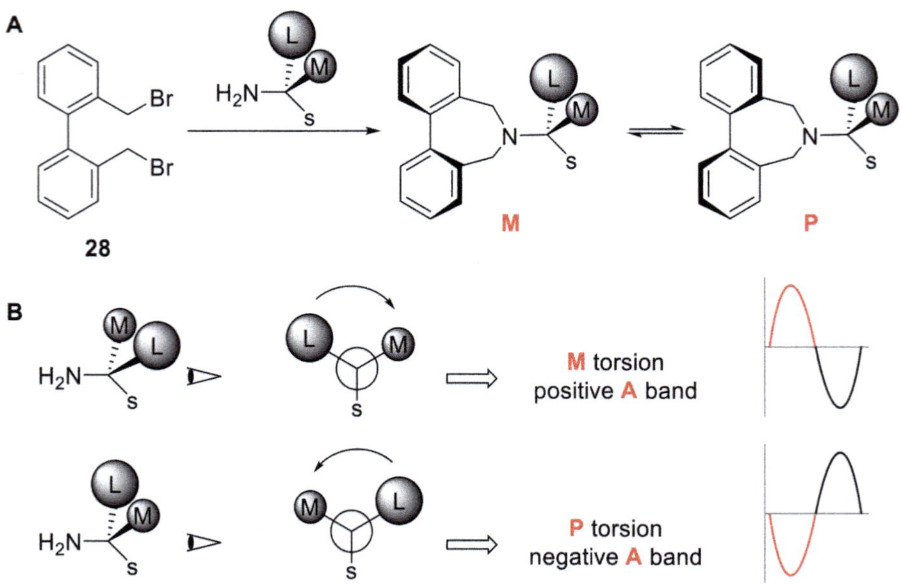

Figure 8.11 (A) Biaryl sensor covalently binding alkyl-substituted chiral amine. (B) Mnemonic to determine absolute configuration of the chiral amine.

electrostatic interactions between the aromatic substituent and nitrogen lone pair, and not to π–π interactions as seen in other biaryl sensors.[53] Biaryl sensors are prolific in UV-vis and fluorescence-based sensors due to the enormous library of BINOL analogs. Consequently, tropos ligands reminiscent of atropisomeric BINOL have been introduced that utilize CD.[54]

8.4.2 Small-molecule and ee Quantitation

In addition to absolute configuration, the ee of analytes can be measured by creating calibration curves relating the magnitude of the Cotton effect and the enantiomeric excess of standard solutions. One of the first applications of this relationship in a tripodal complex was the derivatization of α-amino acids into a tripodal Cu(II) complex, **29**, by Canary, which gave both absolute configuration and ee (Figure 8.12).[55] Inspired by this work, numerous tripodal-type assemblies were designed to accommodate a breadth of functional groups, concentrations, and wavelengths. Sensors developed by Anslyn, Wolf, and Zonta are highly sensitive towards ee and enable its quantitation for a variety of analytes, including amines,[56] alcohols,[57] carboxylates,[58] and many more.[59] The vanadium(V) complex **30** introduced by Zonta and co-workers is particularly suited for ee determination as its CD spectrum exhibits a substantial bathochromic shift upon binding of a chiral analyte. This so-called "g-probe" enables ee determination within 5% margin of error at wavelengths beyond that of common optical interference compounds.

The zinc-based tripodal complex **31** is primarily used for ee determination of alcohols based on the same functional principle as the other tripodal complexes. The covalent incorporation of the chiral alcohol produces a new stereocenter at the hemiaminal ether functionality. The diastereomeric ratio (dr) of this stereocenter is highly influenced by the differing steric pressure imparted by the substituents of the incorporated alcohol. Using Charton parameters as steric descriptors, Anslyn and co-workers were able to predict the maximum CD signal of enantiopure analytes and thus create calibration

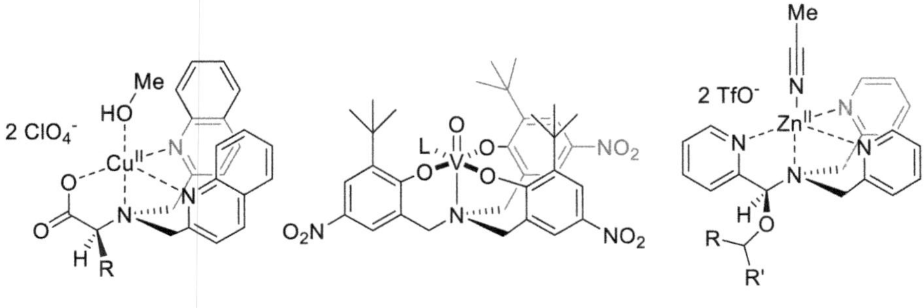

29 **30** **31**

Figure 8.12 Early tripodal Cu complex for ee determination by Canary (**29**). Oxovanadium complex for concentration-independent ee determination by Zonta (**30**). Zn^{II} complex for ee determination of alcohols (**31**).

curves relating CD and ee.[60] This removes the need to create calibration curves *via* tedious calibration experiments and greatly accelerates CD-based ee determination. Further exploration of the free energy relationships towards this application remains scarce despite the potential time savings.

Circular dichroism-based sensors satisfy the requirements for chirality sensing in HTE workflows. Sensing schemes designed on point-to-axial chirality transfer enable ee determination of nanograms of material without the need for chromatographic resolution. The variety of analytes and functional groups tolerated by these sensors demonstrates flexibility without degradation in accuracy, and improvements in limits of detection relative to chromatographic methods. With the advent of HTE, sensors with chromatographic accuracy are the minimum requirements for adoption by the synthetic community. Prediction of CD sensor responses without performing calibration experiments is required for materials that are difficult to isolate, and to assess which sensor is appropriate for the substrate. Further efforts to simplify sensor design and broaden substrate compatibility are the next steps to ease implementation efforts.

8.4.3 Multiple and Distal Stereocenters

Most sensing schemes for both absolute configuration and ee determination target a single stereocenter bearing a reactive functional group, such as an amine, carboxylic acid, or alcohol. Few sensor schemes manage to accurately determine ee at multiple stereocenters and deconvolute diastereomeric excess (de). One such example was introduced by Pu and co-workers, in which four stereoisomers of *N*-Cbz-threonines were fully speciated using an enantiomeric pair of fluorescent BINOL derivatives.[61] A similar system utilizing CD spectroscopy was realized by Anslyn, in which the de of 2-aminocyclohexanol was determined using a dual-sensor strategy. The ee of each stereocenter was measured by using an octahedral iron complex and zinc tripodal complex **31** for the amine and hydroxyl functional groups, respectively. The de is intimately related to the ee of both stereocenters, and a complete speciation can be achieved by deriving the relationship between both values.[62] Wolf and co-workers used a similar strategy to measure the ee and absolute configuration of a mixture of chiral amines and aminoalcohols. The resulting CD spectra can be deconvoluted using latent variable multivariate regression and multiblock principal component analysis to construct multidimensional calibration surfaces. In total, eight structurally similar chiral compounds were analyzed using this method, demonstrating the power of combining statistical methods with the ee sensing schemes.[63]

Most of the sensors previously described in this section probe functional groups with α-stereocenters. In general, the proximity between the stereocenter of interest and the stereodynamic element (sometimes referred to as a "reporter chromophore") is required to maximize point-to-axial chirality transfer. Determination of the ee of distal stereocenters has recently been achieved by derivatizing existing sensors to accommodate β-chiral primary alcohols.[64] Stereocenters further from the point of attachment of a reporter chromophore remain an untapped challenge in the sensing community. Efforts to expand the scope of these sensing techniques are underway and will further improve real-world utility.

8.5 Conclusion

High-throughput screening of chiral compounds is of ever-growing importance. Modern pharmaceuticals are increasingly three-dimensional, with many stereocenters.[65] Optical methods conducive to HTS are well explored and adaptable to many-use cases. Applications that demand high sensitivity are easily approached with fluorescence-based sensors, while detailed structural information is best revealed by circular dichroism spectroscopy. Conversely, the maturity of UV-visible spectroscopy as an optical sensing method lends itself to inexpensive and operator-friendly implementations. As combinatorial and parallel asymmetric syntheses increase in pharmaceutical industry, suitable HTS protocols utilizing the advantages of these sensing methods are of vital importance for the expeditious development of new drugs, their production, and quality control.

References

1. W. H. Brooks, W. C. Guida and K. G. Daniel, The Significance of Chirality in Drug Design and Development, *Curr. Top. Med. Chem.*, 2011, **11**(7), 760–770.
2. S. H. Shabbir, L. A. Joyce, G. M. da Cruz, V. M. Lynch, S. Sorey and E. V. Anslyn, Pattern-Based Recognition for the Rapid Determination of Identity, Concentration, and Enantiomeric Excess of Subtly Different Threo Diols, *J. Am. Chem. Soc.*, 2009, **131**(36), 13125–13131.
3. B. Wang and E. V. Anslyn, *Chemosensors: Principles, Strategies, and Applications*, John Wiley & Sons, 2011, vol. 15.
4. F. Y. Thanzeel, K. Balaraman and C. Wolf, Click Chemistry Enables Quantitative Chiroptical Sensing of Chiral Compounds in Protic Media and Complex Mixtures, *Nat. Commun.*, 2018, **9**(1), 5323.
5. D. J. Cram; R. C. Helgeson; L. R. Sousa; J. M. Timko; M. Newcomb; P. Moreau; F. De Jong; G. W. Gokel; D. H. Hoffman and L. A. Domeier, Chiral Recognition in Complexation of Guests by Designed Host Molecules, in *Organic Synthesis*, Elsevier, 1975, pp. 327–349.
6. E. E. Greciano, R. Rodríguez, K. Maeda and L. Sánchez, Disclosing Chirality in Consecutive Supramolecular Polymerizations: Chiral Induction by Light in N-Annulated Perylenetetracarboxamides, *Chem. Commun.*, 2020, **56**(15), 2244–2247.
7. X. Zhang, J. Yin and J. Yoon, Recent Advances in Development of Chiral Fluorescent and Colorimetric Sensors, *Chem. Rev.*, 2014, **114**(9), 4918–4959.

8. X. Lu, J. Tang, X. Dang, X. Jing, K. Xu, H. Li and B. Liang, Chiral Recognition and Determination of Enantiomeric Excess of Chiral Compounds by UV-Visible-Shortwave near Infrared Diffuse Reflectance Spectroscopy with Chemometrics, *RSC Adv.*, 2017, **7**(22), 13552–13560.
9. L.-L. Wang, Z. Chen, W.-E. Liu, H. Ke, S.-H. Wang and W. Jiang, Molecular Recognition and Chirality Sensing of Epoxides in Water Using Endo-Functionalized Molecular Tubes, *J. Am. Chem. Soc.*, 2017, **139**(25), 8436–8439.
10. V. N. Khose, M. E. John, A. D. Pandey, V. Borovkov and A. V. Karnik, Chiral Heterocycle-Based Receptors for Enantioselective Recognition, *Symmetry*, 2018, **10**(2), 34.
11. J. Wu, B. Kwon, W. Liu, E. V. Anslyn, P. Wang and J. S. Kim, Chromogenic/Fluorogenic Ensemble Chemosensing Systems, *Chem, Rev.*, 2015, **115**(15), 7893–7943.
12. Y. Sasaki, S. Kojima, V. Hamedpour, R. Kubota, S. Takizawa, I. Yoshikawa, H. Houjou, Y. Kubo and T. Minami, Accurate Chiral Pattern Recognition for Amines from Just a Single Chemosensor, *Chem. Sci.*, 2020, **11**(15), 3790–3796.
13. B. T. Herrera, S. L. Pilicer, E. V. Anslyn, L. A. Joyce and C. Wolf, Optical Analysis of Reaction Yield and Enantiomeric Excess: A New Paradigm Ready for Prime Time, *J. Am. Chem. Soc.*, 2018, **140**(33), 10385–10401.
14. D. Leung, S. O. Kang and E. V. Anslyn, Rapid Determination of Enantiomeric Excess: A Focus on Optical Approaches, *Chem. Soc. Rev.*, 2012, **41**(1), 448–479.
15. Y. Kubo, S. Maeda, S. Tokita and M. Kubo, Colorimetric Chiral Recognition by a Molecular Sensor, *Nature*, 1996, **382**(6591), 522–524.
16. K. Tsubaki, M. Nuruzzaman, T. Kusumoto, N. Hayashi, W. Bin-Gui and K. Fuji, Visual Enantiomeric Recognition Using Chiral Phenolphthalein Derivatives, *Org. Lett.*, 2001, **3**(25), 4071–4073.
17. C. Hu, Y. He, Z. Chen and X. Huang, Synthesis and Enantioselective Recognition of an (S)-BINOL-Based Colorimetric Chemosensor for Mandelate Anions, *Tetrahedron: Asymmetry*, 2009, **20**(1), 104–110.
18. K. Tsubaki, K. Mukoyoshi, H. Morikawa, T. Kinoshita and K. Fuji, Enantiomeric Recognition of Amino Acids Using a Chiral Spiropyran Derivative, *Chirality*, 2002, **14**(9), 713–715.
19. A. C. Sedgwick, J. T. Brewster, T. Wu, X. Feng, S. D. Bull, X. Qian, J. L. Sessler, T. D. James, E. V. Anslyn and X. Sun, Indicator Displacement Assays (IDAs): The Past, Present and Future, *Chem. Soc. Rev.*, 2021, **50**(1), 9–38.
20. H. H. Jo, C.-Y. Lin and E. V. Anslyn, Rapid Optical Methods for Enantiomeric Excess Analysis: From Enantioselective Indicator Displacement Assays to Exciton-Coupled Circular Dichroism, *Acc. Chem. Res.*, 2014, **47**(7), 2212–2221.
21. J. F. Folmer-Andersen, V. M. Lynch and E. V. Anslyn, Colorimetric Enantiodiscrimination of α-Amino Acids in Protic Media, *J. Am. Chem. Soc.*, 2005, **127**(22), 7986–7987.

22. J. P. Lorand and J. O. Edwards, Polyol Complexes and Structure of the Benzeneboronate Ion, *J. Org. Chem.*, 1959, **24**(6), 769–774.
23. S. D. Bull, M. G. Davidson, J. M. Van den Elsen, J. S. Fossey, A. T. A. Jenkins, Y.-B. Jiang, Y. Kubo, F. Marken, K. Sakurai and J. Zhao, Exploiting the Reversible Covalent Bonding of Boronic Acids: Recognition, Sensing, and Assembly, *Acc. Chem. Res.*, 2013, **46**(2), 312–326.
24. L. Zhu and E. V. Anslyn, Facile Quantification of Enantiomeric Excess and Concentration with Indicator-Displacement Assays: An Example in the Analyses of α-Hydroxyacids, *J. Am. Chem. Soc.*, 2004, **126**(12), 3676–3677.
25. D. Leung, J. F. Folmer-Andersen, V. M. Lynch and E. V. Anslyn, Using Enantioselective Indicator Displacement Assays to Determine the Enantiomeric Excess of α-Amino Acids, *J. Am. Chem. Soc.*, 2008, **130**(37), 12318–12327.
26. B. Valeur and M. N. Berberan-Santos, A Brief History of Fluorescence and Phosphorescence before the Emergence of Quantum Theory, *J. Chem. Educ.*, 2011, **88**(6), 731–738.
27. A. P. de Silva, H. Q. N. Gunaratne, T. Gunnlaugsson, A. J. M. Huxley, C. P. McCoy, J. T. Rademacher and T. E. Rice, Signaling Recognition Events with Fluorescent Sensors and Switches, *Chem. Rev.*, 1997, **97**(5), 1515–1566.
28. L. Pu, Enantioselective Fluorescent Recognition of Free Amino Acids: Challenges and Opportunities, *Angew. Chem.*, 2020, **59**(49), 21814–21828.
29. K. Wen, S. Yu, Z. Huang, L. Chen, M. Xiao, X. Yu and L. Pu, Rational Design of a Fluorescent Sensor to Simultaneously Determine Both the Enantiomeric Composition and the Concentration of Chiral Functional Amines, *J. Am. Chem. Soc.*, 2015, **137**(13), 4517–4524.
30. S. R. Moor, J. R. Howard, B. T. Herrera and E. V. Anslyn, High-Throughput Screening of α-Chiral-Primary Amines to Determine Yield and Enantiomeric Excess, *Tetrahedron*, 2021, 132315.
31. E. G. Shcherbakova, T. D. James and P. Anzenbacher, High-Throughput Assay for Determining Enantiomeric Excess of Chiral Diols, Amino Alcohols, and Amines and for Direct Asymmetric Reaction Screening, *Nat. Protoc.*, 2020, **15**(7), 2203–2229.
32. K. W. Bentley and C. Wolf, Stereodynamic Chemosensor with Selective Circular Dichroism and Fluorescence Readout for in Situ Determination of Absolute Configuration, Enantiomeric Excess, and Concentration of Chiral Compounds, *J. Am. Chem. Soc.*, 2013, **135**(33), 12200–12203.
33. J. M. Dragna, G. Pescitelli, L. Tran, V. M. Lynch, E. V. Anslyn and L. Di Bari, In Situ Assembly of Octahedral Fe(II) Complexes for the Enantiomeric Excess Determination of Chiral Amines Using Circular Dichroism Spectroscopy, *J. Am. Chem. Soc.*, 2012, **134**(9), 4398–4407.
34. R. Corradini, A. Dossena, R. Marchelli, A. Panagia, G. Sartor, M. Saviano, A. Lombardi and V. Pavone, A Modified Cyclodextrin with a Fully Encapsulated Dansyl Group: Self-Inclusion in the Solid State and in Solution, *Chem. – Eur. J.*, 1996, **2**(4), 373–381.

35. R. Corradini, C. Paganuzzi, R. Marchelli, S. Pagliari, S. Sforza, A. Dossena, G. Galaverna and A. Duchateau, Design and Synthesis of Fluorescent ?-Cyclodextrins for the Enantioselective Sensing of ?-Amino Acids, *Chirality*, 2003, **15**(S1), S30–S39.
36. S. W. Provencher and J. Gloeckner, Estimation of Globular Protein Secondary Structure from Circular Dichroism, *Biochemistry*, 1981, **20**(1), 33–37.
37. P. Manavalan and W. C. Johnson, Variable Selection Method Improves the Prediction of Protein Secondary Structure from Circular Dichroism Spectra, *Anal. Biochem.*, 1987, **167**(1), 76–85.
38. G. D. Fasman, *Circular Dichroism and the Conformational Analysis of Biomolecules*, Springer, New York, NY, 2013.
39. P. Sianglam, S. Kulchat, T. Tuntulani and W. Ngeontae, A Circular Dichroism Sensor for Selective Detection of Cd 2+ and S 2− Based on the In-Situ Generation of Chiral CdS Quantum Dots, *Spectrochim. Acta, Part A*, 2017, **183**, 408–416.
40. J. Nan and X.-P. Yan, A Circular Dichroism Probe for L-Cysteine Based on the Self-Assembly of Chiral Complex Nanoparticles, *Chem. – Eur. J.*, 2010, **16**(2), 423–427.
41. H. Maeda, Y. Bando, K. Shimomura, I. Yamada, M. Naito, K. Nobusawa, H. Tsumatori and T. Kawai, Chemical-Stimuli-Controllable Circularly Polarized Luminescence from Anion-Responsive π-Conjugated Molecules, *J. Am. Chem. Soc.*, 2011, **133**(24), 9266–9269.
42. High-Throughput Analysis in the Pharmaceutical Industry, in *Critical Reviews in Combinatorial Chemistry*, ed. P. G. Wang, CRC Press, Boca Raton, FL, 2009.
43. P. Metola, S. M. Nichols, B. Kahr and E. V. Anslyn, Well Plate Circular Dichroism Reader for the Rapid Determination of Enantiomeric Excess, *Chem. Sci.*, 2014, **5**(11), 4278–4282.
44. IUPAC, *Compendium of Chemical Terminology*, (the "Gold Book"), compiled by A. D. McNaught and A. Wilkinson, Blackwell Scientific Publications, Oxford, 2nd end, 1997, online version (2019-) created by S. J. Chalk, ISBN 0-9678550-9-8, https://doi.org/10.1351/goldbook.
45. H. Eyring, H.-C. Liu and D. Caldwell, Optical Rotatory Dispersion and Circular Dichroism, *Chem. Rev.*, 1968, **68**(5), 525–540.
46. T. A. Keiderling, Protein and Peptide Secondary Structure and Conformational Determination with Vibrational Circular Dichroism, *Curr. Opin. Chem. Biol.*, 2002, **6**(5), 682–688.
47. E. I. Solomon, E. G. Pavel, K. E. Loeb and C. Campochiaro, Magnetic Circular Dichroism Spectroscopy as a Probe of the Geometric and Electronic Structure of Non-Heme Ferrous Enzymes, *Coord. Chem. Rev.*, 1995, **144**, 369–460.
48. V. Person, K. Monde, H. Humpf, N. Berova and K. Nakanishi, A New Approach in Exciton-Coupled Circular Dichroism—(ECCD) Insertion of an Auxiliary Stereogenic Center, *Chirality*, 1995, **7**(3), 128–135.

49. G. Pescitelli, For a Correct Application of the CD Exciton Chirality Method: The Case of Laucysteinamide A, *Mar. Drugs*, 2018, **16**(10), 388.
50. X. Huang, B. Borhan, B. H. Rickman, K. Nakanishi and N. Berova, Zinc Porphyrin Tweezer in Host-Guest Complexation: Determination of Absolute Configurations of Primary Monoamines by Circular Dichroism, *Chem. – Eur. J.*, 2000, **6**(2), 216–224.
51. M. Anyika, H. Gholami, K. D. Ashtekar, R. Acho and B. Borhan, Point-to-Axial Chirality Transfer—A New Probe for "Sensing" the Absolute Configurations of Monoamines, *J. Am. Chem. Soc.*, 2014, **136**(2), 550–553.
52. S. Vergura, L. Pisani, P. Scafato, D. Casarini and S. Superchi, Central-to-Axial Chirality Induction in Biphenyl Chiroptical Probes for the Stereochemical Characterization of Chiral Primary Amines, *Org. Biomol. Chem.*, 2018, **16**(4), 555–565.
53. S. Superchi, R. Bisaccia, D. Casarini, A. Laurita and C. Rosini, Flexible Biphenyl Chromophore as a Circular Dichroism Probe for Assignment of the Absolute Configuration of Carboxylic Acids, *J. Am. Chem. Soc.*, 2006, **128**(21), 6893–6902.
54. K. W. Bentley, Y. G. Nam, J. M. Murphy and C. Wolf, Chirality Sensing of Amines, Diamines, Amino Acids, Amino Alcohols, and α-Hydroxy Acids with a Single Probe, *J. Am. Chem. Soc.*, 2013, **135**(48), 18052–18055.
55. A. E. Holmes, S. Zahn and J. W. Canary, Synthesis and Circular Dichroism Studies of N,N-Bis(2-Quinolylmethyl)Amino Acid Cu(II) Complexes: Determination of Absolute Configuration and Enantiomeric Excess by the Exciton Coupling Method, *Chirality*, 2002, **14**(6), 471–477.
56. J. M. Dragna, G. Pescitelli, L. Tran, V. M. Lynch, E. V. Anslyn and L. Di Bari, In Situ Assembly of Octahedral Fe(II) Complexes for the Enantiomeric Excess Determination of Chiral Amines Using Circular Dichroism Spectroscopy, *J. Am. Chem. Soc.*, 2012, **134**(9), 4398–4407.
57. L. You, G. Pescitelli, E. V. Anslyn and L. Di Bari, An Exciton-Coupled Circular Dichroism Protocol for the Determination of Identity, Chirality, and Enantiomeric Excess of Chiral Secondary Alcohols, *J. Am. Chem. Soc.*, 2012, **134**(16), 7117–7125.
58. L. A. Joyce, M. S. Maynor, J. M. Dragna, G. M. da Cruz, V. M. Lynch, J. W. Canary and E. V. Anslyn, A Simple Method for the Determination of Enantiomeric Excess and Identity of Chiral Carboxylic Acids, *J. Am. Chem. Soc.*, 2011, **133**(34), 13746–13752.
59. H. Gholami, J. Zhang, M. Anyika and B. Borhan, Absolute Stereochemical Determination of Asymmetric Sulfoxides via Central to Axial Induction of Chirality, *Org. Lett.*, 2017, **19**(7), 1722–1725.
60. C.-Y. Lin, S. Lim and E. V. Anslyn, Model Building Using Linear Free Energy Relationship Parameters–Eliminating Calibration Curves for Optical Analysis of Enantiomeric Excess, *J. Am. Chem. Soc.*, 2016, **138**(26), 8045–8047.
61. S. Yu and L. Pu, One Enantiomeric Fluorescent Sensor Pair to Discriminate Four Stereoisomers of Threonines, *Sci. China: Chem.*, 2013, **56**(3), 301–306.

62. B. T. Herrera, S. R. Moor, M. McVeigh, E. K. Roesner, F. Marini and E. V. Anslyn, Rapid Optical Determination of Enantiomeric Excess, Diastereomeric Excess, and Total Concentration Using Dynamic-Covalent Assemblies: A Demonstration Using 2-Aminocyclohexanol and Chemometrics, *J. Am. Chem. Soc.*, 2019, **141**(28), 11151–11160.
63. D. S. Hassan and C. Wolf, Optical Deciphering of Multinary Chiral Compound Mixtures through Organic Reaction Based Chemometric Chirality Sensing, *Nat. Commun.*, 2021, **12**(1), 6451.
64. M. B. Minus, A. L. Featherston, S. Choi, S. C. King, S. J. Miller and E. V. Anslyn, Reengineering a Reversible Covalent-Bonding Assembly to Optically Detect Ee in β-Chiral Primary Alcohols, *Chem*, 2019, **5**(12), 3196–3206.
65. I. P. Silvestri and P. J. J. Colbon, The Growing Importance of Chirality in 3D Chemical Space Exploration and Modern Drug Discovery Approaches for Hit-ID: Topical Innovations, *ACS Med. Chem. Lett.*, 2021, **12**(8), 1220–1229.
66. A. W. Czarnik, *Fluorescent Chemosensors for Ion and Molecule Recognition*, American Chemical Society, Washington, DC, 1993.

CHAPTER 9

Fluorescent Carbon Nanoparticles

MICHAŁ BARTKOWSKI[a] AND SILVIA GIORDANI*[a]

[a]School of Chemical Sciences, Dublin City University, Glasnevin, Dublin, Ireland
*E-mail: silvia.giordani@dcu.ie

9.1 Introduction

Nanoparticles can be crudely defined as zero-dimensional (0D) structures with dimensions below 100 nm.[1] Conversely, nanomaterials include nanoparticles and one-dimensional (1D) and two-dimensional (2D) structures, where two or one axes, respectively, are below 100 nm in size.[2] For instance, a typical carbon nanotube is a 1D nanomaterial; graphene is a 2D nanomaterial; and fullerenes are 0D nanoparticles/nanomaterials. It should also be noted that three-dimensional (3D) structures—such as graphite—are not considered nanomaterials, as all axes are over 100 nm in size. Nanoparticles do not have to be spherical—small graphene flakes under 100 nm in size are considered nanoparticles regardless of their planar morphology.[3] Regarding bulk materials, the EU Commission defines nanoparticles as those in which at least 50% of the particle size distribution measures under 100 nm.[4]

Nanoparticles may possess luminescent properties: particularly, fluorescence. Fluorescence occurs when a molecule absorbs a photon of a specific

Monographs in Supramolecular Chemistry No. 34
Fluorescent Chemosensors
Edited by Luling Wu, Adam C. Sedgwick, Xiao-Peng He and Tony D. James

Published by the Royal Society of Chemistry, www.rsc.org

frequency, which excites a singlet electron in the ground state to a higher electronic state. Subsequently, a photon is emitted as the electron returns to the ground state in a spin-allowed process.[5] The particular characteristics of fluorescence, including the emission wavelength, brightness, and the presence of additional emission bands, may be strongly affected by various chemical and physical factors.[6] These include temperature,[7] viscosity,[8] pH,[9] electric potential,[10] tension,[11] solvent polarity,[12] and the presence of hydrogen bonds,[13] ions,[14] quenchers,[15] and upconversion agents.[16] By carefully controlling the nanoparticle and these factors, it is possible to achieve fluorescent nanoparticles with desired photophysical properties.[17] For this reason, interest in fluorescent nanoparticles has seen a sharp growth between 2000 and 2014, with an average of 3500 studies involving the concept of 'fluorescent nanoparticles' being published per year since 2014 (see Figure 9.1). Notably, the first time the concept of 'fluorescence' was reported in the literature was by John Frederick William Herschel in 1845.[18]

The two fundamental application areas of fluorescent carbon nanoparticles and nanomaterials are imaging and sensing. In 2002, R. Bruce Weisman *et al.* observed that single-walled carbon nanotubes (SWCNTs) exhibit a bright band-gap fluorescence in the near-infrared (NIR) region.[19,20] This observation carries deep significance for the fields of bioimaging and biosensing—NIR light can penetrate human tissues, enabling deep-tissue imaging.[21,22] For instance, in 2018, Daniel A. Heller *et al.* utilised the NIR-emitting properties of SWCNTs to visualise endolysosomal lipid accumulation in Kupffer cells *in vivo*—a breakthrough in the study of non-alcoholic fatty liver disease, as this is the first method that can do this type of dynamic monitoring without the need for a liver biopsy.[23] This type of bioimaging has been enabled by 'hyperspectral microscopy', also developed by the

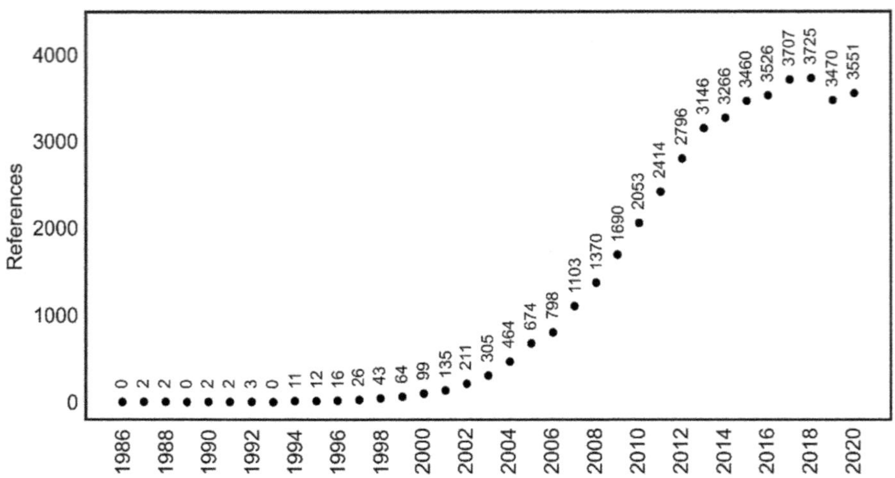

Figure 9.1 Number of publications involving the concept of 'fluorescent nanoparticles' between 1986 and 2020 [*Data collected and processed by the author using SciFinder®.*].

Heller group, which can resolve the NIR fluorescence of CNTs of 17 distinct chiralities (*n*,*m*) (see Figure 9.2).[24]

Aside from their use in NIR bioimaging,[25–28] fluorescent carbon nanoparticles and nanomaterials have also been utilised in analytical biology,[29] fluorescence spectroscopy,[30] bioimaging,[31] clinical diagnosis,[32] biomedical treatment,[33] environmental tracing,[34] and wastewater treatment.[35]

This chapter gives an overview of various fluorescent carbon nanoparticles, supported by selected case examples from the literature, namely on carbon nano-onions, fluorescent carbon dots, and detonation nanodiamonds. Numerous reviews offer an exhaustive discussion of various classes of fluorescent nanoparticles[36–38]—these reviews may interest readers looking for a deeper understanding of the different types of fluorescent nanoparticles.

9.2 Fluorescent Carbon Nanoparticles

9.2.1 Carbon Nano-onions

First seen by Ugarte in 1992, carbon nano-onions (CNOs) are multilayered fullerenes composed of concentric shells of sp^2 carbon.[39,40] Many methods are available to prepare CNOs, including arc discharge,[41] electron beam irradiation,[39] ion implanation,[42,43] laser ablation,[44] pyrolysis,[45] and thermal annealing.[46] These different methods result in CNOs of different shapes, sizes, and morphologies. The approach our group utilises for the preparation of CNOs is the thermal annealing of detonation nanodiamonds (DNDs) under an inert He atmosphere. This approach results in small (approximately 5 nm diameter) CNOs of high purity and low polydispersity (see Figure 9.3A and B).[47] Notably, the biosafety profile of small CNOs is excellent. Their toxicology has been investigated *in vitro, in vivo*, and *ex vivo*. *In vitro* studies were carried out on HeLa[48] and HeLa Kyoto,[49] NIH 3T3,[50] MCF-7,[51] A2780,[52] MDA-MB-231,[52] and KB cells.[48] *In vivo* studies involved *Drosophila melanogaster*,[53] *Hydra vulgaris*,[54] and *Danio rerio*.[55] *Ex vivo* studies were carried out

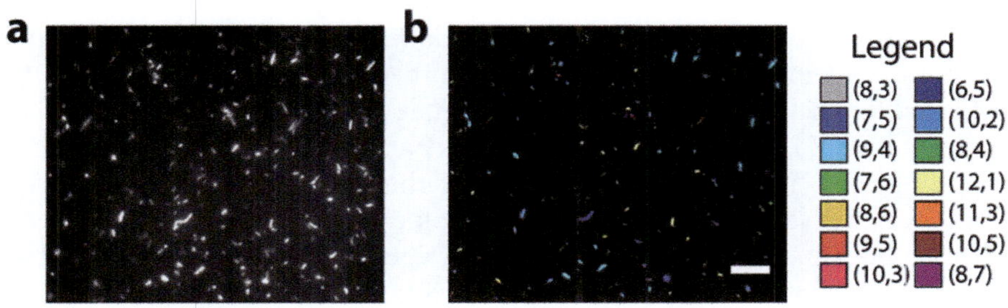

Figure 9.2 (A) Hyperspectral microscopy image of CNTs adsorbed onto a glass surface. (B) False-colour image of (A), coloured according to nanotube chirality (*n*,*m*).[24] Scale bar = 10 μm. Adapted from ref. 24, https://doi.org/10.1038/srep14167, under the terms of the CC BY 4.0 license https://creativecommons.org/licenses/by/4.0/.

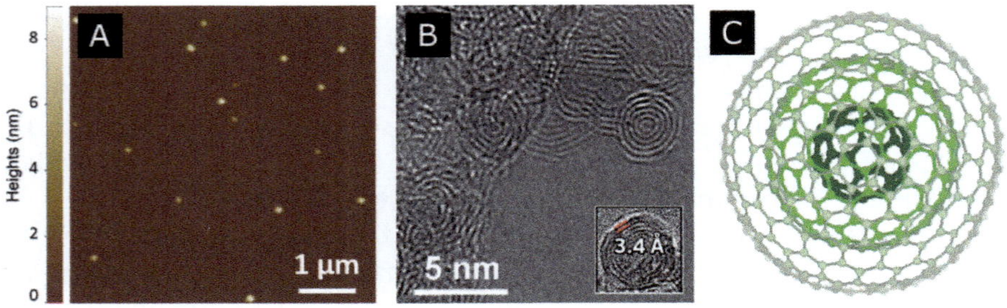

Figure 9.3 (A) Atomic force microscopy image of our pristine CNOs (approximately 5 nm in size).[59] (B) High-resolution transmission electron microscopy (HRTEM) image of our pristine CNOs; inset: the average interlayer distance.[59] (C) 3D representation of a three-layer 2 nm $C_{60}C_{240}C_{500}$ CNO structure. Parts A and B adapted from ref. 59 with permission from the Royal Society of Chemistry.

on murine models.[56–58] In agreement, these studies demonstrate that small CNOs show low systemic toxicity and good biocompatibility. This excellent biosafety profile expands the scope of applicability of small CNOs to the biomedical field. Although CNOs are intrinsically non-luminescent, they are viable scaffolds for coupling with various emitters—the desired photoluminescent properties can be imposed on CNOs through modulation with appropriate fluorophores. Indeed, CNOs have been functionalised with emitters in many instances, thus creating fluorescent nanoparticles capable of emitting from high-energy blue light to low-energy NIR light. Our group reports using aza-BODIPY and BODIPY derivatives to impart green, red, and NIR fluorescence;[49,59–64] fluorescein derivatives to enable green fluorescence;[48,56,65,66] and the tricyclic furo[2,3-*c*]isoquinoline scaffold-based fluorophore to impart deep-blue fluorescence onto CNOs.[67]

In 2014, our group functionalised CNOs with a green-emitting boron dipyrromethane (BODIPY) dye for the first time.[59] The surface of pristine CNOs was covalently functionalised with benzoic acid through a Tour reaction. Subsequently, the BODIPY analogue was covalently attached to the benzoic acid through an EDCl/NHS-mediated amidation (see Figure 9.4, **1**). The emissive properties and high fluorescence afforded to the CNOs by the BODIPY allowed for use of the systems in cellular imaging studies. The BODIPY-CNOs were taken up by MCF-7 cells. Their uptake and distribution were visualised with confocal microscopy—it was found that BODIPY-CNOs are internalised by endocytosis and predominantly colocalised in the lysosome after 48 hours. Overall, the low toxicity and high versatility of BODIPY-CNOs have demonstrated their potential as promising fluorescent nanoparticles for theranostic applications.[59]

In the same year, we developed a red aza-BODIPY-CNO system. In this approach, an alternative BODIPY analogue (see Figure 9.4, **2**) was covalently functionalised onto CNOs to afford their NIR fluorescence.[49] The system was utilised for *in vitro* imaging on HeLa Kyoto cells, showing good fluorescence

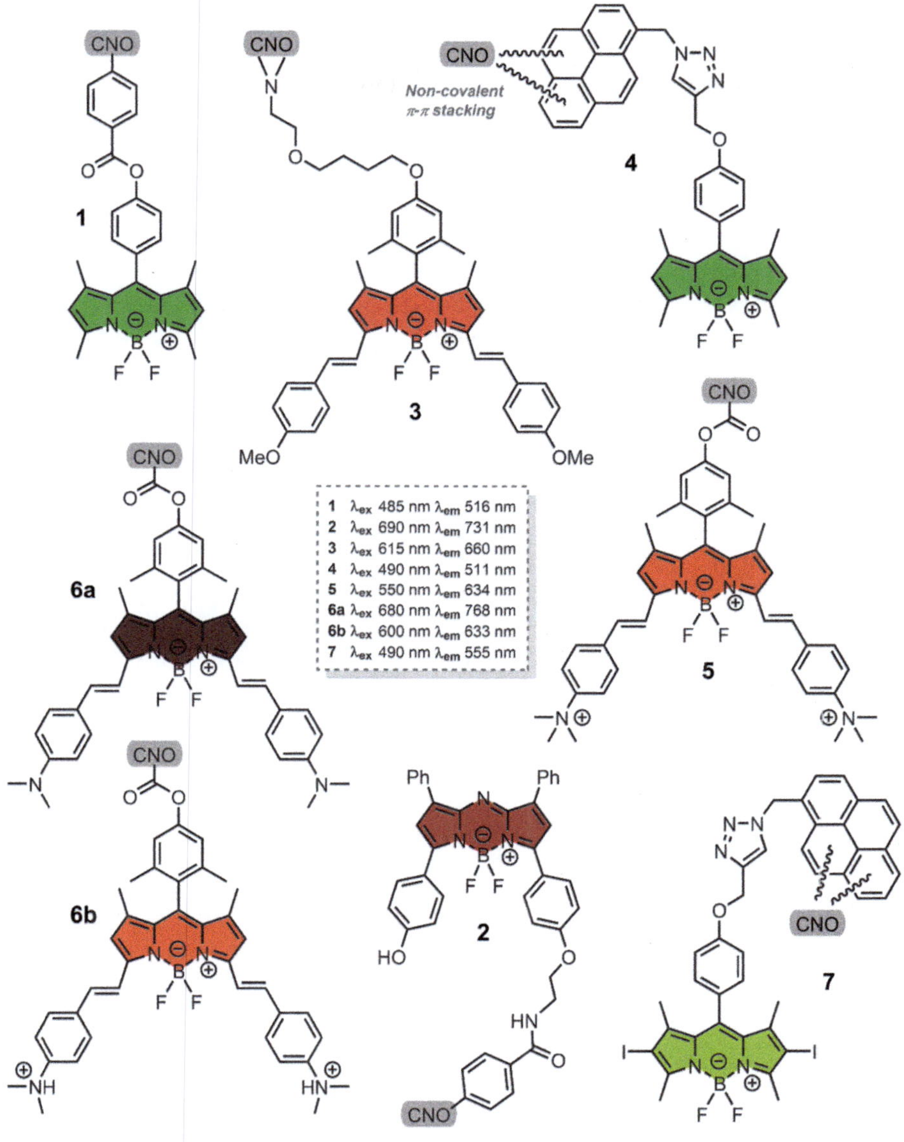

Figure 9.4 The structures of various BODIPY and aza-BODIPY analogues functionalised onto CNOs to create green, red, and NIR fluorescent nanoparticles, as developed by our group: **1**;[59] **2**;[49] **3**;[60] **4**;[61] **5**;[62] **6a** and **6b**;[63] **7**.[64] The photoexcitation properties (in DMSO) have been tabulated in the centre. The fluorescent core has been coloured according to its emission wavelength under the specified conditions.

and the distribution of the systems in the cytosol. We also demonstrated that pH-induced deprotonation of the –OH group of the aza-BODIPY analogue (see Figure 9.4, **2**) results in an effective quenching of fluorescence. Consequently, this reversible on/off fluorescence modulation carries the potential for effective use of this fluorescent nanoparticle in bioimaging and responsive sensing of environmental change.

In an alternative approach, we developed a novel system in which a BODIPY-pyrene derivative was non-covalently attached to the CNO surface (see Figure 9.4, **4**)[61]—a pyrene moiety facilitated a non-covalent interaction with the pristine CNO surface through π–π stacking. Bioimaging studies and a cytotoxicity assay were performed utilising this system—the system showed good cellular uptake and biocompatibility in HeLa cells. Overall, the successful non-covalent functionalisation of CNOs with the BODIPY derivative highlights the potential of CNOs for the cellular delivery of hydrophobic molecules, including various active pharmaceutical ingredients (APIs).[61]

Beyond BODIPY and aza-BODIPY fluorescent CNOs, our group also developed several fluorescein-based CNO systems. In 2015, we covalently surface-functionalised CNOs with fluorescein (see Figure 9.5, **1**). In our approach, we oxidised DND-derived pristine CNOs. Then, we covalently conjugated the fluorophore through an EDCl/NHS-mediated amidation reaction, utilising 4-(dimethylamino)pyridine (DMAP) as a catalyst. Emission spectroscopy in dimethyl sulfoxide (DMSO) confirmed the successful binding of the fluorophore; the CNO-fluorescein nanoparticle had green emission at 518 nm when excited at 490 nm. As a further step in the study, we investigated the cellular viability and imaging potential of the system. Cellular viability studies carried out on HeLa cells showed minimal systemic toxicity. Cellular imaging studies, in which confocal micrographs were taken of HeLa cells incubated with the CNO-fluorescein nanoparticles, showed

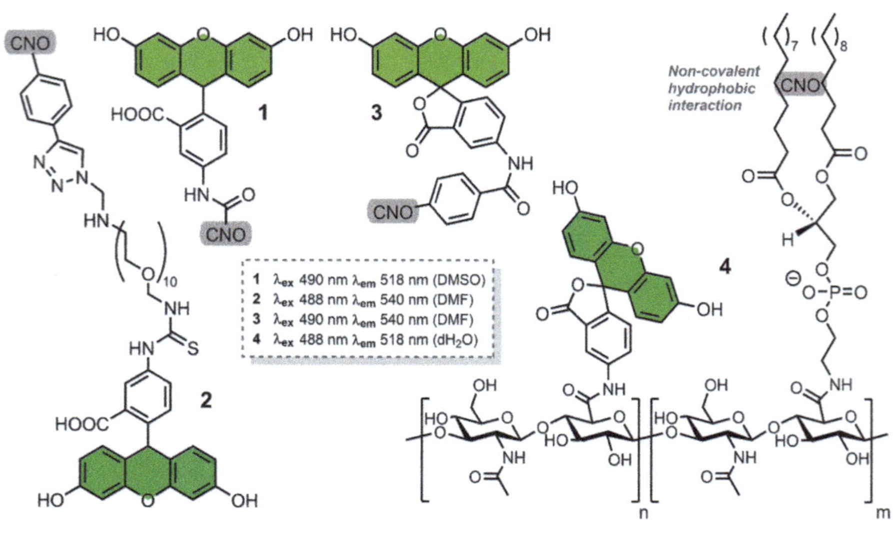

Figure 9.5 The structures of fluorescein-derivative functionalised CNOs developed by our group: 1;[65] 2;[48] 3;[56] 4.[66] The photoexcitation properties have been tabulated in the centre. The fluorescent core has been coloured according to its emission wavelength under the specified conditions.

excellent fluorescence and accumulation of the material in the perinuclear region. Furthermore, the system showed increased dispersibility in water—dispersibility is an important aspect for carbon nanomaterials (CNMs), which typically suffer from low dispersibility in biocompatible solvents due to their hydrophobic nature.[68,69] Overall, the system showed good potential for biomedical and theranostic applications.[65]

In the same year, our group developed another green-fluorescent CNO nanoparticle system.[48] However, in this case, the system was multifunctionalised with both the green-emitting fluorescein and the targeting agent folic acid. The purpose of the folic acid moiety was to impose targetability onto the system towards cancer cells overexpressing the folate receptor. In this system, the fluorophore was attached to CNOs through a polyethylene glycol linker to distance the fluorophore from the CNO surface (see Figure 9.5, 2); folic acid has been coupled similarly. Overall, this CNO system served its purpose of targeting and imaging cancer cells. The system had excellent brightness, stability, and showed rapid, selective uptake in HeLa and KB cancer cell lines, with no significant cytotoxicity.[48] This is an important step towards the utilisation of CNOs as nanocarriers for targeted drug delivery.[70]

Although many red- and green-emitting CNO systems have been designed to date, it was not until 2020 when the first purely blue light-emitting CNO system was developed. Blue light-emitting fluorophores have been coupled with CNOs to afford their blue-emissive properties.[67] The fluorescent molecules in question were heterocyclic luminophores with a tricyclic furo[2,3-*c*]isoquinoline core (see Figure 9.6), prepared through a multicomponent Ugi reaction and a complex Pd-mediated cascade. The fluorophores were covalently conjugated to surface-oxidised CNOs through an EDCl/NHS-mediated amidation. Overall, the three CNO-furo[2,3-*c*]isoquinoline systems showed good deep-blue-emissive properties (see Figure 9.6). A 15–20% quenching of fluorescence was observed for the three systems, as attributed to the intrinsic absorptive properties of CNOs. A possible solution to which, as proposed by the authors, is to electronically distance furo[2,3-*c*]isoquinoline from CNOs

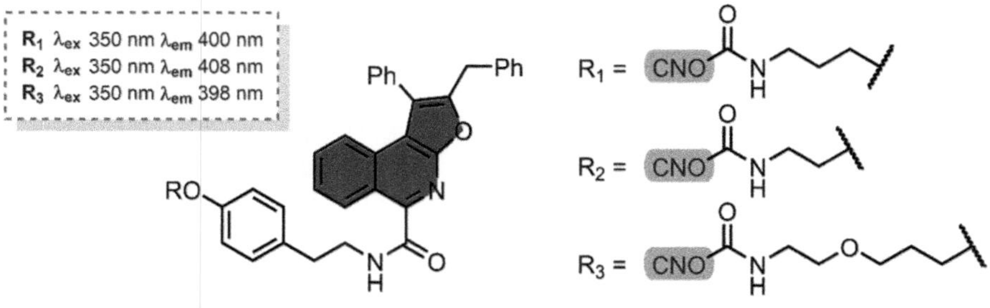

Figure 9.6 The structures of the three CNO-deep-blue emitter dyads reported by Giordani, Riva *et al.* in 2020.[67] The photoexcitation properties (in DMSO) have been tabulated on the left. The fluorescent core has been coloured according to the average (402 nm) of the emissive bands of the three derivatives.

by introducing a longer linker. Overall, CNOs proved to be a very versatile material of choice for deep-blue-emissive nanoparticles. In particular, they were easy to prepare and functionalise, offered excellent stability, and a high surface/volume ratio.[67]

Notably, CNOs have been covalently functionalised with other fluorescent probes. In 2010, our group covalently attached a zinc porphyrin to the surface of CNOs through a 1,3-dipolar cycloaddition (click reaction).[71] Photophysical studies of this porphyrin-CNO system indicated that, when excited at 425 nm, two characteristic emission bands are observed, at 612 and 662 nm. In a further study of this system, X-ray photoelectron spectroscopy and time-of-flight-secondary ion mass spectrometry analyses were performed, and these techniques confirmed with analytical certainty the successful binding of the fluorescent probe to the CNOs.[72]

In summary, fluorescent CNO nanoparticles can be developed through the covalent/non-covalent coupling of carbon nanomaterials with emitters. In addition, the desired photophysical and emissive properties can be obtained by fine-tuning the emitters bonded to the CNO surface. They offer a complementary choice of fluorescent nanoparticles that are emissive from the blue to the NIR light range. Typically, green-, red-, and NIR-emitting CNOs can be obtained by coupling with BODIPY and aza-BODIPY dyes;[49,59–64] fluorescein-based dyes may afford green fluorescence onto CNOs;[48,56,65,66] and deep-blue emission can be obtained through coupling with furo[2,3-*c*] isoquinoline fluorophores.[67] Overall, CNOs show good potential as fluorescent nanoparticles of choice for biomedical and theranostic applications, given their excellent biocompatibility, great cellular penetration ability, ease of manipulation, ease of preparation and chemical functionalisation, and high surface/volume ratio.[48,51,67,73] Notwithstanding, their potential extends further—CNOs can act as starting materials for creating other fluorescent nanoparticles. Such examples are described in the next section on fluorescent carbon dots, where UV-, blue- and green-emissive quantum dots have been made from CNOs.[74,75]

9.2.2 Fluorescent Carbon Dots

There are many fluorescent nanodot (FND) types, although the difference between them can be blurred due to their overlapping properties. In general, they can be broadly classified into inorganic and carbon-based materials. Inorganic FNDs are referred to as semiconductor quantum dots (SQDs), also commonly abbreviated as quantum dots (QDs). Carbon-based FNDs (fluorescent carbon dots) consist of three major categories. The first is graphene quantum dots (GQDs), also commonly abbreviated as graphene dots (GDs); the second being carbon quantum dots (CQDs), also commonly abbreviated as carbon dots (CDs); and the final category is carbon nanodots (CNDs) (see Figure 9.7). In their review, Cayuela *et al.* succinctly summarise the properties and characteristics of these four different classifications of FNDs.[76] Briefly,

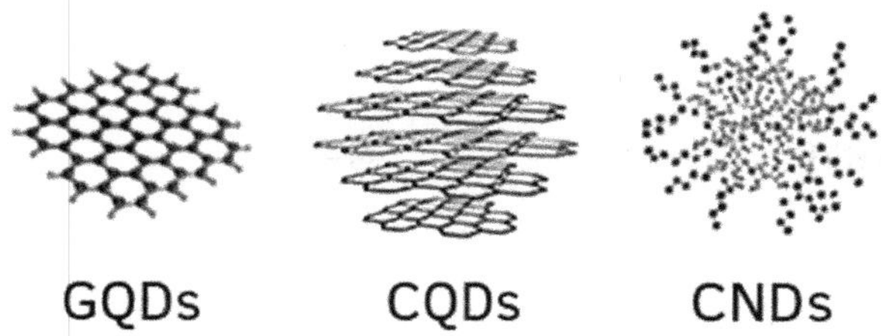

Figure 9.7 The main classes of fluorescent carbon dots.[76] Adapted from ref. 76 with permission from the Royal Society of Chemistry.

SQDs are composed of inorganic elements, GQDs are composed of carbon, and CQDs and CNDs are composed of carbon, with oxygen commonly present. SQDs have a crystalline structure, are spherical, and are usually <6 nm in diameter. Their photoluminescence (PL) is usually composition dependent and excitation independent. GQDs are composed of small discs of single-layer graphite, typically <20 nm in diameter. Their PL is usually excitation and functionalisation dependent. CQDs are quasi-spherical crystalline materials consisting of mainly sp^2-hybridised carbon, typically <10 nm in diameter. Their PL is typically excitation and functionalisation dependent. Lastly, CNDs are quasi-spherical materials with an amorphous carbon core, typically <10 nm in diameter. Their PL characteristics are the same as for CQDs.[76]

As mentioned in Section 9.2.1 on carbon nano-onions, CNOs can serve as a starting material for the preparation of other fluorescent nanoparticles. In 2015, Liu and Kim synthesised UV- and blue-emitting graphene quantum dots from DND-derived CNOs.[74] Typically, GQDs are synthesised through a top-down chemical oxidation of various CNMs, including carbon fibres, graphene sheets, and carbon black. The authors chose CNOs as their starting material, given their small size and high uniformity/low polydispersity. In choosing CNOs, the authors were able to synthesise GQDs of lower size distribution and more energetically uniform emissive properties, as compared to what is possible with the aforementioned CNMs. After dialysis and purification, the authors obtained blue-emissive GQDs (λ_{ex} 330 nm; λ_{em} 450 nm) and UV-emissive GQDs (λ_{ex} 285 nm; λ_{em} 350 nm). The authors note that, unlike blue/green-emissive GQDs, UV-emissive GQDs are very rare, especially when prepared using CNMs in a top-down oxidation approach. This underlines the benefit of utilising CNOs as a precursor for the preparation of QDs. The authors speculate that the blue-emissive GQDs originate from the surface of CNOs and that the UV-emissive fraction originates from the unoxidised core of CNOs. It was also demonstrated that the blue-emissive GQDs could be used to detect Cu^{2+}, given their ultra-sensitivity to this cation.[74]

Similarly, in 2020, our group prepared CDs *via* a top-down oxidation utilising two CNMs as starting materials:[75] namely, CNOs and graphene oxide. The CDs derived from both materials had desirable photoluminescent and physiochemical properties. Graphene oxide-derived CDs had an excitation-dependent fluorescence, with the strongest excitation bands at 314 and 469 nm, and an emission maxima of 533 nm. Interestingly, the fluorescence of CDs prepared using CNOs was not excitation dependent, which relates to the surface functional groups of the material. CNO-derived CDs had an emission of 527 nm, with three notable excitation bands: 340, 400 and 460 nm, the latter giving the highest emission. Notably, both CDs showed excellent water dispersibility, with long-term stability.[75]

9.2.3 Detonation Nanodiamonds

Detonation nanodiamonds (DNDs) are small (approximately 5 nm) diamond particles (crystallites of sp^3-hybridised carbon; see Figure 9.8) originating from a detonation reaction. Typically, they are produced through the detonation of an oxygen-poor TNT/hexogene mixture.[77] DNDs may have multiple fluorescence sources, including fluorescence originating from defect colour centres,[78,79] and from the surface decoration with other emissive species.[80,81] Strictly speaking, the two main types of defect colour centres commonly present in nanodiamonds are nitrogen-vacancy (NV) defects and nitrogen-vacancy-nitrogen (NVN) defects.[82,83] The former is responsible for red/NIR fluorescence, whilst the latter is responsible for the emission of green light. NV and NVN defects have led to various emerging applications of nanodiamonds, ranging from theranostics[84] to nanomagnetometry[82] and nanoscale nuclear magnetic resonance.[82]

Similarly to CNOs, the surface of DNDs may also be functionalised with fluorescent moieties, such as CDs. In 2018, we investigated CD-functionalised DNDs (CD-DNDs) for their biomedical applications.[81] This was the first investigation of the biological viability of CD-DNDs, as this material had not been previously isolated in high yields. Specifically, the fluorescent nanoparticle utilised in this study consisted of single-digit (4–5 nm) DNDs with surface-bound 1–2 nm CDs composed of sp^2 and amorphous carbon. The

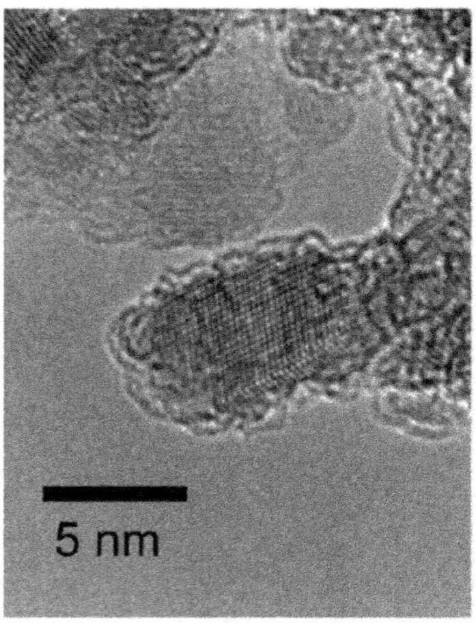

Figure 9.8 HRTEM image of DNDs.[85] Adapted from ref. 85 with permission from AIP Publishing, Copyright 2001.

CD-DNDs were utilised for *in vitro* bioimaging in MDA-MB-231 breast cancer cells—for which the excitation source was a 488 nm Ar laser, resulting in emission in the 500–550 nm range. Confocal imaging results showed that CD-DNDs are sufficiently luminescent for use in bioimaging studies. Furthermore, an *in vivo* biocompatibility study was carried out using *Danio rerio*. This study showed that the nanoparticles exhibit high biocompatibility and no evident toxic biological consequences in that particular animal model. It should also be noted that, in general, DNDs have low colloidal stability, as is the case with the CD-DND nanoparticles reported in this paper. However, we were able to disperse the CD-DNDs indefinitely by coating them with poly(glycerol). Finally, we speculate that CD-DNDs may be an alternative fluorescent nanoparticle to sub-10 nm DNDs containing NV centres, given their cost effectiveness, excellent biocompatibility, and good luminescence.[81]

9.3 Conclusion

Fluorescent nanoparticles are very versatile. Their properties extend well beyond their fluorescence, depending on the type of material/particle used. There is a wide array of various intrinsically fluorescent nanoparticles available, with many available commercially. Non-fluorescent nanoparticles can also be functionalised with emissive moieties, thus creating fluorescent nanoparticle–emitter dyads. This coupling can be carried out by covalent means or through non-covalent interactions. This chapter outlined several fluorescent nanoparticles, namely carbon nano-onions, fluorescent carbon dots, and detonation nanodiamonds. When choosing fluorescent nanoparticles for a given application, their limiting factors need to be considered—these include cytotoxicity, low quantum yield, low dispersibility, and emission outside the biologically transparent NIR optical window. By fine-tuning the nanoparticles, and in certain cases their surface-bound fluorophores, it is possible to develop fluorescent nanoparticles with the desired physicochemical and photoluminescent properties. Such particles are often applied in imaging, sensing, theranostics, and environmental remediation.

Acknowledgements

Financial assistance in the form of a Government of Ireland Postgraduate Scholarship (GOIPG) to MB (GOIPG/2019/1820) from the Irish Research Council (IRC) is gratefully acknowledged.

References

1. M. Auffan, J. Rose, J.-Y. Bottero, G. V. Lowry, J.-P. Jolivet and M. R. Wiesner, *Nat. Nanotechnol.*, 2009, **4**, 634–641.
2. E. A. J. Bleeker, W. H. de Jong, R. E. Geertsma, M. Groenewold, E. H. W. Heugens, M. Koers-Jacquemijns, D. van de Meent, J. R. Popma, A. G. Rietveld, S. W. P. Wijnhoven, F. R. Cassee and A. G. Oomen, *Regul. Toxicol. Pharmacol.*, 2013, **65**, 119–125.

3. Z. M. Markovic, L. M. Harhaji-Trajkovic, B. M. Todorovic-Markovic, D. P. Kepić, K. M. Arsikin, S. P. Jovanović, A. C. Pantovic, M. D. Dramićanin and V. S. Trajkovic, *Biomaterials*, 2011, **32**, 1121–1129.
4. EU-Commission, Comm. EEd Bruss. BE, 2011/696/EU, 2.
5. J. R. Lakowicz, *Principles of Fluorescence Spectroscopy*, Springer Science & Business Media, 2013.
6. A. Chandra, S. Prasad, G. Gigli and L. L. del Mercato, in *Frontiers of Nanoscience*, Elsevier, 2020, vol. 16, pp. 117–149.
7. S. Arai, S.-C. Lee, D. Zhai, M. Suzuki and Y. T. Chang, *Sci. Rep.*, 2014, **4**, 6701.
8. I. López-Duarte, T. Truc Vu, M. Angeles Izquierdo, J. A. Bull and M. K. Kuimova, *Chem. Commun.*, 2014, **50**, 5282–5284.
9. L. Albertazzi, M. Brondi, G. M. Pavan, S. S. Sato, G. Signore, B. Storti, G. M. Ratto and F. Beltram, *PLoS One*, 2011, **6**, 28450.
10. T. D. Fehérvári, Y. Okazaki, H. Sawai and T. Yagi, *PLoS One*, 2015, **10**, 0133853.
11. A. Colom, E. Derivery, S. Soleimanpour, C. Tomba, M. D. Molin, N. Sakai, M. González-Gaitán, S. Matile and A. Roux, *Nat. Chem.*, 2018, **10**, 1118–1125.
12. J. Yin, M. Peng, Y. Ma, R. Guo and W. Lin, *Chem. Commun.*, 2018, **54**, 12093–12096.
13. K. C. Toh, E. A. Stojković, I. H. M. van Stokkum, K. Moffat and J. T. M. Kennis, *Proc. Natl. Acad. Sci. U. S. A.*, 2010, **107**, 9170–9175.
14. W. Jiang, Q. Fu, H. Fan and W. Wang, *Chem. Commun.*, 2008, 259–261.
15. X. Wang, X. Chen, Z. Xie and X. Wang, *Angew. Chem., Int. Ed.*, 2008, **47**, 7450–7453.
16. G. Yi, Y. Peng and Z. Gao, *Chem. Mater.*, 2011, **23**, 2729–2734.
17. H. Piwoński, T. Michinobu and S. Habuchi, *Nat. Commun.*, 2017, **8**, 15256.
18. J. F. W. Herschel, *Philos. Trans. R. Soc. London*, 1845, **135**, 143–145.
19. M. J. O'Connell, S. M. Bachilo, C. B. Huffman, V. C. Moore, M. S. Strano, E. H. Haroz, K. L. Rialon, P. J. Boul, W. H. Noon, C. Kittrell, J. Ma, R. H. Hauge, R. B. Weisman and R. E. Smalley, *Science*, 2002, **297**, 593–596.
20. D. Movia, E. Del Canto and S. Giordani, *Phys. Status Solidi B*, 2009, **246**, 2704–2707.
21. L. A. Sordillo, Y. Pu, S. Pratavieira, Y. Budansky and R. R. Alfano, *J. Biomed. Opt.*, 2014, **19**, 056004.
22. R. G. Aswathy, Y. Yoshida, T. Maekawa and D. S. Kumar, *Anal. Bioanal. Chem.*, 2010, **397**, 1417–1435.
23. T. V. Galassi, P. V. Jena, J. Shah, G. Ao, E. Molitor, Y. Bram, A. Frankel, J. Park, J. Jessurun, D. S. Ory, A. Haimovitz-Friedman, D. Roxbury, J. Mittal, M. Zheng, R. E. Schwartz and D. A. Heller, *Sci. Transl. Med.*, 2018, **10**, 2680.
24. D. Roxbury, P. V. Jena, R. M. Williams, B. Enyedi, P. Niethammer, S. Marcet, M. Verhaegen, S. Blais-Ouellette and D. A. Heller, *Sci. Rep.*, 2015, **5**, 14167.

25. K. Soga, K. Tokuzen, K. Tsuji, T. Yamano, H. Hyodo and H. Kishimoto, *Eur. J. Inorg. Chem.*, 2010, 2673–2677.
26. S. K. Yen, D. Jańczewski, J. L. Lakshmi, S. B. Dolmanan, S. Tripathy, V. H. B. Ho, V. Vijayaragavan, A. Hariharan, P. Padmanabhan, K. K. Bhakoo, T. Sudhaharan, S. Ahmed, Y. Zhang and S. Tamil Selvan, *ACS Nano*, 2013, **7**, 6796–6805.
27. T. Ribeiro, S. Raja, A. S. Rodrigues, F. Fernandes, J. P. S. Farinha and C. Baleizão, *RSC Adv.*, 2013, **3**, 9171–9174.
28. Q. Zhao, K. Li, S. Chen, A. Qin, D. Ding, S. Zhang, Y. Liu, B. Liu, J. Zhi Sun and B. Zhong Tang, *J. Mater. Chem.*, 2012, **22**, 15128–15135.
29. J. Liu, X. Yang, X. He, K. Wang, Q. Wang, Q. Guo, H. Shi, J. Huang and X. Huo, *Sci. China: Chem.*, 2011, **54**, 1157.
30. B. Kang, M. M. Afifi, L. A. Austin and M. A. El-Sayed, *ACS Nano*, 2013, **7**, 7420–7427.
31. D. Liße, V. Wilkens, C. You, K. Busch and J. Piehler, *Angew. Chem., Int. Ed.*, 2011, **50**, 9352–9355.
32. S. Rogalla, K. Flisikowski, D. Gorpas, A. T. Mayer, T. Flisikowska, M. J. Mandella, X. Ma, K. M. Casey, S. A. Felt, D. Saur, V. Ntziachristos, A. Schnieke, C. H. Contag, S. S. Gambhir and S. Harmsen, *Adv. Funct. Mater.*, 2019, **29**, 1904992.
33. E. S. Glazer and S. A. Curley, *Cancer*, 2010, **116**, 3285–3293.
34. F. Tauro, E. Rapiti, J. F. Al-Sharab, L. Ubertini, S. Grimaldi and M. Porfiri, *J. Nanopart. Res.*, 2013, **15**, 1884.
35. L. Otero-González, J. A. Field, I. A. C. Calderon, C. A. Aspinwall, F. Shadman, C. Zeng and R. Sierra-Alvarez, *Water Res.*, 2015, **77**, 170–178.
36. O. S. Wolfbeis, *Chem. Soc. Rev.*, 2015, **44**, 4743–4768.
37. M. J. Ruedas-Rama, J. D. Walters, A. Orte and E. A. H. Hall, *Anal. Chim. Acta*, 2012, **751**, 1–23.
38. S. M. Ng, M. Koneswaran and R. Narayanaswamy, *RSC Adv.*, 2016, **6**, 21624–21661.
39. D. Ugarte, *Nature*, 1992, **359**, 707–709.
40. D. Ugarte, *Carbon*, 1995, **33**, 989–993.
41. R. Borgohain, J. Yang, J. P. Selegue and D. Y. Kim, *Carbon*, 2014, **66**, 272–284.
42. T. Cabioc'h, M. Jaouen, E. Thune, P. Gúerin, C. Fayoux and M. F. Denanot, *Surf. Coat. Technol.*, 2000, **128**, 43–50.
43. E. Thune, T. Cabioc'h, P. Guérin, M. F. Denanot and M. Jaouen, *Mater. Lett.*, 2002, **54**, 222–228.
44. D. Dorobantu, P. M. Bota, I. Boerasu, D. Bojin and M. Enachescu, *Surf. Eng. Appl. Electrochem.*, 2014, **50**, 390–394.
45. S. K. Sonkar, M. Roy, D. G. Babar and S. Sarkar, *Nanoscale*, 2012, **4**, 7670.
46. J. P. Bartolome and A. Fragoso, *Fullerenes, Nanotubes, Carbon Nanostruct.*, 2017, **25**, 327–334.
47. M. Bartkowski and S. Giordani, *Nanoscale*, 2020, **12**, 9352–9358.
48. M. Frasconi, R. Marotta, L. Markey, K. Flavin, V. Spampinato, G. Ceccone, L. Echegoyen, E. M. Scanlan and S. Giordani, *Chem. – Eur. J.*, 2015, **21**, 19071.

49. S. Giordani, J. Bartelmess, M. Frasconi, I. Biondi, S. Cheung, M. Grossi, D. Wu, L. Echegoyen and D. F. O'Shea, *J. Mater. Chem. B*, 2014, **2**, 7459–7463.
50. M. d'Amora, V. Maffeis, R. Brescia, D. Barnes, E. Scanlan and S. Giordani, *Nanomaterials*, 2019, **9**, 1069.
51. J. Bartelmess and S. Giordani, *Beilstein J. Nanotechnol.*, 2014, **5**, 1980–1998.
52. M. d'Amora, A. Camisasca, A. Boarino, S. Arpicco and S. Giordani, *Colloids Surf., B*, 2020, **188**, 110779.
53. M. Ghosh, S. K. Sonkar, M. Saxena and S. Sarkar, *Small*, 2011, **7**, 3170–3177.
54. V. Marchesano, A. Ambrosone, J. Bartelmess, F. Strisciante, A. Tino, L. Echegoyen, C. Tortiglione and S. Giordani, *Nanomaterials*, 2015, **5**, 1331–1350.
55. M. d'Amora, A. Camisasca, S. Lettieri and S. Giordani, *Nanomaterials*, 2017, **7**, 414.
56. M. Yang, K. Flavin, I. Kopf, G. Radics, C. H. A. Hearnden, G. J. McManus, B. Moran, A. Villalta-Cerdas, L. A. Echegoyen, S. Giordani and E. C. Lavelle, *Small*, 2013, **9**, 4194–4206.
57. B. Pakhira, M. Ghosh, A. Allam and S. Sarkar, *RSC Adv.*, 2016, **6**, 29779–29782.
58. M. Trusel, M. Baldrighi, R. Marotta, F. Gatto, M. Pesce, M. Frasconi, T. Catelani, F. Papaleo, P. P. Pompa, R. Tonini and S. Giordani, *ACS Appl. Mater. Interfaces*, 2018, **10**, 16952–16963.
59. J. Bartelmess, E. De Luca, A. Signorelli, M. Baldrighi, M. Becce, R. Brescia, V. Nardone, E. Parisini, L. Echegoyen, P. P. Pompa and S. Giordani, *Nanoscale*, 2014, **6**, 13761–13769.
60. J. Bartelmess, M. Baldrighi, V. Nardone, E. Parisini, D. Buck, L. Echegoyen and S. Giordani, *Chem. - Eur. J.*, 2015, **21**, 9727–9732.
61. J. Bartelmess, M. Frasconi, P. B. Balakrishnan, A. Signorelli, L. Echegoyen, T. Pellegrino and S. Giordani, *RSC Adv.*, 2015, **5**, 50253–50258.
62. S. Lettieri, A. Camisasca, M. D'Amora, A. Diaspro, T. Uchida, Y. Nakajima, K. Yanagisawa, T. Maekawa and S. Giordani, *RSC Adv.*, 2017, **7**, 45676.
63. S. Lettieri, M. d'Amora, A. Camisasca, A. Diaspro and S. Giordani, *Beilstein J. Nanotechnol.*, 2017, **8**, 1878.
64. J. Bartelmess, G. Milcovich, V. Maffeis, M. d'Amora, S. M. Bertozzi and S. Giordani, *Front. Chem.*, 2020, **8**, 573211.
65. M. Frasconi, V. Maffeis, J. Bartelmess, L. Echegoyen and S. Giordani, *Methods Appl. Fluoresc.*, 2015, **3**, 0044005.
66. M. d'Amora, A. Camisasca, A. Boarino, S. Arpicco and S. Giordani, *Colloids Surf., B*, 2020, **188**, 1100779.
67. V. Maffeis, L. Moni, D. Di Stefano, S. Giordani and R. Riva, *Front. Chem. Sci. Eng.*, 2019, **14**, 76–89.
68. S. Giordani, S. Bergin, V. Nicolosi, S. Lebedkin, W. J. Blau and J. N. Coleman, *Phys. Status Solidi B*, 2006, **243**, 3058–3062.
69. J. C. Zuaznabar-Gardona and A. Fragoso, *J. Mol. Liq.*, 2019, **294**, 111646.
70. M. Bartkowski and S. Giordani, *Dalton Trans.*, 2021, **50**, 2300–2309.

71. K. Flavin, M. N. Chaur, L. Echegoyen and S. Giordani, *Org. Lett.*, 2010, **12**, 840–843.
72. V. Spampinato, G. Ceccone and S. Giordani, *Biointerphases*, 2015, **10**, 19006.
73. M. Bartkowski and S. Giordani, *Nanoscale*, 2020, **12**, 9352.
74. Y. Liu and D. Y. Kim, *Chem. Commun.*, 2015, **51**, 4176–4179.
75. A. Ventrella, A. Camisasca, A. Fontana and S. Giordani, *RSC Adv.*, 2020, **10**, 36404–36412.
76. A. Cayuela, M. L. Soriano, C. Carrillo-Carrión and M. Valcárcel, *Chem. Commun.*, 2016, **52**, 1311–1326.
77. K. Iakoubovskii, M. V. Baidakova, B. H. Wouters, A. Stesmans, G. J. Adriaenssens, A. Ya. Vul' and P. J. Grobet, *Diamond Relat. Mater.*, 2000, **9**, 861–865.
78. A. M. Edmonds, U. F. S. D'Haenens-Johansson, R. J. Cruddace, M. E. Newton, K.-M. C. Fu, C. Santori, R. G. Beausoleil, D. J. Twitchen and M. L. Markham, *Phys. Rev. B*, 2012, **86**, 035201.
79. J. O. Orwa, C. Santori, K. M. C. Fu, B. Gibson, D. Simpson, I. Aharonovich, A. Stacey, A. Cimmino, P. Balog, M. Markham, D. Twitchen, A. D. Greentree, R. G. Beausoleil and S. Prawer, *J. Appl. Phys.*, 2011, **109**, 083530.
80. O. Shenderova, S. Hens, I. Vlasov, S. Turner, Y.-G. Lu, G. Van Tendeloo, A. Schrand, S. A. Burikov and T. A. Dolenko, *Part. Part. Syst. Charact.*, 2014, **31**, 580–590.
81. N. Nunn, M. d'Amora, N. Prabhakar, A. M. Panich, N. Froumin, M. D. Torelli, I. Vlasov, P. Reineck, B. Gibson, J. M. Rosenholm, S. Giordani and O. Shenderova, *Methods Appl. Fluoresc.*, 2018, **6**, 35010.
82. R. Schirhagl, K. Chang, M. Loretz and C. L. Degen, *Annu. Rev. Phys. Chem.*, 2014, **65**, 83–105.
83. M. W. Doherty, N. B. Manson, P. Delaney, F. Jelezko, J. Wrachtrup and L. C. L. Hollenberg, *Phys. Rep.*, 2013, **528**, 1–45.
84. V. Vaijayanthimala, D. K. Lee, S. V. Kim, A. Yen, N. Tsai, D. Ho, H.-C. Chang and O. Shenderova, *Expert Opin. Drug Delivery*, 2015, **12**, 735–749.
85. S. Tomita, T. Sakurai, H. Ohta, M. Fujii and S. Hayashi, *J. Chem. Phys.*, 2001, **114**, 7477–7482.

CHAPTER 10

Fluorescent Chemosensors for Phosphates

S. M. BUTLER[a] AND K. A. JOLLIFFE*[a,b]

[a]The University of Sydney, School of Chemistry and Australian Research Council Centre of Excellence for Innovations in Peptide and Protein Science, 2006, NSW, Australia; [b]The University of Sydney Nano Institute (Sydney Nano), The University of Sydney, NSW 2006, Australia
*E-mail: kate.jolliffe@sydney.edu.au

10.1 Introduction

Inorganic phosphate (Pi), oligomers thereof (*e.g.* pyrophosphate, PPi), and phosphorylated biomolecules are essential components in living organisms, playing important roles in a variety of fundamental biological processes.[1] Phosphates form the backbone of DNA and RNA, linking together the sugars and nucleobases to form the basic unit of our genetic material. Phosphate and its derivatives, *e.g.* adenosine triphosphate (ATP), are crucial for energy storage and transduction in biological systems. Phospholipids are a major component of cell membranes, providing structural barriers, in addition to playing roles in membrane transport and cell signalling processes.[2] Phosphorylation and dephosphorylation of proteins provides a key signalling mechanism within cells.[3] Pyrophosphate (PPi)—the dimeric form of phosphate—is generated in numerous biological reactions, including the

Monographs in Supramolecular Chemistry No. 34
Fluorescent Chemosensors
Edited by Luling Wu, Adam C. Sedgwick, Xiao-Peng He and Tony D. James

Published by the Royal Society of Chemistry, www.rsc.org

cellular hydrolysis of ATP and the polymerisation of DNA, and has also been indicated as a biomarker for a variety of diseases, including bone disease and cancer.[4,5]

Inorganic phosphate (Pi) itself plays a key role in bone mineralisation, with approximately 85% of total body inorganic Pi being associated with bone.[6] Only approximately 1% of the inorganic phosphate in the body is present in extracellular fluids,[7] but high or low levels of blood Pi result in serious medical conditions. Excess Pi is normally cleared through the kidneys, but for patients with advanced kidney disease, inability to control blood Pi levels is of serious concern[8] and hyperphosphatemia (abnormally high serum Pi levels) leads to a variety of complications, including soft tissue calcification, cardiovascular issues, and increased morbidity and mortality.[9] In contrast, prolonged phosphate deficiency leads to hypophosphatemia, resulting in impaired bone mineralisation, increased membrane rigidity in erythrocytes, muscle weakness, and dysfunction of the central nervous system.[10] Intracellular levels of inorganic Pi are higher than those in extracellular fluids and vary depending on a range of factors (tissue type, subcellular location, pH, hormones), typically falling in the range of 0.5–5 mM.[11] However, the intracellular levels of phosphorylated biomolecules (*e.g.* ATP, phosphocreatine) can be significantly higher than those of inorganic Pi.

In addition to their importance in living systems, phosphates are of significance in industry and the environment, primarily in the form of fertilisers. Phosphate fertilisers are key to the world's food supply, but their overuse results in significant damage to the environment. Run-off from agricultural use of phosphorus fertilisers is linked to high phosphate levels in waterways, which in turn can result in eutrophication and toxic algal blooms.[12,13]

The importance of phosphates in both the environment and biology has led to widespread interest in the development of methods for sensing phosphates and phosphorylated biomolecules. Initial work in this area commenced as early as 1914, with the development of a colorimetric method for the estimation of inorganic phosphorous in biological material.[14] This method forms the basis for currently used phosphate assays that rely on the formation of a blue phosphomolybdate complex, the absorbance of which is proportional to the level of inorganic phosphate present.[15] However, these assays are time consuming, tedious to perform, and not convenient, even in environmental settings. Alternative methods for sensing of phosphates frequently rely on enzymatic biosensing, but these often suffer from poor stability or require complicated sample preparation.[16]

Synthetic molecular receptors for the detection of phosphate and phosphorylated biomolecules have therefore been extensively investigated.[17–40] Depending on their structures, these can provide readouts through a variety of spectroscopic techniques. Fluorescent chemosensors are frequently favoured because they have a lower limit of detection than those that provide a readout *via* NMR or UV-visible spectroscopy and can also be used *in cellulo* to provide spatiotemporal information, *e.g. via* fluorescence imaging techniques.

The development of methods to selectively sense phosphate and its derivatives is challenging for several reasons. Dihydrogenphosphate is highly hydrophilic with a high Gibbs free energy of hydration (-465 kJ mol^{-1})[41] in comparison to other monovalent anions (see Table 10.1) and this increases significantly as the charge on the anion increases (the hydration energies for HPO_4^{2-} and PO_4^{3-} are -1089 and -2773 kJ mol^{-1}, respectively[42]); this makes selective binding to phosphate in water difficult. Less hydrophilic anions, such as chloride and bicarbonate, are frequent interferants in both biological and environmental settings and are often present at higher concentrations than phosphates. The protonation states of phosphates are pH dependent, *e.g.* the pK_a values of H_3PO_4 are 2.12, 7.20 and 12.36,[43] so at neutral pH both $H_2PO_4^-$ and HPO_4^{2-} are present in aqueous solution in a close to 1 : 1 ratio (see Figure 10.1). Finally, the wide variety of phosphorylated species present in biological systems (see Figure 10.2), together with the differing relative concentrations of each of these at any given location, poses one of the most significant challenges to the development of chemosensors for phosphates for biological applications—the ability to discriminate between these species.

Fluorescent chemosensors for phosphates use a wide variety of sensing mechanisms (*e.g.* photoinduced electron transfer (PeT), Förster resonance energy transfer (FRET), intramolecular charge transfer (ICT), twisted intramolecular charge transfer (TICT), excited-state intramolecular proton transfer (ESIPT), aggregation-induced emission (AIE), indicator displacement assays (IDA), intramolecular indicator displacement assays (IIDA)). They also employ a variety of different functionalities for binding to the phosphate ion, including charge–charge interactions, hydrogen-bonding interactions, and metal–ligand interactions. The examples below are classified by binding interaction and have been selected to illustrate the different sensing mechanisms that can be employed, and how these can be used to discriminate between ions, with a focus on chemosensors where selectivity for phosphates over other anions (halides, carboxylates, *etc.*) has been reported.

10.2 Phosphate-binding Interactions

10.2.1 Charge–Charge Interactions

Early work in the development of chemosensors for phosphate species was influenced by the discovery that polyalkylammonium species, including natural polyamines such as spermine and spermidine that are protonated at neutral pH, bind to adenosine nucleotides (AMP, ADP, and ATP).[44] Thus followed the development of a wide range of cyclic and acyclic polyammonium receptors for phosphate species. Binding of phosphate species was typically determined by careful pH or NMR titrations, and predominantly evaluated in aqueous solution. For the most part, only binding affinities of these receptors towards nucleotides and inorganic phosphate species were examined, and comparisons with other anions that could potentially interact competitively—such as carboxylates—were not reported.

Table 10.1 Gibbs free energies of hydration for monovalent anions.[41]

	ClO_4^-	I^-	CN^-	NO_3^-	Br^-	HSO_4^-	HCO_3^-	Cl^-	AcO^-	F^-	$H_2PO_4^-$
$\Delta G^{\circ}_{\text{hydration}}\ (\text{kJ mol}^{-1})$	−205	−275	−295	−300	−315	−330	−335	−340	−365	−465	−465

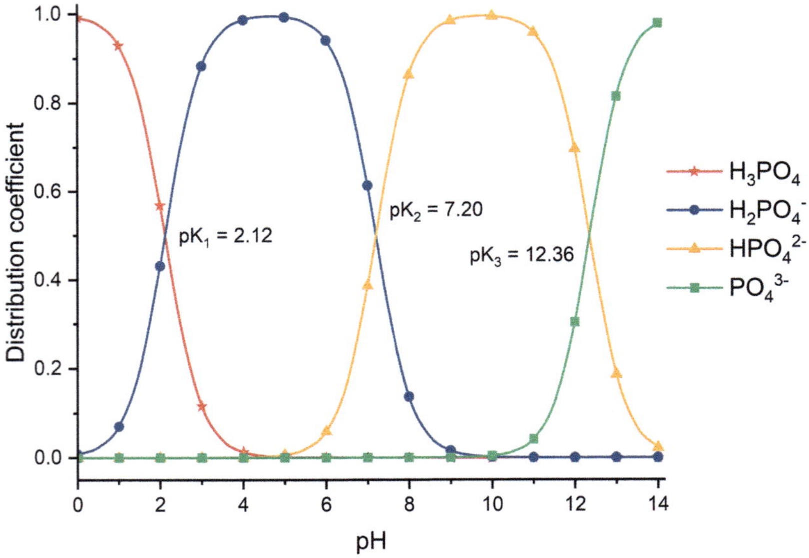

Figure 10.1 Speciation plot for phosphate.

The first receptor to provide a fluorescence response to the binding of phosphate species (or indeed to any anion) was developed by Czarnik and co-workers in 1989.[45] Chemosensor **1** utilized the affinity of polyammonium groups for phosphates, coupled with an anthracene fluorophore to provide a response to binding. A "receptor–spacer–fluorophore" design strategy was used to provide a fluorescent response to binding. PeT, which quenches the anthracene fluorescence, was inhibited in the presence of phosphate species, leading to an increase in fluorescence intensity. The selectivity of this chemosensor and of most related polyammonium-based chemosensors was found to correlate most strongly with the overall charge of the phosphate species: the higher the charge on the phosphate species the greater the affinity. For example, chemosensor **1** was found to bind HPO_4^{2-} with a $\log K_a$ of 0.82, whereas affinity for the trivalent anion ATP was determined to be more than three orders of magnitude greater ($\log K_a$ 4.2). Discrimination based on the size and shape of phosphate species was also possible, and these properties were exploited as chemosensor designs became more advanced and as understanding of the factors influencing selectivity increased.

Given that polyammonium receptors rely on charge–charge interactions (in addition to hydrogen bonds) between an anionic phosphate species and a protonated amine group, they are also highly sensitive to changes in pH. Chemosensor **1** was found to be most effective at pH 6, at which point the chemosensor is triprotonated and $H_2PO_4^-$ is the predominant

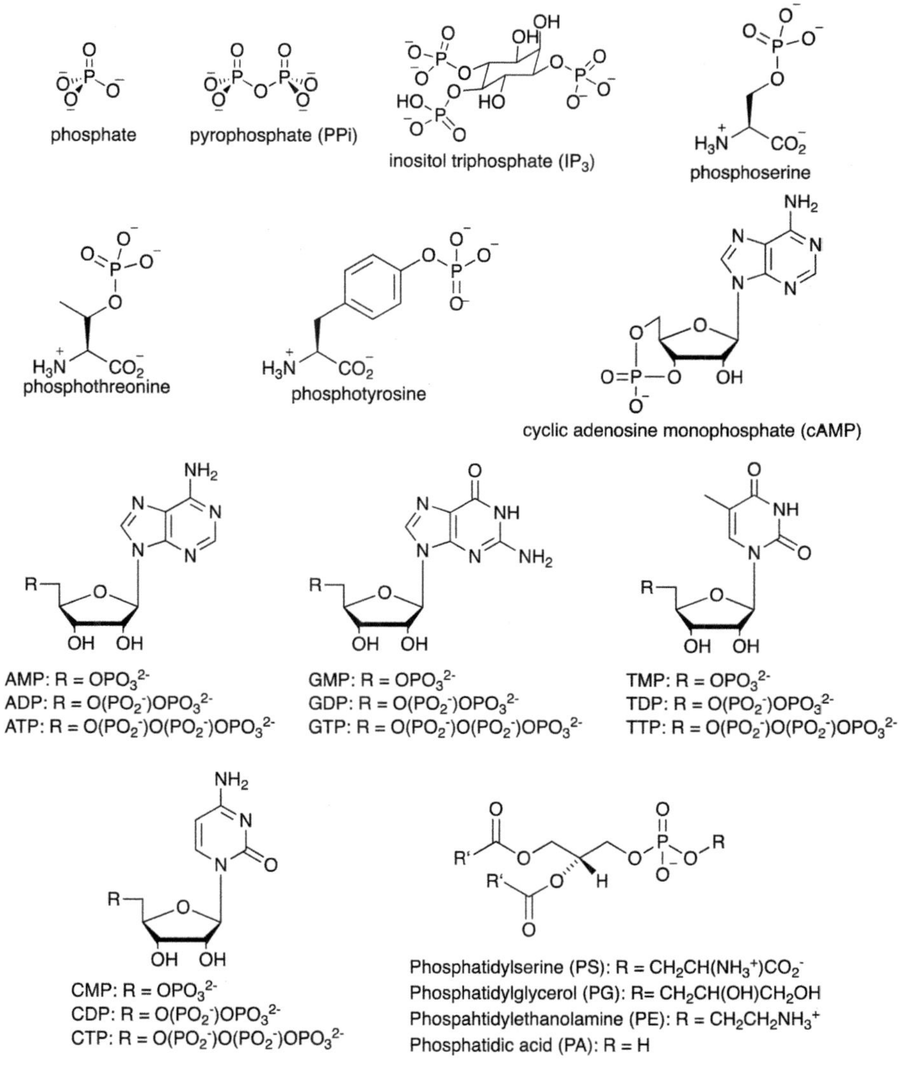

Figure 10.2 Structures of common biological phosphates.

form of inorganic phosphate. A successor compound, double-armed polyamine chemosensor **2** exhibited higher pK_a values and hence was able to be used at pH 7 with each arm triprotonated.[46] This chemosensor exhibited 2200 times greater selectivity for PPi over Pi (log K_a PPi: 5.54; log K_a Pi: 2.20). Although both species are present as dianions at this pH, the structure of PPi is presumed to be better suited to fit between the two polyammonium arms than is Pi.

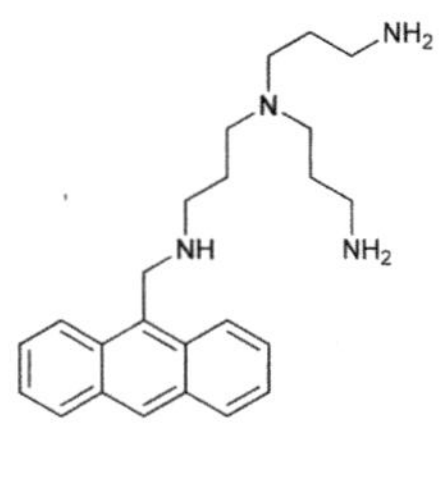

1

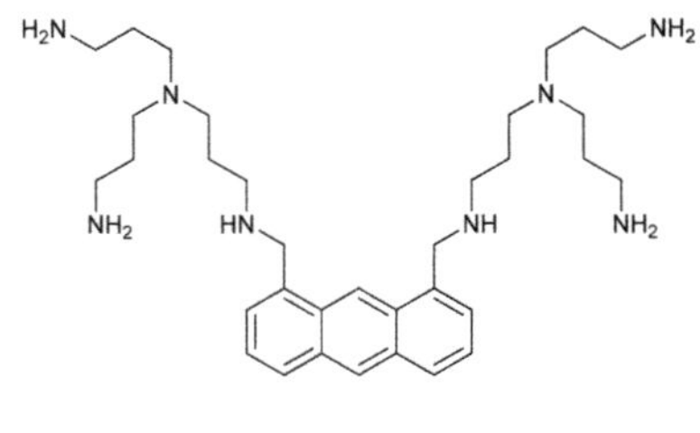

2

The capability of PPi to form such bridging interactions between two charged recognition groups was leveraged by Teramae and co-workers in 1999 to generate what represents the simplest fluorescent chemosensor with selectivity among phosphate species.[47] Guanidinium-based chemosensor **3** binds PPi in a 2 : 1 host : guest stoichiometry in methanol solution. This binding arrangement brings the pyrene groups of two molecules of **3** into proximity, resulting in an increase in excimer emission from the proximal pyrene units. Despite the simplicity of this chemosensor, the response to PPi was significantly greater than the responses to other inorganic phosphate species tested.

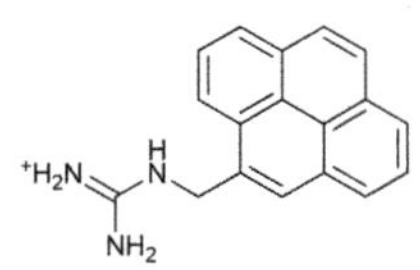

3

Many of the earliest fluorescent chemosensors for phosphate species, including more complex macrocyclic and cage-like polyamine architectures, such as those shown in PeT chemosensors **4–6**, were developed by Lehn and co-workers in the 1990s.[48–51] Lehn's work established the first examples for recognition of nucleotides, and established patterns and challenges in binding selectivity that are still encountered today. Binding affinities of nucleotides, determined at pH 6, were found to increase with increasing negative charge (or with additional phosphate groups).[49] Purine nucleobases, which contain a more extended π-system than their pyrimidine counterparts, were also found to exhibit higher binding affinities due to π–π stacking interactions with the extended π-systems of the fluorophores of these chemosensors. Cage-like chemosensor **5** displayed lower nucleotide-binding affinities than its macrocyclic counterpart with only two acridine units, indicating that the species likely bound partially outside the chemosensor cavity.[50] The introduction of a π-system that is further extended, such as in chemosensor **6**, allowed recognition of two paired nucleotides in a 1 : 2 host : guest arrangement.[51]

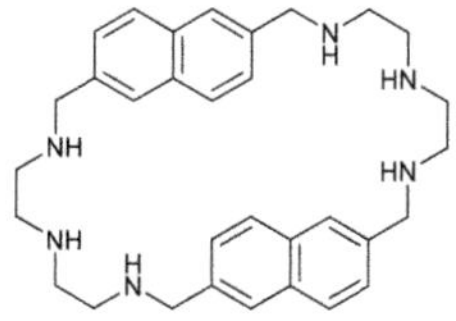

4

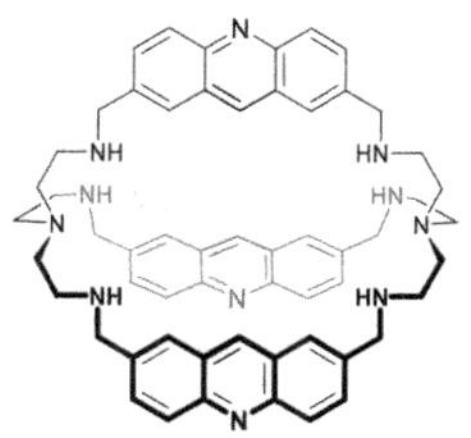

5

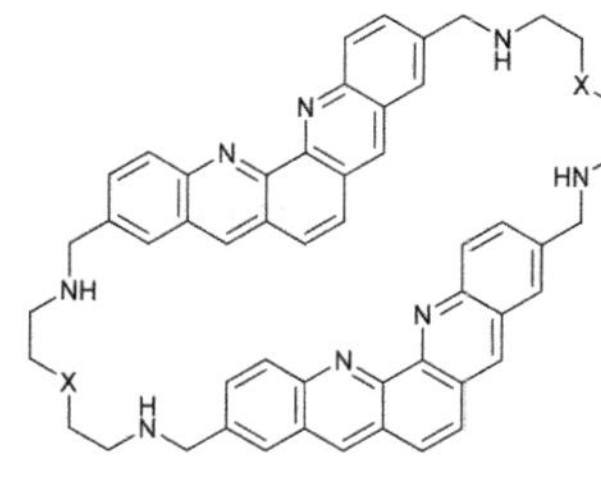

X = NH, O, $CH_2NH^+(CH_3)CH_2$

6

In another key advance in the understanding of nucleotide recognition, the simple phenanthridine receptor **7** investigated by Piantanida and co-workers in 2005 provided curious results, which indicated that increasing the charge on a chemosensor does not necessarily improve affinity.[52] Binding of nucleotides to the tetra-protonated receptor at pH 5 was compared with binding to the di-protonated receptor at pH 7. Binding affinities of nucleotides increased with increasing negative charge; however, the nucleotides were found to bind with greater affinities at pH 7 than at pH 5. Differentiation between nucleobases was also greatest at pH 7. This result was attributed to significant contributions of π–π stacking to binding affinity, which are disrupted when the phenanthridine group is protonated. In results mirroring those obtained by Lehn, purine nucleotides were found to display the strongest binding regardless of pH.

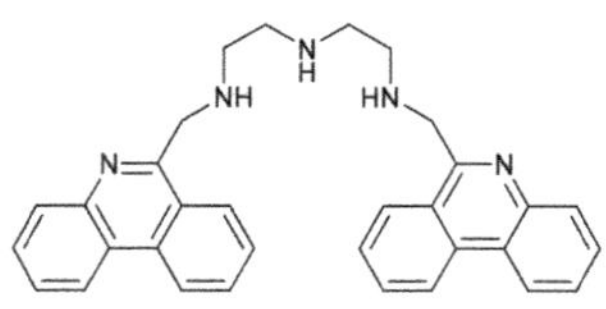

7

Efforts to reduce the pH dependency of polyamine chemosensors for phosphate species have led to the development of chemosensors with a fixed charge. Although some chemosensors employing quaternary amine groups have been developed, those incorporating imidazoliums and related groups have encountered greater success. This difference may be attributed to the screened nature of the charge in quaternary ammonium groups, together with the ability of imidazolium groups to form additional C–H hydrogen-bonding interactions.

Receptor **8**, which possesses a fixed 4+ charge due to both quaternary amine and imidazolium groups was investigated by Yoon and co-workers in 2004 in aqueous HEPES buffer at biological pH (7.4) (see Figure 10.3).[53] This receptor binds nucleotides through a combination of charge–charge, charge-assisted C–H hydrogen bonding, and edge-to-face π–π interactions. Curiously, differential fluorometric responses were observed for adenine and guanine nucleotides, attributed to differences in the strength of the H–π interactions (CH

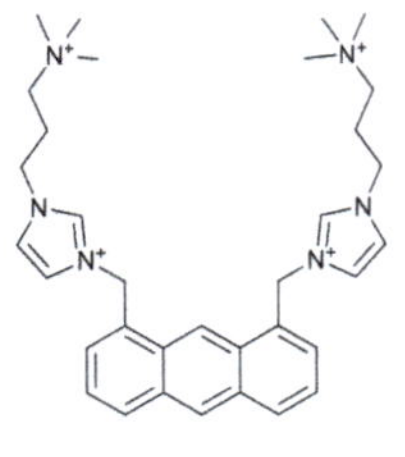

8

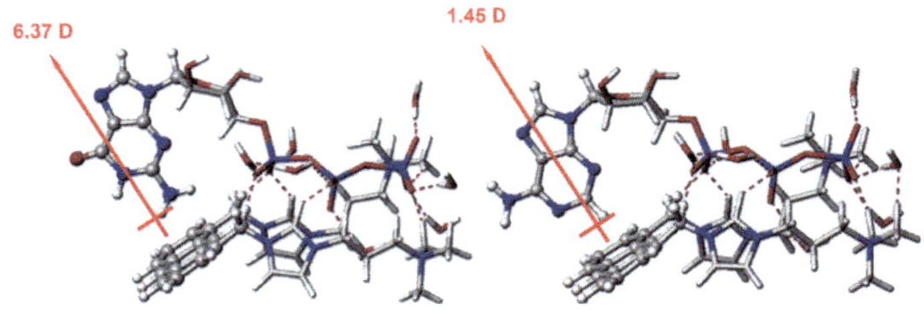

8-GTP 8-ATP

Figure 10.3 Computed geometries of **8**-ATP and **8**-GTP complexes. Adapted from ref. 53 with permission from American Chemical Society, Copyright 2004.

for adenine, NH for guanine) between the nucleobases and the anthracene moiety (see Figure 10.3), together with differences in the dipole moments of the nucleobases. A chelation-enhanced fluorescence (CHEF), *i.e.* turn-on, response to adenine nucleotides was attributed to the low dipole moment of adenine perpendicular to the anthracene ring (1.45 D), which is lower than that of the displaced solvent water (~3 D) and hence enhances the fluorescence transition. The corresponding dipole moment of guanine (6.37 D) is greater than that of water, and hence a chelation-enhanced quenching (CHEQ) response occurs.

Pyrene chemosensors **9** and **10**, also developed by Yoon and co-workers, represent the first examples of imidazolium-based structures exhibiting a fluorescent excimer response.[54,55] Chemosensor **10** was the first ATP-selective ratiometric fluorescent chemosensor to be developed.[55] Interaction with ATP led to an enhancement in monomer emission and quenching of excimer emission, whereas all other nucleotides tested did not affect monomer emission, and Pi and PPi did not influence either. It was postulated that adenine sits between the pyrene units, but other nucleobases interact with only one side of the chemosensor. Chemosensor **10** was successfully used to monitor the activity of apyrase, an enzyme that catalyzes the loss of a phosphate group from ATP and ADP, in aqueous buffer (20 mM HEPES, pH 7.4).

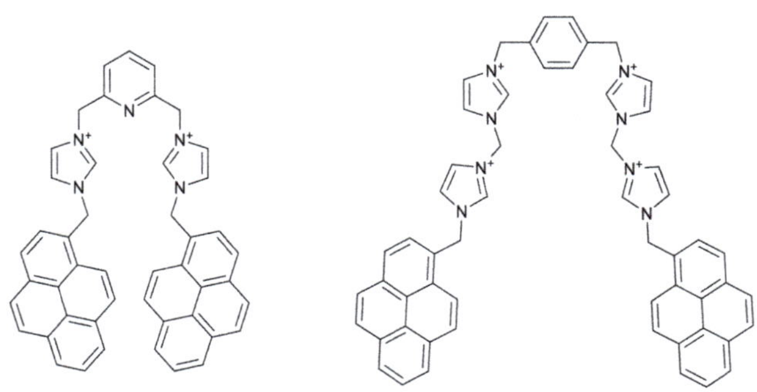

9 10

Use of an IDA approach[56] has also proven to be a fruitful strategy for fluorescent detection of phosphate species. Unlike the examples previously mentioned in this chapter, in which the reporter fluorophore is an integral component of the receptor group, the IDA approach allows the use of unattached fluorophores complexed with fluorescence-inactive receptors. This approach brings a range of benefits, including greater synthetic accessibility, and the ability to easily mix and match receptors and fluorophores to tune the fluorescent response.

Ramaiah and co-workers prepared the macrocyclic receptor **11**, which contains fluorescent anthracene and viologen groups.[57] However, the receptor displayed weak fluorescence ($\Phi_F = 0.0007$) and negligible changes in emission when combined with ATP, despite changes in the absorption spectrum that indicated formation of a host:guest complex. This result was hypothesized to be due to PeT quenching from the excited anthracenes to the viologen units. However, use of an IDA approach rendered the receptor useful as a sensitive fluorescent chemosensor for GTP.[58] Complexation of pyrene fluorophore **12** with receptor **11** quenched the fluorescence of the pyrene group. An increase in fluorescence intensity was then observed when GTP was introduced to the complex and displaced fluorophore **12**. Selective discrimination of GTP over other nucleotides, including ATP, was attributed to well-matched electrostatic and π–π stacking interactions of the guanine nucleobase with the receptor.

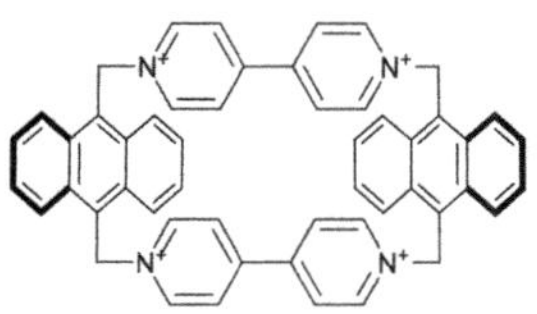

11

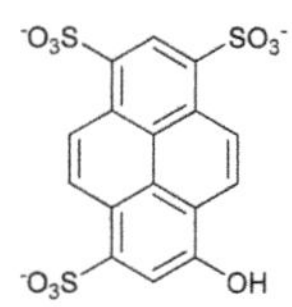

12

In work by Bitter and co-workers, the IDA approach was used to obtain a fluorescence response to nucleotides with receptor pillar[6]arene (**13**), which lacks intrinsic fluorescence properties.[59] Highly charged pillar[6]arene was complexed with solvatochromic fluorophore dapoxyl sodium sulfonate (**14**, DSS). DSS exhibits a turn-on in fluorescence when complexed to **13**, relative to the emission intensity displayed when it is free and unbound in aqueous solution. Introduction of purine nucleotides AMP, ADP, ATP, or GTP was found to displace DSS from pillar[6]arene **13**, resulting in a diminished fluorescence intensity. Selectivity between nucleotides was controlled by charge (K_a ATP > ADP ≫ AMP) and size complementarity with the pillar[6]arene binding pocket (K_a ATP > GTP).

The strengths of an IDA approach to phosphate recognition are further illustrated by work by Guo and co-workers with the guanidinium-based calix[6]arene **15**.[60] In this case, an increase in fluorescence intensity occurred when the reporter fluorophore, fluorescein (**16**), was displaced by a competitive guest molecule. The greatest fluorescence response occurred when the target of this study, the cancer biomarker lysophosphatidic acid, was introduced; however, a similar response to ATP was also noted. The introduction of an additional indicator (aluminium(III) phthalocyanine chloride tetrasulfonic acid, $AlPcS_4$) and analogous calix[4]arene receptor introduced four possible receptor/indicator combinations. Principal component analysis of the fluorescence responses of this small array comprised of the four receptor/indicator pairs resulted in effective discrimination between lysophosphatidic acid, ATP, ADP, and other phosphate species (see reviews for further explanation of array-based sensing).[61,62]

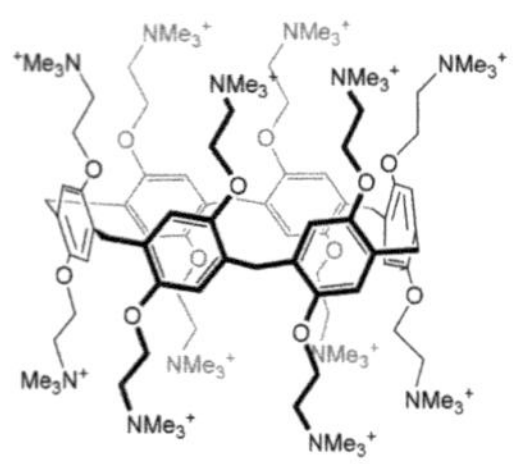

13

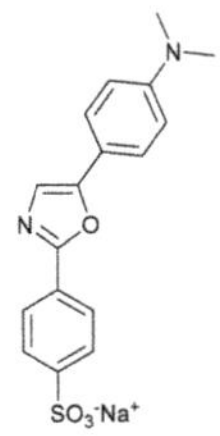

14 (DSS)

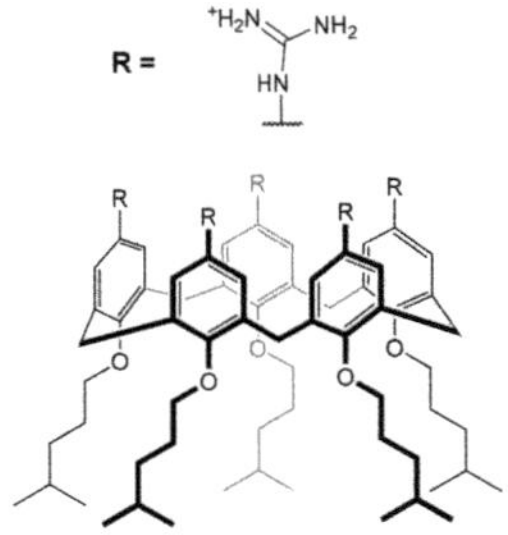

15

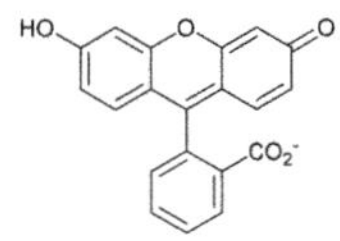

16 (Fluorescein)

Just as natural polyamines inspired the design of the original receptors for phosphate species, other naturally occurring phosphate-binding groups have also proven to be a rich source of inspiration for artificial fluorescent phosphate chemosensors. Anslyn and co-workers took the non-selective receptor **17** and sought to adapt it into a selective chemosensor through the incorporation of peptide arms.[63] Subsequent analysis of over 3000 X-ray crystal structures of phosphate-binding proteins by Hirsch, Fischer, and Diederich has revealed that greater than two-thirds of these proteins rely on charge–charge and hydrogen-bonding interactions from peptide side chains and the amide backbone.[64] In Anslyn's work, combinatorial synthesis was used to generate a library of up to 4913 members with tripeptide arms one of which was appended with either a coumarin or fluorescein fluorophore (see Figure 10.4). Analysis of the library identified a member with Ser-Tyr-Ser arms, which gave a selective turn-on in fluorescence in the presence of ATP and no response to either GTP or AMP. It was hypothesized that binding of the phosphate groups of ATP was predominantly due to interactions with the charged guanidinium groups, while the additional peptide groups provided selectivity for the adenine of ATP over the guanine of GTP through additional hydrogen bonding, π–π stacking, and steric interactions. More recently developed examples, such as Schmuck's peptidic fluorescent chemosensor **18**,[65] have also proven the utility of combining charged binding units with peptide components that provide highly directional interactions and a greater three-dimensional character than many artificial receptors.

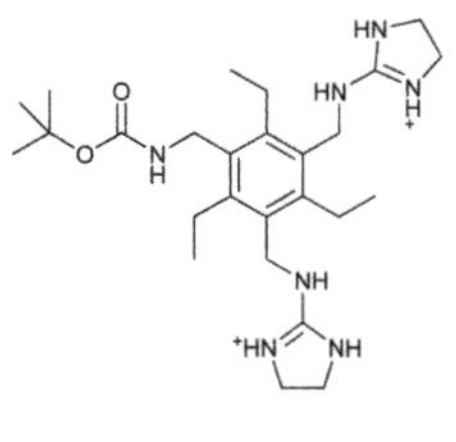

17

Combinatorial library

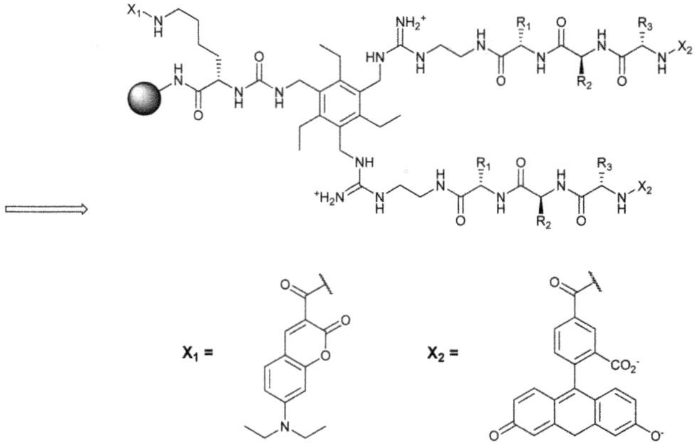

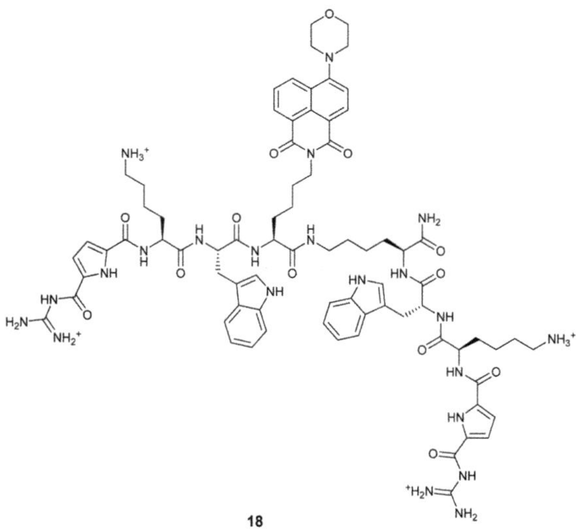

18

Figure 10.4 Structures of peptidic chemosensors.

10.2.2 Hydrogen-bonding Interactions

The use of hydrogen-bonding interactions in chemosensors for phosphates offers several potential advantages over charge–charge interactions. Hydrogen bonds are highly directional, which may lead to greater shape selectivity compared to (non-directional) charge–charge interactions. Dual hydrogen-bond donors (such as ureas, thioureas, and squaramides) and other binding units may offer highly rigid, preorganized structures that are well suited to oxoanion recognition. Hydrogen bonds are also less sensitive to pH than ammonium and guanidinium groups, which are reliant on protonation for the most effective binding to occur.

Additionally, the use of fluorophores that contain hydrogen-bond donors introduces the possibility of a greater range of possible fluorescence response mechanisms than is accessible with charge–charge complexes. Strategies already encountered in this chapter, such as PeT and excimer emission, can also be used in hydrogen-bonding receptors. Furthermore, mechanisms such as ICT and ESIPT—which are either not possible with systems that rely solely on charge–charge interactions or do not lead to significant changes in the emission spectra—can be readily exploited in the generation of hydrogen-bonding receptors for phosphate species.

Despite these advantages, several additional challenges are faced with hydrogen-bonding chemosensors compared to systems reliant on charge–charge interactions. Hydrogen bonds are typically weaker than charge–charge interactions; hence, chemosensors must include a larger number of binding units or be highly preorganized for comparable binding affinities to be obtained. Many hydrogen-bond donor groups also display poor solubility in aqueous media, and so additional solubilizing groups must be used to develop chemosensors appropriate for use in biological and environmental contexts. As a consequence of these two challenges, the majority of hydrogen-bonding chemosensors have only been evaluated in organic solvents or in mixed organic/aqueous solution. These media allow solubilization of the chemosensors, and are less competitive than aqueous solutions, in which weak binding may not be observable.

In contrast to the chemosensors reliant on charge–charge interactions that were outlined in Section 10.2.1, hydrogen-bonding chemosensors have typically been evaluated for their responses to simple inorganic anions, such as phosphate, pyrophosphate, oxoanions (*e.g.* sulfate, carbonate), and halides. The development of hydrogen-bonding chemosensors for larger biological anions, such as nucleotides, is much less common; this is, at least in part, due to the poor solubility of many of these species in organic solvents. Evaluation of selectivity between phosphate species is underexplored.

One of the earliest, and simplest, hydrogen-bonding fluorescent phosphate chemosensors is the thiourea **19** developed by Teramae and co-workers in 1998.[66] This ratiometric chemosensor responds to $H_2PO_4^-$ with an increase in excimer emission and decrease in monomer emission from the pyrene fluorophore. However, unlike the analogous guanidinium compound, chemosensor **3** (see Section 10.2.1), which was utilized in methanol solution,[47] analysis of thiourea **19** was limited to less competitive acetonitrile solutions. Binding of pyrophosphate (the analyte for which chemosensor **3** gave a selective response) was not evaluated, however acetate was found to have a higher binding affinity and gave a larger fluorescence response than $H_2PO_4^-$. This highlights one of the key issues in the development of phosphate chemosensors that operate through hydrogen-bonding interactions: namely, the difficulty in developing chemosensors with high selectivity.

Other simple urea- and thiourea-based chemosensors, such as Gunnlaugsson's PeT chemosensors **20–22**, exhibit similar properties.[67,68] Evaluation of the binding of these chemosensors in DMSO solution indicated that fluoride, acetate, and $H_2PO_4^-$ bind with the highest affinities of the anions tested. This trend correlates loosely with the basicity of the anions: less basic anions, such as chloride and bromide, gave weak responses. These compounds also demonstrated that hydrogen-bonding chemosensors may be finely tuned using different substituents. Increasing the polarization of the thiourea N–H groups with an electron-withdrawing 4-(trifluoromethyl)phenyl group (**22a**) led to increased binding affinities for all anions relative to the phenyl compound (**21a**). While the aryl thioureas **21a** and **22a** exhibited higher binding affinities for acetate than $H_2PO_4^-$, the methyl compound **20** displayed the opposite selectivity (**22a:** AcO^- $\log K_a = 2.55$, $H_2PO_4^-$ $\log K_a = 2.05$; **20:** AcO^- $\log K_a = 1.75$, $H_2PO_4^-$ $\log K_a = 2.05$).

Addition of a second thiourea-binding group, such as in PeT chemosensor **23**, led to the formation of moderate-affinity 1 : 1 complexes with pyrophosphate in DMSO.[69,70] However, affinities for 1 : 1 complexes with dicarboxylates succinate and glutarate were higher, and 1 : 2 host : guest complexes with fluoride and acetate were also observed.

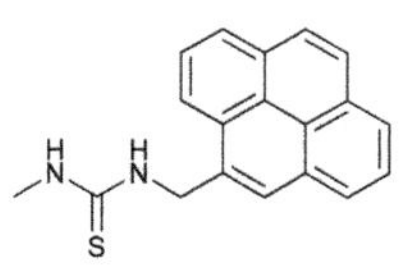

19

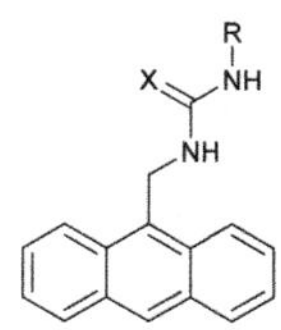

20 R = Me, X = S
21a R = Ph, X = S
21b R = Ph, X = O
22a R = 4-CF$_3$-Ph, X = S
22b R = 4-CF$_3$-Ph, X = O

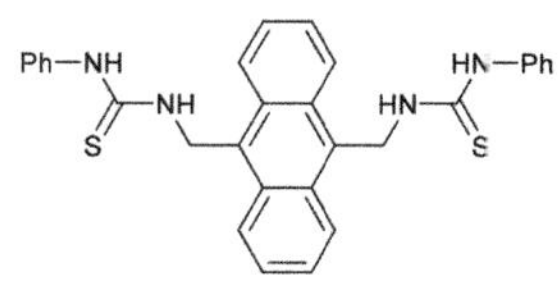

23

In order to improve selectivity for tetrahedral phosphate anions over the geometrically distinct acetate (planar) and fluoride (spherical) anions, a number of tripodal scaffolds have been developed. Tri-urea **24**, developed by Wu and co-workers, displayed an increase in fluorescence in the presence of increasing quantities of $H_2PO_4^-$ in dimethylformamide (DMF).[71] Although this chemosensor contains hydrogen-bond donors in direct conjugation with the fluorescent naphthalene group, no significant shifts in emission wavelength were noted, suggesting that the fluorescence response is largely due to inhibition of PeT, rather than ICT. Selectivity over spherical halide anions and tetrahedral HSO_4^- was observed, but acetate gave a similar response to that observed for $H_2PO_4^-$.

Similarly, tripodal chemosensors **25** and **26** developed by Suzuki and co-workers exhibited selectivity for phosphate over chloride in acetonitrile.[72] Pyrene chemosensor **25** displayed a large increase in excimer emission in the presence of $H_2PO_4^-$, and anthracene chemosensor **26** displayed selective fluorescence quenching. In this case, the binding affinities and fluorescence responses of the tripodal chemosensors were greater for $H_2PO_4^-$ than for acetate. This result is in contrast to simple model systems with only one thiourea, which exhibited a preference for acetate, thus highlighting the importance of the geometrical complementarity between chemosensor and anion.

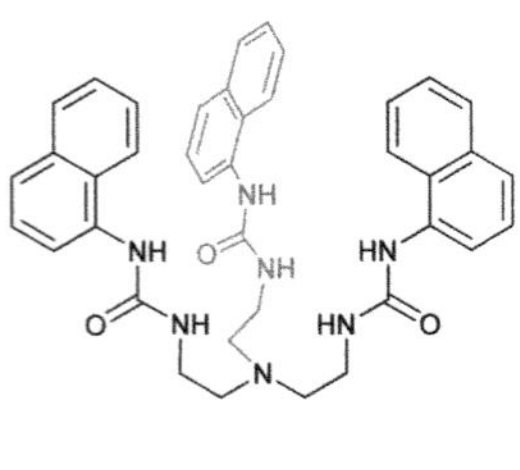

24

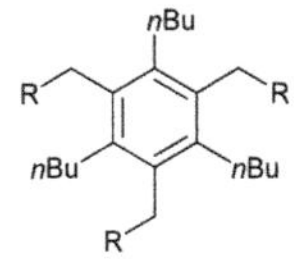

25: R = 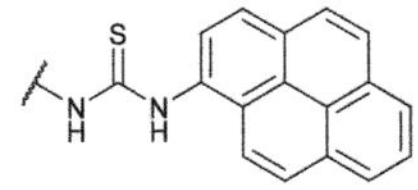

26: R = 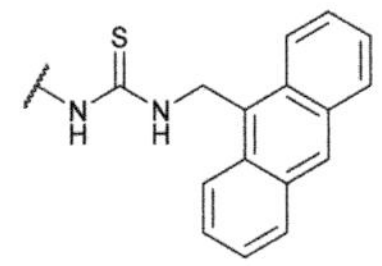

As mentioned previously, the use of hydrogen-bonding interactions in fluorescent phosphate chemosensors also introduces the possibility of a broader range of fluorescence responses *via* alternative mechanisms. Thiourea **27**, developed by Jiang and co-workers, contains a donor–acceptor dimethylaminobenzamide fluorophore.[73,74] This fluorophore displays dual fluorescence, which has been attributed to locally excited (LE) and charge transfer (CT) states with distinct emission wavelengths.[74] In the absence of anions, emission from the CT state predominates. This emission band is quenched and the LE emission is enhanced upon the addition of anions to give a ratiometric response. It was hypothesized that these spectral changes are due to anion-induced rotation around the N–N bond of the chemosensor and disruption of intramolecular hydrogen-bonding interactions in the free receptor. Consistent with the results of other simple thiourea-based chemosensors, the chemosensors of this class displayed strong binding to the basic anions acetate, fluoride, and $H_2PO_4^-$, and a preference for these over other anions.

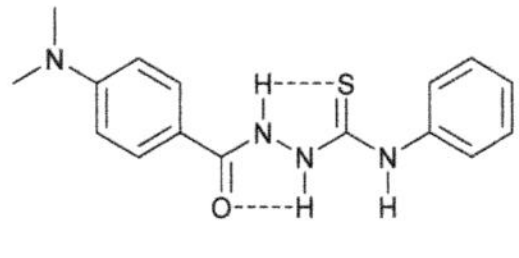

27

Another class of phosphate chemosensors are those that rely on an ESIPT mechanism. Gale's simple acridone chemosensor **28** was shown by ^1H NMR experiments to deprotonate with the addition of a single equivalent of basic anions, such as $H_2PO_4^-$ in DMSO solution, followed by presumed tautomerization of the acridone and subsequent binding interactions upon addition of further equivalents of anion (see Scheme 10.1).[75] In a less basic solvent mixture (96 : 4 acetonitrile : DMSO), chemosensor **28** provided a distinct fluorescence response to $H_2PO_4^-$ anions. Selective quenching of the

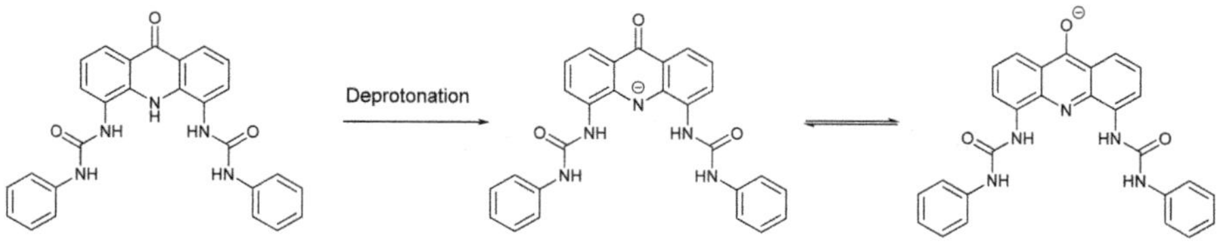

28

Scheme 10.1 Deprotonation states of acridone-based chemosensor **28**.

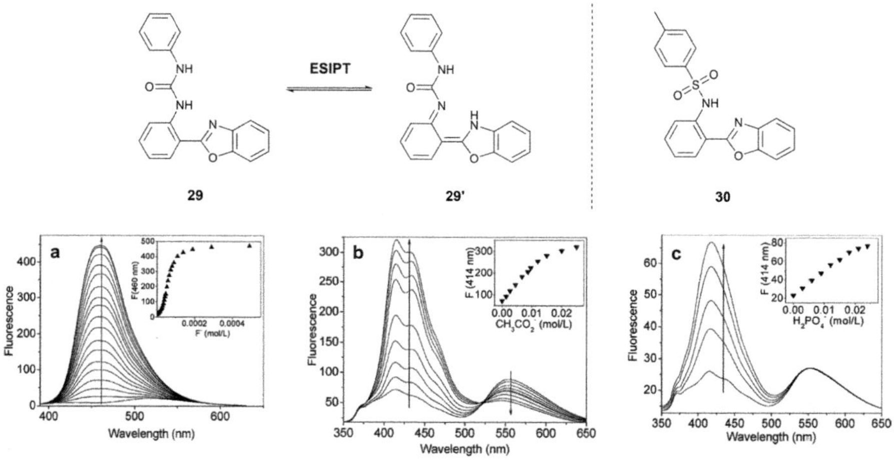

Figure 10.5 ESIPT urea chemosensors **29** and **30**. Changes in fluorescence spectrum (λ_{ex} = 336 nm) of chemosensor **29** (10 μM in DMSO) in the presence of: (a) fluoride; (b) acetate; and (c) dihydrogen phosphate. Reproduced from ref. 76 with permission from American Chemical Society, Copyright 2007.

chemosensor fluorescence was observed for $H_2PO_4^-$ (and to a lesser extent chloride), whereas addition of hydroxide or acetate led to a minor fluorescence enhancement.

ESIPT chemosensor **29**, developed by Peng and co-workers, displayed distinct fluorescence responses depending on the interacting anion (see Figure 10.5).[76] The chemosensor possesses two tautomeric forms, **29** and **29′**, which are interchangeable by ESIPT, and differing fluorescence emissions are observed for each. Chemosensor **29** displays similar binding affinities for fluoride, acetate, and $H_2PO_4^-$, but differing fluorescence responses in the presence of each anion facilitated differentiation between the species. Addition of fluoride to chemosensor **29** in DMSO solution led to deprotonation, and a prominent fluorescence enhancement. Addition of acetate suppressed formation of the tautomer form **29′** and hence the intensity of the corresponding fluorescence band was reduced, accompanied by fluorescence enhancement of the urea form (**29**). Finally, $H_2PO_4^-$ induced spectral changes that were consistent with stabilization of the urea form of the chemosensor, but the fluorescence intensity of the tautomer was unchanged. It was hypothesized that binding to $H_2PO_4^-$ does not suppress the ESIPT process, but rather increases the rigidity of the chemosensor and hence reduces non-radiative energy losses from the urea form. A more acidic chemosensor, sulfonamide **30**, was deprotonated by fluoride, acetate, and $H_2PO_4^-$, and consequently lacked a selective response.

More recently, many researchers have focused efforts on combining the benefits of hydrogen-bonding interactions (relative pH insensitivity, directional interactions) with those of charge–charge interactions (strong binding, high water solubility) by generating fluorescent phosphate chemosensors containing both groups. These chemosensors are often soluble in aqueous media and are suitable for the recognition of nucleotides as well as inorganic phosphate species.

Spirolactam rhodamine derivative **31**, published by Li and co-workers in 2013, utilizes a mixture of charge–charge and hydrogen-bonding interactions to bind the phosphate groups of ATP, and also responds by a fluorescence mechanism that has not yet been discussed in this chapter.[77] This chemosensor equilibrates between a non-fluorescent closed form, and an open form with fluorescence emission at 580 nm. In an important step forward for hydrogen-bonding chemosensors, rhodamine **31** is water soluble and able to function as a fluorescent chemosensor for ATP in biologically relevant conditions. Interaction of the chemosensor with ATP, *via* a combination of hydrogen-bonding, charge–charge, and π–π stacking interactions stabilizes the ring-opened form. A selective turn-on response was obtained for ATP in aqueous buffer at pH 7.2. No response was observed for other adenine nucleotides, inorganic phosphate species, or oxoanions. Li and co-workers further developed rhodamine **31** into a ratiometric FRET chemosensor, naphthalimide **32**.[78] Naphthalimide **32** was evaluated against a greater range of nucleotides and cations—including the triphosphates GTP, CTP, UTP, and TTP—and exhibited a selective colorimetric and fluorescence response for ATP. The chemosensor was successfully used to monitor activity of apyrase, an enzyme that converts ATP into AMP and inorganic phosphate, in HeLa cells.

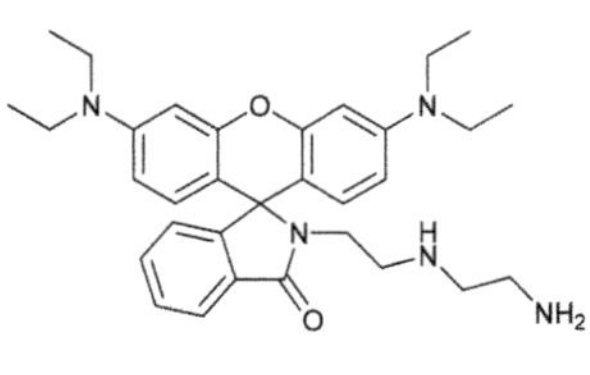

31

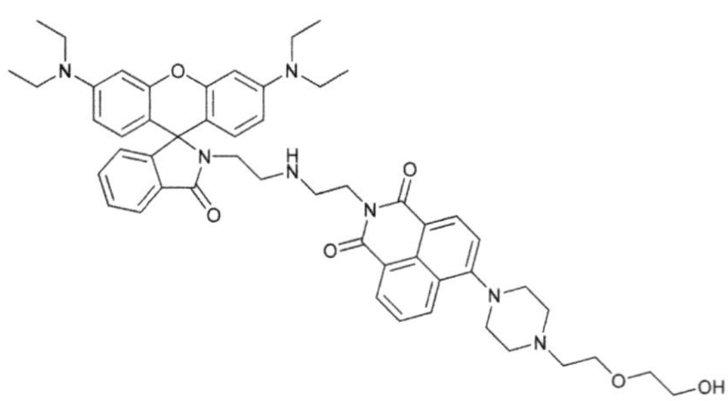

32

Tripodal chemosensor **33**, developed by Ghosh and co-workers in 2015, combines charged pyridinium and ammonium groups with amide hydrogen-bond donors for the recognition of ATP.[79] Chemosensor **33** exhibited an increase in excimer emission from the naphthalene groups in acetonitrile in the presence of $H_2PO_4^-$ but not in the presence of halides, oxoanions, or pyrophosphate. In the more competitive solvent medium of 1:1 water:acetonitrile (10 mM HEPES buffer, pH = 6.5) a broad emission at 390 nm was observed when ATP or ADP were added. This emission is attributed to the formation of an exciplex between the adenine unit of the nucleotide and naphthalene group of the receptor. Chemosensor **33** was successfully used to observe ATP levels in HEK293T cells.

More recently, Kataev and co-workers presented cryptand **34**, which combines ammonium and amide binding groups.[80] This PeT chemosensor provided an increase in fluorescence intensity in the presence of $H_2PO_4^-$ and oxalate in buffered aqueous solution (50 mM Tris, pH 7.2). The fluorescence response to oxalate was greater than that observed for $H_2PO_4^-$, but higher binding affinities were measured for the 1:2 host:guest complex with $H_2PO_4^-$.

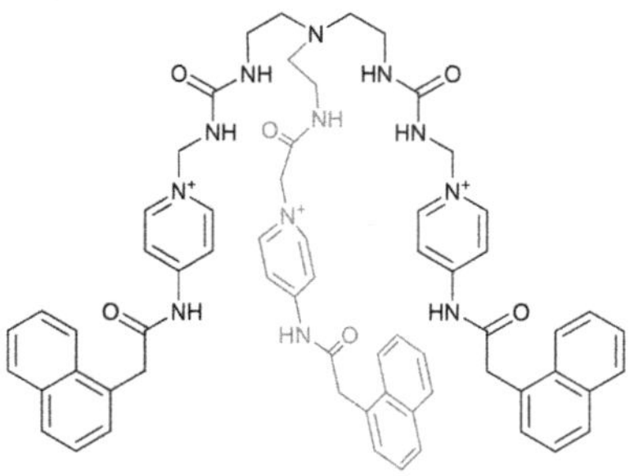

33

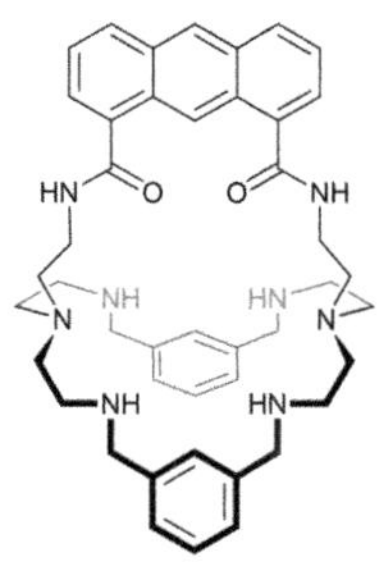

34

The use of arrays with cross-reactive fluorescent chemosensors has also been expanded to hydrogen-bonding receptors. In a 2016 example from Anzenbacher and co-workers, array-based sensing was combined with the use of a polyurethane-based polymer matrix to facilitate the recognition of phosphate species from blood serum samples using chemosensors that are not themselves soluble in aqueous media. Chemosensors **35** and **36** were doped into the polyurethane matrix, which absorbs water and dissolved ions, bringing them in contact with the chemosensors. The combination of this technology with linear discriminant analysis of the responses from the two chemosensors allowed 100% discrimination of adenine nucleotides, oxoanions, and halides from blood serum. This strategy was able to overcome many of the shortcomings associated with hydrogen-bonding fluorescent chemosensors—namely, poor aqueous solubility and affinity in an aqueous environment—without modification of the chemosensors themselves.

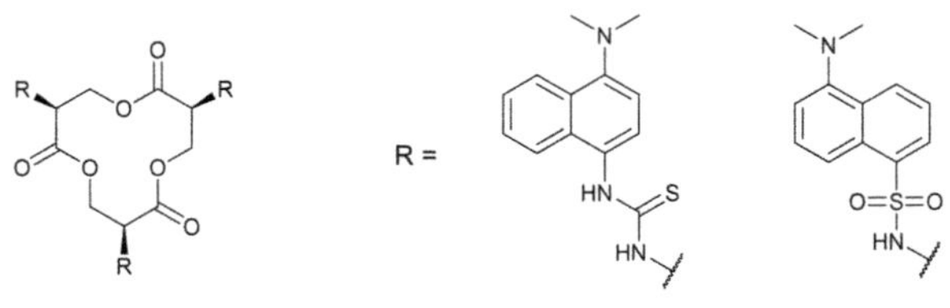

35 36

10.2.3 Metal-cation Coordination

Coordination of phosphate species by a metal cation or cations is the most widely employed interaction trope in the design of fluorescent chemosensors. Phosphate–metal cation interactions provide the highest binding affinities of any class of chemosensors, and such chemosensors are often highly soluble in water. In fact, in many cases the binding affinities of these chemosensors are too high for direct, accurate measurement, and so competition assays must be employed to obtain accurate binding data. Hence, chemosensors involving metal cation-binding groups are highly suited to applications involving the analysis of aqueous samples, such as biological fluids and environmental samples. Furthermore, the strength of metal cation–phosphate interactions is less dependent on pH than charge–charge interactions involving ammonium groups; exceptionally high or low pH is typically necessary to outcompete interaction with the metal centre or eject the metal cation from its coordinating ligand.

The large scope of metal cations available is also a key advantage of this class of phosphate chemosensors. Metal cations vary greatly in their size, charge, and electronic properties. Changing the metal involved in coordination is generally easy to achieve, and may result in varied binding selectivities and fluorescent responses. Thus, small libraries of receptors suitable for use in arrays may be rapidly generated. Lanthanide cations, such as Tb(III) and Eu(III), bring a suite of beneficial properties, including intrinsic luminescence with sharp emission peaks, long fluorescence lifetimes, and large Stokes shifts.

Despite these many advantages, several challenges are faced in the design and optimization of metal cation phosphate chemosensors. First, although high binding affinities are observed between phosphate groups and certain metal cations, selectivity between phosphate species may be more difficult to achieve. In the absence of a well-defined coordination environment, metal cation–phosphate interactions are non-directional, and may display poorly selective interactions. As is the case with chemosensors that apply charge–charge interactions, alternative binding interaction types or complex chemosensor architectures are often necessary to tune the selectivity for the phosphate species of interest. Second, metal cation chemosensors are dependent on coordinative bonds to hold a metal cation in place. Ligands for these metal cations must be selected to ensure that the cation is not too labile; especially in circumstances where large concentrations of competing cations with the potential to displace the metal cation from the chemosensor are present. Metal extrusion chemosensors—which operate by the displacement of a metal cation from a fluorescent ligand by a phosphate species—have been developed; however, these chemosensors are poorly reversible, and great care must be taken in their design and in the analysis of results.

The first metal cation-based fluorescent chemosensors for phosphate species were inspired by Czarnik's polyammonium anthracene chemosensors **1** and **2** (see Section 10.2.1). Hamachi and co-workers appended Zn(II) dipicolylamine (ZnDPA) groups to an anthracene fluorophore to generate chemosensors **37**, **38**, and **39**, which were first reported in 2002.[81] These chemosensors follow the same "receptor–spacer–fluorophore" design model employed by Czarnik and others for the generation of chemosensors that give a change in PeT in response to interaction with a phosphate species.

The use of ZnDPA binding groups is ubiquitous among metal cation phosphate chemosensors. Zn(II) is a d^{10} metal cation, and as such does not undergo excited-state electron transfer processes (at energies relevant to fluorescence spectroscopy) that could interfere with a fluorescence response. Additionally, Zn(II) forms high-affinity interactions with phosphate species that are generally stronger than interactions with any other oxoanion. PeT chemosensors that employ multiple ZnDPA units typically respond with an increase in fluorescence intensity upon interaction with an anion. This is hypothesized to be due to improved coordination of the second Zn(II) group to the DPA ligand.[82]

Chemosensors **37** and **38** were found to respond to phosphotyrosine in aqueous buffer (pH 7.2) with an increase in fluorescence intensity, but compound **39** did not give a fluorescent response.[81] High binding affinity ($\log K_a = 7$) was also recorded for a phosphorylated peptide sequence under similar conditions. Further work with these chemosensors established that binding affinity (and subsequently a fluorescent response) is greatest for negatively charged phosphopeptides and proteins.[81,83] More extensive substrate screening revealed that chemosensors **37** and **38** exhibit the highest binding affinities for ATP and ADP ($\log K_a > 7$ in aqueous buffer), with lesser affinities for other phosphate species, and no response to other oxoanions, including sulfate, carbonate, and nitrate.[82] Chemosensor **37** has been variously used in applications, including the identification of phosphoproteins in SDS–PAGE staining,[83] to monitor dephosphorylation of a model peptide,[82] to monitor the activity of β-1,4-galactotransferase through recognition of UDP,[84] and for the recognition of phosphate species in a hydrogel.[85]

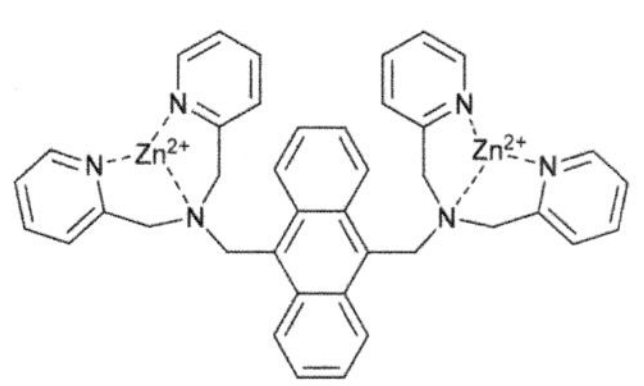

37

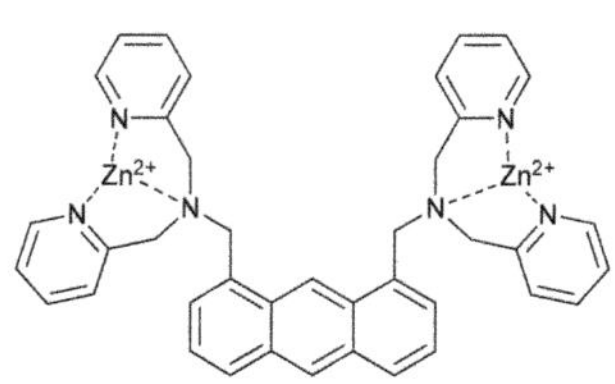

38

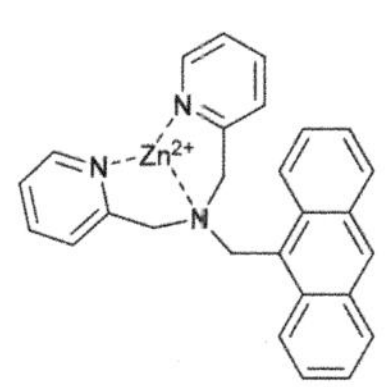

39

The PeT "receptor–spacer–fluorophore" design strategy has been expanded to a wide range of chemosensors employing various metal cations and fluorophores. In a 2007 report, Yoon and co-workers complexed ligand **40**—which contains a long-wavelength dichlorofluorescein fluorophore—with a range of monovalent and divalent metal cations.[86] Of the cations tested, CHEQ was observed upon addition of Hg^{2+}, Cu^{2+}, Ni^{2+}, Ag^{+}, and Mn^{2+} in aqueous buffer at pH 7.4. Of these complexes, the manganese complex of **40** was found to be suitable for the detection of ATP and pyrophosphate, through a fluorescence enhancement observed upon binding these anions. Minimal responses were observed on the addition of other phosphate species, bisulfate, or acetate.

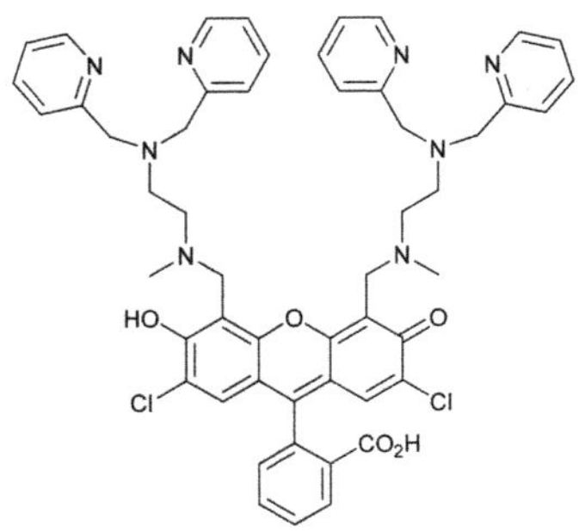

40

Fluorophores that interact directly with the phosphate species or metal cation involved in binding of a phosphate species have also been employed. Chemosensors containing these fluorophores give a fluorescence response as a result of changes in charge density and charge distribution in the fluorophore when a phosphate species is bound. These chemosensors may give a response that is more complex than the changes in fluorescence intensity observed for PeT chemosensors. Acridine chemosensor **41**, which differs from Hamachi's anthracene chemosensor **38** by a single atom, provides differing fluorescence responses to Pi and PPi.[87] Addition of Pi to **41** in aqueous buffer (pH 7.4) led to an increase in fluorescence intensity and slight hypsochromic shift, whereas PPi induced significant fluorescence quenching and a similar hypsochromic shift. No significant change was observed for other anions. The response to phosphate was attributed to hydrogen bonding directly to the acridine nitrogen; pyrophosphate is believed to increase PeT quenching in this case.

ZnDPA-based receptor **42**, developed by Hong and co-workers in 2004, includes an additional phenolate group that also coordinates to the Zn(II) centres.[88] This chemosensor was found to give a pronounced CHEF response to PPi, accompanied by a bathochromic shift in the peak emission. Smaller responses were noted for ATP and ADP, but selectivity for polyphosphates over AMP, Pi, other oxoanions, and halides was observed. The mechanism of response was attributed to weakening of the Zn-phenolate interactions upon anion binding, leading to an increase in the charge character of the oxygen atom.

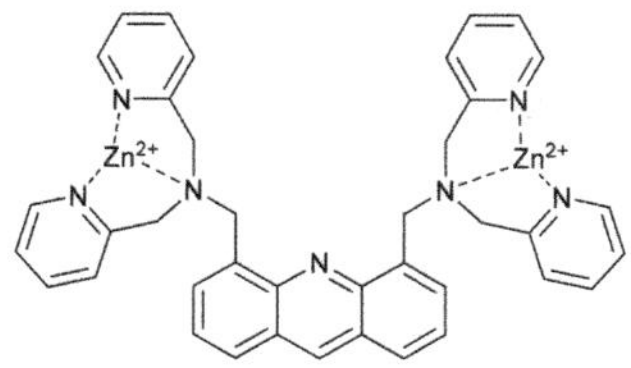

41

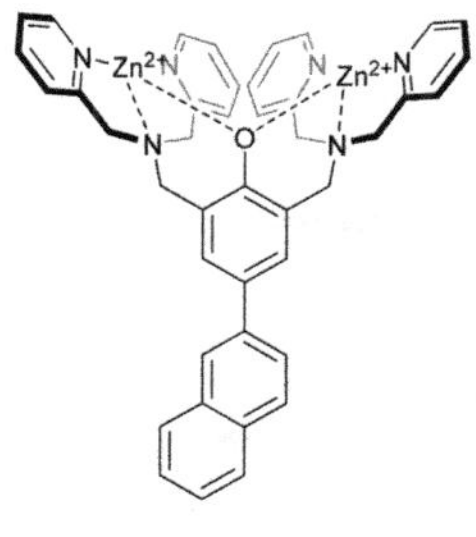

42

Terpyridine-based Zn(II) complex **43**, developed by Tu and co-workers in 2016, is one of the few aggregation-induced emission chemosensors that have been developed for the detection of phosphate species. A notable enhancement in the fluorescence emission intensity was observed upon the addition of 0.25 equivalents of ATP to an aqueous solution of the complex. A less significant response to ADP was also observed, with minimal responses to other phosphate species, halides, or oxoanions. Exogenous ATP could be imaged using confocal fluorescence microscopy in the mitochondria of HeLa cells incubated with complex **43**.

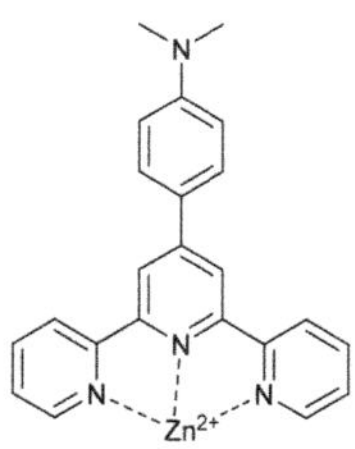

43

With a few exceptions, the simple chemosensors described so far in this section have been most selective for phosphate species such as ATP, ADP, and PPi, which possess a significant negative charge. Many of these chemosensors are also reported to provide a significant response to other highly charged anions, such as citrate. As is the case with other fluorescent phosphate chemosensors reliant on charge–charge and hydrogen-bonding interactions, efforts to increase selectivity have typically involved the generation of complex receptor architectures. Such structures may provide a binding cleft that closely matches the substrate shape and excludes other phosphate species.

Increases in the complexity of the chemosensor scaffold and use of a combination of metal cation and hydrogen-bonding interactions were successfully employed in the development of the ATP-selective chemosensor **44** by Butler in 2014.[89] Addition of Zn(II) to the free ligand resulted in an increase in fluorescence intensity (CHEF). The chemosensor was responsive to the presence of ATP, ADP, and citrate in aqueous buffer (pH 7.4) with an increase in fluorescence intensity at longer wavelength attributed to an excimer emission. Conversely, fluorescence quenching was observed for pyrophosphate, phosphate, AMP, cAMP, and chloride. Although a significant fluorescence response to citrate was observed, the binding affinity for ATP was measured to be more than 100 times higher. The high selectivity of this chemosensor may be attributed to the highly preorganized structure, and combination of high-affinity metal cation interactions with directional hydrogen-bonding interactions. Chemosensor **44** was used to monitor the hydrolysis of ATP by apyrase in a highly competitive aqueous medium at biologically relevant pH.

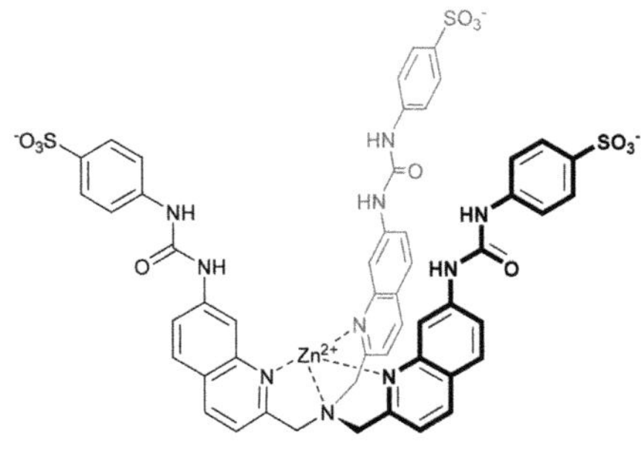

44

Jolliffe and co-workers took inspiration from the lissoclinum family of cyclic peptides to develop highly preorganized chemosensors for phosphate species.[90,91] These structures contain a rigid macrocyclic peptide-derived scaffold that is readily diversified by variation in the selection and functionalization of the amino acid side chains. Binding of phosphate species and oxoanions to ZnDPA macrocycle **45a** was evaluated in aqueous buffer by an indicator displacement approach with fluorescent coumarin derivative **46**.[90,91] High affinity and a prominent fluorescence response were observed for PPi, as well as lesser responses to ATP, ADP, and citrate. The versatility in this scaffold facilitated the determination of the structure–activity relationships.[91] Binding affinities to PPi were able to be increased by decreasing the separation between ZnDPA units (**45** compared with **47**), and the introduction of steric bulk in the non-binding side chains switched selectivity to ATP over PPi (**45a** compared with **45b, 45c**) (see Table 10.2).

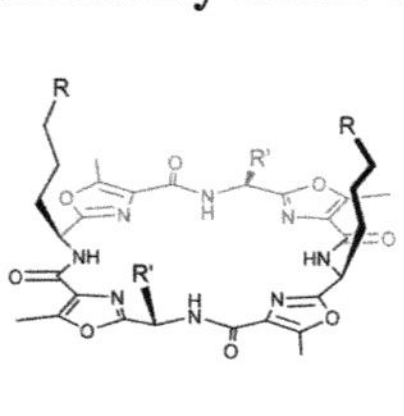

45a: R' = Me
45b: R' = CH_2iPr
45c: R' = CH_2Ph

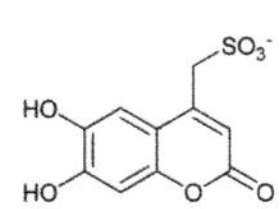

46

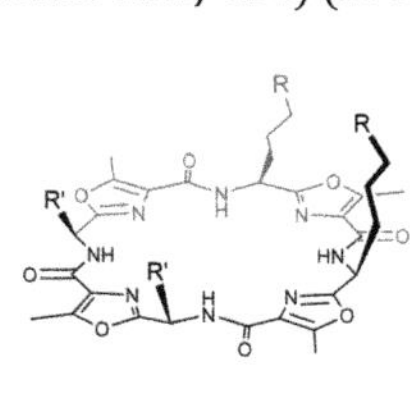

47a: R' = CH_2iPr
47b: R' = CH_2Ph

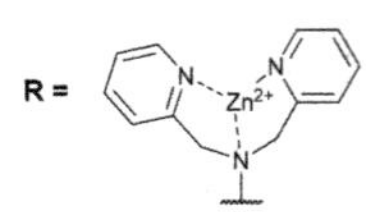

Table 10.2 Apparent association constants ($\log K_a$) determined for receptors **45** and **47** with various polyphosphate anions in aqueous solution (5 mM HEPES, pH 7.4). Each value represents the average of at least two separate titration experiments. Errors < 10%.[91]

Anion	**45a**	**45b**	**45c**	**47a**	**47b**
Indicator 46	6.3	6.2	6.9	6.9	6.8
PPi	7.2	7.4	8.4	8.8	7.9
ATP	6.5	7.6	8.6	8.3	7.2
ADP	6.3	7.4	7.5	7.7	7.0

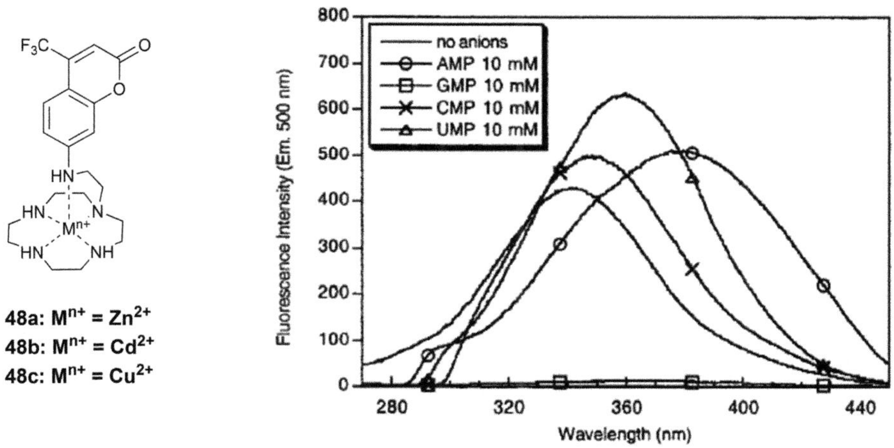

Figure 10.6 Comparison of the excitation spectra of **48b** (5 μM) upon addition of 10 μM of AMP, GMP, CMP, and UMP. Reproduced from ref. 92 with permission from American Chemical Society, Copyright 2002.

Use of an IIDA was demonstrated to be an effective strategy with work by Kikuchi and co-workers published in 2002 with aminocoumarin **48** (see Figure 10.6).[92] Complexes of Zn(II) (**48a**), Cd(II) (**48b**), and Cu(II) (**48c**) were evaluated for the detection of a range of anionic species, and the Cd(II) complex **48b** was found to respond to the presence of PPi, nucleotides, and citrate in aqueous buffer of high ionic strength (100 mM HEPES buffer, pH 7.4). Displacement of the aminocoumarin fluorophore from the Cd(II) metal centre led to distinct changes in the excitation spectra for each phosphate species. Notable responses included a ratiometric change in the excitation spectra in the presence of PPi, and fluorescence quenching for GMP, which was thought to be due to PeT between the guanine nucleobase and the aminocoumarin unit.

Jolliffe and co-workers were able to directly compare inter- and intramolecular indicator displacement strategies for their macrocyclic oxazole scaffold with structures **49** and **50**.[93,94] Covalent attachment of the fluorometric coumarin indicator led to a reduction in binding affinity for PPi, ATP, and

ADP; however, selectivity for PPi was greatly enhanced for **49** and **50** in comparison to the chemosensing ensembles **45a·46** and **47a·46**, respectively. Chemosensor **50** was successfully used to monitor the hydrolysis of PPi by pyrophosphatase in aqueous buffer. Linear peptide **51**, which was designed according to the same principles, was employed as a selective probe to detect cell membranes that are rich in cell-surface phosphatidylserine (PS) in live cell fluorescence imaging studies.[95,96]

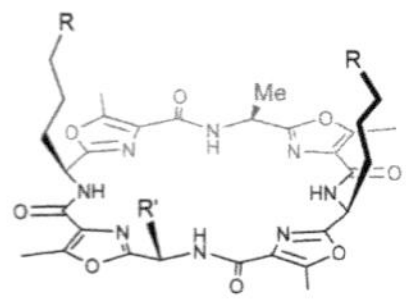

49

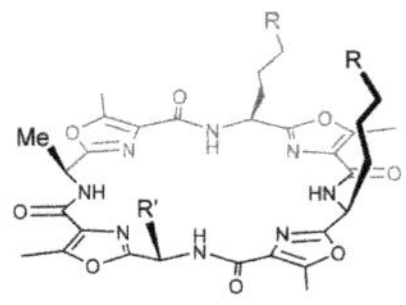

50

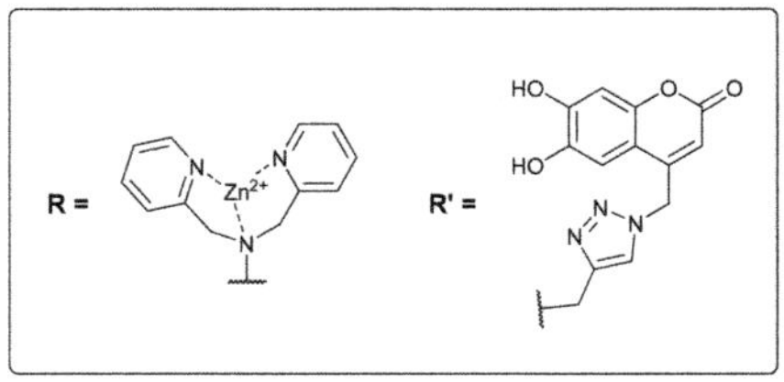

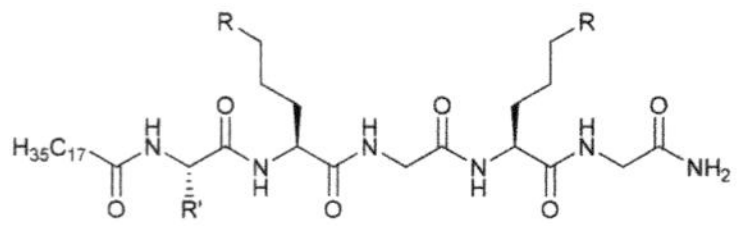

51

10.2.3.1 *Probes for Diphosphorylated Peptides*

The high strength of metal cation–phosphate interactions means that a single metal cation can provide sufficient strength to bind a phosphate group in water. In the example above of the probe for PS (**51**) from Jolliffe and co-workers—as well as in earlier examples by Smith and co-workers[97,98]—one ZnDPA unit is postulated to bind the phosphate group of PS, while the other interacts with the carboxylate of the phospholipid headgroup. Consequently, it is possible to use metal cation-based fluorescent probes for the detection of species with multiple proximal phosphate groups, without the need for extensive chemosensor architectures. Such fluorescent probes are of particular interest for the detection of polyphosphorylated proteins; clusters of proximally phosphorylated amino acids have been implicated in a range of diseases, including cancers, rheumatoid arthritis, and neurodegenerative diseases.[99]

In an extension of their work for the detection of phosphorylated peptides and proteins, Hamachi and co-workers have developed a range of related fluorescent chemosensors for the recognition of peptide sequences with multiple phosphorylation sites.[100–104] These chemosensors differ in the nature, connectivity, and length of the spacer between the two ZnDPA binding units, and operate *via* a variety of fluorescence mechanisms.

Rigid chemosensor **52** proved to be highly selective for diphosphorylated peptide sequences in which the phosphorylated amino acids are separated by four amino acids (i, i + 4),[103,104] and chemosensor **53** displayed significantly higher binding affinities for sequences with adjacent phosphorylated amino acids (i, $i + 1$).[102] The highly flexible nature of chemosensors **54** and **55** led to greater promiscuity in binding. Although chemosensors **54** and **55** contain a methylene spacer unit between the ZnDPA recognition unit and the fluorophore, the changes in fluorescence emission following substrate recognition were more complex than may be explained by a PeT mechanism alone.[100]

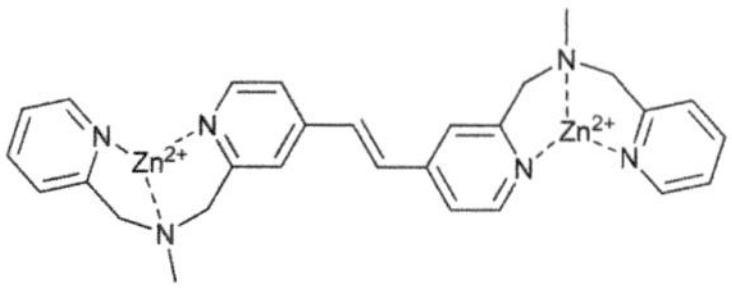

52

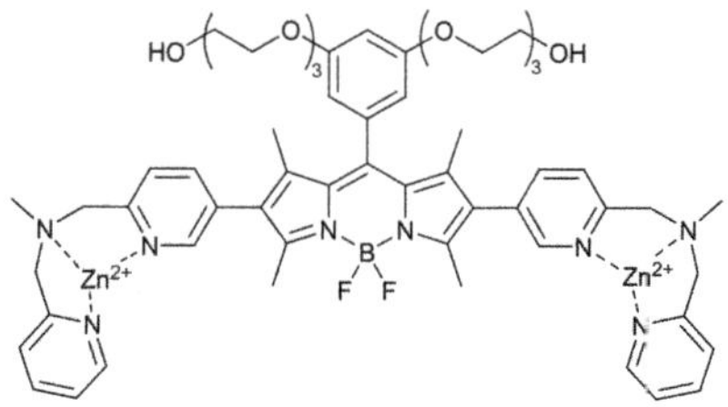

53

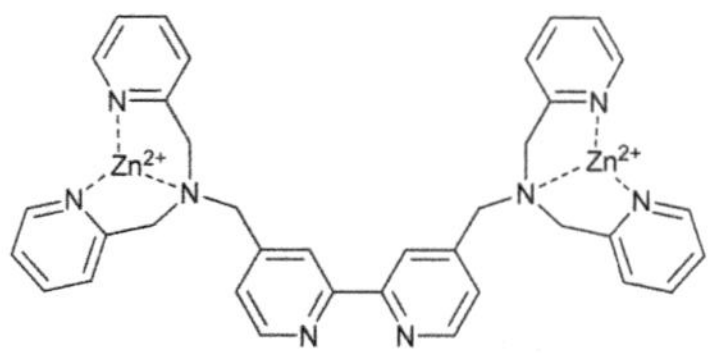

54

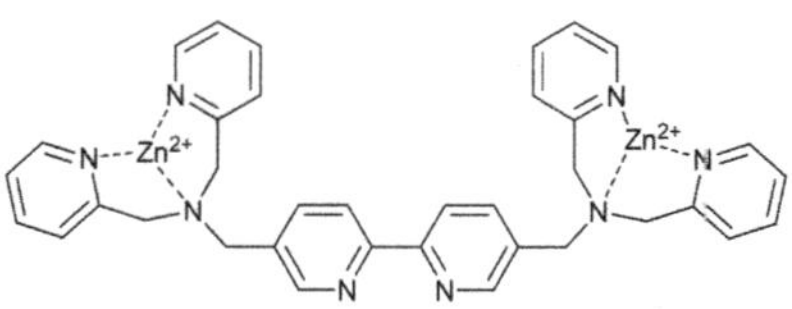

55

Gunning and co-workers adapted the pyrene-based pyrophosphate chemosensors **3** and **19** developed by Teramae and co-workers,[47,66] replacing the charge–charge and hydrogen-bonding interactions in **3** and **19** with metal coordination sites that allow pyrene **56** to function in water.[105–107] Pyrene **56** exhibits an increase in excimer fluorescence intensity in the presence of proximal phosphorylated amino acids, due to association of the pyrene units. This simple chemosensor has been used in a range of applications, including the detection of phosphorylated proteins in solution and by SDS–PAGE,[105,106] and the detection of negatively charged cell membranes rich in PS.[107]

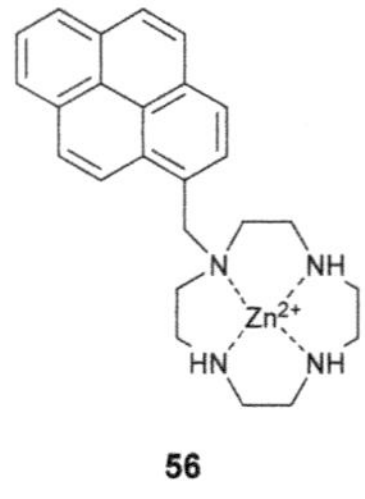

56

10.2.3.2 Chemosensors Employing Lanthanide Cations

As indicated previously, many of the lanthanide metal cations possess intrinsic properties that make them well suited to use in fluorescent chemosensors. In particular, Eu(III) and Tb(III) have been used successfully in a range of chemosensors for phosphate species. These cations possess intrinsic fluorescence that is characterized by large Stokes shifts, long fluorescence lifetimes, and sharp, structured emission peaks. Such properties are of particular interest for the detection of phosphate species in biological milieu, due to the propensity with which background autofluorescence from endogenous biological compounds may be reduced with time-gated fluorescence measurements. Although direct excitation of lanthanides typically results in weak emission due to Laporte forbidden transitions, lanthanide cations are readily excited by energy transfer from a fluorescent ligand. Initial work by Parker and co-workers in 2000 revealed that Eu(III) and Tb(III) complexes of ligand **57** form high-affinity complexes with inorganic phosphate in aqueous media (collidine/HCl buffer, pH 7.4).[108] Further work by the same group indicated that similar lanthanide complexes **58–60** are also responsive to phosphorylated amino acids and peptides.[109]

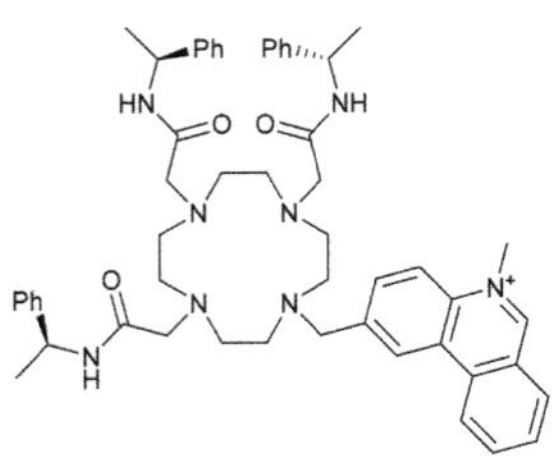

57

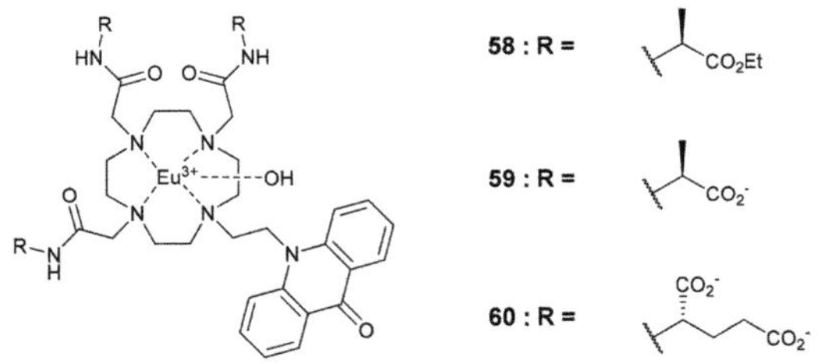

The high affinity of pyrophosphate for lanthanides was utilized by Yuan and co-workers to develop a Eu(III) pyrophosphate-selective chemosensor that operates by a metal extrusion mechanism.[110] Diketonate ligand **61** acts as an antenna for Eu(III) when complexed to the metal cation: excitation of the ligand leads to energy transfer to the metal centre and subsequent emission at a longer wavelength. Addition of PPi to the [Eu-$\mathbf{61}_3$] complex in a cetylpyridinium bromide micellar solution resulted in formation of an Eu-PPi complex through displacement of diketonate ligand **61**, and subsequent reduction in emission intensity. Precise tuning of the ligand affinity was necessary to achieve a selective response to PPi in the presence of potential interferents, including ATP, citrate, and tartrate.

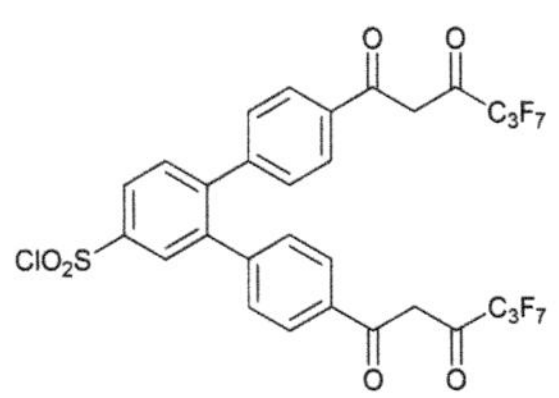

61

One of the primary mechanisms by which emission intensity from lanthanide complexes is reduced is *via* energy losses from labile water molecules in the inner coordination sphere. In 2014, Schäferling and co-workers investigated the effect of the number of coordinated water molecules on selectivity for phosphate species with the series of complexes **62–64**.[111] These complexes differ in the number of coordination sites occupied by the pyridine-based antenna ligand. Complex **62**, which contains two coordinated water molecules, was unresponsive to all anions tested. In this case, increasing the number of coordinated water molecules to three (complex **63**) led to a chemosensor that gave non-selective increases in luminescence intensity in response to ATP, ADP, and phosphate. Complex **64** ($4H_2O$) was selective for ATP—a result that was attributed to the higher number of phosphate units present in ATP relative to ADP and AMP. Displacement of the coordinated water molecules by ATP was accompanied by a 60-fold increase in luminescence intensity, and a three-fold increase in the luminescence lifetime of the complex. Complex **64** was successfully used to monitor the activity of ATPase in aqueous Tris buffer (pH 7.4).

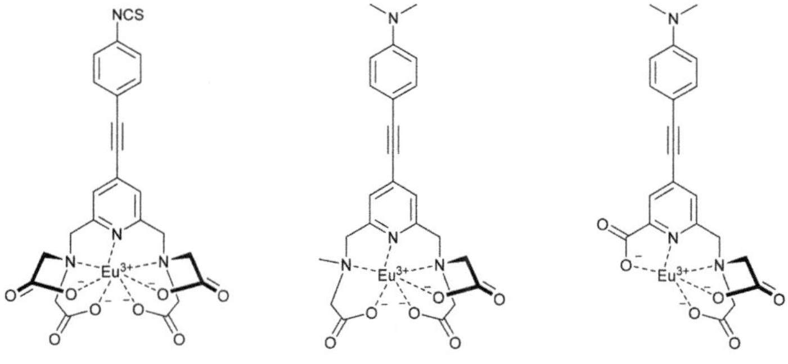

62 **63** **64**

Further refinement of selective lanthanide-based chemosensors for phosphate species has been undertaken by Butler and co-workers.[112–114] Eu(III) complex **65**, originally presented as a luminescent chemosensor for fluoride,[115] was found to bind and respond to nucleotides with an increase in luminescence intensity due to displacement of a coordinated water molecule. Modest selectivity for ADP over ATP was observed in aqueous HEPES buffer; however, addition of a Mg(II) salt enhanced selectivity due to the higher affinity of Mg(II) for ATP.[112] The binding selectivity of complex **65**, coupled with the ability to conduct time-gated luminescence measurements, have made it suitable to monitor the activity of kinases,[112] glycosyltransferases, and phosphodiesterases.[114] The chemosensor was not selective between diphosphorylated nucleotides, but maintained selectivity for diphosphorylated nucleotides over triphosphate species of the same nucleobase. Replacement of negatively charged carboxylate coordinating groups for neutral amides led to increased electropositive character of the Eu(III) centre.[113] Consequentially the affinities for all phosphate species were increased in the resulting chemosensor **66**, and cellular uptake was improved. Although chemosensor **66** was no longer ADP selective, it was successfully used to monitor changes in ATP concentration within the mitochondria of cells.

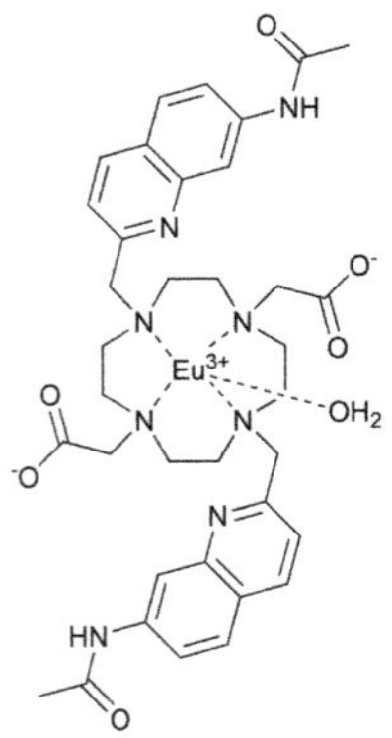

65

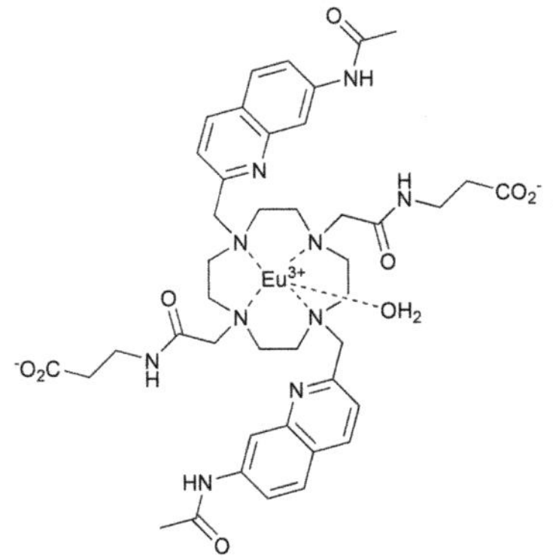

66

10.3 Conclusions

From the seminal report of a fluorescent chemosensor for phosphate species by Czarnik in 1989,[45] improvements in the sensitivity, selectivity, and utility of these chemosensors have continued apace. A wide range of chemical scaffolds and binding interactions have been explored in chemosensors that respond by a range of fluorescence mechanisms, including PeT, ICT, TICT, ESIPT, IDA, IIDA, AIE, and modulation of lanthanide luminescence. This chapter has presented a small subset of the many phosphate chemosensors developed over the last 30 years. The examples outlined above are representative of fundamental advances in the design of fluorescent chemosensors for phosphates, which have facilitated improvements in pH sensitivity, aqueous solubility, binding affinity, and sensitivity. As a result of these advancements, fluorescent chemosensors for phosphates have now been used for a wide variety of applications, including real-time visualization of phosphate species in cells, *in vitro* monitoring of the activity of a range of enzymes, identification of phosphorylated proteins, and analysis of biological samples.

However, despite the impressive volume and scope of work conducted thus far in the field, significant potential for improvement remains. Although many chemosensors are claimed to be selective, serum and cells contain a plethora of phosphate species and other potential anionic interferents (such as dicarboxylate species) that may confound analysis. While not all of these species are present in significant concentrations in every context in which a chemosensor may be used, a comprehensive scope of potential interferents remains underexplored for many chemosensors. Where a large range of phosphate species and dicarboxylates have been examined, functional levels of selectivity between highly related species, such as ATP and GTP, remains rare. There is also currently a dearth of single chemosensors that are selective for phosphate species with lower charge. For example, fluorescent chemosensors reported to be ATP selective are common, whereas chemosensor **65** remains one of only a few chemosensors with selectivity for ADP over ATP;[112] chemosensors with selectivity for AMP are rarer still. The challenge of targeting low-charge phosphate species in water becomes even more significant in the context of analysis of environmental samples. In this circumstance, analyte concentrations are often considerably lower than in biological settings so binding affinities must be significantly greater for effective recognition to occur, requiring chemosensors that combine strong binding interactions and aqueous solubility with well-designed scaffolds to enhance selectivity though complementary geometry to the phosphate of interest.

The development of multi-chemosensor arrays that provide a fingerprint response to differentiate between related phosphate species is a promising advance that may overcome some of the challenges associated with selective phosphate recognition. Although the construction of a chemosensor that displays selectivity through binding strength is a significant obstacle for many phosphate species, the use of multi-chemosensor arrays, such as those presented by Guo (see Section 10.2.1) and Anzenbacher (see Section 10.2.2) may circumvent this difficulty. However, such systems are unlikely to be useful across all applications, *e.g.* real-time imaging of phosphate species in cells, so there remains a need for development of improved selective chemosensors for a variety of phosphate species.

Lastly, the ubiquity of phosphate species in biology in a variety of roles, including as metabolites, signalling molecules, and post-translationally modified proteins, means that the potential applications for the detection of phosphate species are innumerable. There are over 500 human kinases and 200 phosphatases, and many of these enzymes are responsible for the phosphorylation or dephosphorylation of multiple targets.[116] The activity and concentration of each of these enzymes, of their targets, and of a multitude of downstream metabolites are all of potential interest for detection using fluorescent chemosensors. Taken together with the different concentrations of inorganic phosphate species in both biological and environmental settings, this means that further development of new and improved fluorescent phosphate chemosensors tailored for specific applications will remain a fruitful area of research.

Acknowledgements

This work was supported by the Australian Research Council (DP200100848 and CE200100012).

References

1. J. G. Voet and D. Voet, *Biochemistry*, John Wiley and Sons, New York, 4th edn, 2010.
2. T. Harayama and H. Riezman, *Nat. Rev. Mol. Cell Biol.*, 2018, **19**, 281–296.
3. J. D. Graves and E. G. Krebs, *Pharmacol. Ther.*, 1999, **82**, 111–121.
4. A. E. Timms, Y. Zhang, R. G. G. Russell and M. A. Brown, *Rheumatology*, 2002, **41**, 725–729.
5. S. Xu, M. He, H. Yu, X. Cai, X. Tan, B. Lu and B. Shu, *Anal. Biochem.*, 2001, **299**, 188–193.
6. N. M. Hansen, R. Felix, S. Bisaz and H. Fleisch, *Biochim. Biophys. Acta, Gen. Subj.*, 1976, **451**, 549–559.
7. M. Crook and R. Swaminathan, *Ann. Clin. Biochem.*, 1996, **33**, 376–396.
8. G. A. Block, P. S. Klassen, J. M. Lazarus, N. Ofsthun, E. G. Lowrie and G. M. Chertow, *J. Am. Soc. Nephrol.*, 2004, **15**, 2208–2218.
9. R. M. Edwards, *Hyperphosphatemia*, Elsevier, Amsterdam, 2004.
10. K. Hruska and A. Gupta, in *Metabolic Bone Disease and Clinically Related Disorders*, ed. L. V. Avioli and S. M. Krane, Elsevier, Amsterdam, 1998, ch. 7, pp. 207–236.
11. C. Bergwitz and H. Jüppner, *Adv. Chronic Kidney Dis.*, 2011, **18**, 132–144.
12. M. F. Chislock, O. Sarnelle, B. K. Olsen, E. Doster and A. E. Wilson, *Nat. Educ. Knowl.*, 2013, **4**, 10.
13. D. W. Schindler, S. R. Carpenter, S. C. Chapra, R. E. Hecky and D. M. Orihel, *Environ. Sci. Technol.*, 2016, **50**, 8923–8929.
14. A. E. Taylor and C. W. Miller, *J. Biol. Chem.*, 1914, **18**, 215–224.
15. H. Eibl and W. E. M. Lands, *Anal. Biochem.*, 1969, **30**, 51–57.
16. L. S. B. Upadhyay and N. Verma, *Biotechnol. Lett.*, 2015, **37**, 1335–1345.
17. S. Aoki and E. Kimura, *Rev. Mol. Biotechnol.*, 2002, **90**, 129–155.
18. B. Lukasz, R. Ewa, M. Piotr and K. Pawel, *Curr. Org. Chem.*, 2006, **10**, 2285–2306.
19. A. K. Hirsch, F. R. Fischer and F. Diederich, *Angew. Chem., Int. Ed.*, 2007, **46**, 338–352.
20. A. E. Martell, R. J. Motekaitis, Q. Lu and D. A. Nation, *Polyhedron*, 1999, **18**, 3203–3218.
21. A. E. Hargrove, S. Nieto, T. Zhang, J. L. Sessler and E. V. Anslyn, *Chem. Rev.*, 2011, **111**, 6603–6782.
22. J. Wongkongkatep, A. Ojida and I. Hamachi, *Top. Curr. Chem.*, 2017, **375**, 30.
23. S. Lee, K. K. Yuen, K. A. Jolliffe and J. Yoon, *Chem. Soc. Rev.*, 2015, **44**, 1749–1762.
24. H. T. Ngo, X. Liu and K. A. Jolliffe, *Chem. Soc. Rev.*, 2012, **41**, 4928–4965.
25. S. K. Kim, D. H. Lee, J.-I. Hong and J. Yoon, *Acc. Chem. Res.*, 2009, **42**, 23–31.
26. E. A. Katayev, Y. A. Ustynyuk and J. L. Sessler, *Coord. Chem. Rev.*, 2006, **250**, 3004–3037.
27. C. Bazzicalupi, A. Bencini and V. Lippolis, *Chem. Soc. Rev.*, 2010, **39**, 3709–3728.

28. A. T. Law al and S. B. Adeloju, *Talanta*, 2013, **114**, 191–203.
29. C. Warwick, A. Guerreiro and A. Soares, *Biosens. Bioelectron.*, 2013, **41**, 1–11.
30. D. Zhang, J. R. Cochrane, A. Martinez and G. Gao, *RSC Adv.*, 2014, **4**, 29735–29749.
31. A. B. Aletti, D. M. Gillen and T. Gunnlaugsson, *Coord. Chem. Rev.*, 2018, **354**, 98–120.
32. F. P. Schmidtchen and M. Berger, *Chem. Rev.*, 1997, **97**, 1609–1646.
33. Y. W. Jun, S. Sarkar, K. H. Kim and K. H. Ahn, *ChemPhotoChem*, 2019, **3**, 214–219.
34. M. V. Ramakrishnam Raju, S. M. Harris and V. C. Pierre, *Chem. Soc. Rev.*, 2020, **49**, 1090–1108.
35. S. J. Butler and K. A. Jolliffe, *ChemPlusChem*, 2021, **86**, 59–70.
36. A. Nag and S. Das, *Isr. J. Chem.*, 2020, **61**, 169–184.
37. I. Pont, C. Galiana-Rosello, A. Lopera, J. González-García and E. García-España, in *Supramolecular Chemistry in Water*, Wiley-VCH Verlag GmbH & Co. KGaA, Weinheim, 2019, pp. 115–160, DOI: 10.1002/9783527814923.ch4.
38. W. Li, X. Gong, X. Fan, S. Yin, D. Su, X. Zhang and L. Yuan, *Chin. Chem. Lett.*, 2019, **30**, 1775–1790.
39. S. Pal, T. K. Ghosh, R. Ghosh, S. Mondal and P. Ghosh, *Coord. Chem. Rev.*, 2020, **405**, 213128.
40. S.-i. Tamaru and I. Hamachi, in *Recognition of Anions*, Springer Berlin Heidelberg, 2008, pp. 95–125.
41. Y. Marcus, *Biophys. Chem.*, 1994, **51**, 111–127.
42. Y. Marcus, *Ion Properties*, Marcel Dekker, New York, 1997.
43. Z. Zhang, L. Yan, H. Yu, T. Yan and X. Li, *Bioresour. Technol.*, 2019, **284**, 65–71.
44. C. Nakai and W. Glinsmann, *Biochemistry*, 1977, **16**, 5636–5641.
45. M. E. Huston, E. U. Akkaya and A. W. Czarnik, *J. Am. Chem. Soc.*, 1989, **111**, 8735–8737.
46. D. H. Vance and A. W. Czarnik, *J. Am. Chem. Soc.*, 1994, **116**, 9397–9398.
47. S. Nishizawa, Y. Kato and N. Teramae, *J. Am. Chem. Soc.*, 1999, **121**, 9463–9464.
48. M. Dhaenens, J.-M. Lehn and J.-P. Vigneron, *J. Chem. Soc., Perkin Trans. 2*, 1993, 1379–1381.
49. M.-P. Teulade-fichou, J.-P. Vigneron and J.-M. Lehn, *Supramol. Chem.*, 1995, **5**, 139–147.
50. M.-P. Teulade-Fichou, J.-P. Vigneron and J.-M. Lehn, *J. Chem. Soc., Perkin Trans. 2*, 1996, 2169–2175.
51. O. Baudoin, F. Gonnet, M.-P. Teulade-Fichou, J.-P. Vigneron, J.-C. Tabet and J.-M. Lehn, *Chem. – Eur. J.*, 1999, **5**, 2762–2771.
52. G. Malojčić, I. Piantanida, M. Marinić, M. Žinić, M. Marjanović, M. Kralj, K. Pavelić and H.-J. Schneider, *Org. Biomol. Chem.*, 2005, **3**, 4373–4381.
53. J. Y. Kwon, N. J. Singh, H. N. Kim, S. K. Kim, K. S. Kim and J. Yoon, *J. Am. Chem. Soc.*, 2004, **126**, 8892–8893.

54. H. N. Lee, N. J. Singh, S. K. Kim, J. Y. Kwon, Y. Y. Kim, K. S. Kim and J. Yoon, *Tetrahedron Lett.*, 2007, **48**, 169–172.
55. Z. Xu, N. J. Singh, J. Lim, J. Pan, H. N. Kim, S. Park, K. S. Kim and J. Yoon, *J. Am. Chem. Soc.*, 2009, **131**, 15528–15533.
56. A. C. Sedgwick, J. T. Brewster, T. Wu, X. Feng, S. D. Bull, X. Qian, J. L. Sessler, T. D. James, E. V. Anslyn and X. Sun, *Chem. Soc. Rev.*, 2021, **50**, 9–38.
57. P. P. Neelakandan, M. Hariharan and D. Ramaiah, *Org. Lett.*, 2005, **7**, 5765–5768.
58. P. P. Neelakandan, M. Hariharan and D. Ramaiah, *J. Am. Chem. Soc.*, 2006, **128**, 11334–11335.
59. M. Bojtár, J. Kozma, Z. Szakács, D. Hessz, M. Kubinyi and I. Bitter, *Sens. Actuators, B*, 2017, **248**, 305–310.
60. Z. Zheng, W.-C. Geng, J. Gao, Y.-Y. Wang, H. Sun and D.-S. Guo, *Chem. Sci.*, 2018, **9**, 2087–2091.
61. P. Anzenbacher Jr., P. Lubal, P. Bucek, M. A. Palacios and M. E. Kozelkova, *Chem. Soc. Rev.*, 2010, **39**, 3954–3979.
62. D. G. Smith, I. L. Topolnicki, V. E. Zwicker, K. A. Jolliffe and E. J. New, *Analyst*, 2017, **142**, 3549–3563.
63. S. E. Schneider, S. N. O'Nei and E. V. Anslyn, *J. Am. Chem. Soc.*, 2000, **122**, 542–543.
64. A. K. H. Hirsch, F. R. Fischer and F. Diederich, *Angew. Chem., Int. Ed.*, 2007, **46**, 338–352.
65. D. Maity, M. Li, M. Ehlers, A. Gigante and C. Schmuck, *Chem. Commun.*, 2016, **53**, 208–211.
66. S. Nishizawa, H. Kaneda, T. Uchida and N. Teramae, *J. Chem. Soc., Perkin Trans. 2*, 1998, 2325–2327.
67. T. Gunnlaugsson, A. P. Davis and M. Glynn, *Chem. Commun.*, 2001, 2556–2557.
68. C. M. G. Dos Santos, M. Glynn, T. McCabe, J. S. Seixas de Melo, H. D. Burrows and T. Gunnlaugsson, *Supramol. Chem.*, 2009, **20**, 407–418.
69. T. Gunnlaugsson, A. P. Davis, J. E. O'Brien and M. Glynn, *Org. Lett.*, 2002, **4**, 2449–2452.
70. T. Gunnlaugsson, A. P. Davis, J. E. O'Brien and M. Glynn, *Org. Biomol. Chem.*, 2005, **3**, 48–56.
71. H. Xie, S. Yi, X. Yang and S. Wu, *New J. Chem.*, 1999, **23**, 1105–1110.
72. S.-i. Sasaki, D. Citterio, S. Ozawa and K. Suzuki, *J. Chem. Soc., Perkin Trans. 2*, 2001, 2309–2313.
73. F.-Y. Wu, Z. Li, Z.-C. Wen, N. Zhou, Y.-F. Zhao and Y.-B. Jiang, *Org. Lett.*, 2002, **4**, 3203–3205.
74. F.-Y. Wu, Z. Li, L. Guo, X. Wang, M.-H. Lin, Y.-F. Zhao and Y.-B. Jiang, *Org. Biomol. Chem.*, 2006, **4**, 624–630.
75. S. E. García-Garrido, C. Caltagirone, M. E. Light and P. A. Gale, *Chem. Commun.*, 2007, 1450–1452.
76. Y. Wu, X. Peng, J. Fan, S. Gao, M. Tian, J. Zhao and S. Sun, *J. Org. Chem.*, 2007, **72**, 62–70.

77. C.-Y. Li, C.-X. Zou, Y.-F. Li, X.-F. Kong, Y. Zhou, Y.-S. Wu and W.-G. Zhu, *Anal. Chim. Acta*, 2013, **795**, 69–74.
78. J. L. Tang, C. Y. Li, Y. F. Li and C. X. Zou, *Chem. Commun.*, 2014, **50**, 15411–15414.
79. K. Ghosh, D. Tarafdar, A. Samadder and A. R. Khuda-Bukhsh, *RSC Adv.*, 2015, **5**, 35175–35180.
80. B. S. Morozov, S. S. R. Namashivaya, M. A. Zakharko, A. S. Oshchepkov and E. A. Kataev, *ChemistryOpen*, 2020, **9**, 171–175.
81. A. Ojida, Y. Mito-oka, M.-a. Inoue and I. Hamachi, *J. Am. Chem. Soc.*, 2002, **124**, 6256–6258.
82. A. Ojida, Y. Mito-oka, K. Sada and I. Hamachi, *J. Am. Chem. Soc.*, 2004, **126**, 2454–2463.
83. A. Ojida, T. Kohira and I. Hamachi, *Chem. Lett.*, 2004, **33**, 1024–1025.
84. J. Wongkongkatep, Y. Miyahara, A. Ojida and I. Hamachi, *Angew. Chem., Int. Ed.*, 2006, **45**, 665–668.
85. I. Yoshimura, Y. Miyahara, N. Kasagi, H. Yamane, A. Ojida and I. Hamachi, *J. Am. Chem. Soc.*, 2004, **126**, 12204–12205.
86. K. M. K. Swamy, S. K. Kwon, H. N. Lee, S. M. Shantha Kumar, J. S. Kim and J. Yoon, *Tetrahedron Lett.*, 2007, **48**, 8683–8686.
87. H. N. Lee, K. M. K. Swamy, S. K. Kim, J.-Y. Kwon, Y. Kim, S.-J. Kim, Y. J. Yoon and J. Yoon, *Org. Lett.*, 2007, **9**, 243–246.
88. D. H. Lee, S. Y. Kim and J.-I. Hong, *Angew. Chem., Int. Ed.*, 2004, **43**, 4777–4780.
89. S. J. Butler, *Chem. – Eur. J.*, 2014, **20**, 15768–15774.
90. M. J. McDonough, A. J. Reynolds, W. Y. G. Lee and K. A. Jolliffe, *Chem. Commun.*, 2006, 2971–2973.
91. S. J. Butler and K. A. Jolliffe, *Chem. – Asian J.*, 2012, **7**, 2621–2628.
92. S. Mizukami, T. Nagano, Y. Urano, A. Odani and K. Kikuchi, *J. Am. Chem. Soc.*, 2002, **124**, 3920–3925.
93. X. Liu, D. G. Smith and K. A. Jolliffe, *Chem. Commun.*, 2016, **52**, 8463–8466.
94. K. A. Jolliffe, *Acc. Chem. Res.*, 2017, **50**, 2254–2263.
95. V. E. Zwicker, B. L. Oliveira, J. H. Yeo, S. T. Fraser, G. J. L. Bernardes, E. J. New and K. A. Jolliffe, *Angew. Chem., Int. Ed.*, 2019, **58**, 3087–3091.
96. V. E. Zwicker, J. H. Yeo, S. T. Fraser, E. J. New and K. A. Jolliffe, in *Methods in Enzymology*, ed. D. M. Chenoweth, Academic Press, 2020, vol. 640, pp. 291–307.
97. C. Lakshmi, R. G. Hanshaw and B. D. Smith, *Tetrahedron*, 2004, **60**, 11307–11315.
98. R. G. Hanshaw, C. Lakshmi, T. N. Lambert, J. R. Johnson and B. D. Smith, *ChemBioChem*, 2005, **6**, 2214–2220.
99. A. D. Cabral, T. B. Radu, E. D. de Araujo and P. T. Gunning, *RSC Chem. Biol.*, 2021, **2**, 815–829.
100. A. Ojida, M.-A. Inoue, Y. Mito-oka and I. Hamachi, *J. Am. Chem. Soc.*, 2003, **125**, 10184–10185.

101. A. Ojida, M.-A. Inoue, Y. Mito-oka, H. Tsutsumi, K. Sada and I. Hamachi, *J. Am. Chem. Soc.*, 2006, **128**, 2052–2058.
102. Y. Ishida, M.-A. Inoue, T. Inoue, A. Ojida and I. Hamachi, *Chem. Commun.*, 2009, 2848–2850.
103. A. Ojida, T. Sakamoto, M. A. Inoue, S. Fujishima, G. Lippens and I. Hamachi, *J. Am. Chem. Soc.*, 2009, **131**, 6543–6548.
104. T. Sakamoto, M. A. Inoue, A. Ojida and I. Hamachi, *Bioorg. Med. Chem. Lett.*, 2009, **19**, 4175–4177.
105. D. Kraskouskaya, M. Bancerz, H. S. Soor, J. E. Gardiner and P. T. Gunning, *J. Am. Chem. Soc.*, 2014, **136**, 1234–1237.
106. D. Kraskouskaya, A. D. Cabral, R. Fong, M. Bancerz, K. Toutah, D. Rosa, J. E. Gardiner, E. D. de Araujo, E. Duodu, D. Armstrong, U. Fekl and P. T. Gunning, *Analyst*, 2017, **142**, 2451–2459.
107. B. I. Murcar-Evans, A. D. Cabral, K. Toutah, E. D. de Araujo, A. Lai, P. M. Macdonald, A. Berger-Becvar, D. Kraskouskaya and P. T. Gunning, *Analyst*, 2017, **142**, 4511–4521.
108. J. I. Bruce, R. S. Dickins, L. J. Govenlock, T. Gunnlaugsson, S. Lopinski, M. P. Lowe, D. Parker, R. D. Peacock, J. J. B. Perry, S. Aime and M. Botta, *J. Am. Chem. Soc.*, 2000, **122**, 9674–9684.
109. P. Atkinson, Y. Bretonniere and D. Parker, *Chem. Commun.*, 2004, 438–439.
110. N. Shao, J. Jin, G. Wang, Y. Zhang, R. Yang and J. Yuan, *Chem. Commun.*, 2008, 1127–1129.
111. M. Schäferling, T. Ääritalo and T. Soukka, *Chem. - Eur. J.*, 2014, **20**, 5298–5308.
112. S. H. Hewitt, J. Parris, R. Mailhot and S. J. Butler, *Chem. Commun.*, 2017, **53**, 12626–12629.
113. R. Mailhot, T. Traviss-Pollard, R. Pal and S. J. Butler, *Chem. - Eur. J.*, 2018, **24**, 10745–10755.
114. S. H. Hewitt, R. Ali, R. Mailhot, C. R. Antonen, C. A. Dodson and S. J. Butler, *Chem. Sci.*, 2019, **10**, 5373–5381.
115. S. J. Butler, *Chem. Commun.*, 2015, **51**, 10879–10882.
116. F. Sacco, L. Perfetto, L. Castagnoli and G. Cesareni, *FEBS Lett.*, 2012, **586**, 2732–2739.

CHAPTER 11

Fluorescent Sensors for Reactive Oxygen Species

PING LI*[a] AND BO TANG*[a]

[a]College of Chemistry, Chemical Engineering and Materials Science, Collaborative Innovation Center of Functionalized Probes for Chemical Imaging in Universities of Shandong, Key Laboratory of Molecular and Nano Probes, Ministry of Education, Institutes of Biomedical Sciences, Shandong Normal University, Jinan 250014, People's Republic of China
*E-mail: lip@sdnu.edu.cn, tangb@sdnu.edu.cn

11.1 Fluorescent Sensors for $O_2^{\bullet-}$

Superoxide anion radicals ($O_2^{\bullet-}$), as the first reactive oxygen species (ROS) produced in living cells, play a crucial role in molecular events. $O_2^{\bullet-}$ in living cells is an important regulator of cell function, especially that directly involved in cell proliferation, energy supply and signal transduction, as well as synthesis of important biological compounds.[1,2] However, when over produced, $O_2^{\bullet-}$ disrupts the cellular redox balance and its function becomes destructive to the organism. It targets biological macromolecules, such as lipids, proteins and DNA, to perform irreversible oxidative damage, leading to cell death and various diseases.[3] Therefore, in order to accurately investigate the physiological and pathological effects of $O_2^{\bullet-}$, it is essential to develop analytical tools with high sensitivity and specificity.

In situ real-time imaging analysis of $O_2^{\bullet-}$ at the cellular and *in vivo* levels is vital for elucidating the physiological role of $O_2^{\bullet-}$. Therefore, based on

Monographs in Supramolecular Chemistry No. 34
Fluorescent Chemosensors
Edited by Luling Wu, Adam C. Sedgwick, Xiao-Peng He and Tony D. James

Published by the Royal Society of Chemistry, www.rsc.org

the specific hydrogen extraction reaction of $O_2^{\bullet-}$ with catechol derivatives, a series of fluorescent sensors using caffeic acid as a recognition group were constructed and applied to the detection of $O_2^{\bullet-}$ in living organisms (see Figures 11.1–11.4). As illustrated with probe **1**,[4] when it interacts with $O_2^{\bullet-}$, the quinoid reaction product could emanate a brilliant fluorescence around 515 nm. Employing probe **1**, $O_2^{\bullet-}$ levels were imaged in hepatocytes, zebrafish and mice during simulated hepatic ischemia-reperfusion injury processes. The findings prove that reperfusion injury leads to the lethal production of $O_2^{\bullet-}$.

On the basis of caffeic acid derivatives, real-time monitoring of $O_2^{\bullet-}$ in subcellular regions was further achieved. With the assistance of the organelle

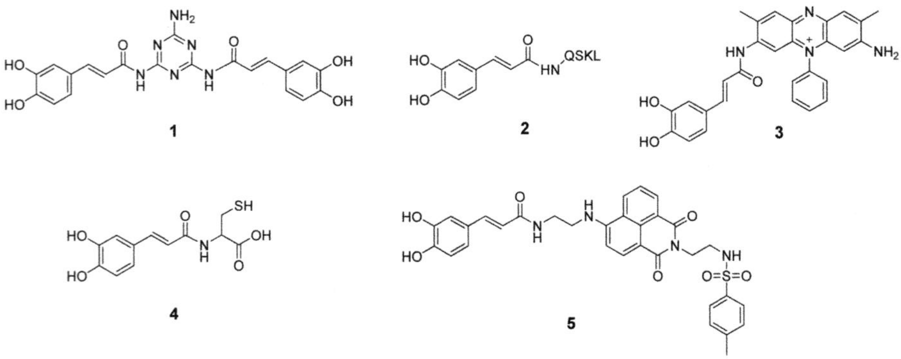

Figure 11.1 Structure of probes **1–5** for fluorescence detection of $O_2^{\bullet-}$ based on caffeic acid derivatives.

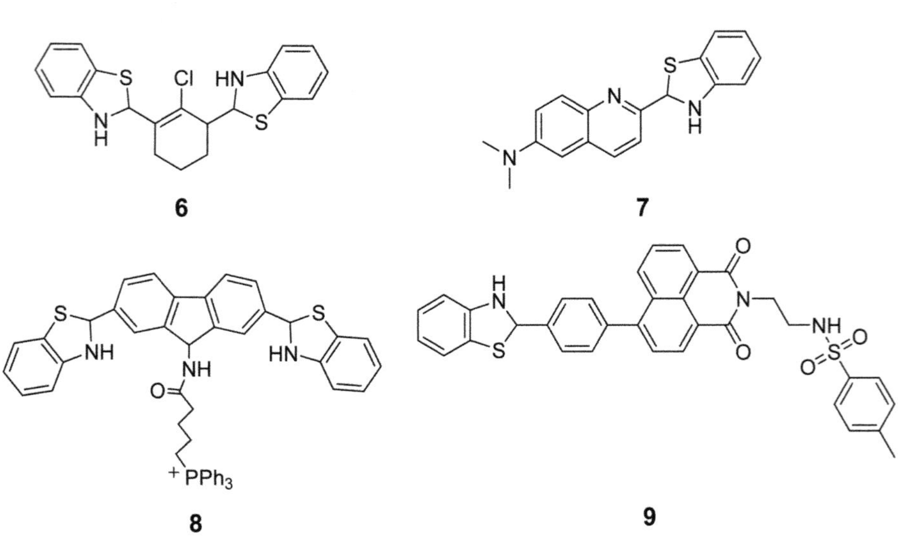

Figure 11.2 Structure of probes **6–9** for fluorescence detection of $O_2^{\bullet-}$ based on benzothiazoline.

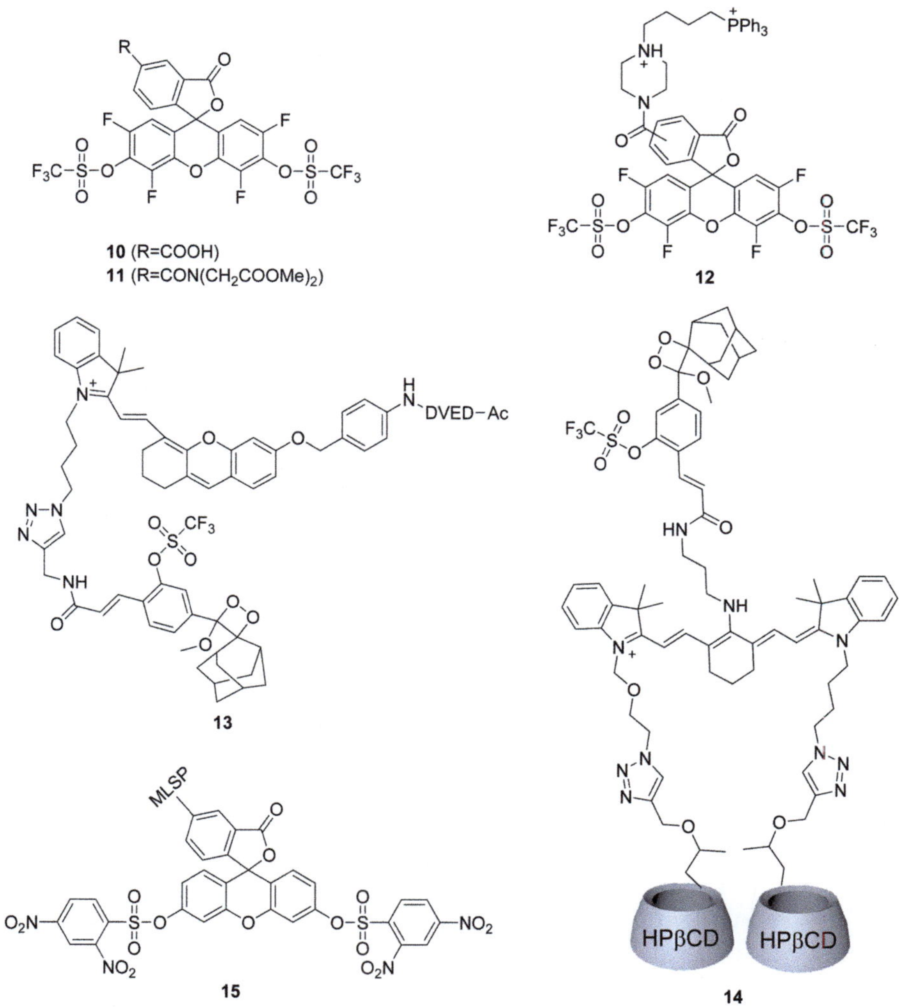

Figure 11.3 Structure of probes **10–15** for fluorescence detection of $O_2^{\bullet-}$ based on the aryl trifluoromethanesulfonate group.

charge and protein distribution characteristics, fluorescent probes were designed for the detection of $O_2^{\bullet-}$ in the peroxisome,[5] mitochondria,[6] Golgi apparatus[7] and endoplasmic reticulum[8] (**2–5**). Probe **2** consists of caffeic acid serving as both the fluorophore and the recognition group for $O_2^{\bullet-}$, and SKL peptide as the peroxisome-targeting group. In the presence of $O_2^{\bullet-}$, the electron-donating catechol of the caffeic acid residue can be oxidized to form electron-withdrawing *o*-benzoquinone, and this reaction can change the electron distribution and increase its emission fluorescence at 495 nm. Probe **2** exhibits excellent selectivity and sensitivity for $O_2^{\bullet-}$. Utilizing probe **2**, peroxisomal $O_2^{\bullet-}$ was successfully observed during oxidative stress in the

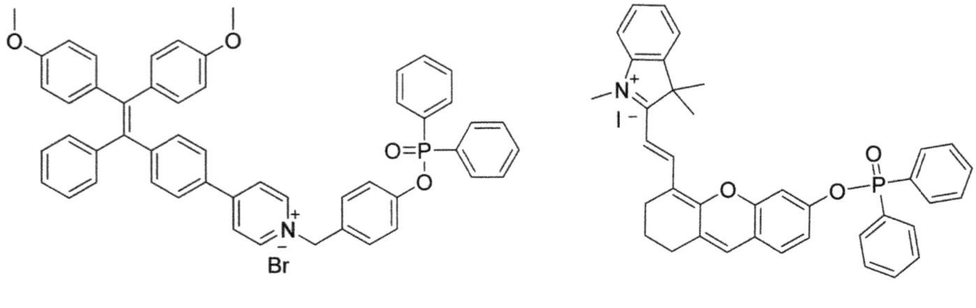

16 17

Figure 11.4 Structure of probes **16** and **17** for fluorescence detection of $O_2^{\bullet-}$ based on the diphenylphosphinyl moiety.

mouse brain by *in vivo* fluorescence imaging, and a signaling pathway mediated by peroxisomal $O_2^{\bullet-}$ in depressed mice was revealed.

To exploit the fluorescence imaging tool for $O_2^{\bullet-}$ location in mitochondria, probe **3** was successfully synthesized. Probe **3** consisted of the $O_2^{\bullet-}$ recognition group caffeic acid and a fluorophore, Safranine T. In addition to serving as the locator group, Safranine T also acted as a reference fluorophore to emit a stable orange fluorescence signal. Based on the change of blue/orange ratio fluorescence signal, real-time quantitative monitoring of $O_2^{\bullet-}$ levels in mitochondria was realized. Using probe **3** and the existing cytoplasmic $O_2^{\bullet-}$ probe dihydroethidium (DHE), $O_2^{\bullet-}$ flux between mitochondria and the cytoplasm during ischemia-reperfusion injury in cells and in living mice were presented, and the signal pathways between mitochondria $O_2^{\bullet-}$ and cytoplasmic $O_2^{\bullet-}$ were also investigated.

To monitor the fluctuations of Golgi $O_2^{\bullet-}$ in real time, the probe **4** was successfully constructed coupling with caffeic acid and L-cysteine. This is because the Golgi apparatus is rich in L-cysteine receptors. With the application of probe **4**, the $O_2^{\bullet-}$ levels of the Golgi were determined and the molecular mechanisms related to the regulation by Golgi $O_2^{\bullet-}$ during hepatic ischemia-reperfusion were elucidated. The fluorescent probe **5** covalently coupled with endoplasmic reticulum (ER) targeting groups (methyl sulfonamide) and caffeic acid was successfully synthesized and exploited for the specific recognition of endoplasmic reticulum $O_2^{\bullet-}$. This probe exhibited extremely high specificity and good biocompatibility and was successfully implemented to detect $O_2^{\bullet-}$ bursts under ER stress.

Inspired by the exclusive dehydrogenation reaction between benzothiazoline and $O_2^{\bullet-}$, a class of probes has been presented for the particular determination of $O_2^{\bullet-}$. Probe **6**, 2-chloro-1,3-dibenzothiazolinecyclohexene, was able to recognize intracellular $O_2^{\bullet-}$ with high selectivity and sensitivity.[9] The imaging analysis of $O_2^{\bullet-}$ was implemented at the cellular level using probe **6**, which revealed an overproduction of $O_2^{\bullet-}$ in RAW264.7 macrophages stimulated with 4β-12-myristate-13-acetate (PMA). Another benzothiazoline-based two-photon (TP) fluorescent probe, **7**, for $O_2^{\bullet-}$ detection was reported.[10] Based on the extension of π-binding and the enhanced intramolecular charge transfer (ICT) process, probe **7** possessed the ability to distinguish $O_2^{\bullet-}$ with a specific "turn-on" fluorescence response. With the help of a two-photon microscope, probe **7** was first employed to visualize endogenous $O_2^{\bullet-}$ in animals with a model of lung inflammation.

Combining the high selectivity of benzothiazoline for $O_2^{\bullet-}$ and its easy synthesis, a variety of organelle-targeted $O_2^{\bullet-}$ probes have been constructed. For instance, the design strategy for probes **8** and **9** was to covalently conjugate the mitochondrial-targeted moiety (triphenylphosphine) and the ER-targeted moiety (methyl sulfonamide), achieving visual detection of $O_2^{\bullet-}$ in the mitochondria and ER.[11,12] Imaging experiments revealed bursts of $O_2^{\bullet-}$ in immune models and ER stress models.

Considering the nucleophilic properties of $O_2^{\bullet-}$, a new class of sensors has been designed by means of nucleophilic substitution reactions between fluorescent probes and $O_2^{\bullet-}$. The aryl trifluoromethanesulfonate group is a promising group for nucleophilic substitution by $O_2^{\bullet-}$.[13,14] Upon reaction with $O_2^{\bullet-}$, the probes **10–12** cleaved off the aryl trifluoromethanesulfonate group while fluorescent-enhanced fluorophores were generated with the phenolic hydroxyl group exposed. In particular, probe **12** combined the advantages of triphenylphosphinyl localization to achieve the detection of mitochondrial $O_2^{\bullet-}$. These probes effectively visualize $O_2^{\bullet-}$ in intact living zebrafish embryos and reveal significant increases in $O_2^{\bullet-}$ levels in models of inflammation and mitochondrial stress. In view of the same recognition mechanism, combined with the advantages of near-infrared fluorophores, deep imaging of $O_2^{\bullet-}$ at the *in vivo* level has been achieved.[15,16] Considering that changes in $O_2^{\bullet-}$ fluctuation is a potential diagnostic parameter for disease, with the aid of probe **13** and **14**, early detection of drug-induced liver/kidney damage was successfully achieved.

In a similar design strategy, 2,4-dinitrobenzenesulfonyl can also be used as a group substituted by the $O_2^{\bullet-}$ nucleophilic reaction. Fluorescent probe **15** for mitochondrial $O_2^{\bullet-}$ was designed and synthesized by bonding the coupled mitochondrial localization peptide.[17] At the living cell level, applying probe **15** could allow observation of $O_2^{\bullet-}$ overproduction during single-walled carbon nanotube invasion.

The diphenylphosphinyl moiety also represents an excellent substrate for $O_2^{\bullet-}$ nucleophilic substitution reactions. Upon reaction with $O_2^{\bullet-}$, the diphenylphosphinyl moiety of the probe is cleaved off while exposing the fluorophore containing the phenolic oxygen negative ion, generating a bright fluorescence. Probe **16** emitted dual-color fluorescence due to changes in the donor–acceptor structure of the $O_2^{\bullet-}$ reaction product, which resulted in aggregate-induced shifts in emission wavelength from red to green.[18] The probe **16** displayed good selectivity for applications in imaging analysis of $O_2^{\bullet-}$ fluctuations in living cells. On the basis of the recognition of $O_2^{\bullet-}$ by the diphenylphosphinyl moiety, coupling the indole-xanthene derivative, probe **17**, achieved the near-infrared fluorescence detection of $O_2^{\bullet-}$.[19] Furthermore, probe **17** was successfully employed for $O_2^{\bullet-}$ fluctuation imaging of living cells, zebrafish and mice.

11.2 Fluorescent Sensors for H_2O_2

Hydrogen peroxide (H_2O_2) is the main ROS produced in the metabolic processes of organisms, and plays a vital role in a variety of physiological processes. This makes it a common marker of oxidative stress. Therefore, it is very important to develop fluorescent probes with fast response, high sensitivity and low cell autofluorescence interference for monitoring H_2O_2 levels *in vivo*. In recent years, many fluorescent probes for the detection of intracellular H_2O_2 have emerged (see Figure 11.5).

A large number of studies have shown that oxidative stress plays an important role in the pathogenesis of hypertension by disturbing the

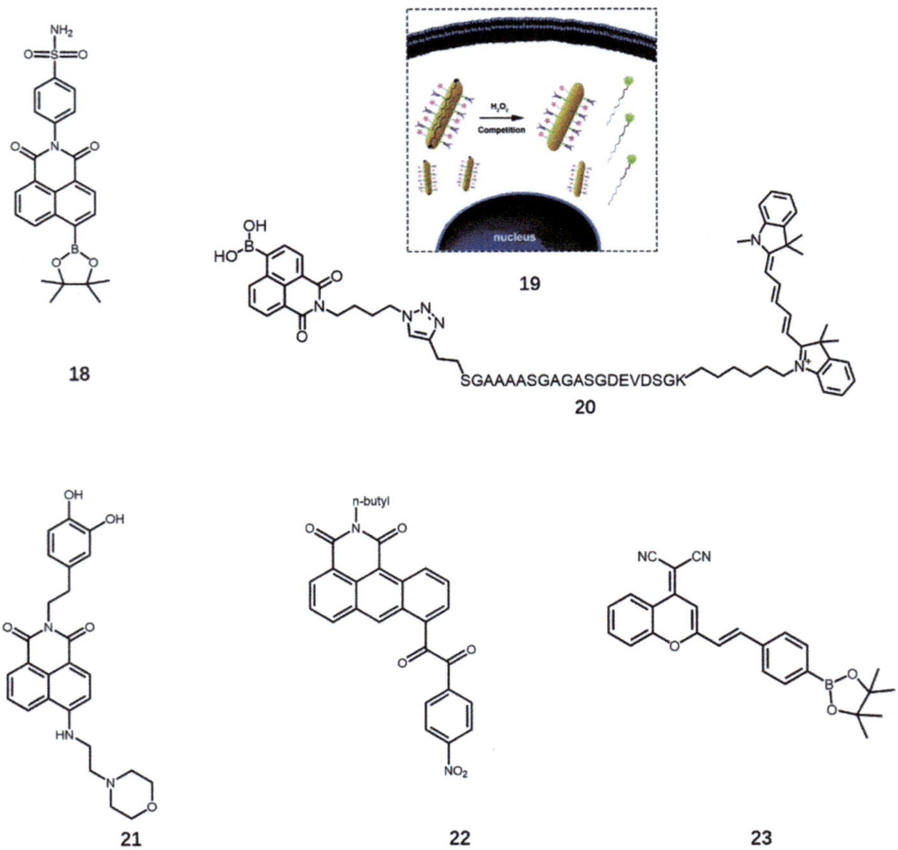

Figure 11.5 Structure of probes **18–23** for fluorescence detection of H_2O_2. Reproduced from ref. 21 with permission from the Royal Society of Chemistry.

balance between ROS and antioxidant defences. Notably, the concentration of H_2O_2 is proportional to the degree of oxidative stress in the Golgi apparatus. Although many TP fluorescent probes have been developed to visualize and image H_2O_2 in various organelles, *in situ* bioimaging of Golgi H_2O_2 is still scarce. Thus, Tang and colleagues designed a TP ratiometric fluorescent probe, **18** (Np-Golgi), with a phenyl sulfonamide moiety as the Golgi-targeting group, 1,8-naphthalimide as a fluorescent reporter and borate as a specific H_2O_2-recognition group.[20] The boronate ester reacts with H_2O_2 and is converted into a hydroxyl group, which promotes the push–pull electronic effect of the conjugated system, thus producing strong fluorescence emission. Np-Golgi can effectively accumulate in the Golgi to detect H_2O_2 with high selectivity and sensitivity both *in vivo* and *in vitro*, and Np-Golgi has been successfully applied to detect the Golgi H_2O_2 level in hypertensive mice.

The rupture of atherosclerotic plaques triggered by oxidative stress is a well-recognized consequence in the life-threatening conditions of myocardial infarction and stroke. Early quantification of plaque-specific oxidation states and more accurate assessment of plaque inflammatory activity and vulnerability *in vivo* are urgent requirements. H_2O_2 is a member of the ROS family. It can help to accelerate plaque progression, and can be used as an important marker to detect changes of oxidation status in atherosclerotic plaques. The Tang group developed a selective probe for H_2O_2, **19**.[21] Because the binding ability between Ce^{4+} ions and H_2O_2 is stronger than that between phosphate groups in DNA, when it encounters H_2O_2, **19** (CeO_2-NW) recovered the fluorescence signal by dissociation of DNA from the surface. This fluorescent nanosensor can achieve non-invasive imaging of H_2O_2 *in vitro* and *in vivo* with good sensitivity and selectivity, and so can accurately assess the level of inflammation and provide early warning of plaque vulnerability.

The DNA damage caused by ROS will promote the activation of programmed cell apoptosis, if not repaired in time. Apoptosis requires activation of the initiator cysteine protease, and caspase 3 has been recognized as a biomarker of apoptosis. In order to understand the relationship between ROS and apoptosis, it is urgent to monitor these two substances simultaneously. The group of Yi and Xiao developed a fluorescent probe **20** that can simultaneously detect H_2O_2 and caspase 3. This probe connects a H_2O_2 fluorescent reporter and Cy5 fluorescent chromophore through a caspase 3-specific recognition peptide.[22] In living cells, the accumulation of H_2O_2 will induce red fluorescence in living cells due to the presence of Förster resonance energy transfer (FRET). In contrast, because the overexpressed caspase 3 cleaves the peptide spacer, FRET is inhibited in apoptotic cells and emits blue fluorescence when exposed to H_2O_2. Therefore, **20** (pep4-NP1) successfully displayed different fluorescent colors when detecting H_2O_2 in living cells and apoptotic cells, indicating that the generation of endogenous ROS is not always accompanied by apoptosis.

The H_2O_2 inside the lysosome reacts with a labile iron pool to produce toxic hydroxyl free radicals (·OH), promote the production of other ROS and, finally, induce cell apoptosis. Therefore, the detection and imaging of hydrogen peroxide in lysosomes is very important. The group of Kumar and Kaur developed a lysosomal-targeted naphthalimide-based fluorescent probe, **21**.[23] The addition of hydrogen peroxide to the solution of **21** (LyNC), the catechol moiety is converted into an *o*-quinone part, thus inhibiting the photoinduced electron transfer (PeT), which has been successfully applied to high-sensitivity imaging of endogenous hydrogen peroxide in living cells and brain tissue.

Drug abuse can cause serious organ damage, especially drug-induced liver damage. This pathological process is accompanied by a rapid increase in the level of some physiological substances. Therefore, tracking H_2O_2 levels *in vivo* may be a way to determine organ damage and help evaluate drug-induced pathological processes. The Wu and Zeng groups designed and synthesized a ratiometric fluorescent nanoprobe **22**, which introduced nitrophenyldicarbonyl (H_2O_2 recognition moiety) onto the sixth position of the anthracene carboximide scaffold.[24] After reacting with H_2O_2, the recognition part is converted to a carboxyl group, thereby giving ratiometric detection of H_2O_2. By embedding probe molecules into the phospholipid bilayer, the nanoprobe has excellent biocompatibility, so **22** (NPS-A) can be successfully applied to the detection of endogenous H_2O_2 in living cells and can track the H_2O_2 level in the zebrafish APAP-induced organ injury model.

Peng and colleagues designed and synthesized a new borate-based two-photon fluorescent probe **23**, which includes dicyanomethylene-4*H*-pyran (chromophore) and boronate (H_2O_2 response group).[25] When the boronate is deprotected by H_2O_2 to the corresponding hydroxyl compound (TPNR-OH), near-infrared fluorescence emission is obtained due to the ICT effect. This two-photon fluorescent probe exhibits excellent photostability and may develop into a good hydrogen peroxide probe for chemical and biological applications.

11.3 Fluorescent Sensors for $^{\bullet}OH$

Hydroxyl radical ($^{\bullet}OH$) is the most oxidative reactive oxygen species in biological systems, and can attack biological molecules such as DNA, proteins and lipids, causing severe oxidative stress damage. The $^{\bullet}OH$ radical is associated with a variety of physiological and pathological processes, such as inflammation, cancer and psychiatric disorders. However, due its short lifespan and low concentration, it is difficult to accurately assess $^{\bullet}OH$ *in vivo*. In recent years, researchers have developed a series of fluorescent probes to detect $^{\bullet}OH$ (see Figure 11.6).

Based on the coumarin-merocyanine platform, the Lin group developed a ratiometric detection probe for $^{\bullet}OH$ (probe **24**).[26] In the absence of $^{\bullet}OH$, the central merocyanine structure of the probe was not conjugated with the coumarin structure, so it showed the typical emission of coumarin dye (usually 500 nm). However, after the reaction with $^{\bullet}OH$, the merocyanine structure was conjugated with the coumarin structure, and the emission was obviously red shifted, which realized the ratiometric detection of $^{\bullet}OH$. The probe had

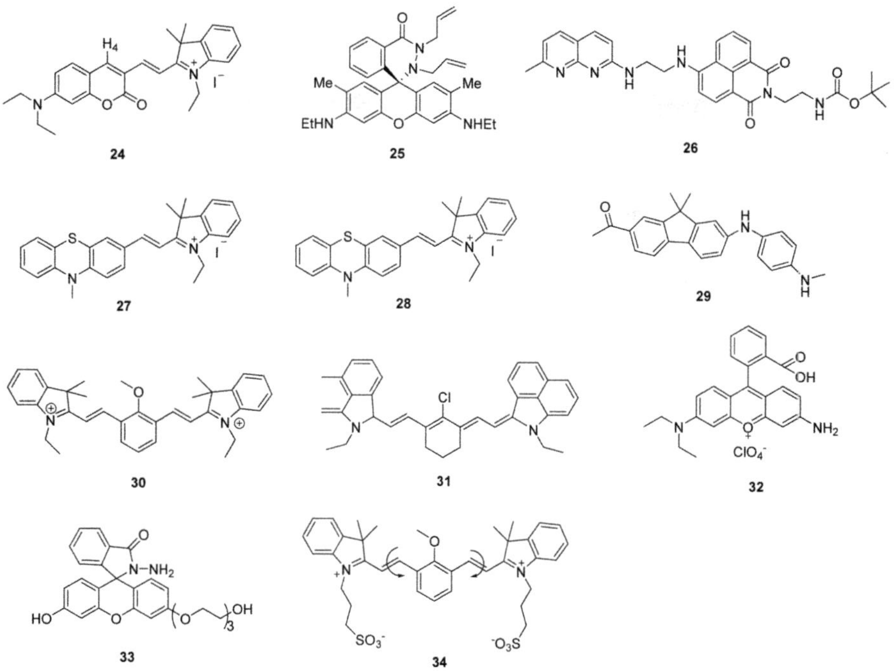

Figure 11.6 Structure of probes **24–34** for fluorescence detection of •OH.

the advantages of high selectivity to •OH, large Stocks shift (156 nm), high ratiometric signal change (210 times), near-infrared emission, good optical stability, excitation in the visible light region, good membrane permeability and so on. These excellent properties meant that the probe realized ratiometric fluorescence imaging of •OH in living cells for the first time. The research strategy provides a new design concept for the development of ratiometric fluorescent probes.

Inspired by rhodamine 6G, Jinsung Tae *et al.* developed a new fluorescent probe for detecting •OH by using the C–H bond oxidative dehydrogenation of rhodamine cyclic hydrazide (probe **25**).[27] The probe was highly selective for •OH and was almost free from interference from other ROS/RNS species. The probe was successfully used for fluorescence imaging of •OH in A549 and RAW264.7 cells. This practical imaging tool has broad applications for biological research related to •OH.

Through ethylenediamine linking position 4 of the naphthalenediimide group with position 2 of the naphthyridine group, a new naphthalimide–naphthyridine derivative for the ratiometric fluorescence imaging analysis of •OH was designed and synthesized by the Yi group (probe **26**).[28] The probe was highly selective for •OH and can effectively distinguish •OH from other ROS. The detection limit of the probe was 0.2 μM, and it showed a highly sensitive response to •OH. The probe achieved a rapid response to •OH in water and dimethylformamide (DMF) solutions in less than 2 minutes. In addition, the probe also had good photostability, low cytotoxicity and high biocompatibility. Using this probe $^{\bullet}OH$ was successfully visualized in RAW264.7 cells.

Sulfur atoms on phenothiazine have high electron cloud density and can be oxidized to sulfoxides. Inspired by this property of phenothiazine, the Sun group developed a new fluorescent probe based on the phenothiazine-cyanine platform, which can be used to detect $^{\bullet}OH$ in mitochondria of living cells and its distribution *in vivo* (probe **27**).[29] The changes in probe absorption and fluorescence signal were attributed to the oxidation of sulfur atoms in the Schiff base. The probe can be utilized as a ratiometric absorbance and a "turn-on" type fluorescence sensor to detect $^{\bullet}OH$. The probe had good selectivity for $^{\bullet}OH$ and can detect endogenous $^{\bullet}OH$ in living cells and the mitochondria of various bacteria. In addition, the probe can also be used to evaluate the ability TiO_2 nanoparticles of different diameter to produce $^{\bullet}OH$ under light irradiation, and to track $^{\bullet}OH$ in living zebrafish. The probe can be further applied to the detection of $^{\bullet}OH$ in cells, bacteria and other organisms.

The Yang group developed a highly sensitive and selective probe for $^{\bullet}OH$ detection based on the selective *O*-deacetylation of $^{\bullet}OH$ by 2,6-diiodophenol (probe **28**).[30] Two iodine atoms were introduced into the *ortho*-position of phenol hydroxyl to form diiodophenol. While quenching the fluorescence of fluorescein or 2′, 7′-dichlorofluorescein, the steric hindrance of the two iodine atoms would prevent the oxidation of other strong oxidants, such as HOCl or $ONOO^-$, thus improving the selectivity of the probe. The $^{\bullet}OH$ can react with diiodiophenol to generate monophenol free radicals, resulting in an electron transfer reaction, and eventually leading to the decomposition of aromatic compounds. At this point in the reaction, luciferin or 2′,7′-dichloroluciferin will be released, resulting in enhanced fluorescence. Using this probe, the authors calibrated and quantitatively analyzed the scavenging ability of a series of reported $^{\bullet}OH$ scavengers, and analyzed the endogenous $^{\bullet}OH$ produced by RAW264.7 cells treated with PMA using fluorescence confocal imaging and flow cytometry. Therefore, this probe has a good application prospects in exploring the biological processes related to $^{\bullet}OH$.

The Liu group reported the first two-photon small-molecule fluorescent probe with high sensitivity and specificity for tracing $^{\bullet}OH$ fluctuation *in vivo* (probe **29**).[31] Based on the *N*-dealkylation reaction mechanism mediated by $^{\bullet}OH$, the probe exhibited high sensitivity and high selective response to $^{\bullet}OH$ (1 equivalent of $^{\bullet}OH$ gave a fluorescence enhancement of 200 times). The probe was successfully applied to reveal the change of $^{\bullet}OH$ in the superoxide dismutase 1 (SOD1)-involved biological process, which confirmed that the inhibition of SOD1 activity and the downregulation of SOD1 expression would lead to an increase in $^{\bullet}OH$ level in cells. The relationship between SOD1 and $^{\bullet}OH$ levels was visually revealed for the first time. Changes in the fluorescence intensity of the probe can reflect vital information *in vivo* and help to elucidate the functions and effects of $^{\bullet}OH$-related biology.

Inspired by the unique aromatic hydroxylation reaction and the electrophilicity of $^{\bullet}OH$, the Ma group designed a high-sensitivity fluorescent probe for monitoring $^{\bullet}OH$ (probe **30**),[32] which enhanced the capture ability of $^{\bullet}OH$ by introducing a strong electron-donor methoxy group into the fluorochromic cyanine group. At the same time, the selective hydroxylation of the probe and the subsequent deprotonation reaction resulted in a larger π-conjugated system and a near-infrared fluorescence response. The probe had good biocompatibility and can be used for imaging $^{\bullet}OH$ in living cells under different conditions. Notably, the probe had high sensitivity and can monitor the formation of trace $^{\bullet}OH$ in the process of iron self-oxidation. This work contributes to a further understanding of the physiological and pathological processes associated with iron oxidation.

By directly destroying/restoring the conjugated system and rigid plane structure of organic fluorescent molecules, the Liu group prepared the first near-infrared region II fluorescent probe for $^{\bullet}OH$ detection (probe **31**).[33] After reaction with $^{\bullet}OH$, the probe emitted bright fluorescence in the range 1000–1400 nm. The probe had good sensitivity (limit of detection (LOD) = 0.5 nM) and high selectivity for $^{\bullet}OH$. Using this probe, Liu and colleagues successfully visualized the increase of $^{\bullet}OH$ in mouse abdominal cavity stimulated by lipopolysaccharide. More importantly, the probe can trace the subtle changes of $^{\bullet}OH$ in liver during acetaminophen-induced liver injury, and can provide high contrast and clear images.

Based on the breaking of the azo bond with $^{\bullet}OH$, the Lin group developed a novel mitochondrial-targeting fluorescent probe for monitoring $^{\bullet}OH$ in living cells and zebrafish (probe **32**).[34] The probe consisted of two parts: the anthracene group containing a lipophilic cation serves as the mitochondrial-targeting group and fluorophore, and the azo group serves as the recognition group and fluorescence-quenching group for $^{\bullet}OH$. In the absence of $^{\bullet}OH$, the probe had almost no fluorescence emission at 550 nm. With increasing $^{\bullet}OH$ concentration, the fluorescence emission at 550 nm gradually increased, the fluorescence intensity increased by 50 times and the detection limit of the probe was as low as 8 nM. Selectivity experiments showed that the probe possessed good selectivity to $^{\bullet}OH$ in all ROS/RNS. In addition, the probe has also been successfully applied to image $^{\bullet}OH$ in living cells and zebrafish, which provides a strong theoretical basis and technical support for the study of $^{\bullet}OH$ involved in physiological and pathological effects.

The Zhang group designed a fluorescent probe with dual reaction sites to recognize and analyze $^{\bullet}OH$ and HClO through the reconstruction and modification of common fluorescent dyes (probe **33**).[35] By introducing a hydrazide five-membered heterocycle into the fluorescein matrix, the dual reaction sites for $^{\bullet}OH$ and HClO were constructed at the same time as quenching of the fluorescein fluorescence. In addition, a triethylene glycol side chain attached to the fluorescein matrix greatly increased the ability of the probe to penetrate various biological barriers. After reacting with $^{\bullet}OH$ and HClO, the probe produced cyan and green fluorescence emissions, respectively, which could realize the real-time identification and quantitative analysis of two kinds of ROS in the mitochondria of living cells. Based on the advantages of dual-channel imaging, the probe can clearly show the distribution of $^{\bullet}OH$ and HClO in the intestine and liver of zebrafish during gut-blood-organ/tissue-kidney-excretion. In addition, the probe was applied to observe the production of $^{\bullet}OH$ in fresh wounds of zebrafish for the first time. These successful and unique applications make this multifunctional probe a promising tool for exploring ROS *in vivo*.

The Ma group designed a bifunctional fluorescent probe to visualize changes in intracellular $^{\bullet}OH$ and viscosity during ferroptosis (probe **34**).[36] The probe had a typical molecular rotor structure, in which there are two symmetric small π-conjugated structures, the indole part (electron acceptor) and the anisole part (electron donor), which are connected by a rotatable vinyl bond to form a D–π-A structure. With increase of the microenvironment viscosity, the intramolecular rotation is limited, thus inhibiting the distorted ICT, leading to enhanced fluorescence. The specific hydroxylation of aromatic compounds was utilized in the detection of $^{\bullet}OH$ by the probe. The strong electron-donor methoxy group was introduced to improve the capture ability of $^{\bullet}OH$, and thus improve the detection sensitivity. The probe showed high selectivity and sensitivity to both $^{\bullet}OH$ and viscosity in two completely different spectral regions, and realized the separate and simultaneous selective detection of both $^{\bullet}OH$ and viscosity. Using the probe, it was found that the process of ferroptosis was accompanied by the formation of $^{\bullet}OH$, increase of cytoplasmic viscosity and acceleration of lipid droplet formation. This fluorescent probe has broad application prospects in the study of viscosity and $^{\bullet}OH$ in cells and *in vivo*.

11.4 Fluorescent Sensors for HOCl

Hypochlorous acid (HOCl) is mainly generated from the peroxidation reaction between H_2O_2 and chloride ions (Cl^-) catalysed by myeloperoxidase (MPO) in living systems. As an essential oxidant produced in phagolysosomes, HOCl functions in the elimination of pathogens in the immune system. Furthermore, endogenous excessive HOCl has been implicated in many pathological processes, including rheumatoid arthritis, cardiovascular disease, atherosclerosis and Alzheimer's disease. Hence, tracing and detecting intracellular HOCl fluctuation precisely is of vital significance for demonstrating its biological function. In the past few years, plenty of fluorescent probes have been reported for the detection of HOCl *in vivo* based on the strongly oxidative property of HOCl (see Figure 11.7).

The first TP HOCl-responsive mitochondria- (probe **35**) and lysosome- (probe **36**) targeted fluorescence probes were reported by Chang.[37] Acedan, with strong "push–pull" structure, and 2-mercaptoethanol were employed as the TP fluorophore and recognition site for HOCl, respectively. In the presence of HOCl (20 equiv.), 670-fold fluorescence enhancement at 500 nm in PBS/EtOH (1:1, pH 7.4) solution was observed due to the recovery of the "push–pull" structure of acedan. The probe displayed high sensitivity, pH stability and excellent optical response towards HOCl in the

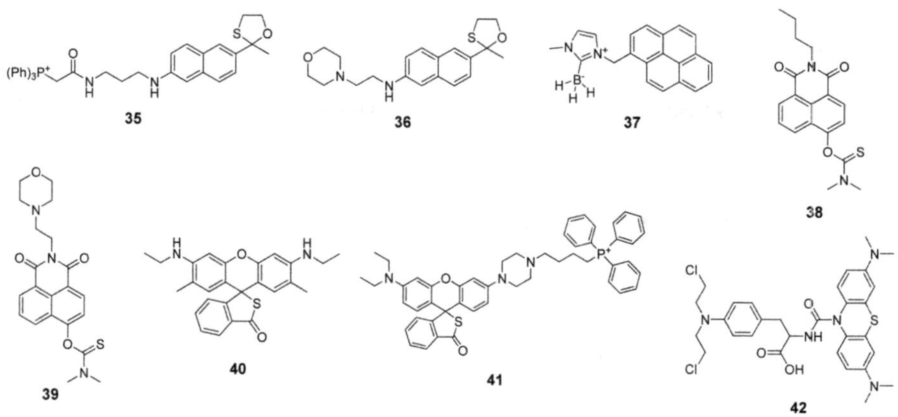

Figure 11.7 Structure of probes **35–42** for fluorescence detection of HOCl.

MPO/H_2O_2/Cl^- system. By using probes **35** and **36** in RAW 264.7 macrophages and inflamed mouse stimulated by lipopolysaccharides, elevated HOCl levels were confirmed.

A novel two-photon fluorescent probe (probe **37**) has been constructed by Yoon for TP ratiometric imaging of HOCl in living cells and tissues based on highly specific oxidation of the B–H bond.[38] The authors chose *N*-heterocyclic carbene boranes and pyrene as the HOCl recognition site and fluorescence reporter, respectively, to construct the probe skeleton. Upon titration of HOCl, a gradual fluorescence decrease at 477 nm and remarkable fluorescence increase at 374 nm were detected. The detection limit for probe **37** was calculated to be 3 μM, indicating its high sensitivity for HOCl. Facilitated by probe **37**, the authors found enhanced HOCl fluctuations in phorbol myristate acetate (PMA) pretreated RAW 264.7 cells and fresh rat hippocampal slices by TP fluorescence imaging, demonstrating its potential for tracing HOCl levels in living systems.

An ICT-based TP fluorescent probe (probe **38**) was reported by Tang to monitor HOCl, and which possesses strong sensitivity and selectivity.[39] *N*-Butyl-1,8-naphthalimide with typical "push–pull" structure and *N*,*N*-dimethylthiocarbamate (DMTC) were chosen as the fluorophore and recognition moiety, respectively. Due to the presence of DMTC, probe **38** is almost non-fluorescent in PBS/ethanol buffer solution (1 : 1, v/v, pH 7.4). When HOCl was added (0–1500.0 pM), an obvious fluorescence enhancement at 547 nm was detected under 410 nm excitation. The detection limit was determined as 7.6 pM for probe **38** towards HOCl. Probe **39**, containing the alkylmorpholine group, was developed to monitor lysosomal HClO levels in living cells. TP fluorescence images of live macrophages demonstrate that probe **38** and probe **39** are capable of fluorescence imaging of endogenous HOCl fluctuation. Furthermore, probe **38** and probe **39** could successfully achieve discrimination of tumors and normal tissues on the basis of HOCl levels, through TP tissue fluorescence imaging.

A rhodamine-based HOCl fluorescent probe (probe **40**) was developed to monitor endogenous HOCl in live *Drosophila* and neutrophils in a mouse model with infection.[40] Probe **40** was easily synthesized with two steps. After titration of HOCl, probe **40** displayed a 60-fold fluorescence enhancement at 545 nm in a KH_2PO_4 buffer. Furthermore, probe **40** remained active in the detection of HOCl in an acidic environment (pH 4–6), indicating that it could be applied to monitor HOCl generation in phagosomes. More importantly, endogenous HOCl production induced by *Drosophila* dual oxidase was visualized by probe **40**. Probe **40** served as a robust tool for tracing HOCl generation in neutrophils in the infection mouse model, demonstrating its capability in therapeutic drug screening for inhibiting HOCl production.

A rhodamine-based fluorescent probe (probe **41**) was designed for the detection of mitochondrial HOCl.[41] To monitor mitochondrial HOCl in macrophages, probe **41** was implemented by using a triphenylphosphonium moiety as the mitochondria-targeted group and a thiolactone unit as the response moiety. HOCl triggered a remarkable fluorescence enhancement of more than 200-fold with probe **41** at 580 nm. In addition, probe **41** exhibited pH stability, rapid response, high sensitivity and desirable specificity. Probe **41** could target mitochondria in macrophages specifically, with a high Pearson's coefficient of 0.92. Imaging results for probe **41**-incubated RAW264.7 cells suggest that HOCl exists in mitochondria of macrophages during bacterial infection.

A series of methylene blue-based probes that were triggered by HOCl were reported by Yi.[42] Detection, imaging and therapeutic functions of the probes enable quantitative monitoring of HOCl and release of prodrugs. When HOCl (15 μM) was added, probe **42** presented an excellent fluorescence response and inertia to other ROS/RNS species, as shown through detection in the near-infrared and by naked eye. Probe **42**-containing anticancer drug melphalan was applied to destroy human promyelocytic leukemia (HL-60) cells. After treatment with probe **42**, the HL-60 cells displayed reduced cell viability, indicating the therapeutic function.

11.5 Fluorescent Sensors for HOBr

As a kind of ROS, hypobromous acid (HOBr) is produced from the heme peroxidase-catalysed peroxidation reaction between H_2O_2 and bromide anion (Br^-). HOBr displays effective antibacterial action *in vivo*, however, excessive amounts of HOBr have been shown to be involved in many diseases. Many HOBr-responsive fluorescent probes have been designed to elucidate physiological and pathological functions of HOBr in living systems (see Figure 11.8).

Tang reported an ultrasensitive HOBr fluorescent probe (probe **43**), which was based on sulfilimine.[43] HOBr caused a cyclization reaction between the amino group and *S*-methyl group, resulting in a red-shifted fluorescence

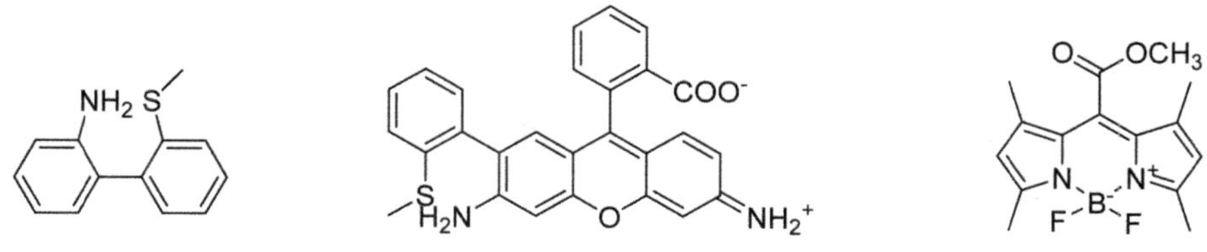

43 44 45

Figure 11.8 Structure of probes **43–45** for fluorescence detection of HOBr.

emission (*ca.* 100 nm). Probe **43** showed high specificity and ultra-sensitivity towards HOBr, with a detection limit of 17 pM. Probe **43** also presented other advantages, including broad pH tolerance (from 2 to 12), fast response towards HOBr, low cytotoxicity and good photostability. Using probe **43**, native HOBr levels were observed in HepG2 cells without Br^- stimulation. Moreover, it has been demonstrated in zebrafish that HOBr could be generated from Br^-, which suggests that probe **43** will be a powerful tool for monitoring HOBr in living systems.

A rhodamine 110-containing fluorescent probe (probe **44**) was designed by Tang for visualizing mitochondrial HOBr levels in HepG2 cells and zebrafish based on the cyclization reaction mechanism triggered by HOBr.[44] With addition of HOBr (0–100 nM) in HEPES 7.4 buffer, a cyclization reaction between the amino group and *S*-methyl group occurred, resulting in remarkable fluorescence enhancement of probe **44** at 663 nm. In addition, the detection limit of probe **44** was calculated to be 20 pM. Applying probe **44** in HepG2 cells, intracellular enhanced HOBr levels stimulated by Br^- and HOBr were detected under excitation at 633 nm. Endogenous HOBr upregulation has been verified by fluorescence imaging in zebrafish embryos.

Probe **45** contains a BODIPY chromophore and acts as a fluorescent probe for the detection of HOBr produced by EPO (eosinophil peroxidase).[45] The highly emissive J-aggregates of the dibrominated reaction product render probe **45** specifically responsive to HOBr, without interference from HOCl. In the presence of 2 equiv. of HOBr, an approximately 22-fold fluorescence enhancement at 616 nm was detected. Probe **45** exhibited desirable attributes, including a detection limit as low as 3.8 nM, excellent selectivity toward HOBr and pH stability of J-aggregates. Intraperitoneal injection of *Salmonella* leads to acute infection in mouse, and HOBr upregulation has been confirmed using probe **45**.

11.6 Fluorescent Sensors for $^1O_2$

Singlet oxygen ($^1O_2$) is the lowest excited electronic state of molecular oxygen, which functions in cell signal transmission and gene expression. As a strong oxidant, $^1O_2$ could oxidize biomolecules in living systems, including proteins, lipids and DNA, which may lead to many diseases. Due to the high

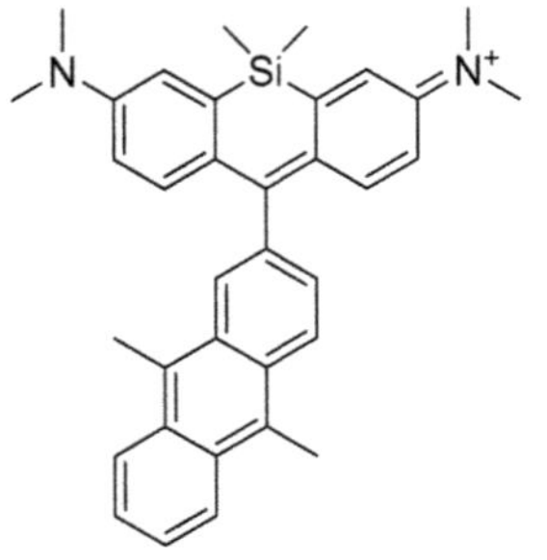

46

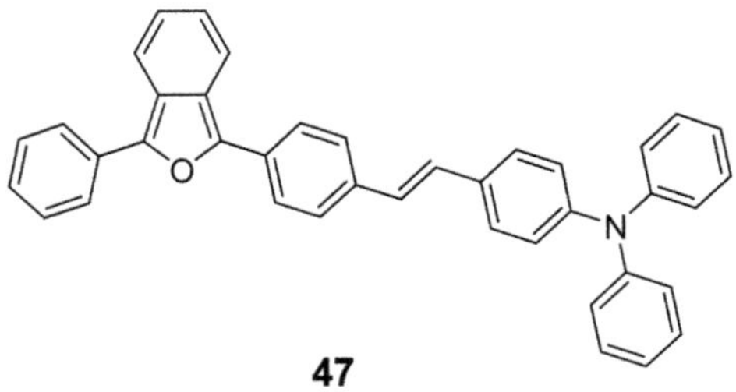

47

Figure 11.9 Structure of probes **46** and **47** for fluorescence detection of $^1O_2$.

cytotoxicity of $^1O_2$, the application of $^1O_2$ in photodynamic therapy (PDT) has received a great deal of attention.

Over the past several decades, many fluorescent probes have been reported for $^1O_2$ recognition (see Figure 11.9). Majima synthesized a far-red probe, probe **46**, for detecting $^1O_2$.[46] Silicon-containing rhodamine and anthracene moieties were employed as fluorescence reporter and recognition sites for $^1O_2$, respectively. Endoperoxide formation triggered by $^1O_2$ occurred in the anthracene moiety, resulting in 18-fold fluorescence enhancement of probe **46** in MeOH. Intracellular $^1O_2$ produced from aminolevulinic acid-derived protoporphyrin IX during PDT was mapped with probe **46**.

A series of 1,3-diphenylisobenzofuran (DPBF)-based ratiometric fluorescent probes for $^1O_2$ were developed (probe **47**).[47] The optimized probe exhibited high photoluminescence quantum yields (6.4–39%). The optical ratiometric response between probe **47** and $^1O_2$ was investigated in tetrahydrofuran (THF) solution containing IrF photosensitizer. The TP microscope images of PMA-treated RAW 264.7 cells displayed stronger green fluorescence (377–481 nm) under 750 nm TP excitation.

11.7 Fluorescent Sensors for O_3

Ozone (O_3) is closely involved in lung dysfunction, depression, *etc.* Recently, a number of well-designed fluorescent probes specific for O_3 have been created (see Figure 11.10). A near-infrared "turn-on" fluorescent probe (probe **48**) was designed for visualizing O_3 in brains of mouse.[48] It is expected that the cycloaddition reaction of the terminal olefin in probe **48** with O_3 would release the quinone derivative QCy7, which results in a more than 80-fold fluorescence enhancement at 690 nm under 570 nm excitation. Elevated O_3 levels in the brain tissue of mice with depression were monitored for the first time with probe **48**. More importantly, the authors disclosed that O_3 in the brain promotes depression through generating excessive interleukin-8.

Tang reported a near-infrared fluorescent probe (probe **49**) developed by linking the tricarbocyanine scaffold and L-tryptophan recognition moiety,

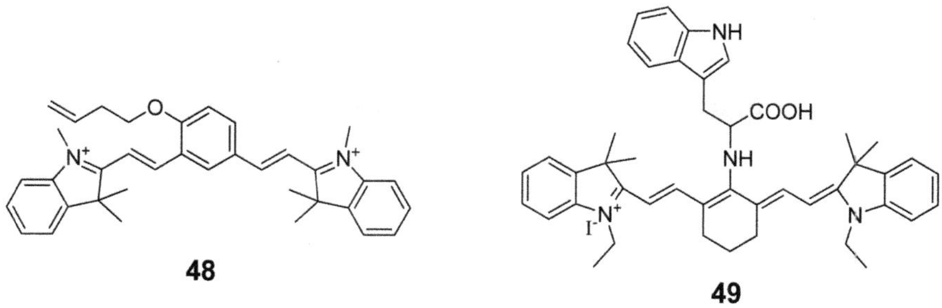

Figure 11.10 Structure of probes **48** and **49** for fluorescence detection of O_3.

which could quickly detect O_3 levels in living cells.[49] Probe **49** displayed a linear response between fluorescence intensity at 770 nm and O_3 concentration (0.05–7.0 μM). The detection limit of probe **49** was calculated to be 17 nM. Probe **49**, in conjunction with confocal fluorescence microscopy, enabled tracing endogenous O_3 levels in RAW 264.7 cells.

11.8 Fluorescent Sensors for L• and LOO•

Polyunsaturated fatty acids (PUFAs) are prone to oxidative modification by ROS, leading to lipid peroxidation.[50] Lipid peroxidation could destroy the structure of lipid membrane, which eventually causes serious cell damage.[51] Classical products of lipid peroxidation, such as lipid radical (L•), lipid peroxy radical (LOO•) and lipid hydroperoxides (LOOH), can cause protein destruction or other fatal gene mutations, leading to related diseases.[52,53] Therefore, it is essential to exploit tools for *in situ* and real-time tracking of these early lipid peroxidation products, for exploring the related physiological and pathological mechanisms.[54] Herein, recent research of fluorescent probes that respond to lipid peroxidation products in organisms is introduced (see Figure 11.11).

Yamada *et al.* designed the first fluorescent probe for L• (probe **50**).[55] The α-substituted nitroxide was the recognition group and has a high affinity for L•, realizing specific recognition of L•. The fluorescence of probe **50** was quenched by α-substituted nitroxide, and the fluorescence was restored when the α-substituted nitroxide reacted with L•. Probe **50** could directly capture L• in living cells and in living animal models. More importantly, the nitroxide portion of probe **50** showed potential for reducing inflammation in the early stages of carcinogenesis.

Considering the hydrophobic properties of lipid membranes, fluorescent probes for monitoring LOO• generally mimic the structure of PUFAs, and are designed to the lipophilic structure with unsaturated double bonds. Hudson *et al.* developed the first fluorescent probes for the detection of LOO• in the early 1970s (probe **51**).[56] They selected the natural polyunsaturated fatty acid *cis*-parinaric acid as the fluorescent probe for LOO•. When the double bond

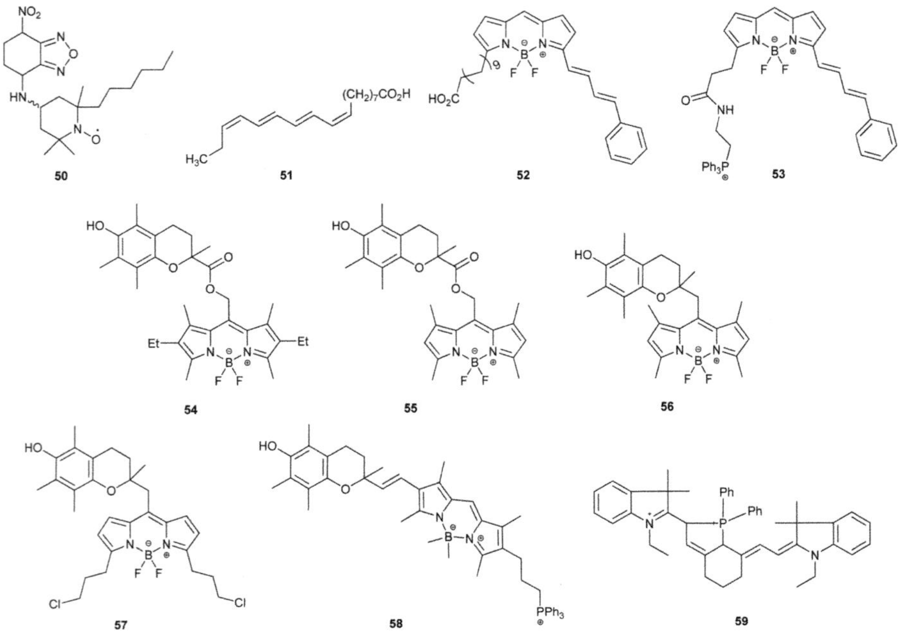

Figure 11.11 Structure of probes **50–59** for fluorescence detection of L$^{\bullet}$ and LOO$^{\bullet}$.

of the probe was oxidized by LOO$^{\bullet}$, the fluorescence of the compound was blue shifted due to the destruction of the conjugated structure. However, the absorption in the UV area and photoinstability of the probe limited its application in biological systems.

After that, a number of similar probes were developed based on the above detection mechanism of *cis*-parinaric acid on LOO$^{\bullet}$. Taking probe **52** as an example,[57] the probe itself was highly conjugated and emitted fluoresce at 592 nm. After being oxidized by LOO$^{\bullet}$, the conjugation of the probe was destroyed and the fluorescence blue shifted to 520 nm. By measuring the ratiometric emission intensity of the red/green channel, probe **52** realized the quantitative monitoring of LOO$^{\bullet}$, effectively reducing the error caused by false signal and instrument factors. Later, Murphy and colleagues further modified the structure of probe **52** by fusing the mitochondrial locator triphenylphosphonium lipophilic cation into the probe, and created probe **53**.[58] Probe **53** made it possible to quantitatively determine the LOO$^{\bullet}$ in mitochondria. However, the selectivity of these probes is not perfect because of their reactivity with other free radicals. In addition, the optical stability of these imaging tools is deficient.

As a scavenger of LOO$^{\bullet}$, vitamin E, especially α-tocopherol, can effectively stop the autooxidation of lipid chains. In recent years, Cosa *et al.* have simulated the structure and activity of α-tocopherol and developed a series of fluorescent probes for LOO$^{\bullet}$ detection (probes **54–58**). The chromanol moiety (Trolox) of the first-generated probe (probe **54**) owned similar activity to α-tocopherol, which was used to specifically recognize LOO˙. Probe **54** identified LOO˙ by an intramolecular PeT mechanism.[59] The fluorescence of the BODIPY moiety was quenched by the phenolic hydroxyl group in Trolox, before the reaction with LOO˙. After the reaction with LOO˙, the phenolic hydroxyl group was converted to quinone and the fluorescence was turned on, achieving the capture of LOO˙. In addition, the lipophilic BODIPY fluorophore in probe **54** could ensure targeting of the lipid membrane, thus improving the sensitivity of the probe. However, compared with homogeneous solution, the sensitivity of probe **54** in the lipid membrane was noticeably reduced.

In order to improve the sensitivity of the first-generated probe, Cosa *et al.* modulated the structure of probe **54** and exploited the second-generation probe (probe **55**).[60] By changing the C2 and C6 substituents of the BODIPY core to electron-withdrawing groups, the highest occupied molecular orbital level of the fluorophore was reduced, achieving more electron transfer. Thus, the fluorescence emission of probe **55** was enhanced about 40-fold, and the sensitivity was significantly elevated. For further improving the sensitivity of the probe to LOO˙, Cosa's laboratory constructed the third-generation probe (probe **56**).[60] Compared with the previous probe, the linker in probe **56** was changed from ester to methylene, thereby enhancing the phenolic hydroxyl due to the electron-withdrawing effect. The modification of this structure significantly increased the reaction ability of the chromanol moiety towards LOO˙, and the fluorescence emission enhancement of probe **56** was elevated 100-fold, thus realizing higher sensitivity.

After that, to further mimic the lipid solubility of α-tocopherol, the fourth-generation probe, probe **57**, was developed in subsequent years.[61] In probe **57**, Cosa *et al.* replaced the methyl groups at positions 3 and 5 of the BODIPY core with propyl-chlorides, and the fluorescence emission of probe **57** was enhanced 1000-fold. More importantly, the amelioration of probe **57** made it extremely sensitive to LOO˙ at the cell level. Using this well-designed probe **57**, the authors achieved real-time and non-invasive visualization of endogenous LOO˙ in the cellular lipid membrane. In addition, Cosa's laboratory continued to functionalize these kinds of probes. Over the next few years, probe **58** was designed and synthesized.[62] By introducing triphenylphosphine cations, probe **58** could selectively target the mitochondrial inner lipid membrane (IMM) and track LOO˙ on the mitochondrial membrane in real time.

Tang and co-workers developed a near-infrared fluorescent probe **59** for detecting LOOH in living cells.[63] The probe was almost non-fluorescent due to its non-conjugation, whereas upon reacting with LOOH it turned into the conjugated structure, and the fluorescence signal was strongly restored. Probe **59** exhibited rapid response, specific selectivity and high sensitivity to LOOH, and the LOD is 38 pM. What is more, the near-infrared excitation and emission of probe **59** could effectively avoid background fluorescence interference that was emitted from the biological organism. This probe, with low cytotoxicity, provides a potential tool in tracing LOOH in living biological systems.

11.9 Conclusion

We have presented the impressive advances in the fluorescent sensors for ROS in living cells and *in vivo*. Although fluorescent sensors that are specific for ROS have been widely developed, several efforts need to be improved.

1. Short emission wavelengths and the unstable structures of the fluorescent sensors are still the main factors that limit the depth of tissue penetration. Thus, fluorescent sensors with longer wavelengths and more rigid conjugated structures need to be developed. What is more, the development of fluorescent sensors that combine with multimodal imaging techniques, such as photoacoustic imaging and magnetic resonance imaging, could also improve the depth of tissue penetration.
2. Due to the short lifetime and high reactivity of ROS, it is urgent to design more fluorescent sensors that can provide reversible, highly sensitive, real-time and quantitative monitoring of ROS.
3. For further exploration of the signaling pathways between ROS in living cells and *in vivo*, developing fluorescent sensors for simultaneous detection of multiple ROS is essential.

In conclusion, developing functional fluorescence sensors for ROS in living cells and *in vivo* is still a future development direction in the field of oxidative stress. We firmly believe that the physiological and pathological mechanisms of ROS will be gradually interpreted and understood.

Acknowledgements

The authors would like to thank Prof. Wei Zhang, Dr Wen Zhang, Qi Ding, Jihong Liu and Di Su for their invaluable support during the drafting of this chapter.

References

1. B. Halliwell and M. Whiteman, *Br. J. Pharmacol.*, 2004, **142**, 231–255.
2. A. Gomes, E. Fernandes and J. L. Lima, *J. Biochem. Biophys. Methods*, 2005, **65**, 45–80.
3. X. Chen, X. Tian, I. Shin and J. Yoon, *Chem. Soc. Rev.*, 2011, **40**, 4783–4804.
4. W. Zhang, P. Li, F. Yang, X. Hu, C. Sun, W. Zhang, D. Chen and B. Tang, *J. Am. Chem. Soc.*, 2013, **135**, 14956–14959.
5. Q. Ding, Y. Tian, X. Wang, P. Li, D. Su, C. Wu, W. Zhang and B. Tang, *J. Am. Chem. Soc.*, 2020, **142**, 20735–20743.
6. W. Zhang, J. Liu, P. Li, X. Wang, S. Bi, J. Zhang, W. Zhang, H. Wang and B. Tang, *Biomaterials*, 2019, **225**, 119499.
7. W. Zhang, J. Zhang, P. Li, J. Liu, D. Su and B. Tang, *Chem. Sci.*, 2019, **10**, 879–883.
8. H. Xiao, C. Wu, P. Li and B. Tang, *Anal. Chem.*, 2018, **90**, 6081–6088.

9. J. J. Gao, K. H. Xu, B. Tang, L. L. Yin, G. W. Yang and L. G. An, *FEBS J.*, 2007, **274**, 1725–1733.
10. R. Q. Li, Z. Q. Mao, L. Rong, N. Wu, Q. Lei, J. Y. Zhu, L. Zhuang, X. Z. Zhang and Z. H. Liu, *Biosens. Bioelectron.*, 2017, **87**, 73–80.
11. P. Li, W. Zhang, K. Li, X. Liu, H. Xiao, W. Zhang and B. Tang, *Anal. Chem.*, 2013, **85**, 9877–9881.
12. H. Xiao, X. Liu, C. Wu, Y. Wu, P. Li, X. Guo and B. Tang, *Biosens. Bioelectron.*, 2017, **91**, 449–455.
13. J. J. Hu, N.-K. Wong, S. Ye, X. Chen, M.-Y. Lu, A. Q. Zhao, Y. Guo, A. C.-H. Ma, A. Y.-H. Leung, J. Shen and D. Yang, *J. Am. Chem. Soc.*, 2015, **137**, 6837–6843.
14. D. Q. Lu, L. Y. Zhou, R. W. Wang, X. B. Zhang, L. He, J. Zhang, X. X. Hu and W. H. Tan, *Sens. Actuators, B*, 2017, **250**, 259–266.
15. J. Huang, J. Li, Y. Lyu, Q. Miao and K. Pu, *Nat. Mater.*, 2019, **18**, 1133–1143.
16. P. Cheng, Q. Miao, J. Li, J. Huang, C. Xie and K. Pu, *J. Am. Chem. Soc.*, 2019, **141**, 10581–10584.
17. F. Si, Y. Liu, K. Yan and W. Zhong, *Chem. Commun.*, 2015, **51**, 7931–7934.
18. X. Gao, G. Feng, P. N. Manghnani, F. Hu, N. Jiang, J. Liu, B. Liu, J. Z. Sun and B. Z. Tang, *Chem. Commun.*, 2017, **53**, 1653–1656.
19. J. Zhang, C. Li, R. Zhang, F. Zhang, W. Liu, X. Liu, S. M.-Y. Lee and H. Zhang, *Chem. Commun.*, 2016, **52**, 2679–2682.
20. H. Wang, Z. He, Y. Yang, J. Zhu, W. Zhang, W. Zhang, P. Li and B. Tang, *Chem. Sci.*, 2019, **10**, 10876–10880.
21. Z. Liu, Y. Cao, X. Zhang, H. Yang, Y. Zhao, W. Gao and B. Tang, *J. Mater. Chem. B.*, 2020, **8**, 3502–3505.
22. Y. Wen, F. Xue, H. Lan, Z. Li, S. Xiao and T. Yi, *Biosens. Bioelectron.*, 2017, **91**, 115–121.
23. S. I. Reja, M. Gupta, N. Gupta, V. Bhalla, P. Ohri, G. Kaur and M. Kumar, *Chem. Commun.*, 2017, **53**, 3701–3704.
24. J. Peng, X. Hou, F. Zeng and S. Wu, *Biosens. Bioelectron.*, 2017, **94**, 278–285.
25. H. Li, Q. Yao, J. Fan, J. Du, J. Wang and X. Peng, *Biosens. Bioelectron.*, 2017, **94**, 536–543.
26. Y. Lin, L. Weiying and S. Jizeng, *Chem. Commun.*, 2010, **46**, 7930.
27. K. Minjeong, K. Sung-Kyun, K. Hyemi, S. Injae and T. Jinsung, *Chem. Commun.*, 2013, **49**, 7959.
28. M. Luyan, W. Yongquan and Y. Tao, *Chem. Commun.*, 2014, **50**, 4843.
29. L. Fei, D. Juan, S. Da, X. Meiying and S. Guoping, *Chem. Commun.*, 2016, **52**, 4636.
30. B. Xiaoyu, H. Yueyang, L. Mingyang and Y. Dan, *Angew. Chem., Int. Ed.*, 2017, **56**, 12873.
31. Z. Lingyu, X. Tian, H. Wei, C. Shiyu, C. Siyu, L. Yidi and L. Zhihong, *Anal. Chem.*, 2018, **90**, 1317.
32. L. Hongyu, L. Xiaohua, S. Wen, X. Yanhui and M. Huimin, *Angew. Chem., Int. Ed.*, 2018, **130**, 13012.
33. F. Wenqi, Z. Yuying, L. Zhen, Z. Shuyang, L. Weijie and L. Zhihong, *Anal. Chem.*, 2019, **91**, 15757.

34. Y. Gangqiang, D. Haiyuan, S. Hongyan, Z. Liyi and L. Qinlu, *Sens. Actuators, B*, 2019, **296**, 126706.
35. Z. Ruilong, Z. Jun, H. Guangmei, L. Zhengjie, L. Cui, Z. Cheng, L. Bianhua, J. Changlong, L. Renyong, Z. Tingting, H. Ming-Yong and Z. Zhongping, *J. Am. Chem. Soc.*, 2016, **138**, 3769.
36. L. Hongyu, S. Wen, L. Xiaohua, H. Yiming, F. Yu and M. Huimin, *J. Am. Chem. Soc.*, 2019, **141**, 18301.
37. L. Yuan, L. Wang, B. K. Agrawalla, S.-J. Park, H. Zhu, B. Sivaraman, J. Peng, Q.-H. Xu and Y.-T. Chang, *J. Am. Chem. Soc.*, 2015, **137**, 5930–5938.
38. Y. L. Pak, S. J. Park, D. Wu, B. Cheon, H. M. Kim, J. Bouffard and J. Yoon, *Angew. Chem., Int. Ed.*, 2018, **57**, 1567–1571.
39. B. Zhu, P. Li, W. Shu, X. Wang, C. Liu, Y. Wang, Z. Wang, Y. Wang and B. Tang, *Anal. Chem.*, 2016, **88**, 12532–12538.
40. X. Chen, K.-A. Lee, X. Ren, J.-C. Ryu, G. Kim, J.-H. Ryu, W.-J. Lee and J. Yoon, *Nat. Protoc.*, 2016, **11**, 1219–1228.
41. J. Zhou, L. Li, W. Shi, X. Gao, X. Li and H. Ma, *Chem. Sci.*, 2015, **6**, 4884–4888.
42. P. Wei, L. Liu, Y. Wen, G. Zhao, F. Xue, W. Yuan, R. Li, Y. Zhong, M. Zhang and T. Yi, *Angew. Chem., Int. Ed.*, 2019, **58**, 4547–4551.
43. K. Xu, D. Luan, X. Wang, B. Hu, X. Liu, F. Kong and B. Tang, *Angew. Chem., Int. Ed.*, 2016, **55**, 12751–12754.
44. X. Liu, A. Zheng, D. Luan, X. Wang, F. Kong, L. Tong, K. Xu and B. Tang, *Anal. Chem.*, 2017, **89**, 1787–1792.
45. T.-I. Kim, B. Hwang, B. Lee, J. Bae and Y. Kim, *J. Am. Chem. Soc.*, 2018, **140**, 11771–11776.
46. S. Kim, T. Tachikawa, M. Fujitsuka and T. Majima, *J. Am. Chem. Soc.*, 2014, **136**, 11707–11715.
47. D. Song, S. Cho, Y. Han, Y. You and W. Nam, *Org. Lett.*, 2013, **15**, 3582–3585.
48. P. Li, J. Wang, X. Wang, Q. Ding, X. Bai, Y. Zhang, D. Su, W. Zhang, W. Zhang and B. Tang, *Chem. Sci.*, 2019, **10**, 2805–2810.
49. K. Xu, S. Sun, J. Li, L. Li, M. Qiang and B. Tang, *Chem. Commun.*, 2012, **48**, 684–686.
50. K. J. Barnham, C. L. Masters and A. I. Bush, *Nat. Rev. Drug Discovery*, 2004, **3**, 205–214.
51. C. Chatgilialoglu and C. Ferreri, *Acc. Chem. Res.*, 2005, **38**, 441–448.
52. B. Kalyanaraman, *Redox Biol.*, 2013, **1**, 244–257.
53. K. Krumova and G. Cosa, *Photochemistry*, 2013, **41**, 279.
54. X. Wang, P. Li, Q. Ding, C. Wu, W. Zhang and B. Tang, *J. Am. Chem. Soc.*, 2019, **141**, 2061–2068.
55. K. Yamada, F. Mito, Y. Matsuoka, S. Ide, K. Shikimachi, A. Fujiki, D. Kusakabe, Y. Ishida, M. Enoki, A. Tada, M. Ariyoshi, T. Yamasaki and M. Yamato, *Nat. Chem. Biol.*, 2016, **12**, 608–613.
56. L. A. Sklar, B. S. Hudson and R. D. Simoni, *Proc. Natl. Acad. Sci. U. S. A.*, 1975, **72**, 1649–1653.
57. G. P. Drummen, L. C. van Liebergen, J. A. Op den Kamp and J. A. Post, *Free Radical Biol. Med.*, 2002, **33**, 473–490.

58. T. A. Prime, M. Forkink, A. Logan, P. G. Finichiu, J. McLachlan, P. B. Li Pun, W. J. Koopman, L. Larsen, M. J. Latter, R. A. Smith and M. P. Murphy, *Free Radical Biol. Med.*, 2012, **53**, 544–553.
59. P. Oleynik, Y. Ishihara and G. Cosa, *J. Am. Chem. Soc.*, 2007, **129**, 1842.
60. K. Krumova, S. Friedland and G. Cosa, *J. Am. Chem. Soc.*, 2012, **134**, 10102–10113.
61. L. E. Greene, R. Lincoln and G. Cosa, *J. Am. Chem. Soc.*, 2017, **139**, 15801–15811.
62. K. Krumova, L. E. Greene and G. Cosa, *J. Am. Chem. Soc.*, 2013, **135**, 17135–17143.
63. P. Li, B. Tang, Y. Xing, P. Li, G. Yang and L. Zhang, *Analyst*, 2008, **133**, 1409–1415.

CHAPTER 12

Subcellular Compartment-targeting Fluorescent Zn^{2+} Probes

TOSHIYUKI KOWADA[a] AND SHIN MIZUKAMI*[a]

[a]Institute of Multidisciplinary Research for Advanced Materials, Tohoku University, 2-1-1 Katahira, Aoba-ku, Sendai, Miyagi 980–8577, Japan
*E-mail: shin.mizukami@tohoku.ac.jp

12.1 Introduction

Since fluorescence imaging can visualize various biological events in cells in real time because of its high spatiotemporal resolution and sensitivity, it is an indispensable tool in biological research. To date, many fluorescent probes have been developed and have contributed to the evolution of this field. Among them, Fura-2 is widely used in neuroscience as it can monitor changes in [Ca^{2+}], which is responsible for intracellular signaling.[1] Recent studies have revealed that Zn^{2+} functions as a signaling mediator and is involved in intracellular signaling.[2] Intracellular Zn^{2+} is mostly bound to proteins, such as metallothionein.[3] The transport of labile Zn^{2+}, which exists at very low concentrations, into and out of cells and organelles is regulated by two families of zinc transporters, ZIPs and

Monographs in Supramolecular Chemistry No. 34
Fluorescent Chemosensors
Edited by Luling Wu, Adam C. Sedgwick, Xiao-Peng He and Tony D. James

Published by the Royal Society of Chemistry, www.rsc.org

ZnTs.[4] Dysfunction of these zinc transporters leads to disruption of zinc homeostasis, which can cause various diseases. To date, more than 20 types of zinc transporters are known in mammals, and the location of their expression varies depending on their subtypes. To analyze the function of each zinc transporter, it is necessary to image the Zn^{2+} dynamics in each organelle.

Organelle-selective detection and quantification of labile Zn^{2+} have been performed by fusing signal sequences for organelle localization to protein-based Zn^{2+} sensors using genetic engineering techniques, which are summarized in other reviews.[5,6] Small-molecule Zn^{2+} probes can be targeted to various organelles by attaching a functional group that responds to each organellar environment, such as acidic pH, negative membrane potential, or a high-affinity structure toward organelle-specific proteins (Figure 12.1).[7] As another methodology, there have been increasing reports of hybrid probes that combine the advantages of proteins and small molecules. This chapter summarizes the research on these small-molecule and hybrid fluorescent Zn^{2+} probes (Table 12.1) and discusses future challenges.

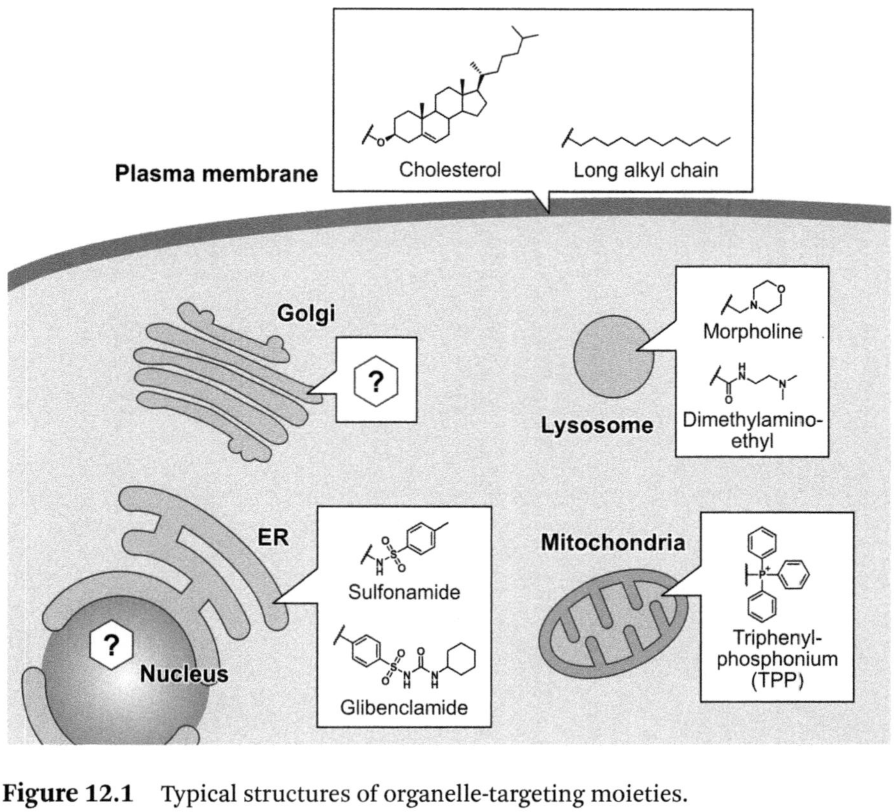

Figure 12.1 Typical structures of organelle-targeting moieties.

Table 12.1 Photophysical properties of Zn^{2+} probes.

	Absorption				Emission						
Probe	λ_{free} [nm]	ε_{free} [$\times10^4$ M^{-1} cm^{-1}]	λ_{bound} [nm]	ε_{bound} [$\times10^4$ M^{-1} cm^{-1}]	λ_{free} [nm]	Φ_{free}	λ_{bound} [nm]	Φ_{bound}	K_d (pH)	Cell line	Reference
Cytosol											
Zinbo-5	337	2.3	376	1.9	407	0.02	443	0.10	2.2 nM (7.2)	L(TK)[−]	16
FluoZin-3	494	7.2	494	—	515	<0.005	515	0.43	15 nM (7.4)	HT-29	15 and 17
BZn-Cyto	367	2.4	367	—	470	0.025	470	—	1.8 nM (7.2)[d]	HeLa	51
Plasma membrane											
LF-Chol[a]	470	—	470	—	517	<0.005	517	0.18	0.13 µM (7.2)	HeLa	21
ZIMIR[b]	493	7.3	493	—	515	0.0032	515	0.23	0.45 µM (7.5)	INS-1, MIN6	22
ZIMIR-Ex4(9–39)	490	—	490	—	515	0.02	515	0.21	0.9 µM (7.5)	MIN6	23
Palm-ZP1	517	6.6	506	7.4	534	0.16	529	0.79	82 pM (7)	HeLa	24
Palm-ZQ	336	0.39	360	0.36	483	≤0.014	483	0.35	20 nM (7)	RWPE-1	24
FZn-mem[c]	400	—	—	—	506	0.19	—	—	19 nM[d]	HeLa	25
ER											
ZBR1	478	1.9	530	2.6	625	0.067	628	0.41	0.69 nM (7.0)	HeLa, RAW 264.7, NSCs	28
1	346	—	346	—	414	0.05	414	0.28	2.8 nM (7.2)	HeLa	30
2	346	—	346	—	414	0.041	414	0.25	3.5 nM (7.2)	HeLa, MCF-7, EC23, HepG2	32
3	346	—	346	—	414	0.013	414	0.041	4.7 nM (7.2)	HeLa, MCF-7, EC23, HepG2	32
Mitochondria											
RhodZin-3	545	—	545	—	575	—	575	—	65 nM (7.0)	Neurons	42 and 53
ZS5	497	3.3	490	4.2	522	0.36	517	0.80	1.5 µM (7)	HeLa	44
ZBR4	569	2.8	573	3.7	629	0.036	631	0.50	3.3 nM (7.0)	HeLa	45
4[e]	347 (D), 602 (A)	8.1 (D), 2.3 (A)	363 (D), 602 (A)	6.7 (D), 2.1 (A)	633	0.28	633	0.29	9.1 µM (–)	HeLa S3	46
DA-ZP1-TPP	—	—	510	9.2[f]	—	<0.001	529	0.75	0.60 nM (7)[f]	HeLa, RWPE-1, RWPE-2, PC-3	47
SZn-Mito	388	1.9	375	2.3	500	0.15	493	0.92	3.1 nM (7.2)	HeLa	48
SZn2-Mito	413	1.9	395	2.1	536	0.0048	536	0.33	1.4 nM (7.2)[d]	HeLa	50

FZn-Mito	392	1.3	392	—	559	0.023	559	—	17 nM (7.2)[d]	HeLa	51
DQZn2	420	—	370	—	550	0.11	504	0.22	0.45 nM (7.4), 0.15 nM (8.0)	NIH3T3	52
Mito-ST	453	0.62	391	0.42	577	—	542	0.03	8.2 nM (7.2)	HeLa, COC2, MCF-7	54
Lysosome											
SpiroZin2	—	—	518	3.1	—	—	645	0.001	3.6 nM (7)	HeLa, MIN6	56 and 57
DQZn4	405	1.8	405	1.8	536	0.07	495	0.20	0.2 nM (7.4), 16 nM (5.2)	NIH3T3	58
LysoZn-1	597	14	568	13	647	—	578	0.23	68 μM (7.2), 81 μM (4.7)	HeLa, MCF-7, NSCs	59
5[g]	435	1.6	444	1.8	528	0.03	536	0.23	8.5 μM (5.0)	NIH 3T3	60
Lys-NBD-TPEA	—	—	—	—	536	—	536	—	—	A549	61
LysoDPP-C4	453	—	453	—	512	0.092[g]	517	0.37[g]	1.9 nM (7.2)	HeLa, PC3, DU145, RWPE1	62
6	350		359		504	0.32	504	0.33	0.17 nM^2 (7.3)	HeLa S3	63
Golgi											
BPSQ	302	—	364	0.47	532	0.015	532	0.055	1.8 pM (7.2)	HeLa	64
NBD-TPEA	496	2.1	478	1.4	542	0.003	550	0.046	2 nM (7.2)	HeLa	65
Zinpyr-1	515	8.0	507	8.4	529	0.39	525	0.87	0.7 nM (7.0)	COS-7	14
SZnC	388	—	393	—	499	0.12	499	0.93	1.5 nM (7.2)[d]	HeLa, A431, RAW 264.7, HepG2	66
Hybrid probes											
ZP1BG[h]	490	—	490	—	532	—	532	—	—	HeLa	68
ZP1-12Cl[h]	518	8.1	512	8.5	532	0.27	531	0.61	0.11 nM (7.0)	HeLa	70
ZIMIR-HaloTag	—	—	—	—	—	—	515	—	0.13 μM (7.5)	INS-1, MIN6	72
ZnDA-1H[h]	403	2.2	438	2.8	516	0.11	507	0.91	0.54 μM (6.5)	HeLa	73 and 74

[a]A methyl derivative, LF-Me, was used for the measurements instead of the cholesterol derivative, LF-Chol.
[b]An ethyl derivative was used for the measurements instead of the dodecyl derivative, ZIMIR.
[c]Measured in MOPS (pH 7.2)/DPPC (1 : 1).
[d]Measured using a two-photon process.
[e]D and A represent the donor and acceptor parts, respectively. Measured in MeCN.
[f]Obtained for the hydrolyzed form, ZP1-TPP.
[g]Measured at pH 5.0.
[h]Measured in the protein-tag-bound state.

12.2 Small-molecule Probes With Organelle-localizable Properties

12.2.1 Nucleus

A Zn^{2+}-responsive transcription factor, metal-responsive transcription factor 1 (MTF-1), binds to Zn^{2+} in the cytosol and translocates into the nucleus.[8] Thus, investigating the distribution of [Zn^{2+}] and Zn^{2+} dynamics in the cytosol and nucleus is critical for studying the functions of these proteins in the nucleus. Although nuclear localization of probes has been achieved using nuclear localization signal (NLS)-fused protein-based probes,[9] to the best of our knowledge, no nuclear-localizable small-molecule Zn^{2+} probes have been reported so far. Recent studies have shown that various fluorescent dyes can be localized in the nucleus using Hoechst as a localization motif.[10–12] This method can be applied to the development of nuclear-localizable Zn^{2+} probes.

12.2.2 Cytosol

As a result of intensive studies in the early 2000s, many fluorescent Zn^{2+} probes, such as ZnAF-2,[13] Zinpyr-1,[14] and FluoZin-3,[15] have been developed, many of which are commercially available. Zinbo-5 and FluoZin-3 (Figure 12.2) were used for the quantitative analysis of cytosolic labile Zn^{2+}. In 2004, O'Halloran *et al.* developed a benzoxazole-based ratiometric probe, Zinbo-5, whose absorption and emission maxima were red shifted upon binding to Zn^{2+} by approximately 40 nm.[16] The K_d value for Zn^{2+} was estimated to be 2.2 nM at pH 7.2. Zinbo-5 showed sufficient cell membrane permeability and visualized changes in [Zn^{2+}] induced by adding zinc pyrithione (ZPT) and *N*, *N*, *N'*, *N'*-tetrakis (2-pyridylmethyl)-ethyenediamine (TPEN) in the cytosol of fibroblast [L(TK)$^-$] cells. The ratio of emission intensities at 445 and 402 nm in the initial state was significantly higher than that after TPEN addition, suggesting that approximately 1 nM of Zn^{2+} is present in the cytosol of [L(TK)$^-$] cells, although this was not mentioned in the paper.

The concentration of intracellular labile Zn^{2+} in HT-29 cells was quantified using FluoZin-3,[17] and the K_d for Zn^{2+} was reported to be 15 nM.[15] The probe was incorporated into the cells using an acetoxymethyl (AM) ester derivative.

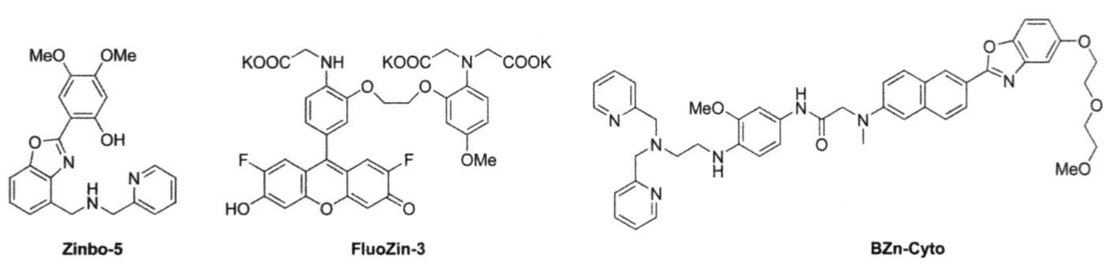

Figure 12.2 Structures of cytosol-localizable Zn^{2+} probes.

The [Zn^{2+}] in the absence of the probe was estimated by extrapolating the labile [Zn^{2+}] calculated from the fractional saturation of FluoZin-3 at various intracellular probe concentrations (0.35–5 μM). As a result, the labile [Zn^{2+}] of resting, proliferating, and differentiated cells were estimated to be 614 pM, 784 pM, and 1.25 nM, respectively.

Palmer *et al.* speculated that the significant differences in reported [Zn^{2+}] might be due to differences in cell types, probes, and measurement conditions.[18] They clarified the causes by directly comparing the protein-based probe, ZapCY2 (K_d = 811 pM), and small-molecule probe, FluoZin-3 (K_d = 15 nM). The advantage of protein-based probes is their accurate subcellular localization by fusing with a localization signal peptide, which allows the evaluation of changes in [Zn^{2+}] in different organelles. ZapCY2 could be localized to the cytosol and nucleus by fusion with the nuclear export signal or NLS, although ZapCY2 without a signal sequence was used to quantify labile [Zn^{2+}] in this study. In contrast, FluoZin-3 showed strong fluorescence in the Golgi apparatus of HeLa cells (Pearson's correlation coefficient (PCC) = 0.60 with a Golgi marker, GalT-mCherry). Although this was not the case in MIN6 or cortical neurons (PCC = 0.37, 0.41), it seemed to be preferentially localized in the synaptic vesicles of neurons (PCC = 0.66 with VAMP2-mCherry). In the quantification, as in previous reports,[17] different [Zn^{2+}] values ranging from 5–14 nM were calculated using FluoZin-3 depending on the probe concentration, that is, lower concentration probes gave higher quantitative values of labile [Zn^{2+}]. This result indicates that the intracellular Zn^{2+}-buffering system was affected by FluoZin-3 because the probe was treated just before the imaging experiments. In contrast, ZapCY2 provided a constant value, regardless of the intracellular protein probe concentration. This result implies that ZapCY2 was incorporated into the intracellular zinc-buffering system as the imaging experiments were performed after sufficient time from protein expression. Finally, the labile [Zn^{2+}] in HeLa, RWPE1, LNCaP, MIN6, and rat cortical neuron cells were calculated using ZapCY2 to be 83, 55, 272, 110, and 80 pM, respectively.

12.2.3 Plasma Membrane

The synaptic vesicles of glutamatergic neurons contain large amounts of Zn^{2+}, and labile Zn^{2+} is released with glutamate during neuronal excitation.[19] In addition, in pancreatic beta cells, Zn^{2+} exists mainly as a complex with insulin in the insulin granules and is released from the cells together with insulin.[20] Zn^{2+} probes localized on the cell membranes enable a practical analysis of such physiological phenomena, and cholesterol and long-chain alkyl groups have been used as localization motifs (Figure 12.3).

Taki *et al.* developed a fluorescein-based probe, LF-Chol, with a cholesterol moiety as a membrane anchor.[21] Owing to the negative charge of the Zn^{2+} chelator and the cholesterol interacting with the lipid membrane, LF-Chol can be immobilized on the extracellular side of the cell membrane. Upon binding to Zn^{2+}, its fluorescence intensity was enhanced 45-fold at pH 7.2,

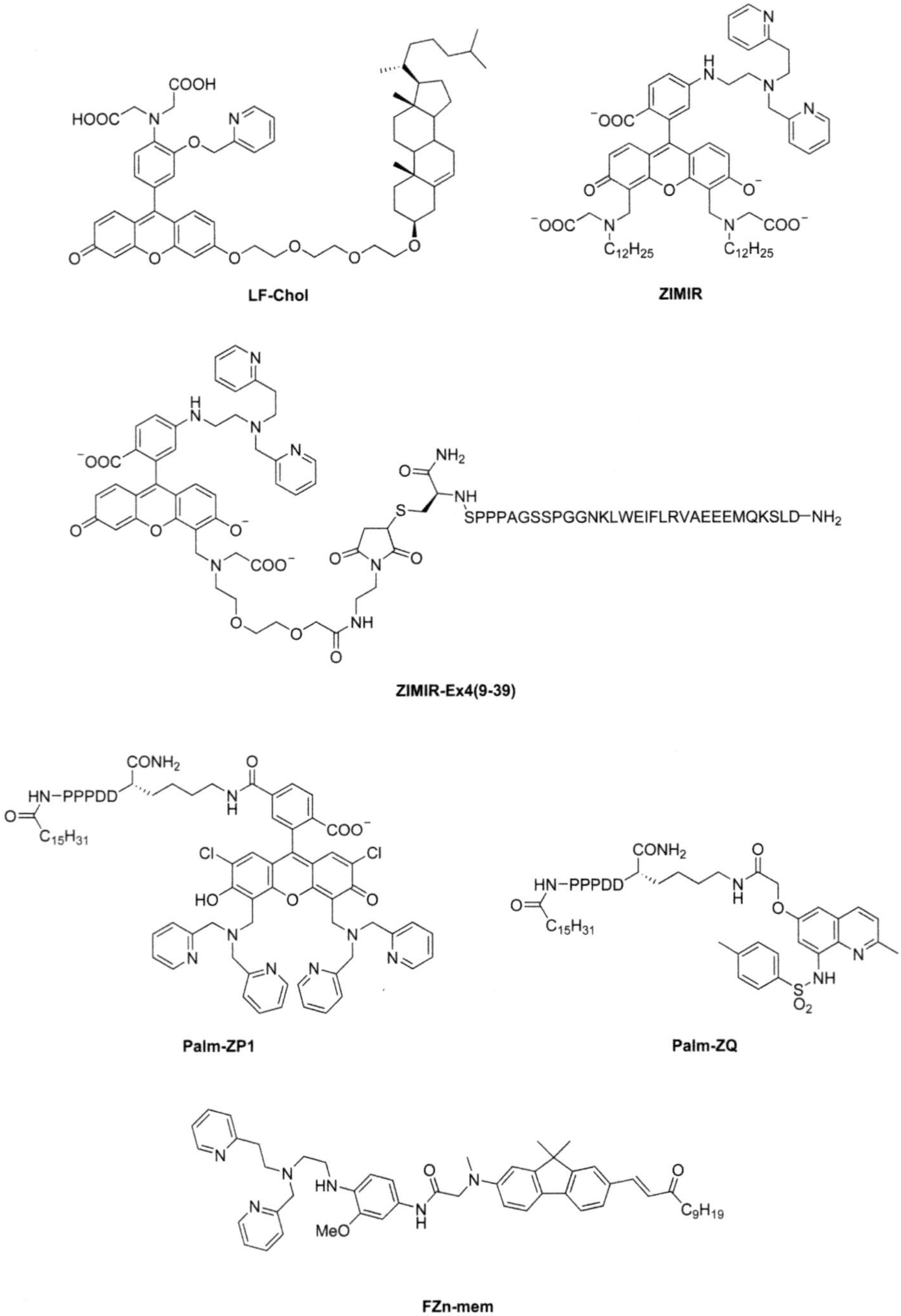

Figure 12.3 Structures of plasma-membrane-localizable probes.

as the fluorescence quenching by photoinduced electron transfer (PeT) is canceled. By adding 2.5 μM LF-Chol to HeLa cells, a faint fluorescence was observed from the membranes, and the fluorescence signal was increased by adding 20 μM Zn^{2+} and 100 μM ethylenediaminetetraacetic acid (EDTA).

Insulin granules contain high concentrations of Zn^{2+}, which are released from the cell by exocytosis during insulin release. Li *et al.* developed a membrane-anchoring Zn^{2+} probe, ZIMIR, to observe this phenomenon.[22] ZIMIR possesses a dodecyl group for membrane anchoring, and such a long-chain alkyl group causes aggregation in aqueous solution, which hampers the evaluation of photophysical properties. Thus, the derivative with an ethyl group was used for the evaluation in this study, which showed a 70-fold fluorescence enhancement upon Zn^{2+} binding. The K_d value for Zn^{2+} was estimated to be 0.45 μM at pH 7.5. At physiological pH, ZIMIR is negatively charged and firmly anchored to the outer leaflet of the plasma membrane by a combination of long alkyl chains. Once the probe was anchored to the membrane, intense fluorescence remained even after washing the cells. On the other hand, the derivative with nonyl groups, instead of dodecyl groups, was readily removed by washing. After adding the probe to MIN6 cells, the fluorescence intensity on the membrane gradually increased over 20 min, followed by probe uptake into the cells by endocytosis. The fluorescence signal reversibly changed by modifying [Zn^{2+}] in the extracellular fluid. In addition, the fluorescence of ZIMIR on the membrane increased due to insulin release upon stimulation with high concentrations of glucose and KCl. No such fluorescence increase was observed when using beta cells from ZnT8 knockout mice. ZnT8 is a zinc transporter that is expressed specifically in the insulin granules. ZIMIR can be used to assess insulin/Zn^{2+} release in pancreatic islets.

The same group reported the usefulness of exendin-4 (Ex4) for probe localization on the membrane of MIN6 cells.[23] Ex4 is a 39-amino-acid peptide that can bind to the glucagon-like peptide-1 receptor (GLP-1R), which is highly expressed on the surface of pancreatic beta cells. The C-terminal side of Ex4 contributes to the high affinity for GLP-1R, while its N-terminal side is crucial for the activation of GLP-1R. Thus, the anchoring of ZIMIR on the plasma membrane was accomplished by linking the N-terminal truncated Ex4(9–39).

Lippard *et al.* utilized solid-phase synthesis to develop membrane-anchoring probes, Palm-ZP1 and Palm-ZQ, consisting of an N-terminal palmitoyl group, a negatively charged rigid linker peptide PPPDDK, and a Zn^{2+} sensor.[24] Palm-ZP1 was localized to the membrane of HeLa cells, showing good colocalization with CellMask Orange (PCC = 0.70) and responded well to changes in [Zn^{2+}] in the extracellular fluid. A strong fluorescence signal of Palm-ZP1 was observed from the membrane for 2 h and then internalized into the cells. Similarly, Palm-ZQ localized on the plasma membranes of RWPE-1 cells and responded to changes in extracellular Zn^{2+}. However, the low brightness of Palm-ZQ makes it unsuitable for imaging compared to Palm-ZP1.

Another membrane-localized Zn^{2+} probe, FZn-mem, is a push–pull-type fluorene-based probe with good two-photon absorption properties. It was tethered to the plasma membrane by inserting a terminal dodecenyl group into the membrane.[25] In the absence of Zn^{2+}, its fluorescence was quenched by PeT and increased five-fold upon binding to Zn^{2+}. Taking advantage of its two-photon absorption properties, FZn-mem was applied to the imaging of rat hippocampal slices.

12.2.4 Endoplasmic Reticulum

The endoplasmic reticulum (ER) is considered to be an intracellular zinc store. For example, the protein kinase CK2 phosphorylates and activates the zinc transporter ZIP7 on the ER membrane, promoting Zn^{2+} release from the ER to the cytosol. This Zn^{2+} release induces phosphorylation of kinases associated with cell signaling, which causes proliferation and cell migration.[26] In mast cells, antigen stimulation causes Zn^{2+} release from the ER, which is known as a zinc wave.[27] However, since cytosol-retention probes, such as Zinquin and Newport Green DCF, were used for these analyses of intracellular zinc signaling and Zn^{2+} concentration changes in the ER were not directly monitored, it cannot be concluded that Zn^{2+} was released from the ER. Therefore, the development of ER-localizable Zn^{2+} probes is required to analyze Zn^{2+} dynamics in the ER (Figure 12.4).

Most fluorescent Zn^{2+} probes are based on fluorescein and BODIPY with green fluorescence, and there are limited examples of red fluorescent probes.[28] Under such circumstances, Lippard *et al.* developed a benzoresorufin-based red fluorescent probe, ZBR1. ZBR1 exhibited PeT-induced fluorescence quenching in the free-base form. However, because the pK_a values of the tertiary amino group in the Zn^{2+} chelator and the phenolic hydroxy group of the benzoresorufin skeleton are 8.15 and 5.35, ZBR1 emits relatively strong fluorescence at pH 7 because of the protonation of the tertiary amino group, which cancels the PeT-based quenching. Zn^{2+}binding also increased by eight-fold the fluorescence intensity of ZBR1. The K_d of ZBR1 for Zn^{2+} at pH 7.0 was determined to be 0.69 nM. The subcellular localization

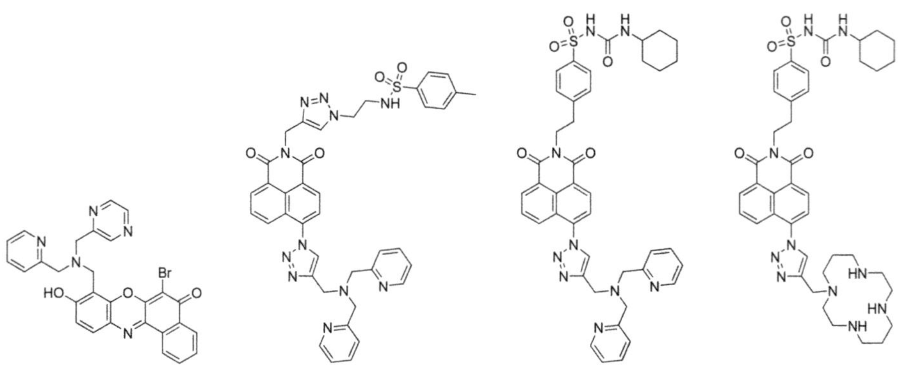

ZBR1 1 2 3

Figure 12.4 Structures of ER-localizable probes.

was determined in HeLa cells, RAW 264.7, and neuronal stem cells (NSCs), where the PCC values with ER-Tracker Blue-White DPX were determined to be 0.65, 0.58, and 0.70, respectively. When neurons are exposed to high concentrations of exogenous NO, peroxynitrite is formed to release Zn^{2+} from intracellular stores. In addition, high concentrations of NO and peroxynitrite are known to induce ER stress.[29] To assess the cellular response to NO, NSCs were incubated with ZBR1 and subsequently treated with 3-morpholinosydnonimine, which generates NO and superoxide. As a result, the ZBR1 fluorescence gradually decreased. It was concluded that this was due to Zn^{2+} release caused by peroxynitrite-induced ER stress.

For efficient probe synthesis and derivatization, Watkinson *et al.* utilized two sequential azide-alkyne click reactions to introduce an organelle-targeting group and tris(2-pyridylmethyl)amine, a Zn^{2+} chelator, into a naphthalimide fluorophore.[30] They adopted a *p*-toluenesulfonamide moiety for ER targeting and developed compound **1**. The triazole formed by the click reaction was also involved in the coordination of Zn^{2+} to form a 1 : 1 complex, which was confirmed by ^{1}H NMR and density functional theory calculations. Compound **1** showed pH-insensitive fluorescence enhancement upon binding to Zn^{2+} between pH 5 and 10, and its K_d was determined to be 2.8 nM for Zn^{2+} at pH 7.4. To evaluate the subcellular localization in HeLa cells, compound **1** was co-incubated with ER-Tracker Red, LysoTracker Red, and MitoTracker Red, resulting in PCC values of 0.88, 0.50, and 0.73, respectively, indicating low organelle specificity.

To improve the organelle-targeting ability of probes, a glibenclamide moiety, which is an inhibitor of ER-resident ATP-sensitive K^+ channels[31] and used in the structures of ER-Tracker Red and Green, was incorporated in compounds **2** and **3**.[32] Both compounds formed a 1 : 1 complex with Zn^{2+} and showed a 16- and 4-fold increase in fluorescence, respectively. Similar to compound **1**, the triazole moieties of compounds **2** and **3** are likely involved in the Zn^{2+} coordination, and the K_d values were determined to be 3.5 and 4.7 nM, respectively. Because of the organelle targetability of the glibenclamide moiety, both compounds were explicitly localized in the ER of various cells, such as HeLa, MCF-7, EC23, and HepG2. In addition, these probes could be used for the assessment of ER stress induced by tunicamycin or thapsigargin.

12.2.5 Mitochondria

It has been suggested that mitochondria may also store labile Zn^{2+}.[33] However, the presence of zinc transporters in the mitochondria is under debate, and the mechanisms by which Zn^{2+} release and influx occur are not yet understood.[34] Moreover, the quantitative values of labile [Zn^{2+}] in the mitochondria have been reported differently, ranging from relatively high (0.23 nM in HeLa)[35] to extremely low (0.14 pM in HeLa).[36] Therefore, it is important to establish a more reliable Zn^{2+} quantification method to clarify the Zn^{2+} dynamics in the mitochondria.

Mitochondrial zinc uptake was studied using FluoZin-3 AM in the mitochondria isolated from the rat brain, and two mechanisms were proposed for Zn^{2+} import into mitochondria.[37] One is dependent on the mitochondrial membrane potential, and the other is mediated by a calcium uniporter.[38] Although Zn^{2+} efflux was monitored by the change in FluoZin-3 fluorescence intensity, the efflux was uniporter-independent and not affected by the loss of mitochondrial membrane potential. As mentioned by the authors, a mitochondria-localized fluorescent probe that can be used in living cells is critical to verify whether this mechanism occurs in living cells. FluoZin-3 AM was used to study the oxidative stress response of cultured hippocampal neurons, and it did not localize to the mitochondria.[39] When hippocampal neurons were exposed to oxidative stress by H_2O_2 treatment, the fluorescence intensity of FluoZin-3 in the lysosome, cytosol, and nucleus increased. Oxidative stress likely weakens the integrity of the membrane, and FluoZin-3 localizes to the lysosome, or the localization is different in different cell types. To avoid confusion caused by such ambiguous subcellular localization of probes, it is important to endow the probes with accurate localization properties.

To date, many mitochondria-targeting fluorescent probes have been developed (Figure 12.5).[40] Since the mitochondrial matrix is negatively charged, lipophilic and cationic compounds tend to localize in the mitochondria.[41] Rhodamine, which has lipophilicity and a cationic chromophore, is a representative fluorophore that tends to localize in the mitochondria, and a rhodamine-based probe, RhodZin-3, was developed in 2003.[42] RhodZin-3 is a pH-independent orange fluorescent probe that does not respond to Ca^{2+} in the physiological concentration range (≤40 μM). The fluorescence increased 75-fold at pH 7.0 upon binding to Zn^{2+}, and its K_d for Zn^{2+} is 65 nM.

When cytosolic [Zn^{2+}] is excessively elevated, Zn^{2+} is taken up by the mitochondria to maintain intracellular zinc homeostasis.[43] However, increased [Zn^{2+}] in the mitochondria promotes loss of the mitochondrial membrane potential, induces the production of reactive oxygen species (ROS), and releases pro-apoptotic factors, which cause neurotoxicity. The AM ester derivative of RhodZin-3 was used to visualize Zn^{2+} in the mitochondria of cultured neurons, and its localization in the mitochondria was confirmed by colocalization with MitoTracker Green. When cortical neurons were depolarized with 60 mM K^+ in the presence of 50 μM Zn^{2+}, the fluorescence signal of RhodZin-3 in the mitochondria increased by approximately 45% with an increase in cytosolic [Zn^{2+}].

Lippard *et al.* developed ZS5 by changing one picolyl group of each 2,2′-dipicolylamine (DPA) group of Zinpyr-1 to a 2-thiophenylmethyl group.[44] The fact that thiophenes are not involved in coordination to Zn^{2+} was verified by X-ray crystallographic analysis of a model compound. ZS5 formed a 1 : 1 complex in the low concentration range of Zn^{2+} and a 1 : 2 complex in the high concentration range. However, only $K_{d,1}$ was calculated to be 1.5 μM because the fluorescence intensity hardly changed in the second binding step. Due to the protonation effect on the tertiary amines and the phenolic OH, ZS5 showed pH-dependent fluorescence ($\Phi = 0.36$, pH 7; $\Phi = 0.04$, pH 9) in the absence of Zn^{2+}. In HeLa cells, the intense fluorescence signal of ZS5 was unexpectedly

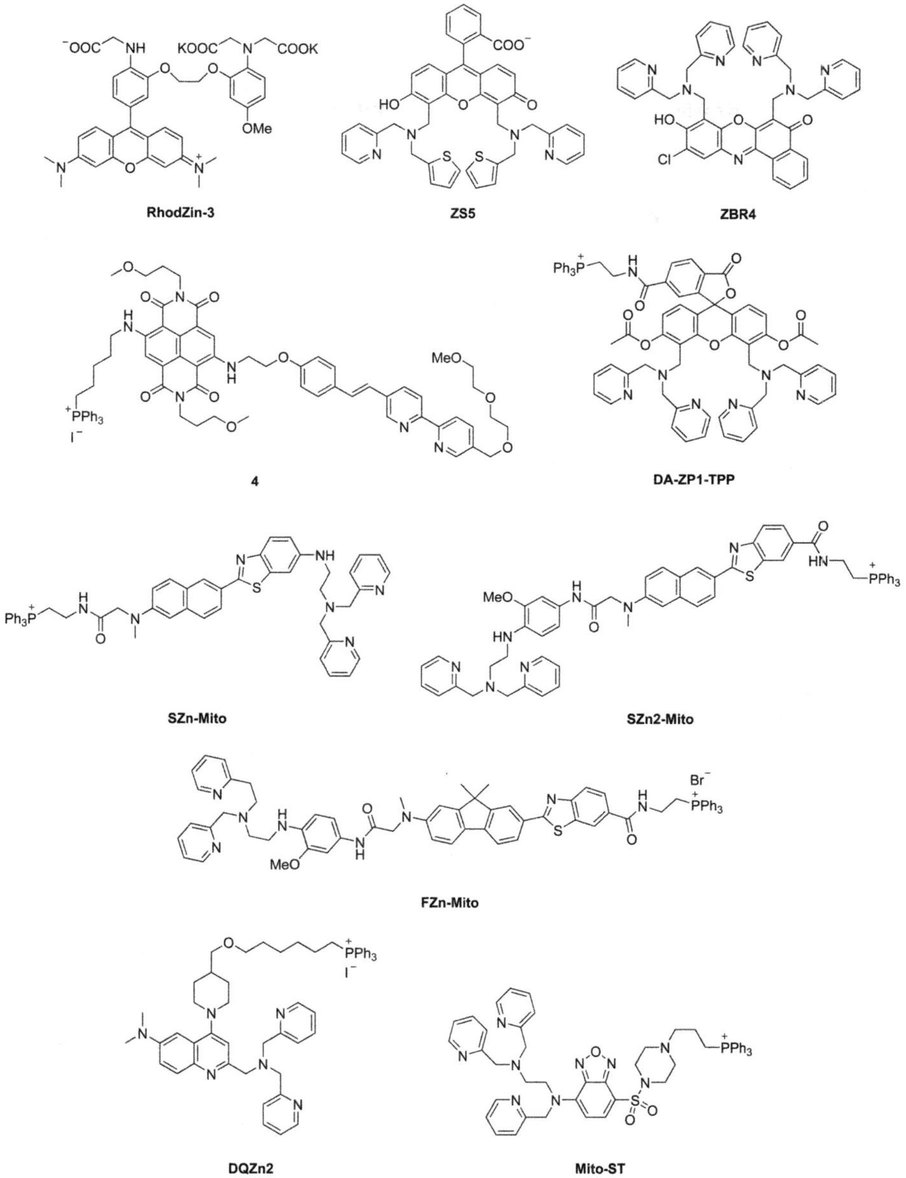

Figure 12.5 Structures of mitochondria-localizable probes.

observed in the mitochondria, whereas the fluorescence signal was observed in the whole-cell region and some puncta in COS-7 and HEK293T cells.

Lippard *et al.* also developed a benzoresorufin-based probe, ZBR4, by introducing another DPA structure to ZBR1, which caused a subcellular localization change from the ER to the mitochondria in HeLa cells.[45] The presence of two DPA groups positively affected fluorescence quenching in the absence of

Zn^{2+}, resulting in a 14-fold fluorescence enhancement upon binding to Zn^{2+}. The pK_a values for the two tertiary amines in Zn^{2+} chelators were 6.3 and 3.2, suggesting that PeT efficiently occurs under neutral conditions. In addition, ZBR4 is membrane permeable, without masking the phenolic hydroxy group.

As an alternative strategy for mitochondria targeting, a lipophilic and cationic moiety, such as triphenylphosphonium (TPP), has been attached to the probes. Compound **4** is a mitochondria-targeting TPP-attached fluorescence resonance energy transfer (FRET) probe composed of a 2,2′-bipyridine derivative and a diamino-substituted naphthalenediimide (NDI) as a FRET donor and acceptor, respectively.[46] The photophysical properties were evaluated using a derivative without a TPP group in acetonitrile. Upon excitation at 400 nm, fluorescence at 630 nm was observed owing to energy transfer from the bipyridine to the NDI moieties. The fluorescence at 630 nm increased 12-fold with the addition of Zn^{2+}. Direct excitation of the NDI moiety at 600 nm also showed Zn^{2+}-insensitive fluorescence at 630 nm. This result suggests that ratiometric Zn^{2+} imaging using 400 nm and 600 nm excitation might be possible, although this was not demonstrated in this study. Compound **4** showed good colocalization with mCerulean3 TOMM20 expressed in the mitochondria of HeLa S3 cells.

The membrane permeability and subcellular localization of small-molecule fluorescent probes can be significantly affected by minor structural differences. Regardless of the TPP moiety, ZP1-TPP did not localize to the mitochondria (PCC = −0.15) but was partially localized to the endosomes and lysosomes (PCC = 0.45 with LysoTracker Red) in HeLa cells.[47] In addition, ZP1-TPP hardly showed Zn^{2+} responsiveness in acidic environments because of the intrinsic pH sensitivity of ZP1. On the other hand, since the phenolic hydroxy groups were masked by the acetyl groups, DA-ZP1-TPP was nonfluorescent and relatively stable against intracellular esterases for more than 2 h. However, Zn^{2+}-mediated cleavage of these acetyl groups occurred rapidly ($t_{1/2}$ = 8.1 s), leading to a 140-fold fluorescence enhancement. Therefore, although the fluorescence signal of DA-ZP1-TPP was hardly observed just after incubation with HeLa cells, the zinc pyrithione (ZPT) addition caused a 12-fold fluorescence enhancement, and the probes were colocalized with MitoTracker Red (PCC = 0.64).

Cho *et al.* reported a series of two-photon probes localized in mitochondria using a TPP moiety. The first example is SZn-Mito, which has two-photon absorption at 760 nm with a cross-section of 75 GM and emission at 493 nm at pH 7.2 in the presence of Zn^{2+}.[48] Upon addition of Zn^{2+}, SZn-Mito showed a seven-fold fluorescence enhancement. In the pH range of 5–8, the fluorescence properties of SZn-Mito in the presence and absence of Zn^{2+} were scarcely affected by pH changes. SZn-Mito formed a 1:1 complex with Zn^{2+}, which was confirmed by a Job plot, and the K_d was determined to be 3.1 nM for both one-photon and two-photon excitation. Two-photon imaging revealed that SZn-Mito colocalized well with MitoTracker Red FM (PCC = 0.89) in HeLa cells. The fluorescence signal was increased by the addition of ZPT or 2,2′-dithiodipyridine (DTDP), which oxidizes thiols in Zn^{2+}-binding proteins to promote Zn^{2+} release,[49] and decreased by the addition of TPEN or CCCP, which induce Zn^{2+} release by loss of mitochondrial membrane potential. SZn-Mito was also applied to rat hippocampal slice imaging and provided intense fluorescence in the CA3 and dentate gyrus regions. SZn-2-Mito, which has the same π-backbone as SZn-Mito, was designed to improve sensitivity by introducing push–pull substituents.[50] As expected, owing to the increased intramolecular charge transfer (ICT) nature of the probe, SZn2-Mito showed weaker fluorescence in the absence of Zn^{2+} than SZn-Mito, resulting in a 70-fold increase in fluorescence. SZn2-Mito was also localized in the mitochondria (PCC = 0.85 with MitoTracker Red FM) and showed a Zn^{2+}-dependent fluorescence response in living cells. In the imaging of rat hippocampal tissue, brighter and clearer images were obtained because the large fluorescence turn-on ratio suppressed the background signals.

Since the elevation of [Zn^{2+}] in the cytosol causes cytotoxicity, excess Zn^{2+} is excreted from the cell or sequestered in subcellular compartments, such as the ER and mitochondria.[51] To validate this process, it is necessary to detect Zn^{2+} in the cytosol and mitochondria simultaneously and discriminately. BZn-Cyto and FZn-Mito were designed for this purpose and showed 28-fold and 16-fold Zn^{2+}-dependent fluorescence enhancement, respectively, in two-photon excitation. The GM values of these probes were 103 and 80 GM, and the K_d values were 1.8 and 17 nM. In HeLa cells, FZn-Mito localized in the mitochondria (PCC = 0.86), and the authors claimed that BZn-Cyto localized in the cytosol (Figure 12.2). However, the fluorescent signal of BZn-Cyto did not spread throughout the cell, but appeared to be lacking in the nuclear region, suggesting that it localized in particular subcellular compartments. Both probes were similarly distributed in the CA2 and CA3 regions in two-photon microscopy imaging of rat hippocampal slices. In addition, DTDP and CCCP induced a fluorescence response in both HeLa cells and hippocampal slices.

Jiang *et al.* reported DQZn2, which is one of the few small-molecule probes capable of quantifying intracellular [Zn^{2+}].[52] At pH 7.4, the nitrogen in quinoline is protonated cooperatively with the Zn^{2+} ligand DPA, leading to the possible formation of a 1,4-dihydroquinoline-type resonance structure and effective ICT from the electron-donating amino group at position 4 of the quinoline. This resonance structure formation was inhibited by coordination to Zn^{2+}, resulting in a 46 nm blue shift in the fluorescence maxima from 550 nm ($\Phi = 0.11$) to 504 nm ($\Phi = 0.22$) with an iso-emissive point at 567 nm. The fluorescence intensity ratio at 504 nm and 550 nm (F_{504}/F_{550}) changed from 0.39 to 1.94, and the K_d was determined to be 0.45 nM at pH 7.4. Since the pH in the mitochondria is maintained slightly basic, the K_d value at pH 8.0 was also measured and was 0.15 nM. Regardless of the presence or absence of Zn^{2+}, the fluorescence ratio values were almost constant in the pH range of 5–8. The mitochondrial localization of DQZn2 was confirmed using NIH3T3 cells (PCC = 0.86, MitoTracker Deep Red). The intracellular fluorescence ratio (F_{Green}/F_{Red}) was calculated to be 1.17 by acquiring fluorescence microscopy images for 430–490 nm (green) and 520–580 nm (red) upon 405 nm excitation. By intracellular calibration, the K_d of the probe was determined to be 77 pM, and the steady-state labile [Zn^{2+}] in the mitochondria of NIH3T3 cells was calculated to be 72 pM. It was also demonstrated that NO-induced Zn^{2+} release occurred from the mitochondria to the cytosol. After the addition of *S*-nitrosocystein, an NO donor, the fluorescence ratio values in the mitochondria increased for 10 min and immediately decreased after the addition of TPEN. The authors concluded that Zn^{2+} released into the cytosol is likely taken up rapidly into the mitochondria.

Mitochondria are thought to function as a zinc pool in neurons.[53] An ICT-based ratiometric probe, Mito-ST, was developed by He and Guo *et al.* to monitor mitochondrial Zn^{2+} release induced by oxidative stress.[54] The aromatic amine in the chelator structure acts as an electron donor. Thus, when monitoring the fluorescence at 577 nm, the excitation wavelength was hypsochromically shifted from 466 to 397 nm upon binding with Zn^{2+}. Mito-ST formed a 1 : 1 complex with Zn^{2+}, and its K_d was determined to be 8.2 nM. The fluorescence intensity ratio at 397 nm and 488 nm changed significantly from 0.36–6.2 depending on [Zn^{2+}], whereas it hardly changed between pH 6 and 8 in the absence of Zn^{2+}. Owing to the TPP moiety, Mito-ST was well localized in the mitochondria of HeLa, COC2, and MCF-7 cells. The response to oxidative stress in MCF-7 cells was investigated by ratiometric imaging using excitation light at 405 nm and 488 nm. Before stimulation, the ratio value was 0.03, and [Zn^{2+}] was estimated to be 0.6 nM. After treatment with 10 mM H_2O_2, the ratio changed to 0.34, which corresponds to 1.6 nM Zn^{2+}. When the cells were incubated with 10 mM SNOC, a precursor of the NO radical, there was a lag phase of 15 min, at which time the ratio value increased to a level comparable to that of 10 mM H_2O_2 treatment. The ratio increased rapidly over the next 5 min, and the mitochondrial [Zn^{2+}] reached 76 nM. At this point, the probe leaked from the mitochondria, and the cell membrane ruptured.

12.2.6 Lysosome

Lysosomes contain various hydrolases and are the center of biomolecule degradation, in which the pH is kept acidic (pH 4.5–5.5).[55] Due to these severe conditions in the lysosome, to the best of our knowledge, there is no protein-based Zn^{2+} probe that functions in lysosomes. To take advantage of the acidic environment in the lysosome, aliphatic amines, which are protonated in the lysosome and lose their membrane permeability, have been used to deliver small-molecule probes (Figure 12.6). However, the fluorescence property or the Zn^{2+} affinity of small-molecule Zn^{2+} probes are affected by acidic pH because Zn^{2+} chelators generally comprise amines or nitrogen-containing heterocycles. Thus, the imaging of Zn^{2+} in lysosomes is more challenging than in other organelles.

In many Zn^{2+} probes, fluorescence quenching is caused by PeT from a tertiary amine in the Zn^{2+} chelator. However, in this mechanism, fluorescence enhancement can also occur by protonation of the chelator in an acidic

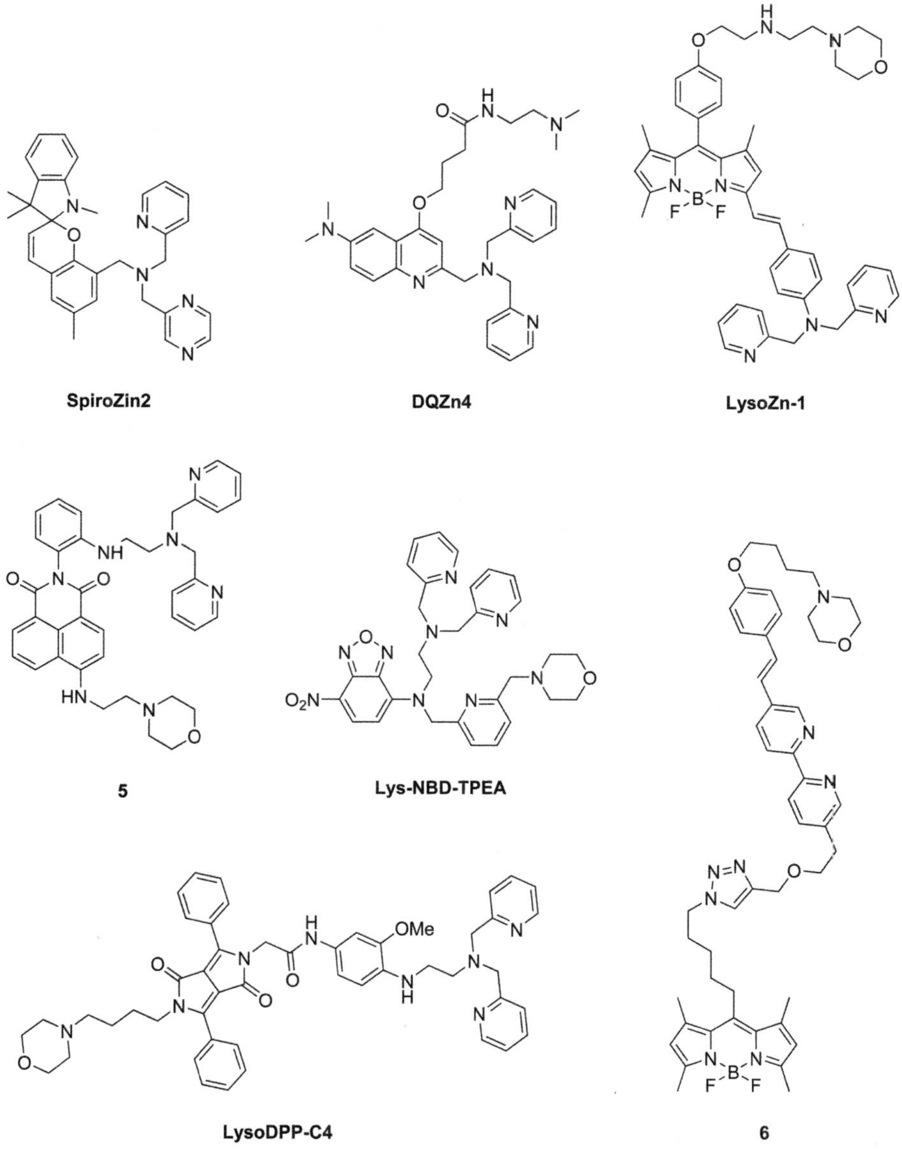

Figure 12.6 Structures of lysosome-localizable probes.

environment. To avoid this, Lippard *et al.* designed SpiroZin2 to emit fluorescence when the spiropyran ring opens into a merocyanine structure upon coordination to Zn^{2+}.[56] The K_d of SpiroZin2 for Zn^{2+} was determined to be 3.6 nM at pH 7. In the absence of Zn^{2+}, the fluorescence intensity of SpiroZin2 was weak and did not change significantly in the pH range of 3–7, whereas SpiroZin2 showed Zn^{2+}-dependent fluorescence enhancement in this pH range. SpiroZin2 localized in the lysosomes of HeLa cells (PCC = 0.80) and in the insulin granules of MIN6 cells (PCC = 0.60 with LysoTracker Green), although the mechanism for subcellular localization is unclear. SpiroZin2 was also applied to monitor changes in lysosomal Zn^{2+} levels during the differentiation of HC11 cells.[57]

A dimethylaminoethyl group, a common structure in LysoTracker Green and Red, was utilized to design a lysosome-targeting ratiometric probe, DQZn4.[58] At physiological pH, the amino group was protonated, and fluorescence quenching by PeT did not occur. Upon binding to Zn^{2+}, DQZn4 showed a blue shift in emission maxima from 542 nm to 495 nm at pH 5.2, which was induced by the change in ICT properties upon binding to Zn^{2+}, similar to DQZn2. The fluorescence ratio at 495 nm and 592 nm (F_{495}/F_{592}) changed from 0.37 to 1.88 depending on [Zn^{2+}]. The K_d values for Zn^{2+} were determined to be 0.2 nM at pH 7.4 and 16 nM at pH 5.2. In the absence of Zn^{2+}, the fluorescence intensity of DQZn4 was affected by the pH change and reached its maximum at pH 6 (pK_{a1} = 7.19, pK_{a2} = 4.52). DQZn4 localized in the lysosome in NIH3T3 cells (PCC = 0.82 with LysoTracker Red DND-99), whereas DQZn3, its derivative without the lysosome-targeting group, showed a lower tendency to localize in the lysosome (PCC = 0.52). The ratiometric images were constructed by acquiring images for emission at 430–510 nm (green) and 520–620 nm (red) upon 405 nm excitation, and the ratio value (F_{Green}/F_{Red}) at the steady state was calculated to be 0.43. Although intracellular calibration was performed using ZPT and TPEN, the ratio value after 50 μM TPEN treatment was higher than that at the steady state, indicating that the quantification of labile [Zn^{2+}] was difficult under these conditions.

Lysosomal membrane permeabilization induced by excessive Zn^{2+} incorporation into lysosomes is involved in oxidative neuronal death. To investigate this phenomenon, Peng *et al.* developed a BODIPY-based Zn^{2+} probe, LysoZn-1, with a morpholine moiety as a lysosome-targeting group.[59] The fluorescence intensity of LysoZn-1 was almost constant between pH 4.5–7.0, regardless of the presence of Zn^{2+}. Weak and broad emission at 647 nm derived from the ICT state was observed before the Zn^{2+} addition, whereas strong and sharp emission at 578 nm from the locally excited state was observed after Zn^{2+} addition. From the fluorescence ratio at 578 nm and 680 nm (F_{578}/F_{680}), the K_d at pH 7.2 and pH 4.7 were determined to be 123 μM and 160 μM, respectively. These results indicate that LysoZn-1 could detect Zn^{2+} in the vesicles or lysosomes of neurons, where more than 100 μM Zn^{2+} is thought to be present, without significant influence from pH variation. In both NSCs and MCF-7 cells, LysoZn-1 colocalized well with LysoSensor Green DND-189 (PCC = 0.98 and 0.97). Zn^{2+} addition into the culture medium induced a fluorescence signal increase both for LyzoZn-1-labeled NSCs and MCF-7 cells, and the fluorescence signal decreased to the initial levels with TPEN addition. However, fluorescence ratio enhancement induced by oxidative stress with H_2O_2 was observed only in NSCs, and MCF-7 cells did not show such a response. This result suggests that the amount of Zn^{2+} stored in cancer cells is probably lower than that in neuronal cells.

A naphthalimide-based two-photon probe **5** was localized in the lysosome of NIH3T3 cells (PCC = 0.87 with LysoTracker Deep Red) and was used for imaging of mouse brain tissue.[60] Although compound **5** showed pH-dependent fluorescence both in the absence and presence of Zn^{2+}, fluorescence quenching by PeT hardly occurred at pH 5.0, because of the protonation of the morpholine group. It was thought that the lower fluorescence intensity of compound **5** at neutral pH could be an advantage for imaging Zn^{2+} in the lysosome because of the lower background signals from other compartments.

Zn^{2+} release induced by oxidative stress was investigated using Lys-NBD-TPEA, which has a morpholine moiety for lysosome targeting.[61] Lys-NBD-TPEA showed a 4.5-fold enhancement in fluorescence at pH 5.0. However, the degree of amplification of fluorescence intensity varied with pH, and the degree at pH 5.0, was half of that at pH 7.4. Lysosomal localization of Lys-NBD-TPEA was confirmed in A549 and RAW 264.7 cells. After adding H_2O_2 to A549 cells, the probe fluorescence increased, likely due to Zn^{2+} release by ROS-induced oxidation of cysteine residues in proteins, and Zn^{2+} was sequestered into the lysosome. In addition, lipopolysaccharide-induced inflammation in RAW 264.7 cells also caused the probe fluorescence enhancement, which was inhibited by *N*-acetyl-L-cysteine, a ROS scavenger.

Sessler *et al.* developed an AND-logic diketopyrrolopyrrole-based probe, LysoDPP-C4, to detect both acidic pH and Zn^{2+}.[62] Fluorescence quenching by PeT occurred from both the Zn^{2+} chelator and the morpholine moiety. Therefore, LysoDPP-C4 showed intense fluorescence only when morpholine was protonated and Zn^{2+} was bound to the chelator. However, probe fluorescence was not completely quenched under acidic conditions (pH < 6.0). The K_d value at pH 7.2 was determined to be 1.9 nM. The lysosomal localization of the probe was evaluated in HeLa cells (PCC = 0.68 with LysoTracker Red DND-99). After incubation of the probe in prostate cancer cell lines (PC3 and DU145) and normal prostate epithelial cell line (RWPE1), Zn^{2+} was added to the culture medium, resulting in fluorescence signal enhancement only in RWPE1 cells. The reason for this phenomenon is that the uptake of extracellular Zn^{2+} is reduced in the cancer cell line due to downregulation of the zinc transporter, ZIP1. When LysoDPP-C4 was administered to nude mice through the tail vein, an intense fluorescence signal was observed in the prostate region, where 10–100 mM Zn^{2+} was thought to be present. In *ex vivo* experiments, the prostate and seminal vesicles showed strong fluorescence. When LysoDPP-C4 was used on human tissue sections from a prostate cancer patient, cancerous tissues showing weaker fluorescence were distinguished from normal tissues.

When a fluorophore functions as a Zn^{2+} chelator, the probe often shows a broad fluorescence spectrum and low photostability. To overcome this, probe **6** using BODIPY, which shows sharp and strong fluorescence, as a FRET acceptor was developed.[63] By using FRET, the donor quickly returns to the ground state from the excited state, thus suppressing the photobleaching of the donor. The fluorescence intensity derived from BODIPY upon donor excitation at 405 nm increased in acidic solutions. The pK_a was determined

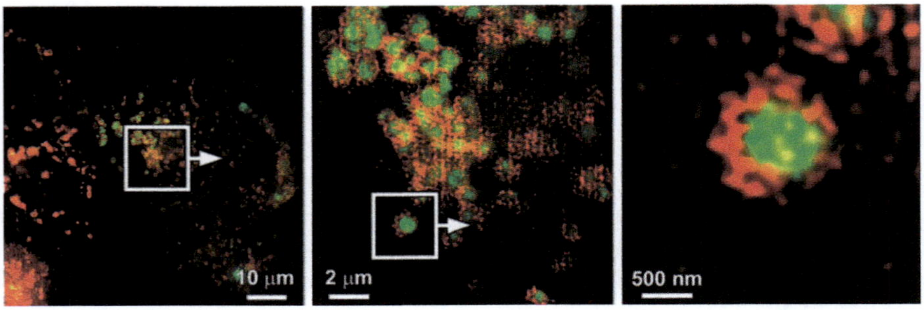

Figure 12.7 Live-cell multicolor SIM imaging of compound **6** (green) with the fluorescent protein FusionRed (red) fused to a lysosome-associated membrane protein (LAMP) in HeLa S3 cells. An enlarged image of the white square area is shown in the next panel indicated by white arrows. Reproduced from ref. 63 with permission from John Wiley & Sons, Copyright © 2015 WILEY-VCH Verlag GmbH & Co. KGaA, Weinheim.

to be 2.7, suggesting that the effect of pH variation is negligible under physiological conditions. In addition, compound **6** also showed comparable Zn^{2+}-dependent fluorescence enhancement at pH 4.8 compared to pH 7.3. It formed a 2:1 complex with Zn^{2+}, and K_a was determined to be 6×10^9 M^{-2}. Because of the morpholino group, compound **6** was localized in the lysosomes of HeLa S3 cells (PCC = 0.67 with FusionRed-LAMP). A detailed analysis using structured illumination microscopy (SIM) revealed that the green fluorescence signals of compound **6** were encapsulated in red fluorescent circles corresponding to FusionRed-LAMP (Figure 12.7).

12.2.7 Golgi Apparatus

Multiple types of ZnTs and ZIPs are present in the Golgi apparatus.[4] To understand the physiological functions of these transporters and the Zn^{2+} homeostasis mechanism, it is essential to monitor Zn^{2+} dynamics in the Golgi apparatus. However, no known small-molecule motifs enable localization to the Golgi lumen, and a rational design strategy for Golgi-targeting small-molecule probes has not yet been established (Figure 12.8).

Guo *et al.* reported an 8-sulfonamidoquinoline-based probe, BPSQ, localized preferentially in the Golgi apparatus of HeLa cells and partially in the mitochondria.[64] Its absorption maximum shifted from 302 nm to 360 nm upon Zn^{2+} addition at pH 7.2. In the absence of Zn^{2+}, the probe fluorescence was quenched by PeT between pH 5.5–10 and enhanced 9.3-fold upon Zn^{2+} binding. The formation of a 1:1 complex with Zn^{2+} was confirmed by X-ray crystallography and ^1H NMR titration. However, BPSQ was not bright, even after binding to Zn^{2+}, because the molar absorption coefficient at 364 nm was 4680 M^{-1} cm^{-1}, and the fluorescence quantum yield was 0.055.

NBD-TPEA was designed based on the ICT fluorophore, 4-amino-7-nitro-2,1,3-benzoxadiazole, and can be excited by visible light despite its small π-conjugated system.[65] NBD-TPEA showed an absorption maximum at

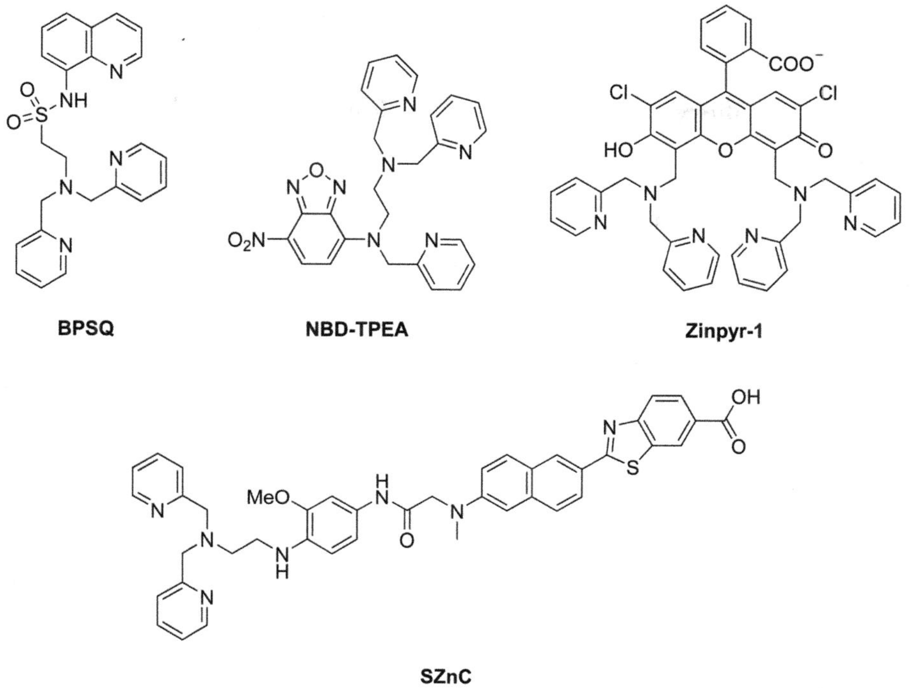

Figure 12.8 Structures of Golgi-localizable probes.

469 nm, which blue shifted to 478 nm upon complexation with Zn^{2+}. The NBD-TPEA fluorescence was quenched by PeT in the range of pH 7.1–10.1 and increased with acidification due to the protonation of the Zn^{2+} chelator. The fluorescence maximum shifted slightly from 550 nm to 534 nm upon Zn^{2+} addition with 14-fold fluorescence enhancement. The K_d at pH 7.2 was determined to be about 2 nM. The subcellular localization of NBD-TPEA was investigated in HeLa cells using LysoTracker Red DND-99, MitoTracker Red CMXRos, and BODIPY TR ceramide. As a result, it was confirmed that NBD-TPEA localized preferentially in both the lysosomes and Golgi apparatus. Finally, NBD-TPEA was applied to zebrafish larva imaging, and bright punctate signals were observed between the eyes.

Zinpyr-1, a 2′,7′-dichlorofluorescein with two DPA moieties on the 4′- and 5′-positions, was developed by Lippard and Tsien *et al.* and is one of the most famous probes for Zn^{2+} imaging.[14] Zinpyr-1 showed an absorption maximum at 515 nm and an emission maximum at 525 nm at pH 7.0. The fluorescence quantum yield was 0.39 without Zn^{2+}, indicating that PeT quenching did not occur efficiently because of the protonation of the tertiary amines in the DPA moieties at pH 7.0. Upon binding to Zn^{2+}, Zinpyr-1 showed approximately 3.1-fold fluorescence enhancement, and its K_d was determined to be 0.7 nM. Zinpyr-1 showed cell membrane permeability without masking the phenolic hydroxy groups, and colocalized with the Golgi marker GalT-ECFP in COS-7 cells.

A two-photon probe, SZnC, was localized in the Golgi apparatus of HeLa, A431, RAW 264.7, and HepG2 cells and was applied to rat hippocampal slice imaging.[66] Kim *et al.* proposed that the mechanism of Golgi localization is probe lipophilicity and a weakly acidic environment in the Golgi apparatus (pH 6–6.7). Theoretical predictions suggest that molecules with a log P value of 3–5 are more likely to be retained in the Golgi.[67] In addition, the weakly basic Zn^{2+} chelator may also function positively for localization in the Golgi. The log P of SZnC was 2.9. The photophysical properties were evaluated in a 1:1 mixture of MOPS buffer (pH 7.2) and ethanol. SZnC showed absorption and emission maxima at 384 and 518 nm, respectively. The fluorescence intensity of SZnC did not change between pH 5.5 and 8 in the absence of Zn^{2+} and was enhanced 8.5-fold by Zn^{2+} addition. The K_d was determined to be 1.7 nM.

12.3 Protein-tag–Small-molecule Hybrid Probes

One of the advantages of protein-based probes is that various localization motifs specific for each cellular compartment are available.[5] By introducing the gene encoding the amino acid sequence of the fusion protein of the probe and localization motif into cells, the probe can be precisely localized to the target organelle. On the other hand, the advantage of small-molecule probes is that the probe properties, such as spectral properties, Zn^{2+} affinity, and resistance to various environmental factors, can be fine-tuned through rational molecular design. Hybrid probes combine these characteristics with self-labeling protein tags (Figure 12.9).

Lippard *et al.* developed a Zinpyr-1 derivative, ZP1BG, with a benzylguanine structure,[68] which is a ligand for human O^6-alkylguanine transferase (AGT), also known as SNAP-tag (Figure 12.10).[69] To evaluate the fluorescence properties, a recombinant AGT was labeled with ZP1BG *in vitro* to form a conjugate ZP1-AGT, which showed approximately 2.7-fold fluorescence enhancement in response to Zn^{2+}, comparable to Zinpyr-1. The absorption and emission wavelengths were also unchanged, indicating that labeling with AGT had little effect on the fluorescence properties. On the other hand, the unlabeled ZP1BG showed little increase in fluorescence intensity with Zn^{2+} addition, which might be due to fluorescence quenching by guanine. To express AGT specifically in the Golgi apparatus and mitochondria, AGT fusion proteins with GalT and COX8 were used as localization motifs, respectively. By incubation of ZP1BG with HeLa cells expressing these fusion proteins, ZP1BG could be specifically localized to the Golgi apparatus or mitochondria.

Lippard *et al.* also developed a Zinpyr-1 derivative, ZP1-12Cl,[70] which covalently labels HaloTag (Figure 12.11).[71] After binding to HaloTag, ZP1-12Cl showed similar photophysical properties as the free probe, such as approximately three-fold fluorescence enhancement by Zn^{2+} binding and a K_d of 0.11 nM. For HaloTag localization in the nucleus and to the cytosolic face of the ER and mitochondrial membrane, NLS, the N-terminal 29 amino acids of cytochrome P450 2C1, and hFis1 were used. In addition, ratiometric Zn^{2+}

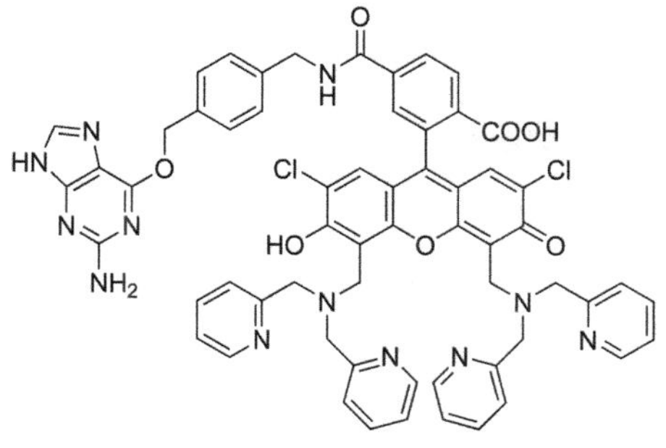

ZP1BG

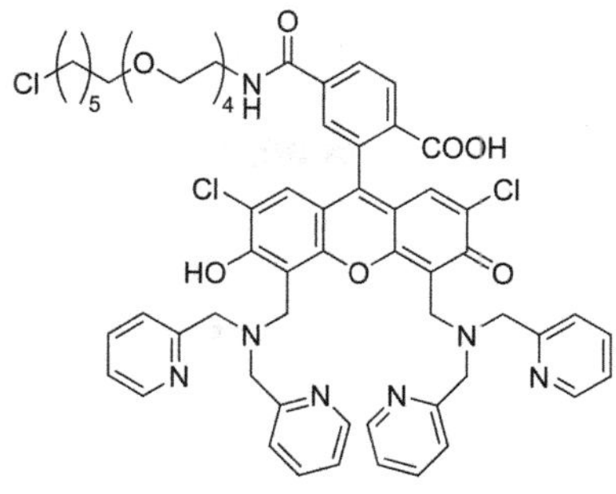

ZP1-12Cl

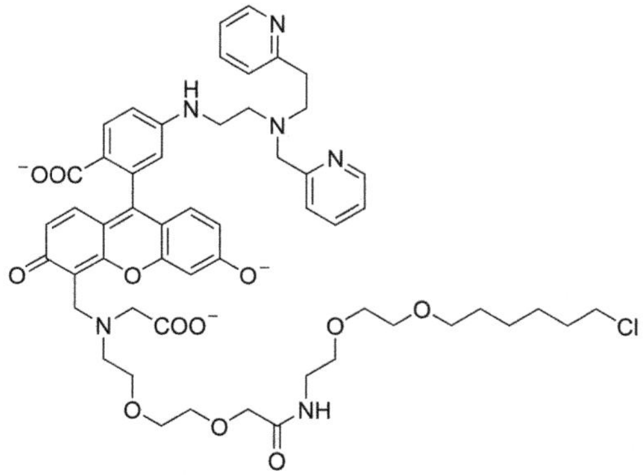

ZIMIR-HaloTag

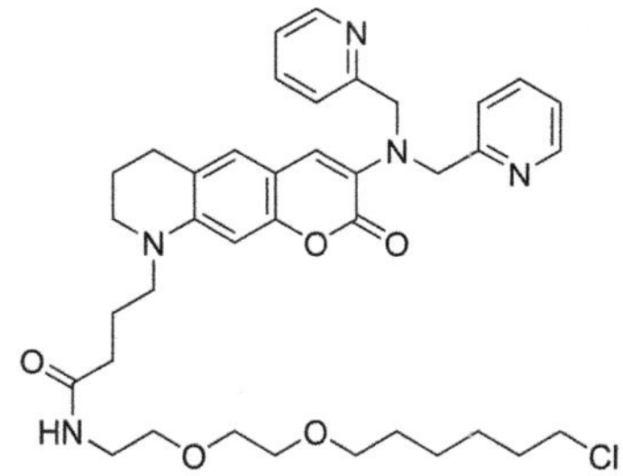

ZnDA-1H

Figure 12.9 Structures of Zn^{2+} probes with a ligand moiety for a protein tag.

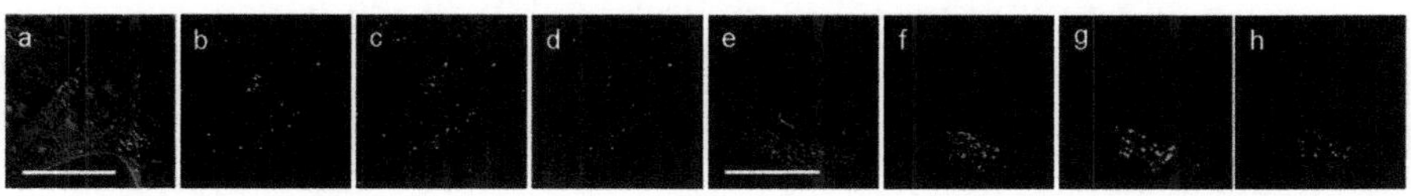

Figure 12.10 HeLa cells expressing AGT in the mitochondria (a–d) and Golgi apparatus (e–h) were labeled with ZP1BG (10 μM). The organelles were marked in red by staining with Mitotracker Red (b) or by expression of mCherry in the Golgi apparatus (f). Scale bar: 25 μm. Reproduced from ref. 68 with permission from American Chemical Society, Copyright 2008.

detection was achieved by using HaloTag proteins fused to mCherry. As a future issue to be solved, free ZP1-12Cl showed a tendency to accumulate in the Golgi apparatus, similar to Zinpyr-1. This drawback may require the complete removal of unlabeled ZP1-12Cl for detailed analysis, such as quantifying [Zn^{2+}] in the Golgi.

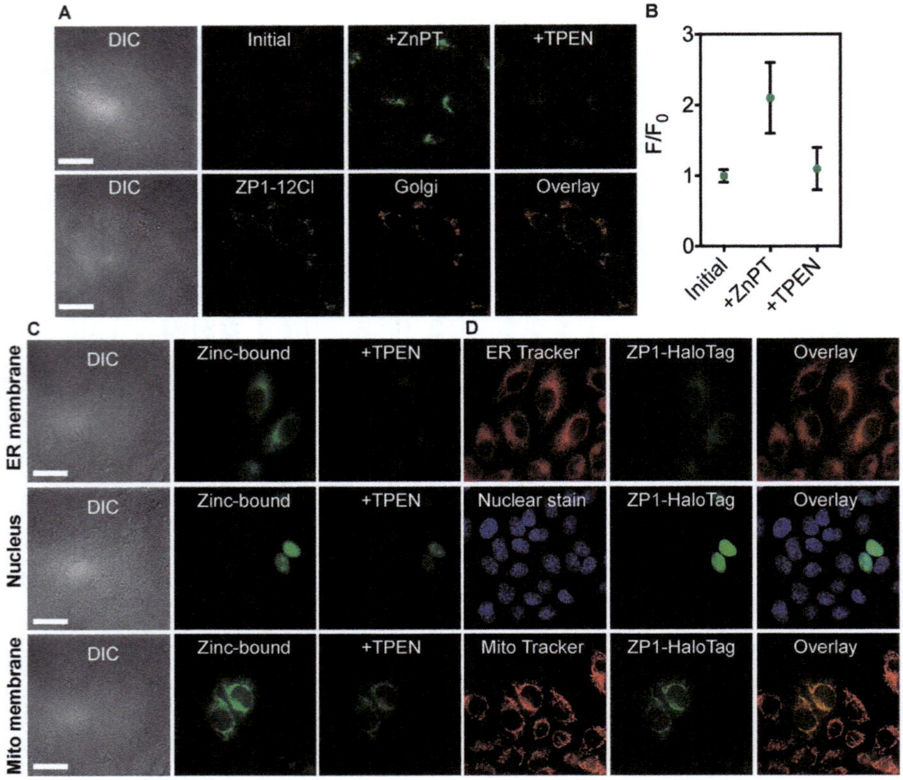

Figure 12.11 Fluorescence microscopy images of live non-transfected (A) and HaloTag-transfected HeLa cells (C, D) pretreated with 5 μM ZP1-12Cl. (B) Quantification of the change in green fluorescence signal intensity from the top row of (A). Reproduced from ref. 70 with permission from American Chemical Society, Copyright 2020.

Li *et al.* achieved the localization of ZIMIR on the outer leaflet of the plasma membrane of pancreatic beta cells (INS-1 and MIN6) through covalent labeling with HaloTag, which was expressed using a pDisplay vector (Figure 12.12). ZIMIR-HaloTag was designed to have an ethylene glycol linker because the labeling efficiency was poor in the derivative without the linker between the fluorophore and the HaloTag ligand.[72] This probe showed 15-fold fluorescence enhancement by Zn^{2+} addition, and K_d was determined to be 0.13 μM. Intracellular Ca^{2+} influx and insulin/Zn^{2+} release *via* KCl addition and channelrhodopsin activation were investigated by simultaneous use of the red fluorescent calcium sensor, R-GECO1.2, which is expressed in the inner leaflet of the plasma membrane or cytosol.

We recently reported an aminocoumarin-based Zn^{2+} probe, ZnDA-1H,[73,74] specifically localized in several organelles, such as the nucleus, ER, mitochondria, and Golgi apparatus, through labeling to HaloTag (Figure 12.13). The photophysical properties of the HaloTag conjugate Halo-ZnDA-1H were

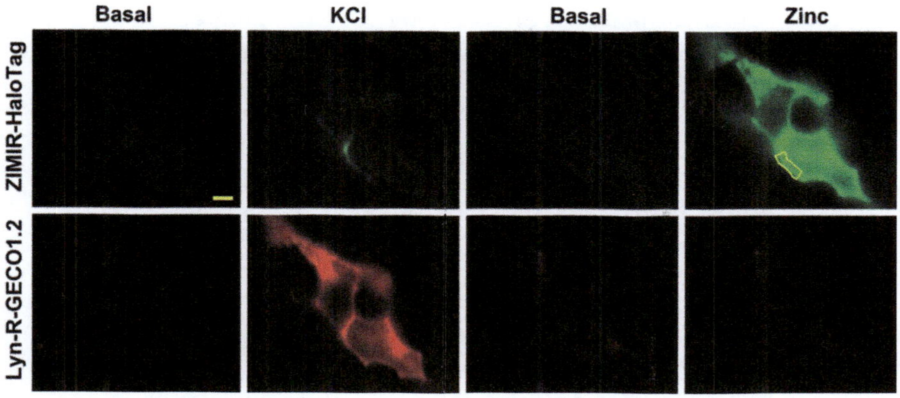

Figure 12.12 Live-cell imaging of Zn^{2+}/insulin release with ZIMIR-HaloTag in MIN6 beta cells cotransfected with HaloTag (pDisplay) and Lyn-R-GECO1.2. Scale bar = 5 μm. Reproduced from ref. 72 with permission from American Chemical Society, Copyright 2015.

evaluated. The fluorescence intensity at 507 nm upon excitation at 440 nm increased about 30-fold after Zn^{2+} addition in a neutral buffer solution (Φ = 0.91, pH 7.5). A similar level of fluorescence enhancement was observed at pH 6–8, and the K_d values at pH 6.5, 7.0, and 7.4 were determined to be 0.54, 0.35, and 0.30 μM, respectively. These results indicate the low pH sensitivity of the probe, which would be an advantage for the accurate detection of Zn^{2+} dynamics in cells without the influence of pH variation. To quantify labile [Zn^{2+}] in a cell, we proposed an *in situ* standard quantification method, in which ZnDA-1H and a Zn^{2+}-insensitive red fluorescent HaloTag ligand bearing tetramethyl rhodamine, HTL-TMR, were simultaneously used for HaloTag labeling (Figure 12.14). By calculating the fluorescence signal ratio, we could monitor changes in [Zn^{2+}] by minimizing the influence of focal-plane drift or cell morphology changes during microscopic observation. As a result, we found that ZnDA-1H had a suitable Zn^{2+} affinity to detect labile Zn^{2+} in the Golgi apparatus and achieved quantification of a labile [Zn^{2+}] of 25 nM.

12.4 Conclusions and Outlook

With the progress of research on the functions of zinc transporters and the role of intracellular labile Zn^{2+} as a signaling mediator, there has been a growing demand for the detailed analysis of Zn^{2+} dynamics in individual compartments rather than the entirety of the cell. At the same time, small-molecule fluorescent probes, which are sufficient to have cell membrane permeability during their emergence, have evolved to facilitate imaging at the organelle level. In this chapter, we have outlined small-molecule probes with targeting properties for each organelle and described the hybrid probes that are robust

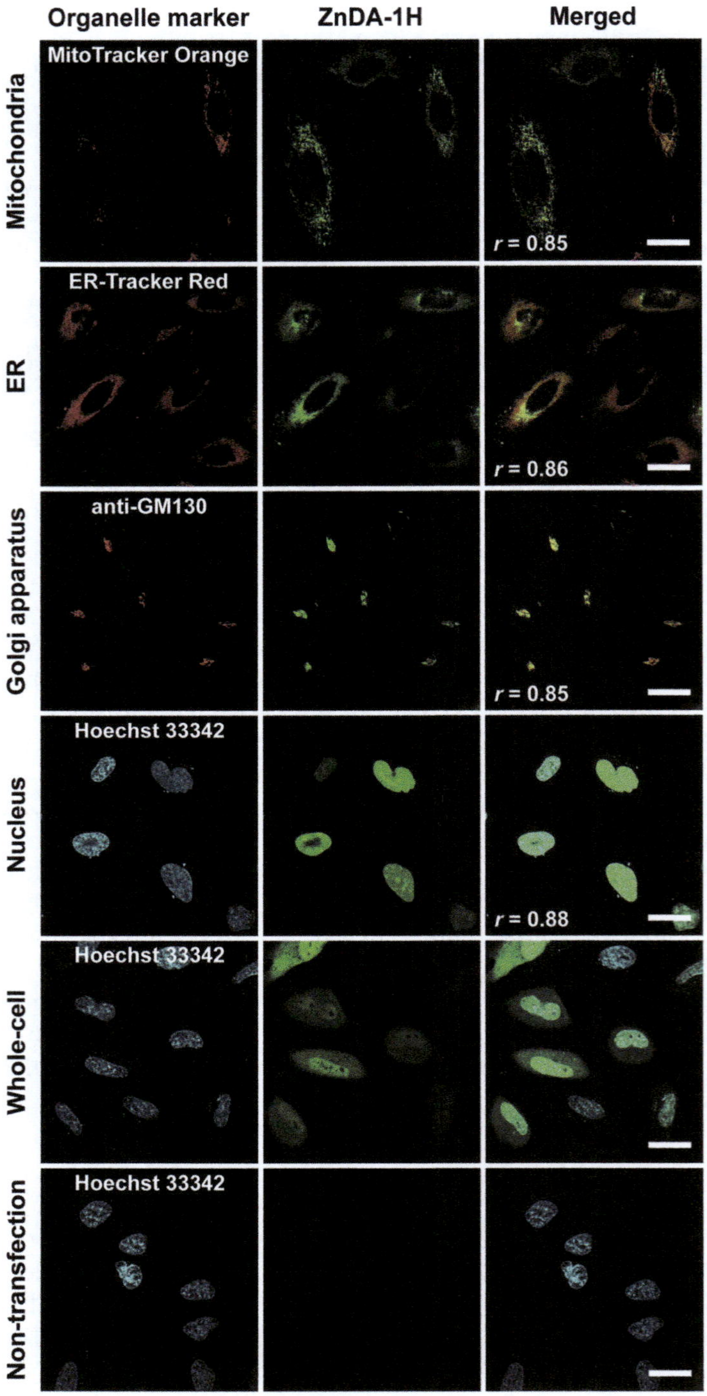

Figure 12.13 Confocal microscopy images of HaloTag-expressing HeLa cells using ZnDA-1H and organelle markers. Reproduced from ref. 73 with permission from Elsevier, Copyright 2020.

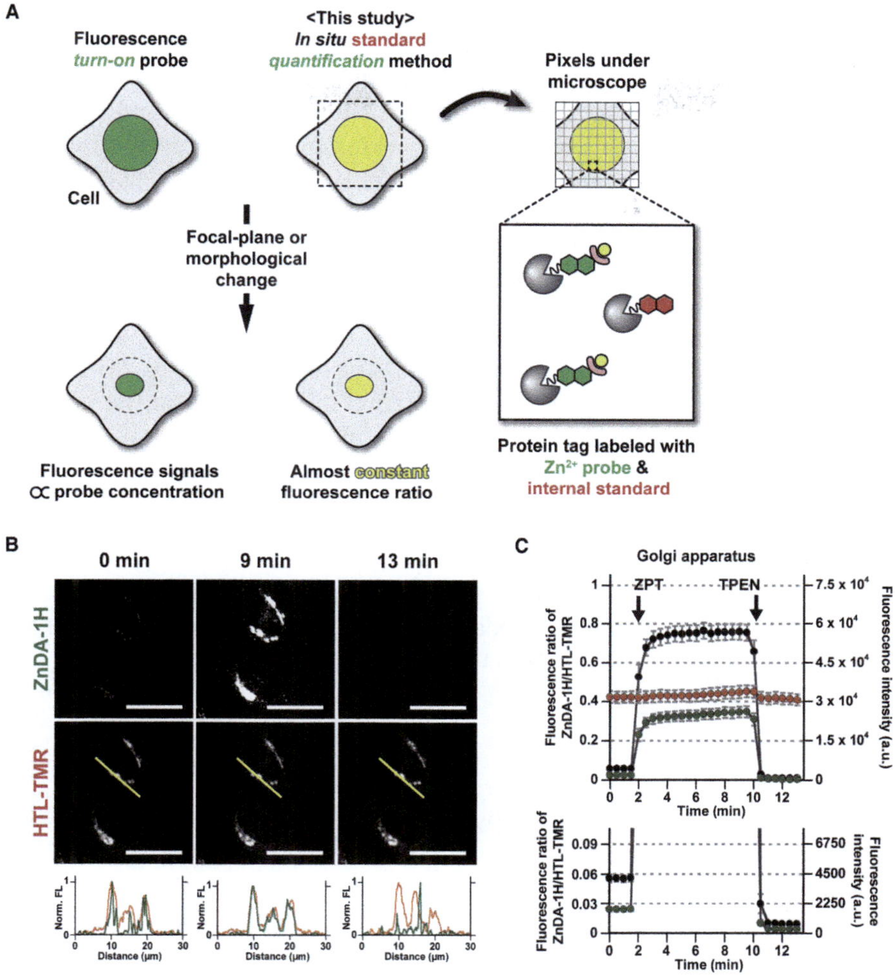

Figure 12.14 (A) Schematic illustration of the *in situ* standard quantification method for monitoring [Zn^{2+}] changes using ZnDA-1H and HTL-TMR. (B) Confocal microscopy images of ZnDA-1H and HTL-TMR targeted into the Golgi apparatus at steady state (0 min) and under the conditions of labile Zn^{2+} saturation (9 min) and depletion (13 min). (C) Time course of the fluorescence ratios (black) and intensities of ZnDA-1H (green) and HTL-TMR (orange), in response to Zn^{2+}. Reproduced from ref. 73 with permission from Elsevier, Copyright 2020.

to changes in the intracellular environment and have the potential to enable reliable observation of Zn^{2+} dynamics.

However, the currently reported small-molecule Zn^{2+} probes still face some challenges. First, there are few color options for fluorescent emissions. Since most probes consist of green fluorophores, such as fluorescein, they cannot be used together with GFP or other green-fluorescent probes. Second, localization motifs are not known for some organelles. In particular, it is desirable to develop effective functional groups for the localization of probes to the nucleus and Golgi lumen. On the other hand, effective localization motifs, such as glibenclamide and TPP, are known for the ER and mitochondria. However, since glibenclamide is a common structure for K^+ channel inhibitors,[31] there is a concern that it may perturb the physiological function of the cell depending on the amount of probe added. For probes with a TPP group, the interpretation of the results depends on whether the Zn^{2+} chelator part is in the mitochondrial matrix or inserted into the membrane.

For further progress in zinc biology, it will be essential to understand the maintenance mechanism of zinc homeostasis and elucidate the disease mechanisms of zinc-related diseases, such as Alzheimer's disease and diabetes, by quantitative and dynamic observation of labile Zn^{2+} at the organelle level. We hope that novel small-molecule and hybrid probes will be further developed to contribute to this research field.

Acknowledgements

This work was supported by JSPS KAKENHI (Nos. JP19K22241, JP20K05702, and 21H05252), and by the Takeda Science Foundation, the Nakatani Foundation, and AMED-CREST (21gm1410006h0001).

References

1. C. Grienberger and A. Konnerth, *Neuron*, 2012, **73**, 862.
2. T. Fukada, S. Yamasaki, K. Nishida, M. Murakami and T. Hirano, *J. Biol. Inorg. Chem.*, 2011, **16**, 1123.
3. T. Kochańczyk, A. Drozd and A. Krężel, *Metallomics*, 2015, **7**, 244.
4. T. Kambe, T. Tsuji, A. Hashimoto and N. Itsumura, *Physiol. Rev.*, 2015, **95**, 749.
5. K. P. Carter, A. M. Young and A. E. Palmer, *Chem. Rev.*, 2014, **114**, 4564.
6. A. M. Hessels and M. Merkx, *Metallomics*, 2015, **7**, 258.
7. L. Fang and M. Watkinson, *Chem. Sci.*, 2020, **11**, 11366.
8. I. Kukic, J. K. Lee, J. Coblentz, S. L. Kelleher and K. Kiselyov, *Biochem. J.*, 2013, **451**, 155.
9. M. N. Lo, L. J. Damon, J. W. Tay, S. Jia and A. E. Palmer, *eLife*, 2020, **9**, e51107.
10. A. Nakamura, K. Takigawa, Y. Kurishita, K. Kuwata, M. Ishida, Y. Shimoda, I. Hamachi and S. Tsukiji, *Chem. Commun.*, 2014, **50**, 6149.
11. G. Lukinavičius, C. Blaukopf, E. Pershagen, A. Schena, L. Reymond, E. Derivery, M. Gonzalez-Gaitan, E. D'Este, S. W. Hell, D. W. Gerlich and K. Johnsson, *Nat. Commun.*, 2015, **6**, 8497.
12. A. Nakamura and S. Tsukiji, *Bioorg. Med. Chem. Lett.*, 2017, **27**, 3127.
13. K. Komatsu, K. Kikuchi, H. Kojima, Y. Urano and T. Nagano, *J. Am. Chem. Soc.*, 2005, **127**, 10197.

14. G. K. Walkup, S. C. Burdette, S. J. Lippard and R. Y. Tsien, *J. Am. Chem. Soc.*, 2000, **122**, 5644.
15. K. R. Gee, Z.-L. Zhou, W.-J. Qian and R. Kennedy, *J. Am. Chem. Soc.*, 2002, **124**, 776.
16. M. Taki, J. L. Wolford and T. V. O'Halloran, *J. Am. Chem. Soc.*, 2004, **126**, 712.
17. A. Krężel and W. Maret, *J. Biol. Inorg. Chem.*, 2006, **11**, 1049.
18. Y. Qin, J. G. Miranda, C. I. Stoddard, K. M. Dean, D. F. Galati and A. E. Palmer, *ACS Chem. Biol.*, 2013, **8**, 2366.
19. C. J. Frederickson, J.-Y. Koh and A. I. Bush, *Nat. Rev. Neurosci.*, 2005, **6**, 449.
20. P. Chabosseau and G. A. Rutter, *Arch. Biochem. Biophys.*, 2016, **611**, 79.
21. S. Iyoshi, M. Taki and Y. Yamamoto, *Org. Lett.*, 2011, **13**, 4558.
22. D. Li, S. Chen, E. A. Bellomo, A. I. Tarasov, C. Kaut, G. A. Rutter and W.-h. Li, *Proc. Natl. Acad. Sci. U. S. A.*, 2011, **108**, 21063.
23. D. Li, Z. Huang, S. Chen, Z. Hu and W.-h. Li, *Bioconjugate Chem.*, 2015, **26**, 1443.
24. R. J. Radford, W. Chyan and S. J. Lippard, *Chem. Sci.*, 2013, **4**, 3080.
25. K. Rathore, C. S. Lim, Y. Lee, H. J. Park and B. R. Cho, *Asian J. Org. Chem.*, 2014, **3**, 1070.
26. K. M. Taylor, S. Hiscox, R. I. Nicholson, C. Hogstrand and P. Kille, *Sci. Signaling*, 2012, **5**, ra11.
27. S. Yamasaki, K. Sakata-Sogawa, A. Hasegawa, T. Suzuki, K. Kabu, E. Sato, T. Kurosaki, S. Yamashita, M. Tokunaga, K. Nishida and T. Hirano, *J. Cell Biol.*, 2007, **177**, 637.
28. W. Lin, D. Buccella and S. J. Lippard, *J. Am. Chem. Soc.*, 2013, **135**, 13512.
29. M. T. Forrester, M. Benhar and J. S. Stamler, *ACS Chem. Biol.*, 2006, **1**, 355.
30. L. Fang, G. Trigiante, R. Crespo-Otero, M. P. Philpott, C. R. Jones and M. Watkinson, *Org. Biomol. Chem.*, 2019, **17**, 10013.
31. Y. Shi, S. Wang, J. Wu, X. Jin and J. You, *J. Controlled Release*, 2021, **329**, 337.
32. L. Fang, G. Trigiante, R. Crespo-Otero, C. S. Hawes, M. P. Philpott, C. R. Jones and M. Watkinson, *Chem. Sci.*, 2019, **10**, 10881.
33. R. A. Colvin, W. R. Holmes, C. P. Fontaine and W. Maret, *Metallomics*, 2010, **2**, 306.
34. P. Chabosseau, J. Woodier, R. Cheung and G. A. Rutter, *Metallomics*, 2018, **10**, 229.
35. P. Chabosseau, E. Tuncay, G. Meur, E. A. Bellomo, A. Hessels, S. Hughes, P. R. V. Johnson, M. Bugliani, P. Marchetti, B. Turan, A. R. Lyon, M. Merkx and G. A. Rutter, *ACS Chem. Biol.*, 2014, **9**, 2111.
36. J. G. Park, Y. Qin, D. F. Galati and A. E. Palmer, *ACS Chem. Biol.*, 2012, **7**, 1636.
37. L. M. Malaiyandi, O. Vergun, K. E. Dineley and I. J. Reynolds, *J. Neurochem.*, 2005, **93**, 1242.
38. S. G. Ji, Y. V. Medvedeva and J. H. Weiss, *Exp. Neurol.*, 2020, **325**, 113161.

39. J. J. Hwang, S.-J. Lee, T.-Y. Kim, J.-H. Cho and J.-Y. Koh, *J. Neurosci.*, 2008, **28**, 3114.
40. Z. Xu and L. Xu, *Chem. Commun.*, 2016, **52**, 1094.
41. L. F. Yousif, K. M. Stewart and S. O. Kelley, *ChemBioChem*, 2009, **10**, 1939.
42. S. L. Sensi, D. Ton-That, J. H. Weiss, A. Rothe and K. R. Gee, *Cell Calcium*, 2003, **34**, 281.
43. S. L. Sensi, H. Z. Yin and J. H. Weiss, *Eur. J. Neurosci.*, 2000, **12**, 3813.
44. E. M. Nolan, J. W. Ryu, J. Jaworski, R. P. Feazell, M. Sheng and S. J. Lippard, *J. Am. Chem. Soc.*, 2006, **128**, 15517.
45. A. Loas, R. J. Radford and S. J. Lippard, *Inorg. Chem.*, 2014, **53**, 6491.
46. K. Sreenath, J. R. Allen, M. W. Davidson and L. Zhu, *Chem. Commun.*, 2011, **47**, 11730.
47. W. Chyan, D. Y. Zhang, S. J. Lippard and R. J. Radford, *Proc. Natl. Acad. Sci. U. S. A.*, 2014, **111**, 143.
48. G. Masanta, C. S. Lim, H. J. Kim, J. H. Han, H. M. Kim and B. R. Cho, *J. Am. Chem. Soc.*, 2011, **133**, 5698.
49. E. Aizenman, A. K. Stout, K. A. Hartnett, K. E. Dineley, B. McLaughlin and I. J. Reynolds, *J. Neurochem.*, 2000, **75**, 1878.
50. N. Y. Baek, C. H. Heo, C. S. Lim, G. Masanta, B. R. Cho and H. M. Kim, *Chem. Commun.*, 2012, **48**, 4546.
51. K. Rathore, C. S. Lim, Y. Lee and B. R. Cho, *Org. Biomol. Chem.*, 2014, **12**, 3406.
52. L. Xue, G. Li, C. Yu and H. Jiang, *Chem. – Eur. J.*, 2012, **18**, 1050.
53. S. L. Sensi, D. Ton-That, P. G. Sullivan, E. A. Jonas, K. R. Gee, L. K. Kaczmarek and J. H. Weiss, *Proc. Natl. Acad. Sci. U. S. A.*, 2003, **100**, 6157.
54. Z. Liu, C. Zhang, Y. Chen, W. He and Z. Guo, *Chem. Commun.*, 2012, **48**, 8365.
55. H. Appelqvist, P. Wäster, K. Kågedal and K. Öllinger, *J. Mol. Cell Biol.*, 2013, **5**, 214.
56. P. Rivera-Fuentes, A. T. Wrobel, M. L. Zastrow, M. Khan, J. Georgiou, T. T. Luyben, J. C. Roder, K. Okamoto and S. J. Lippard, *Chem. Sci.*, 2015, **6**, 1944.
57. Y. Han, J. M. Goldberg, S. J. Lippard and A. E. Palmer, *Sci. Rep.*, 2018, **8**, 15034.
58. L. Xue, G. Li, D. Zhu, Q. Liu and H. Jiang, *Inorg. Chem.*, 2012, **51**, 10842.
59. H. Zhu, J. Fan, S. Zhang, J. Cao, K. Song, D. Ge, H. Dong, J. Wang and X. Peng, *Biomater. Sci.*, 2014, **2**, 89.
60. H.-J. Lee, C.-W. Cho, H. Seo, S. Singha, Y. W. Jun, K.-H. Lee, Y. Jung, K.-T. Kim, S. Park, S. C. Bae and K. H. Ahn, *Chem. Commun.*, 2016, **52**, 124.
61. Z. Hu, G. Yang, J. Hu, H. Wang, P. Eriksson, R. Zhang, Z. Zhang and K. Uvdal, *Sens. Actuators, B*, 2018, **264**, 419.
62. C. Du, S. Fu, X. Wang, A. C. Sedgwick, W. Zhen, M. Li, X. Li, J. Zhou, Z. Wang, H. Wang and J. L. Sessler, *Chem. Sci.*, 2019, **10**, 5699.
63. K. Sreenath, Z. Yuan, J. R. Allen, M. W. Davidson and L. Zhu, *Chem. – Eur. J.*, 2015, **21**, 867.

64. C. Zhang, Y. Zhang, Y. Chen, Z. Xie, Z. Liu, X. Dong, W. He, C. Shen and Z. Guo, *Inorg. Chem. Commun.*, 2011, **14**, 304.
65. F. Qian, C. Zhang, Y. Zhang, W. He, X. Gao, P. Hu and Z. Guo, *J. Am. Chem. Soc.*, 2009, **131**, 1460.
66. H. Singh, H. W. Lee, C. H. Heo, J. W. Byun, A. R. Sarkar and H. M. Kim, *Chem. Commun.*, 2015, **51**, 12099.
67. R. W. Horobin, J. C. Stockert and H. Zhang, *Color. Technol.*, 2022, **138**, 3.
68. E. Tomat, E. M. Nolan, J. Jaworski and S. J. Lippard, *J. Am. Chem. Soc.*, 2008, **130**, 15776.
69. A. Keppler, S. Gendreizig, T. Gronemeyer, H. Pick, H. Vogel and K. Johnsson, *Nat. Biotechnol.*, 2003, **21**, 86.
70. M. L. Zastrow, Z. Huang and S. J. Lippard, *ACS Chem. Biol.*, 2020, **15**, 396.
71. G. V. Los, L. P. Encell, M. G. McDougall, D. D. Hartzell, N. Karassina, C. Zimprich, M. G. Wood, R. Learish, R. F. Ohana, M. Urh, D. Simpson, J. Mendez, K. Zimmerman, P. Otto, G. Vidugiris, J. Zhu, A. Darzins, D. H. Klaubert, R. F. Bulleit and K. V. Wood, *ACS Chem. Biol.*, 2008, **3**, 373.
72. D. Li, L. Liu and W.-H. Li, *ACS Chem. Biol.*, 2015, **10**, 1054.
73. T. Kowada, T. Watanabe, Y. Amagai, R. Liu, M. Yamada, H. Takahashi, T. Matsui, K. Inaba and S. Mizukami, *Cell Chem. Biol.*, 2020, **27**, 1521.
74. T. Kowada, T. Watanabe, R. Liu and S. Mizukami, *STAR Protoc.*, 2021, **2**, 100395.

CHAPTER 13

Molecular Fluorescent Probes for the Detection and Imaging of Sulfane Sulfur and Reactive Selenium Species

ZHENKAI WANG[†a,b], FEIFEI YU[†a,b], YANLONG XING[a,b], RUI WANG[a,b], HENG LIU[a,b], ZIYI CHENG[a,b], JIANFENG JIN[a,b], LINLU ZHAO*[a,b] AND FABIAO YU*[a,b]

[a]Key Laboratory of Hainan Trauma and Disaster Rescue, Department of Neurology, The First Affiliated Hospital of Hainan Medical University, Hainan Medical University, Haikou 571199, China; [b]Key Laboratory of Emergency and Trauma, Ministry of Education, Key Laboratory of Hainan Functional Materials and Molecular Imaging, College of Pharmacy, College of Emergency and Trauma, Hainan Medical University, Haikou 571199, China
*E-mail: zhaolinlu@hainmc.edu.cn, yufabiao@hainmc.edu.cn

13.1 Introduction

Since the concept of oxidative stress was added to the field of redox biology and medicine in 1985,[1] the issues of oxidative stress have been evolving in physiology and pathophysiology research, including chemistry, biochemistry, cell biology, as well as molecular biology in the medical field. Oxidative stress refers to the dysfunction between the antioxidant system and reactive

[†]These authors contributed equally.

Monographs in Supramolecular Chemistry No. 34
Fluorescent Chemosensors
Edited by Luling Wu, Adam C. Sedgwick, Xiao-Peng He and Tony D. James

Published by the Royal Society of Chemistry, www.rsc.org

oxygen species (ROS) in an organism. Under normal conditions, the intracellular milieu maintains oxidation–reduction (redox) homeostasis. ROS produced in a steady-state redox balance play important roles in biomolecule modification *via* redox reactions and then contribute to redox signalling. However, chronic abnormal oxidative stress has been shown to be involved in diseases, including cancer, Parkinson's disease, Alzheimer's disease, atherosclerosis, autoimmune diseases, inflammation, reproductive system diseases, neurodegenerative diseases, and so on.[2–4] Currently, the investigation of molecular redox switches manipulating oxidative stress responses has been elegantly established.[5] Antioxidant defense is crucial for maintaining intracellular redox homeostasis. Although the major role in antioxidant defense is directly performed through antioxidant enzymes, small-molecular antioxidant compounds, as indispensable members of the defense, are involved in the elimination of reactive oxygen species *via* special processes. To prevent damage to organ systems from oxidative stress, low molecular weight compounds, such as ascorbic acid, vitamin A, vitamin E, carotenoids, coenzyme Q, α-tocopherol, uric acid, melatonin, aminoindoles, flavonoids, and polyphenols, behave interactively and synergistically to remove excess reactive oxygen species. In addition, another type of non-enzymatic antioxidant has attracted special attention in the field of redox homeostasis research, that is, reactive chalcogenide species, which mainly contain reactive sulfur species and reactive selenium species, such as glutathione (GSH), cysteine (Cys), homocysteine (Hcy), hydrogen sulfide (H_2S), α-lipoic acid, sulfite (SO_3^{2-}), sulfane sulfurs, selenocysteine (Sec), hydrogen selenide (H_2Se), and selenite (SeO_3^{2-}). As a class of antioxidants with potential clinical biomarker value, the intracellular level and distribution of reactive chalcogenide species can directly demonstrate the dynamic state of oxidative stress, which may reveal the differences and discriminate between physiological and pathological conditions.[6,7]

As reactive chalcogenide species are vital to the health of living organisms, it is crucial to evaluate their intracellular concentration accurately and sensitively. At present, detection methods such as high-performance liquid chromatography (HPLC), capillary electrophoresis, spectrophotometry, mass spectrometry (MS)/HPLC-MS have been elegantly established.[8,9] However, these detection methods require expensive instruments, invasive methods, tedious sample pretreatment procedures, and are susceptible to environmental interference. In particular, they cannot satisfy the *in situ* and real-time inspection of analytes that are highly reactive, unstable in their properties, and impossible to quickly separate from biological samples.

It is desirable that the visualization of cells and organism is accomplished with non-invasive technology, meanwhile, avoiding damage to the body or disturbing the distribution of intracellular constituents and isolating the intracellular analytes. Therefore, techniques for imaging the level and distribution changes of biologically reactive species under physiological or pathophysiological states have become increasingly important in the biomedical field. Since fluorophores can be excited from a light source and their fluorescence emission detected without direct contact, their use means that the whole measurement procedure can be operated non-invasively, *in situ*, and in real time. Fluorescence bioimaging technology based on fluorescent probes has several distinguishing features due to outstanding advantages in terms of high sensitivity, good selectivity, short response time, real-time monitoring, *in situ* observation, low cost, simple operation, high temporal and spatial resolution, and non-invasive measurement.[8,10–12] There have been many summaries and discussions on biological reactive species, such as reactive oxygen species, reactive sulfur species, anions and cations, biological enzymes, and so on.[13–15] As far as we know, there are many studies of fluorescent probes related to imaging and detection of sulfane sulfur and reactive selenium species that have been reported, but there are few dedicated articles that introduce them. Herein, we summarize the current fluorescent probes for detection of sulfane sulfur and reactive selenium species. These paradigms illustrate that fluorescent probes can serve as powerful tools for selectively detecting sulfane sulfur and reactive selenium species in cells and *in vivo*, therefore providing a promising approach for inspecting physiological and pathological mechanisms. We expect to point out a path centered on fluorescence imaging for the detection of these highly biological reactive species in living systems.

13.2 Fluorescent Probes for Sulfane Sulfur

Sulfane sulfurs refer to a type of biospecies containing a sulfur atom with six valence electrons but no charge (designated as S^0). Compared with other biomercaptans, sulfane sulfurs exhibit both nucleophilic and electrophilic properties.[16,17] The dominant existing forms of sulfane sulfurs in living systems mainly include hydrogen polysulfides (H_2S_n, $n > 1$), persulfides (R–S_n–SH, $n > 1$), polysulfides (R–S–S_n–S–R, $n \geq 1$), and protein-bound elemental sulfur (S_8) simultaneously accompanying their thiosulfoxide tautomers. Sulfane sulfurs can transfer the electrophilic sulfur sulfide (S^0) to protein thiols to generate protein-SSH, which affects the function of some proteins and protects proteins from oxidative stress, thereby adjusting the redox homeostasis of the body.[18–20] Sulfane sulfurs exhibit unique physiological functions, such as participation in sulfur-containing vitamins, regulation of enzyme function, and synthesis of cofactors. The significance of sulfane sulfurs is emerging, but because of its reactive chemical properties, the detection technology *in vivo* remains challenging.

The traditional method for the detection of sulfane sulfurs was based on the reaction of cyanide ion (CN^-) to form thiocyanate (SCN^-), which is then measured as iron thiocyanate (SCN^-) (Figure 13.1A).[21] Using this knowledge, a series of fluorescent probes have been developed for the detection of sulfane sulfurs. Xian's group pioneered the fluorescent turn-on strategy for detecting sulfane sulfur.[22] Sulfane sulfurs undergo a nucleophilic reaction with the fluorescent probe to form an intermediate containing –SH. Conceiving a pseudo-fluorescent group (shown in Figure 13.1B) existed as a potential

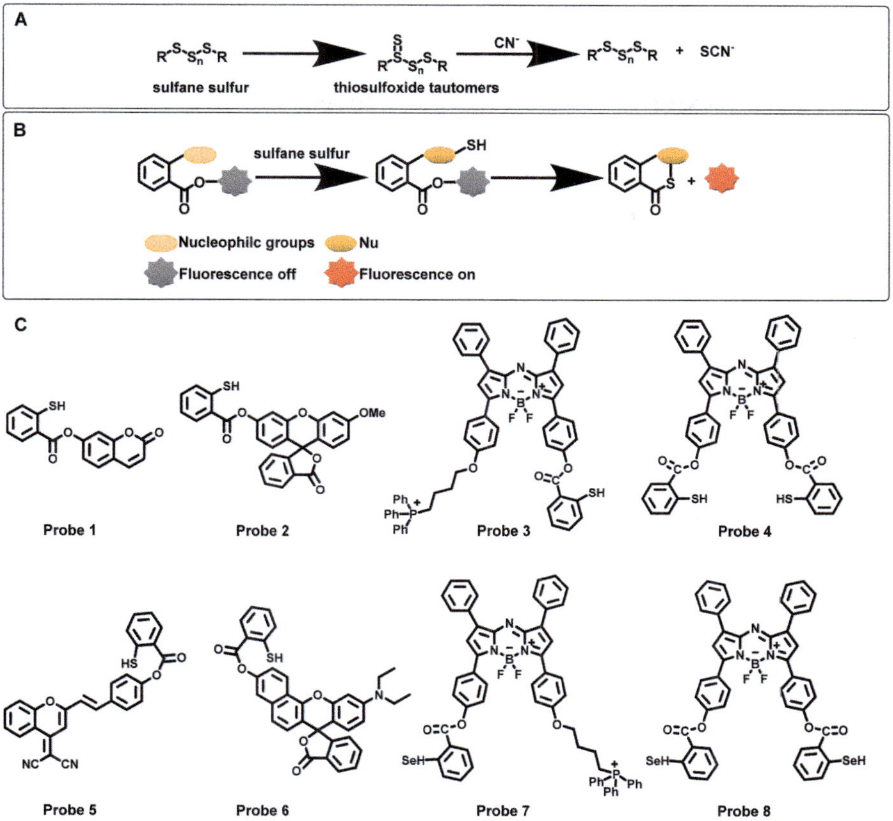

Figure 13.1 (A) Traditional sulfane sulfur detection method based on cyanide ion (CN^-). (B) Fluorescence turn-on strategy for the detection of sulfane sulfurs. (C) Representative fluorescent probes 1–8 for sulfane sulfur detection.

electrophile in a suitable location, the –SH then rapidly and spontaneously cyclizes to free the fluorophore and release a fluorescence emission. Based on the above reaction mechanism, two fluorescent probes (probe 1 and probe 2) were designed that gave faint fluorescence quantum yields, with that of probe 1 being 0.06 and that of probe 2 being 0.05. The probes had high selectivity for sulfane sulfurs. The excitation of probe 1 was at 380 nm and the emission was at 458 nm, while probe 2 was excited at 482 nm and the emission was at 518 nm. When responding to sulfane sulfurs, such as H_2S_2, cysteine hydrosulfide (Cys-SSH), and elemental sulfur, the fluorescence intensity of the two probes significantly increased (25 times for probe 1 and 50 times for probe 2). The maximum fluorescence intensity was reached within 5 minutes. The detection limits of probe 1 and probe 2 were 73 nM and 32 nM, respectively. The probe 2 was an effective tool for detecting exogenous sulfane sulfur in H9c2 and HeLa cells.

The near-infrared absorption and emission profiles can maximize tissue penetration while minimizing the absorbance of heme in hemoglobin and myoglobin, water, and lipids. Yu *et al.* reported a series of near-infrared fluorescent probes for sulfane sulfur detection.[20] These probes selected azo-BODIPY (boron dipyrromethene) as the near-infrared fluorophore and 2-mercaptobenzoate as the recognition unit. The mitochondrial-targeted fluorescent probe 3 possessed a mitochondrial-targeting moiety lipophilic triphenylphosphonium cation.[19] The probe offered good selectivity, excellent sensitivity, and fast fluorescence response to sulfane sulfur with an absorption centered at 679 nm and fluorescence emission at 723 nm. The detection limit was determined as 73 nM. The probe could be used to detect the dynamic changes of sulfane sulfur in hypoxic glucose-deprived cell models. The *in vivo* experiment results illustrated that acute ischemia in mice can cause a large amount of ROS production in the hippocampus, and sulfane sulfur plays a protective role by directly eliminating ROS. Probe 4 had low cytotoxicity and good permeability.[20] The probe could quickly detect sulfane sulfur within 2 min. The detection limit was determined as 10 nM. The quantum yield of probe 4 changed from 0.016 to 0.29, and probe 4 localized in the cytoplasm for intracellular sulfane sulfur detection with a larger detection capacity. Probe 4 could image the production of endogenous sulfane sulfur in living SMMC-7721 cells, in the isolated organs, and in mouse hippocampal tissue.

Liu *et al.* reported a near-infrared fluorescent probe (probe 5) based on the fluorophore of 2-(2-(4-hydroxystyrene)-4*H*-chromene-4-methylene) malononitrile (DCPO).[23] The detection of sulfane sulfur would switch the emission at 680 nm with a response time 1 min with excitation at 560 nm. Probe 5 could detect endogenous sulfane sulfur in living HeLa cells and in the living root tissue of *Arabidopsis thaliana*. Xian's group reported a fluorescent probe (probe 6) using a smartphone for sulfane sulfur detection.[24] The fluorophore of probe 6 was a rhodium derivative with excitation wavelength at 582 nm and emission wavelength at 634 nm. The probe had selectivity and sensitivity to sulfane sulfur species and avoiding interference from other biologically related sulfur species. The proposed probe 6 with smartphone (S4A) system proved to be an effective method for the fast and high-precision detection of sulfane sulfur.

The strong nucleophilic property of the selenol group (–SeH) group may provide superior selectivity and sensitivity for sulfane sulfur detection. Yu *et al.* developed a new type of near-infrared fluorescent probe (probes 7 and 8) for detecting changes of sulfane sulfur in cells and *in vivo* under hypoxic stress.[25,26] The mitochondria-targeted fluorescent probe 7 consisted of three moieties: –SeH as a sulfur acceptor; near-infrared azo-BODIPY as fluorophore; and a lipophilic alkyltriphenylphosphonium cation as a mitochondrial localization group.[25] Probe 7 showed a fluorescence emission at 720 nm when excited at 680 nm. The calculated detection limit was 3.1 nM. The probe could detect in real time the changes of mitochondrial sulfane sulfur induced by hypoxia in vascular smooth-muscle cells. Hypoxic stress stimulated mitochondrial ROS bursts and induced apoptosis, but sulfane sulfur in cells could prevent hypoxia-induced ROS damage. Therefore, sulfane sulfur has significant antioxidant and cytoprotective effects. The results of isolated organs clarified the relationship between sulfane sulfur and ROS in hypoxic mice. The utilization of probe 7 verified that there were significant differences in the expression level of sulfane sulfur in different cell lines, and the probe was used to image sulfane sulfur in three-dimensional (3D) multicellular spheroids and mouse hippocampus in a hypoxia model. The overall level of sulfane sulfur was affected by the degree and duration of hypoxic stress. Based on the above strategy, Yu *et al.* developed another new near-infrared fluorescent probe (probe 8), which was used to image the changes of sulfane sulfur in cells and *in vivo* under hypoxic stress.[26] The probe consists of two parts: a strong nucleophilic phenylselenol group (–SeH) and an azo-BODIPY fluorophore with near-infrared emission characteristics. The quantum yield of the probe was determined to be 0.002. After reacted with sulfane sulfur, a new absorption peak appeared at 707 nm, and the maximum emission was at 737 nm. Probe 8 could track the dynamic changes of endogenous sulfane sulfur based on the rapid and spontaneous intramolecular cyclization reaction. The probe verified that there were significant differences in the expression level of sulfane sulfur in A549 cells, SH-SY5Y cells, RAW 264.7 cells, HeLa cells, HEK 293 cells, HepG2 cells, and SMMC7721 cells, and the probe was used to image sulfane sulfur in 3D multicellular spheroids, mouse hippocampus, and zebrafish in a hypoxia model. The results showed that the overall level of sulfane sulfur was affected by the degree and duration of hypoxic stress, revealing that there was a close relationship between hypoxia and sulfane sulfur in living cells and the body.

13.2.1 Fluorescent Probe for the Detection of H_2S_n

The third endogenous gasotransmitter, H_2S, is involved in cytoprotection, regulation of neuromodulation, vascular tone regulation, inflammation regulation, oxygen sensing, and cell growth. As a direct redox couple, H_2S_n and H_2S coexist and jointly regulate sulfur-related redox homeostasis in living systems. However, recent research suggests that H_2S_n may be the signalling molecule rather than H_2S. The traditional methods for detecting H_2S_n, such as UV-visible spectroscopy and MS, cannot meet the requirement for *in situ* detection in organisms.[27,28]

13.2.1.1 Strategy Based on Nitro-activated Fluorobenzoate

As illustrated in Figure 13.2A, nucleophilic aromatic substitution reactions of hydrogen polysulfide (H_2S_n, $n > 1$) occur with activated fluorobenzoic acid and form a corresponding sulfurized intermediate, I, which was cysteinated to produce benzodithiones and releases the fluorophore. Therefore, the hydroxyl (–OH)-sensitive fluorophores can be introduced into the fluorobenzoic acid moiety to quench fluorescence, and removal of the quencher

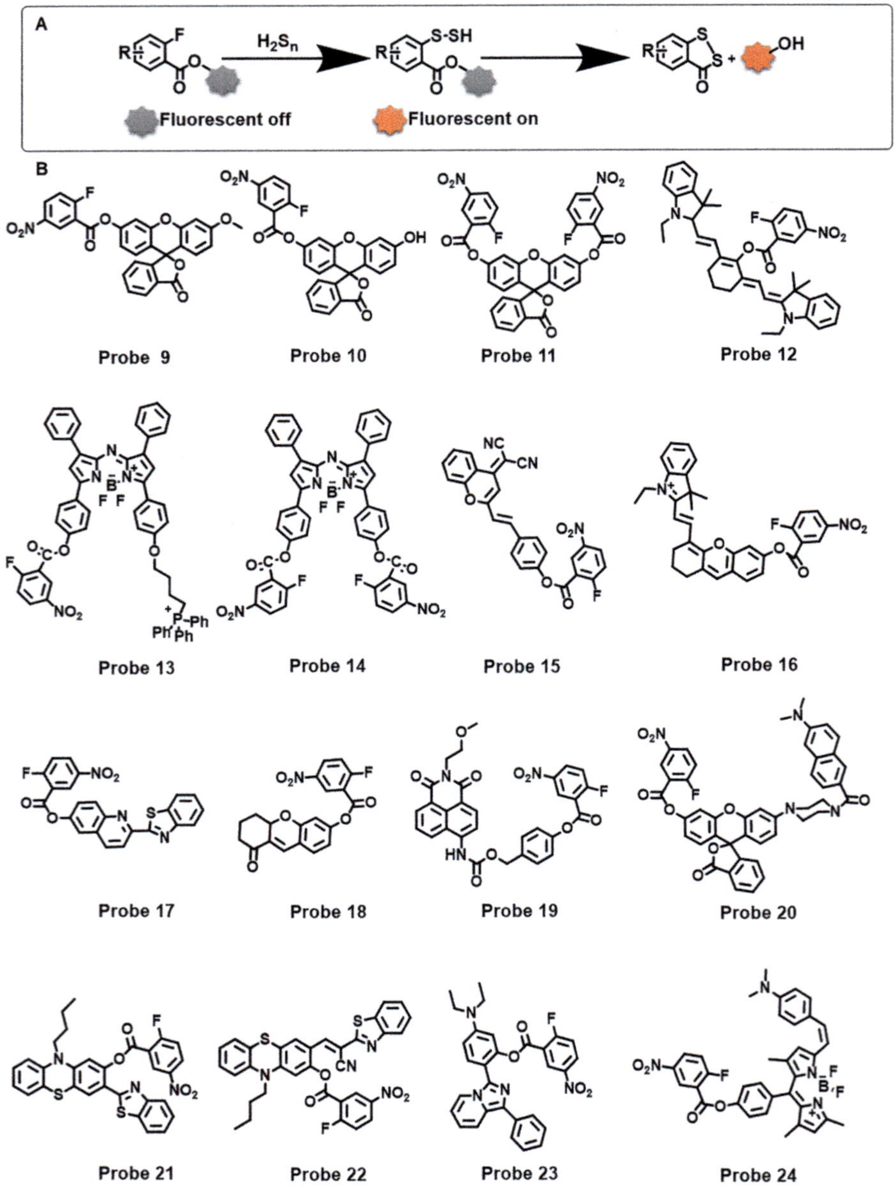

Figure 13.2 (A) Design strategy of fluorescent probes based on nitro-activated fluorobenzoate. (B) Representative fluorescent probes 9–24 for the detection of H_2S_n based on nitro-activated fluorobenzoate.

will restore fluorescence. Xian's group designed and synthesized a series of fluorescent probes (probes 9–11) to capture H_2S_n.[29] However, probe 10 had a strong background fluorescence, and the fluorescence response of probe 9 to H_2S_n was lower than probe 11 (57 times *vs.* 137 times), so probe 11 was the first choice. Probe 11 could detect H_2S_n generated by H_2S and ROS *in situ*.

Its maximum emission intensity at 515 nm was reached within 5 min, and the limit of detection was around 71 nM. Importantly, probe 11 could be used to detect H_2S_n in HeLa cells.

Sulfane sulfur can be derived from the reaction of $O_2^{\bullet-}$ and H_2S. Therefore, it is particularly important to study the relationship between $O_2^{\bullet-}$ and H_2S. Based on the hydrogen abstraction reaction of $O_2^{\bullet-}$ and the amphiphilic reaction of H_2S_n, Yu's group developed a near-infrared mitochondrial-targeting probe (probe 12) that can continuously respond to $O_2^{\bullet-}$ and H_2S_n.[30] The imine cation of near-infrared heptamethine cyanine could be reduced to hydrocyanine for $O_2^{\bullet-}$ detection. Next, the bis-electrophilic nitro-activated fluorobenzoate was used to capture H_2S_n. When detecting $O_2^{\bullet-}$, the absorption wavelength of the probe 12 was 775 nm, the emission wavelength was 794 nm, and the fluorescence quantum yield was 0.0203. Following the detection of H_2S_n, the absorption wavelength shifted to 535 nm, the emission wavelength was at 625 nm, and the fluorescence quantum yield was 0.361. Fluorescence images in RAW264.7 cells and in mice demonstrated that the probe could be used to study the process of $O_2^{\bullet-}$ burst and H_2S_n production *in situ* in real time. The results confirmed that H_2S_n could be derived from H_2S in the presence of $O_2^{\bullet-}$, which was considered a potential direct biosynthetic pathway of H_2S_n in cells.

Yu *et al.* reported a probe (probe 13) for mitochondrial H_2S_n detection through a photoinduced electron transfer (PeT) process from the excited fluorophore to a strong electron-withdrawing group (d-PeT) process.[31] The absorption peak of probe 13 was at 660 nm. When it reacted with H_2S_n, the peak shifted to 688 nm. When treated with Na_2S_2, the detection response time was within 25 s. The detection limit was calculated to be 25 nM. The rapid response of the probe facilitated its use for real-time bioimaging in cells and *in vivo*. Probe 13 could detect H_2S_n in RAW264.7 cells, U87 cells, human umbilical vein endothelial cells (HUVEC), and A549 cells. When the probe was injected into BALB/c mice, there was a 180-fold fluorescence enhancement after 20 minutes following injection of Na_2S_2, suggesting that probe 13 responded to Na_2S_2 in mice. The near-infrared fluorescence probe 14 showed an absorption peak at 660 nm. After reaction with H_2S_n, the absorption peak shifted to 707 nm, and the experimental detection limit of probe 14 for H_2S_n was 50 nM. This probe was found to be sensitive enough to determine H_2S_n in serum. Probe 14 detected that the mouse macrophage cell line RAW264.7 cells could convert excess H_2S to H_2S_n.[32]

Han's group reported a near-infrared fluorescent probe (probe 15).[33] The probe was synthesized *via* esterification of dicyanomethylene-benzopyran (DCMB) dye with 2-fluoro-5-nitrobenzoic acid. The emission wavelength of probe 15 was 682 nm and the absorption peak of the probe itself was 410 nm. When detecting Na_2S_2, the absorption peak shifted to 535 nm, and the limit of detection for H_2S_2 was determined to be 8.2 nM. Probe 15 could be used to detect endogenous H_2S_n in living MCF-7 cells. Based on the semi-heptamethine derivative, Peng *et al.* reported a near-infrared fluorescent probe (probe 16) for polysulfides.[34] The nitrobenzene-benzoate group could react with H_2S_n and release the Cy-OH fluorophore. The probe emitted fluorescence at 720 nm excitation. The detection limit of probe 16 was 2.2×10^{-8} M, and the response time was 5 min. Probe 16 could selectively detect exogenous H_2S_n in living RAW264.7 cells and in mice. Liu *et al.* designed a two-photon fluorescent probe (probe 17) for H_2S_n detection.[35] 2-Benzothiazol-2-yl-quinolinyl-6-ol was the fluorophore because the quinoline core exhibits two-photon absorption capacity. Probe 17 exhibited absorption maxima at 368 nm and 362 nm, and fluorescence maxima were at 536 nm ($\Phi = 0.01$) and 534 nm ($\Phi = 0.57$), respectively. Probe 17 exhibited a 57-fold quantum enhancement. The maximal two-photon action cross-section of probe 17 was 50 GM at 730 nm. The detection limit was 0.5 μM. Probe 17 exhibited the ability to detect endogenous H_2S_n in living HeLa cells and in zebrafish embryo. Lin *et al.* developed a two-photon fluorescent probe (probe 18) to capture H_2S_n.[36] The authors chose GCTPOC (Lin *et al.* previously studied a unique family of similar rigid GFP chromophores, which they called GCTPOC)[37] as the fluorophore. When different concentrations of H_2S_2 were added, probe 18 showed a dramatic fluorescence increase at 512 nm upon excitation at 450 nm. The reaction product of the probe gave a large two-photon cross-section of ~500 GM. The calculated detection limit of probe 18 was 1.52×10^{-7} M. Probe 18 could be used to detect H_2S_n in living MCF-7 cells and in the liver tissue of Kunming mice. Liu's group reported a two-photon ratio fluorescent probe (probe 19) for the detection of H_2S_n in cells and tissues.[38] The fluorophore 1,8-naphthalimide exhibited highly efficient intramolecular charge transfer (ICT). The absorption spectrum of probe 19 shifted from 380 to 432 nm with regulation of H_2S_n, and at the same time the excitation wavelength shifted from 372 nm to 425 nm. This ratiometric probe could avoid the interference from loading concentration, photobleaching, and illumination in biological samples. The probe could quickly detect H_2S_n, within 5 min. The detection limit was measured as 0.1 μM. The two-photon absorption (TPA) cross-section of probe 19 was 155 GM at 780 nm; however, when H_2S_n was added, the TPA cross-section of probe 19 was 113 GM at 790 nm. The ratiometric images indicated the fluorescent probe could quantitatively detect H_2S_n in HeLa cells and in mouse tumor tissue. Zhang's group synthesized a ratiometric fluorescent probe (probe 20) based on the Förster resonance energy transfer (FRET) principle.[39] They chose a two-photon fluorophore naphthalene derivative as the energy donor and the *p*-aminophenol derivative fluorophore as the acceptor. When H_2S_n was added, the deprotection of probe 20 resulted in the fluorescent signal decreasing at 448 nm and increasing at 541 nm, when it was excited at 511 nm, with a limit of detection of 0.7 μM. The probe could be used to detect H_2S_n in HeLa cells, in rat liver tissue sections, and in male C57BL/6 mice for evaluating lipopolysaccharide-induced organ damage. Chen *et al.* developed a fluorescent probe (probe 21) based on excited-state intramolecular proton-transfer inhibition quenching and PeT quenching.[40] The fluorophore was 3-(benzo[*d*]thiazol-2-yl)-10-butyl-10*H*-phenothiazin-2-ol. When the probe interacted with H_2S_n, a maximum absorption peak appeared at 397 nm, and it emitted fluorescence at 534 nm, with a quantum yield of 0.27. The probe exhibited a significant Stokes shift of 137 nm. The detection limit was 26 nM. The probe was used to image H_2S_n in live A549 cells and zebrafish. Zhao *et al.* prepared a fluorescent probe (probe 22) based on phenothiazine coumarin dye for H_2S_n detection.[41] The probe emitted fluorescence centred at 620 nm with a large Stokes shift of 130 nm under excitation at 480 nm. The detection limit of the probe was estimated to be about 1 nM, which was lower than most probes for detecting H_2S_n. The probe could be used to analyze exogenous and endogenous H_2S_n in living RAW-264.7 cells and zebrafish. Zha *et al.* reported a lysosomal targeted "turn-on" fluorescent probe (probe 23) for monitoring endogenous H_2S_n.[42] The probe could respond to H_2S_n with fluorescence emission at 520 nm within 6 min when the probe was excited at 305 nm, and the fluorescence quantum yield increased from 0.00074 to 0.21. The Stokes shift was 215 nm. The limit of detection was 84 nM. Probe 23 was used for the detection of endogenous H_2S_n in living A549 cells. Li *et al.* designed a BODIPY-based near-infrared fluorescent probe (probe 24) for the detection of H_2S_n.[43] Probe 24 responded to H_2S_n with the a pseudo-first-order rate constant of 0.095 min^{-1}. When H_2S_n was added, the absorption maximum shifted from 613 to 605 nm, and the emission was blue shifted from 696 nm to 684 nm. The detection limit was 12 nM. Probe 24 was used to visualize H_2S_n in HeLa cells and in mice. Liu *et al.* designed a fluorescent probe (probe 25 in Figure 13.3) for ratiometric response towards H_2S_n.[44] The reaction with H_2S_n resulted in the absorption gradually decreasing at 563 nm and increasing at 576 nm. The fluorescence emission at 574 nm weakened, while the intensities at 593 nm and 650 nm increased. The quasi-first-order rate constant was 0.071 min^{-1}, and the detection limit was 57 nM. The probe was used to detect H_2S_n in HeLa cells. Zhou *et al.* designed a two-photon turn-on fluorescent probe (probe 26) for H_2S_n detection.[45] In the absence of H_2S_2, an absorption peak was observed at the maximum absorption at 350 nm. After probe 26 reacted with H_2S_n, two absorption peaks were observed at the maximum absorption wavelengths of 350 and 475 nm and the fluorescence intensity at 550 nm increased by more than 80-fold. The detection limit of H_2S_n was determined to be 33 nM. Probe 26 was used for the detection of H_2S_n in living HeLa cells.

Compared with common coumarin, carotene-coumarin exhibits higher fluorescence quantum yield and longer emission wavelength. Song *et al.* synthesized a fluorescent probe (probe 27) for faster response to H_2S_n.[46] After reaction with H_2S_n, the maximum absorption of the probe was blue shifted from 493 nm to 472 nm and the fluorescence enhanced 52-fold at 543 nm. The fluorescence quantum yield increased from 0.09 to 0.79. The detection limit for H_2S_n was 98.3 nM. The probe was used to detect H_2S_n in living MCF-7 cells and zebrafish. Liu *et al.* designed a lysosome-targeting ratiometric two-photon H_2S_n fluorescent probe (probe 28).[47] After reacting with H_2S_n, the absorption peak was red shifted from 384 nm to 432 nm, the emission peak was red shifted from 472 nm to 548 nm, and the limit of detection was 10 nM. The two-photon absorption cross-section of probe 28 was 324 GM at 790 nm, and after treatment with H_2S_n, the reaction product of probe 28 showed

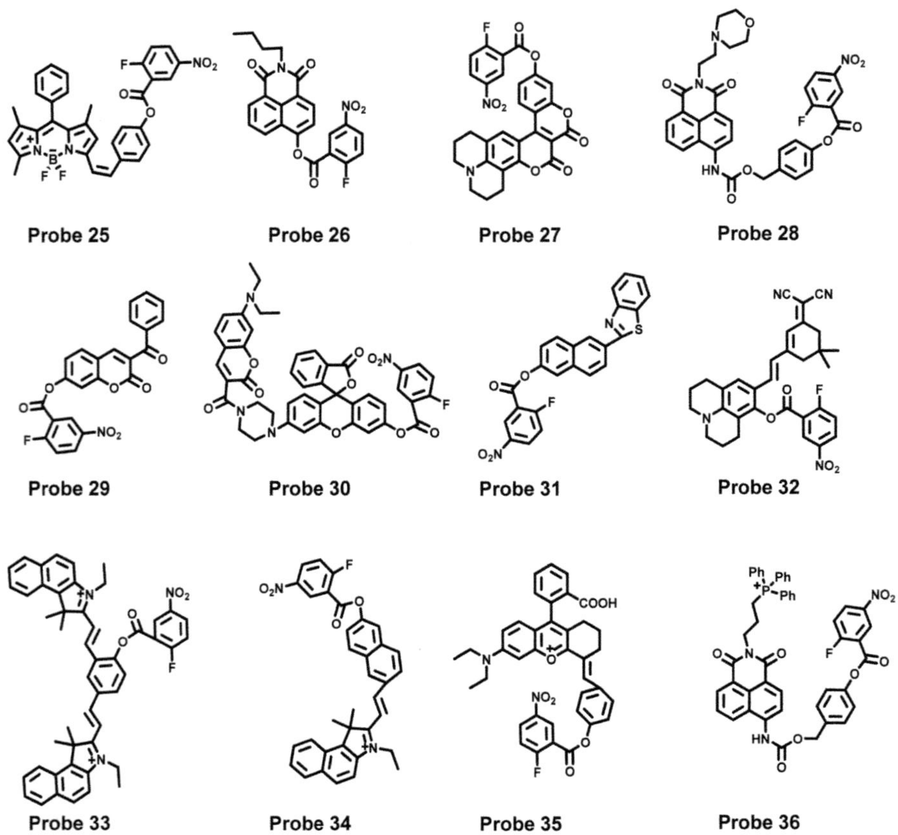

Figure 13.3 The structures of representative fluorescent probes 25–36 based on nitro-activated fluorobenzoate for the detection of H_2S_n.

a maximum two-photon absorption cross-section value of 421 GM at 810 nm, and the effective two-photon-induced excitation range was from 725 nm to 950 nm. Probe 28 was used to detect H_2S_n in BHK cells, HeLa cells, and fresh kidney slices.

Pu *et al.* synthesized a probe (probe 29) for H_2S_n detection.[48] The recognition of H_2S_n led to colour changes from colourless to yellow following the decrease of the original peaks at 291 nm and 323 nm, and a new absorption peak appeared at 429 nm. The detection limit of probe 29 for H_2S_n was 0.03 μM. The cell experiments showed that the probe was suitable for the detection and imaging of exogenous H_2S_n in living HeLa cells. Based on the FRET process, Wu *et al.* designed a proportional fluorescent probe (probe 30) for the detection of H_2S_n.[49] When H_2S_n was detected the probe was excited at 418 nm, and the emission intensity at 537 nm increased and the emission at 460 nm decreased. The detection limit of probe 30 for H_2S_n was 8.0×10^{-7} mol L^{-1}, and it has been successfully used in dual emission channels for confocal imaging of H_2S_n in HepG2 cells. Zhang *et al.* designed and synthesized a fluorescent probe (probe 31) based on the H_2S_n-mediated aromatic substitution-cyclization reaction of naphthalene derivatives to detect H_2S_n. With the addition of H_2S_n, a peak appeared at 338 nm, and the fluorescence emission intensity at 502 nm was observed to increase approximately 18 times. The limit of detection was 5×10^{-7} mol L^{-1}, and it had been successfully applied to the confocal imaging of hydrogen polysulfide in HepG2 cells.[50]

Song *et al.* reported an open fluorescent probe (probe 32) with high sensitivity and good selectivity for H_2S_n, and realized monitoring of H_2S_n in living cells and zebrafish cells.[51] When treated with H_2S_n, the absorption of the probe was red shifted from 521 nm to 578 nm, accompanied by a clear colour change from pink to blue-violet, and the fluorescence was significantly enhanced at 701 nm. The fluorescence quantum yield increased from less than 0.01 to 0.08. The detection limit of the probe was 34 nM, and it has been used to detect the expression of endogenous H_2S_n in live HeLa cells and zebrafish. Chen's group designed a near-infrared fluorescent probe (probe 33) to detect H_2S_n. In the presence of H_2S_n, the probe gradually showed two characteristic absorption peaks at 499 nm and 653 nm, the fluorescence emission increased at 727 nm, and the limit of detection was 46 nM.[52] The probe observed thionitrous acid (HSNO)-mediated H_2S_n production in ZF_4 cells, zebrafish, and BALB/c mice. Qi *et al.* reported a two-photon excitation fluorescent probe (probe 34) for H_2S_n and H_2S. When detecting H_2S_n, and excited at 410 nm, the fluorescence emission at 606 nm was weakened, and the fluorescence emission at 468 nm was enhanced. The ratio of emission intensity at 468 nm and 606 nm increased by about 194 times, and the maximum fluorescence intensity was reached within 175 s. The limit of detection for H_2S_n was 21 nM. However, when detecting H_2S, the emission intensity decreased at 606 nm, and a new emission peak was formed at 519 nm. The ratio of emission intensity at 519 and 606 nm increased by about 37 times and the limit of detection for H_2S was 21 nM. The probe was used to visualize endogenous and exogenous H_2S_n and H_2S in RAW264.7 cells, liver tissue, and zebrafish.[53] Xue *et al.* designed a novel near-infrared fluorescent probe (probe 35), which can be used to detect H_2S_n in living cells and *in vivo*.[54] When detecting H_2S_n, the maximum absorption of the probe changed from 552 nm to 570 nm, and the fluorescence intensity at 670 nm was increased by 20 times. The limit of detection was 18 nM. The probe has been applied to visualize H_2S_n in living MCF-7 cells and mice. Wang *et al.* reported a ratiometric fluorescent probe (probe 36), which detects H_2S_n in mitochondria by integrating triphenylphosphonium and 2-fluoro-5-nitrobenzoic ester groups with a 1,8-naphthalimide-based fluorophore.[55] In the presence of H_2S_n, under excitation at 405 nm, the fluorescence emission intensity of the probe at 485 nm was significantly reduced, and at the same time a new emission peak was formed at about 550 nm, which reaches equilibrium within 10 minutes. In addition, the fluorescence colour changes from blue to yellow. The limit of detection for the probe was 20 nM. The probe was used to detect H_2S_n in living HeLa cells and zebrafish.

13.2.1.2 Strategy Based on Protection by Thioester

The probes reported so far could potentially respond to thiols other than hydrogen polysulfide, as shown in Figure 13.4A.[29] In order to solve this problem, a new strategy has emerged. The group was protected with a thioester, and after deprotection with intracellular nucleophilic species, such as

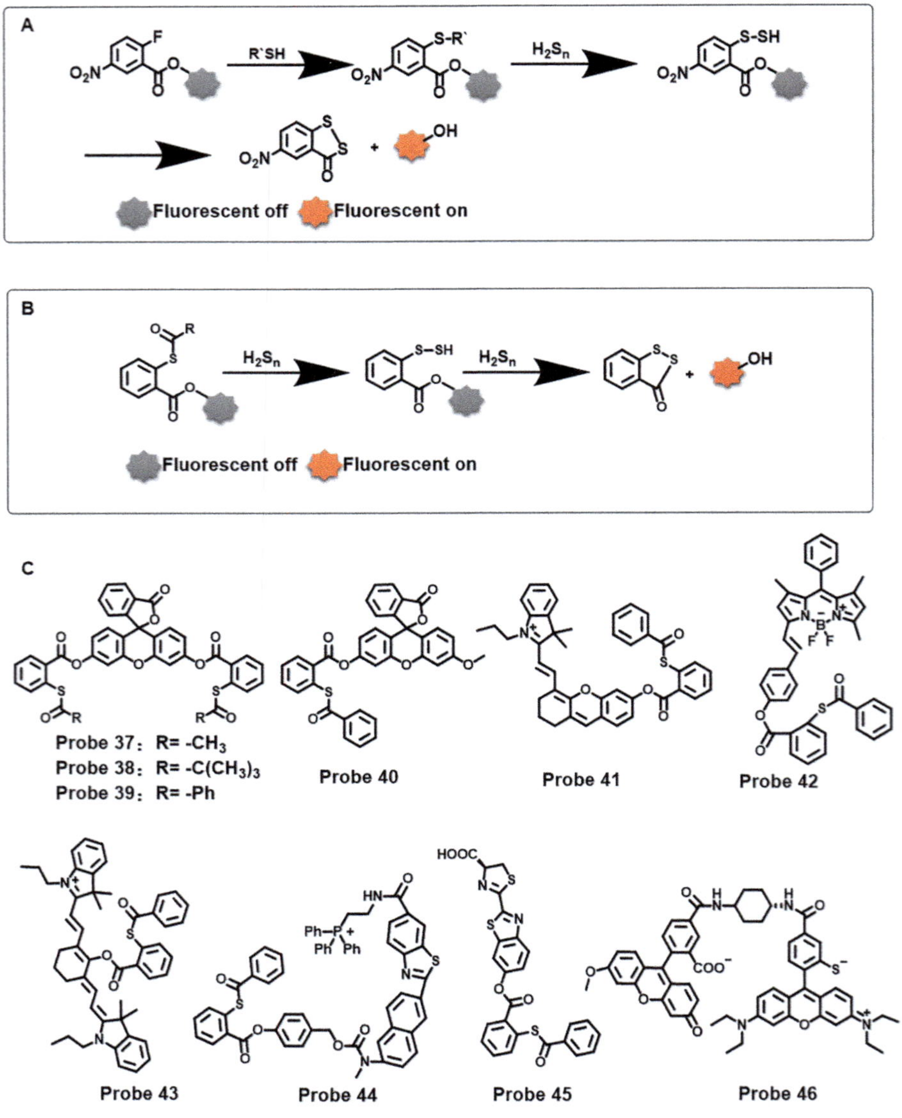

Figure 13.4 (A) Previous reaction strategy for fluorescent probes protected with other thiols. (B) Design strategy of fluorescent probes protected with thioester. (C) Representative fluorescent probes 37–46 for the detection of H_2S_n based on the thioester as a protecting group.

biothiols, the response moiety spontaneously reacts with H_2S_n, and subsequent cyclization would release the fluorophore, as shown in Figure 13.4B. Xian *et al.* reported fluorescent probes (probes 37–39) for sensing *in situ* endogenous H_2S_n formed from H_2S and ROS in cells. Probe 39 offered a desirable off/on fluorescence response to H_2S_n. The absorption of probe 39 was concentrated at 490 nm, and the emission at 515 nm. The maximum emission of probe 39 was achieved within 10 minutes. The detection limit was calculated to be 3 nm. The probe could be used to monitor endogenous H_2S_n in HeLa, RAW264.7, and Vero cells.[56] Xu *et al.* reported a fluorescent lifetime probe (probe 40) for sensing ultra-trace endogenous H_2S_n in native biological systems.[57] The probe initially exhibited a weak fluorescence emission and a short lifetime $\tau = 0.75$ ns. When it reacted with H_2S_n, the fluorescein was released to a "bright" state with long lifetime $\tau = 3.47$ ns. The detection limit of probe 40 was about 2 nM. The probe was also capable of quantification of the average concentration of H_2S_n in native HeLa cells, which was approximately 65 nM. The endogenous H_2S_n of zebrafish was mainly distributed in the abdomen region and had a wide range of 0–160 nM, with an average value of ~70–80 nM.

Near-infrared fluorescent probes have apparent advantages in animal imaging, such as deep tissue penetration and minimal background autofluorescence. Ma *et al.* reported a near-infrared off–on probe (probe 41) for imaging of H_2S_n.[58] The fluorophore contained a hemicyanine skeleton. The probe's maximum absorption was at 550 nm but almost no fluorescence was observed. After reaction with H_2S_n, the maximum absorption was red shifted to 680 nm, accompanied by a fluorescence increase at 708 nm. The detection limit was 35 nM. Probe 41 could be used to image the variations of H_2S_n in HeLa cells and in mice. Das *et al.* reported a probe (probe 42) that could locate at the endoplasmic reticulum.[59] The absorption spectrum of probe 42 was concentrated at 570 nm. When the probe reacted with Na_2S_2 to release the fluorophore, the fluorescence intensity at 584 nm increased significantly, and the quantum yield increased from 0.007 to 0.125. The calculated detection limit of probe 42 was 26.01 nM. Probe 42 could detect H_2S_n in living RAW 264.7 cells. Xiao *et al.* developed a novel red-emitting mitochondrial-targeting fluorescent probe (probe 43) for the detection of hydrogen polysulfide in living cells and tissues that did not react with H_2S or biothiols. The probe was constructed by linking the phenyl 2-(benzoylthio) benzoate group to a heptamethine cyanine (HQO) *via* an ester-linkage. In the presence of hydrogen polysulfide, the excitation of the probe 43 was at 520 nm and the emission at 633 nm, and the fluorescence intensity of the probe increased by 86 times. Probe 43 had fast response, high sensitivity, and good selectivity, after reacting with hydrogen polysulfide (20 μM). The fluorescence intensity of probe 43 (20 μM) reached the maximum within 7 minutes, and the limit of detection was 95.2 nM. The probe was successfully used to image H_2S_n in A549 cells, and it was also successfully applied to monitor the dynamic changes of H_2S_n in mitochondrial lung tissue.[60] Kim *et al.* reported a two-photon ratiometric fluorescent probe (probe 44) for detecting H_2S_n changes in Parkinson's disease.[61]

The two-photon fluorophore was 6-(benzo[*d*]thiazol-2′-yl)-2-(methylamino) naphthalene, and triphenylphosphonium salt was the mitochondrial-targeting moiety. The linkage between the electron-donating amines of the fluorophore and H_2S_n receptor resulted in an internal charge transfer detection mechanism. Probe 44 exhibited a maximum absorption at 346 nm and an emission maximum at 430 nm ($\Phi = 0.86$). On reaction with H_2S_n, the emission spectrum gradually increased at 506 nm, and the two-photon cross-section of probe 44 changed from 12 GM to 110 GM at 740 nm and 759 nm, respectively. The detection limit of probe 44 was 0.1 μM. Probe 44 could measure mitochondrial H_2S_n level changes in HeLa cells and SH-SY5Y cells. The results indicated that a H_2S_n level increase occurred in the brain tissues of a heterozygous and homozygous transgenic Parkinson's disease mouse model.

Zhang *et al.* proposed a bioluminescent probe (probe 45) for detecting H_2S_n *in vitro* and *in vivo*.[62] Probe 45 was constructed by installing an H_2S_n response unit into D-luciferin. The limit of detection was calculated to be 30 nM. The addition of H_2S_n resulted in a red shift of the absorption to 345 nm, and a strong bioluminescence signal, which could be used to monitor the changes of H_2S_n in living MDA-MB-231-luc cells. Probe 45 could be used to detect endogenous H_2S_n during bacterial infection in living mice. The reversible fluorescent probes are suitable for real-time and quantitatively visualizing dynamic changes of intracellular sulfane sulfur. Urano *et al.* reported a reversible fluorescent probe (probe 46) triggered by sulfane sulfur and GSH.[63] Based on a FRET fluorescence regulation mechanism, sulfane sulfur exhibited the unique ability to reversibly combine with the 2-thio-rhodamine B switching moiety, and GSH reduced the intramolecular disulfide bond of the spiro ring with a 15-fold fluorescence increase. The probe exhibited maximum absorption peaks at 480 nm and 560 nm. After adding sulfane sulfur, the absorbance at 560 nm decreases within 1 min, and the fluorescence at 525 nm increases, and the apparent fluorescence quantum yield increased from 0.015 to 0.20. The probe exhibits good selectivity for sulfane sulfur and the detection limit was 0.42 μM. The probe was capable of reversibly detecting sulfane sulfur in living A549 cells and in primary cultured hippocampal astrocytes.

13.2.2 Fluorescent Probes with Thioester Response Site

13.2.2.1 Strategy Based on Nucleophilic Reaction for Double-response Fluorescent Probes

Development of dual-channel fluorescent probes for detecting H_2S and H_2S_n are helpful for understanding the interrelationship between H_2S and H_2S_n. Xian's group reported a dual-detection probe (probe 47 in Figure 13.5) that could distinguish H_2S_n from H_2S.[64] Phenyl 2-(benzoylthio) benzoate and H_2S-mediated azide acted as the reaction sites for H_2S_n and H_2S, respectively. The rigid piperazine linker bridged two fluorophores, rhodamine and coumarin, and the detection mechanism was FRET. The detection of H_2S_n

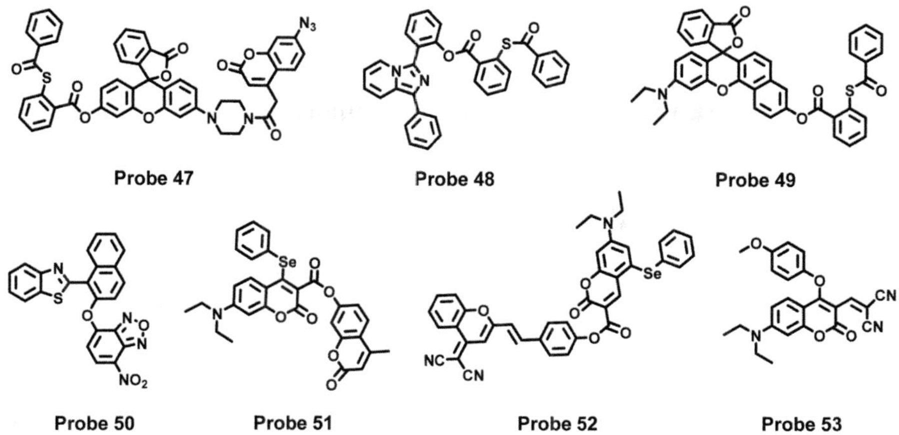

Figure 13.5 Fluorescent probes 47–53 for the detection of thioester based on dual-responsive nucleophilic reaction.

switched the emission at 542 nm, while H_2S simultaneously triggered two distinct emissions at 452 nm and 542 nm. The detection limit for H_2S_n was calculated to be 24 nM. This probe could be used to analyze *in situ* H_2S_n generation. The results indicated that probe 47 could detect H_2S and H_2S_n in HeLa cells with different emission channels. Yuan *et al.* designed a new type of fluorescent probe (probe 48) to detect endogenous H_2S_n. When the probe was used to detect H_2S_n, it was excited at 300 nm and the fluorescence at 478 nm increased significantly, and the quantum yield changed from 0.0132 to 0.12549. The limit of detection was 1 nM. The probe was used to detect the expression of H_2S_n in A549 cells, tumor tissues and mice.[65] Yu *et al.* constructed a semi-naphthalenetetrafluoro-based red-emitting fluorescent probe (probe 49), which could selectively detect H_2S_n. When used to detect H_2S_n, the probe was excited at 580 nm, and the fluorescence emission at 640 nm increased significantly, and reached saturation in 10 minutes. The limit of detection was 6.7 nM. The probe was used to detect the expression of H_2S_n in live A549 cells and zebrafish.[66]

Guo *et al.* reported a design strategy involving the conjugation of 1-(benzo[*d*]thiazol-2-yl) naphthalen-2-oxy and 7-nitrobenzo-2-oxa-1,3-diazole.[67] The probe 50 could simultaneously discriminate Cys/Hcy, GSH, and H_2S_n. In the presence of H_2S_n, the absorption at 561 nm was enhanced, and the colour of the solution changed from colourless to purple. The absorption spectrum in the presence of Cys/Hcy decreased at 375 nm and increased at 480 nm, fluorescence appeared at 548 nm, and the absolute fluorescence quantum yields were 0.127 and 0.138, respectively, corresponding to the colour change from colourless to orange. In the presence of GSH, the absorption band weakened at 375 nm and increased at 430 nm. The colour changed from colourless to yellow. The limit of detection was 0.17 μM, 0.30 μM, 0.32 μM, and 0.30 μM for Cys, Hcy, GSH, and H_2S_n, respectively. The probe was used to detect intracellular Cys/Hcy and GSH/H_2S_n, simultaneously, from different emission channels.

Sheng *et al.* reported a dual response probe (probe 51) for the detection of H_2S_n and GSH, respectively.[68] 7-Diethylaminocoumarin and 7-hydroxycoumarin were used as fluorophores. Due to their well-separated emission spectra, the detection of GSH and H_2S_n could be performed using different emission channels. The phenyl selenide site was selected to respond to GSH. The 3-position ester group of 7-diethylaminocoumarin was chosen to react with H_2S_n. The trigger of H_2S_n led to a strong emission centred at 465 nm, while the detection of GSH resulted in a fluorescence increase at 465 nm. The detection limit for H_2S_n and GSH were calculated to be 40 nM and 56 nM, respectively. The results showed that probe 51 could detect endogenous H_2S_n in RAW264.7 cells. Liu *et al.* reported a probe (probe 52) that could also be used to detect GSH and H_2S_n.[69] After the addition of GSH, probe 52 emitted green fluorescence at 530 nm. In the presence of H_2S_n, the probe showed a fluorescence enhancement at 680 nm. The probe exhibited a low detection limit 0.12 μM for GSH and 0.19 μM for H_2S_n. Probe 52 could be used to sense endogenous GSH and H_2S_n in living MGC-803 cells and RAW264.7 cells through dual-colour fluorescence imaging. Sheng *et al.* developed a fluorescent probe (probe 53) for H_2S_n detection by combining the 4-methoxyphenol moiety on the 4-position of the coumarin malononitrile derivative with a simple ether bond.[70] The cyclization of H_2S_n caused a significant increase in fluorescence at 576 nm within 6 minutes. Probe 53 showed a limit of detection of 20 nM, and could be used to image H_2S_n in living MGC-803 cells.

13.2.2.2 *Michael Addition/Cyclization Cascade Reactions*

Acrylate ester containing a α,β-unsaturated carbonyl unit is an effective response template for Hcy and Cys *via* conjugate addition/cyclization sequences. This reaction mechanism also satisfies the requirements of H_2S_n detection because of the dual nucleophilic features. Yang *et al.* utilized H_2S_n-mediated Michael addition/cyclization cascade of acrylate analogues to construct H_2S_n-specific fluorescent probes (probes 54–57 in Figure 13.6).[71] The fluorescent scaffold 3-benzothiazole-7-hydroxy-marin was modified with four different acrylate derivatives. After the detection of H_2S_n, probe 57 produced an emission peak at 490 nm, and the fluorescence quantum yield changed from 0.006 to 0.50. Probe 57 exhibited almost no response to related biothiols, such as GSH and Cys. The detection limit for probe 57 was 50 nM. The results indicated that probe 57 was suitable for monitoring H_2S_n in HeLa and RAW264.7 cells. Yang *et al.* reported a series of fluorescent probes (probes 58–63, Figure 13.6) based on the fluorophores *N*,*N*-diethylrhodyl and 3-(trifluoromethyl) cinnamoyl.[72] Probe 60 could respond to H_2S_n with low background interference. To meet the requirement of intracellular imaging, another desirable probe, 64, was developed. The calculated detection limit was 0.4 μM. When the probe reacted with H_2S_n, the fluorescence

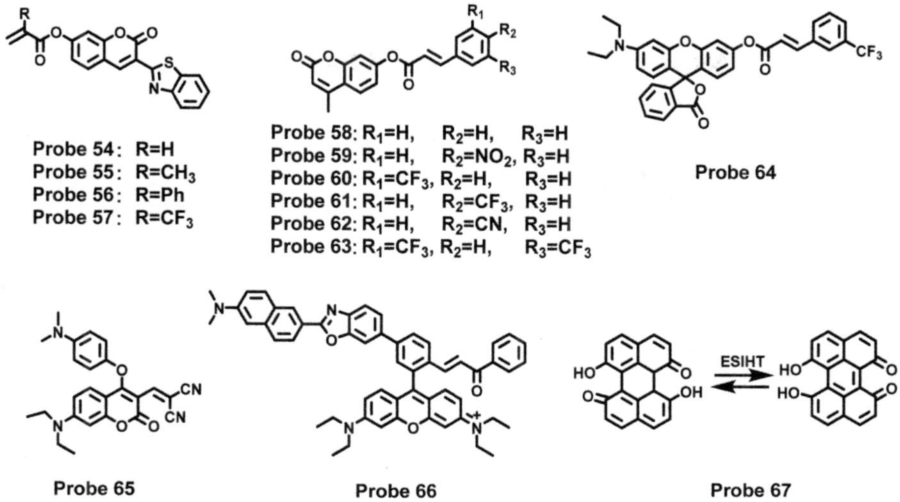

Figure 13.6 Fluorescent probes 54–67 for the detection of thioester based on Michael addition/cyclization cascade and ring-opening reactions.

at 548 nm gradually increased. The probe could be applied to detect H_2S_n in living SMMC-7721 cells.

Song's group designed a fluorescent probe (probe 65) for detection of H_2S, H_2S_n, and biological thiols using different emission channels.[73] When treated with H_2S, H_2S_n, and biothiols, probe 65 could emit fluorescence at 508, 576, and 469 nm, respectively. Probe 65 was selected for the potential identification of H_2S, H_2S_n, and biological thiols in RAW 264.7 cells. Yuan's group adopted a strategy including recognition site screening, probe construction, and nanoparticle self-assembly to construct a ratiometric probe (probe 66) that selectively responds to H_2S_n.[74] When the probe detected H_2S_n, it showed two well-separated absorbances at 355 nm and 565 nm, the fluorescence at 486 nm was reduced, and the fluorescence at 591 nm was increased, resulting in a fluorescence ratio of 59 times (I_{486}/I_{591}). The limit of detection was 9.4 nM. The probe could effectively avoid interaction with biothiols and persulfides. The probe was applied to reveal the H_2S_n generation pathway in L02 cells and RAW264.7 cells. Zeng *et al.* synthesized a reversible fluorescent probe (probe 67) for the detection of H_2S_n in solvents and living cells.[75] The probe was based on the tautomeric reduction of 6,7-dihydroxypentene-1,12-anthracene (DHPQ). After reduction of the carbonyl group at the side chain by H_2S_n, probe 67 released intense fluorescence. In the presence of H_2S_n solution, when the probe was excited at 400 nm, it emitted maximum green fluorescence at 462 nm and 490 nm, with a high quantum yield of 0.858. The turn-on intensity of fluorescence was significantly enhanced, by nearly 600 times. The fluorescence intensity reached the maximum after about 7 minutes, and the detection limit was 1.82×10^{-8} M. The reductant probe could be oxidized by hypochlorous acid (HClO). This reversible cycle could be used to redox cycle between H_2S_n and HClO in living HeLa cells.

13.2.2.3 Strategy Based on Ring-opening

The strong nucleophilicity of H_2S_n under physiological conditions can be recognized by activated aziridinyl chemical sensors. Xian *et al.* designed a mediated aziridine fluorescent probe (probe 68 in Figure 13.7) for H_2S_n.[76] The aziridine and fluorescent dansyl were used as recognition units and signal units, respectively. The probe provided a green fluorescence at 530 nm with excitation at 350 nm, and maximum intensity could be reached in 15 minutes. The detection limit was calculated as 0.3 μM. The probe also exhibited excellent two-photon photophysical and solid-state fluorescence properties ($\Phi = 1.00$).

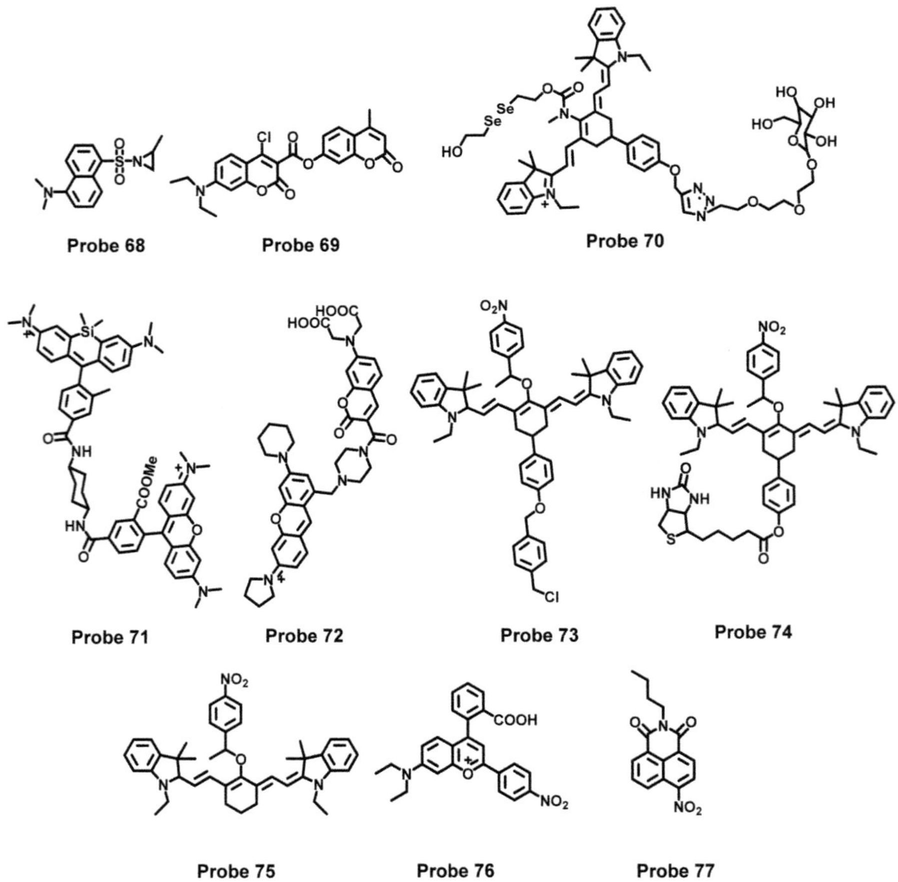

Figure 13.7 Fluorescent probes 68–77 for the detection of H_2S_n based on ring-opening reactions.

Chen *et al.* synthesized a ratiometric fluorescent probe (probe 69) based on H_2S_n-mediated coumarindithiolone formation.[77] The fluorophore was 4-chloro-7-diethylamino-coumarin ester (CDCE). The detection of H_2S_n would cause a reduction of the fluorescence at 510 nm and an increase in fluorescence at 461 nm, when excited at 385 nm. The calculated detection limit was 65 nM. The probe could recognize H_2S_n in living RAW264.7 cells.

Based on selenium–sulfur exchange reactions, Yu *et al.* reported a ratiometric near-infrared fluorescent probe (probe 70) for quantitative determination of Cys-SSH.[78] The probe contained bis(2-hydroxyethyl) diselenide (DiSe) as response modifier; methine cyanine (Cy) as a signal sensor; D-galactose as targeting group. When detecting Cys-SSH, the maximum absorption changed from 790 nm to 614 nm. And the maximum emission wavelength was shifted from 797 nm ($\Phi_{797\ \mathrm{nm}} = 0.05$) to 749 nm ($\Phi_{749\ \mathrm{nm}} = 0.11$). The experimental detection limit was 0.12 μM, and the theoretical detection limit was calculated to be 57 nM. The probe was used for quantification of Cys-SSH in living HL-7702 cells, HepG2 cells, and primary mouse hepatocyte cells, and the results were 1.85 ± 0.2 μM, 0.80 ± 0.1 μM, and 2.51 ± 0.4 μM, respectively, which were verified by flow cytometry and LCMS/MS. The probe could localize in liver cells, and visualize Cys-SSH in hepatic carcinoma models.

Utilizing the nucleophilic ability of hydropolysulfides (R-S_nSSH, $n > 1$), Urano *et al.* synthesized a reversible probe (probe 71) to distinguish R-S_nSSH from other sulfane sulfurs.[79] The probe was excited at 646 nm, the emission peak was at 667 nm, and the absolute quantum yield was 0.31, and the probe showed a K_d value of 10 μM towards R-S_nSSH over GSH and other sulfane sulfur species. The probe could achieve orthogonal and selective imaging of R-S_nSH in A549 cells. Ojida *et al.* reported a ratiometric fluorescent probe (probe 72) for sensing R-S_nSSH based on the FRET mechanism.[80] The probe contained coumarin as the FRET donor and xanthene as the FRET acceptor. When excited at 410 nm, the probe showed two different emissions at 479 nm and 584 nm. The reaction with R-S_nSSH led to a reduction in emission at 584 nm and increase at 479 nm. The detection limit ($3\sigma/k$) was determined to be 1.0 μM. The probe could detect changes of R-S_nSSH caused by enzymes in A549 cells.

To clarify the relationship between reactive oxygen species and H_2S_n, Yu's group synthesized a series of multi-response probes for the sequential detection of $O_2^{\bullet-}$ and H_2S_n in cells and *in vivo*. Probes 73 and 74 used the near-infrared heptamethine cyanine dyes as fluorophores.[30] The detection mechanism for $O_2^{\bullet-}$ was the hydrogen abstraction reaction, and the nitro group was reduced to an amino group by H_2S_n. The absorption and emission spectra of the two probes were basically the same. After reacting with $O_2^{\bullet-}$ and H_2S_n, probe 73 exhibited the maximum absorption at 770 nm and emission spectra centred at 780 nm. For probe 73, the calculated detection limit for $O_2^{\bullet-}$ was 50 nM, and the detection limit obtained by the experiment was 0.1 μM. The detection limit of H_2S_n was 80 nM, and the experimental detection limit was 0.2 μM. The probes 73 and 74 could locate in the mitochondrion in RAW264.7 cells, HUVEC cells, and could target tumor tissues for the detection of exogenous and endogenous changes of $O_2^{\bullet-}$ and H_2S_n. Yu *et al.* further developed a nitrobenzene derivative functional fluorescent probe (probe 75) for the continuous detection of $O_2^{\bullet-}$ and H_2S_n in cells and mouse hypoxic stress models.[81] The near-infrared fluorescence intensity gradually increased at 785 nm. Probe 75 was used to detect H_2S_n, and a strong fluorescence emission appeared at 635 nm though the 1,6-rearrangement–elimination to release the cyanine fluorophore. Probe 75 could simultaneously capture changes of $O_2^{\bullet-}$ and H_2S_n under continuous hypoxia and intermittent hypoxia to reveal the relationship between H_2S_n and $O_2^{\bullet-}$ in living SH-SY5Y cells and in an acute peritonitis mouse model. Based on the ICT process caused by the reduction of the nitro group to an amino group, Yang *et al.* designed a turn-on fluorescent probe (probe 76) for H_2S_n detection.[82] The reaction of H_2S_n led to an absorption shift from 544 nm to 568 nm. When used to detect H_2S_n, the fluorescence intensity of the probe at 620 nm increased significantly, by 87 times, and the fluorescence quantum yield increased from 0.001 to 0.028. The limit of detection was 75 nM. Probe 76 could detect H_2S_n in living SMMC-7721 cells and in biological samples. Zhou *et al.* synthesized an endoplasmic reticulum-targeting H_2S_n fluorescent probe (probe 77).[83] The probe exhibited an absorption peak at 356 nm. After treatment with H_2S_n, the peak shifted to 369 nm, and a shoulder peak appeared at 430 nm. At the same time, fluorescence emission appeared at 540 nm. The detection limit was 26 nM. The probe could be used to observe H_2S_n changes in the endoplasmic reticulum during inflammation and anti-inflammatory processes.

13.3 Fluorescent Probes for the Detection of Reactive Selenium Species

Selenium is an essential trace element for the 21st amino acid selenocysteine (Sec), which is mainly attributed to maintaining the intracellular redox homeostasis.[84] Selenoproteins are a family of proteins containing Sec, including GSH peroxidase and thioredoxin reductase (TrxR).[85] A total of 25 known selenoproteins have been identified and are related to the health of humans. As the main reactive site, Sec is an important functional component of selenoprotein. It can affect the protein structure, regulation of substance transport, anti-oxidation, anti-inflammatory response, and regulation of genes, and is linked with a variety of diseases, including cancer, sub-optimal health, decline in immunity.[86] H_2Se is the most reactive form of selenium in organisms. Under physiological conditions, it exists in the form of HSe^- and is always available for the biosynthesis of selenium-containing molecules.[87] Although biological selenols play a pivotal role in cells, their high reactivity and low concentration in living organisms result in great challenges for its detection in living systems. As such, fluorescent probes have become extremely advantageous tools for monitoring biological selenols in living systems due to their low cost, high sensitivity, high specificity, and non-invasive characteristics.[8–12]

13.3.1 Fluorescent Probes for the Detection of Selenocysteine

The detection of bioselenols mainly depends on their nucleophilic properties. There exist quite distinct pK_a differences between biothiols and bioselenols under physiological conditions; for example, the pK_a of Sec is ~5.2, the pK_a of Cys is ~8.29, and the pK_a of GSH is ~8.0. The levels of biothiols (GSH as a representative) are ~mM and the concentrations of bioselenols (Sec as a representative) are μM in certain cells. As such, the concentration of deprotonated anion of biothiols (GS^-) must be much higher than that of bioselenols (RSe^-) in living cells. Therefore, the biggest obstacle for the selective detection of bioselenols in living systems is to avoid interference from biothiols.

13.3.1.1 Nucleophilic Substitution Reaction of 2,4-Dinitrobenzenesulfonate Ester

Taking advantage of the special features of the pK_a and nucleophilic substitution properties of selenols (Figure 13.8A),[88] Maeda *et al.* reported a fluorescent probe (probe 78) for selenol detection, which was based on nucleophilic reaction with 2,4-dinitrobenzenesulfonate ester.[89,90] The fluorophore was dimethyl-fluorescein with characteristic fluorescence emission at 580 nm. The probe could detect selenols at pH = 5.8. The detection limit was 0.8 pmol/well. Wang *et al.* established a fluorescence probe (probe 79) for Sec detection based on tetralin-xanthene fused dye.[91] The maximum absorption of probe 79 was 550 nm, which red shifted to 584 nm after reaction of Sec. The reaction was complete within 2 min and resulted in a 124-fold fluorescence enhancement. The probe could be used to selectively detect Sec in a physiological environment, in HeLa cells. Feng's group reported a colorimetric and fluorescence turn-on probe (probe 80) that responded to Sec under physiological pH conditions.[92] The fluorophore was 3-benzothiazolyl-7-hydroxycoumarin (BTHC) with high fluorescence quantum yields up to 0.65. The addition of Sec resulted in a significant turn-on fluorescence signal at 493 nm. The detection limit was determined to be about 47 nM. The probe could detect Sec in serum samples and in living HeLa cells. Jiang *et al.* designed a turn-on fluorescent probe (probe 81) to detect Sec.[93] The fluorophore structure was modified by inserting a benzene ring into the xanthene fluorophore to stabilize the positive ions. After treatment with Sec, a new fluorescence peak arose at 614 nm. The limit of detection was calculated as 11.2 nM. The probe could image Sec in living A549 cells.

Tan's research group reported on a two-photon ratiometric fluorescent nanoprobe (probe 82 in Figure 13.9) with the FRET mechanism.[94] The probe contained a FRET donor–acceptor pair: two-photon fluorophore 2-naphthalenecarboxylic acid and fluorescein fluorophore in a silica nanocarrier. Upon reaction with Sec, the emission at 450 nm decreased distinctly and a strong emission at 520 nm developed. The limit of detection towards Sec was determined to be 0.5 μM. The probe could detect Sec in living RAW264.7 cells and inflamed liver tissues, suggesting promise for the diagnosis of Sec-relevant

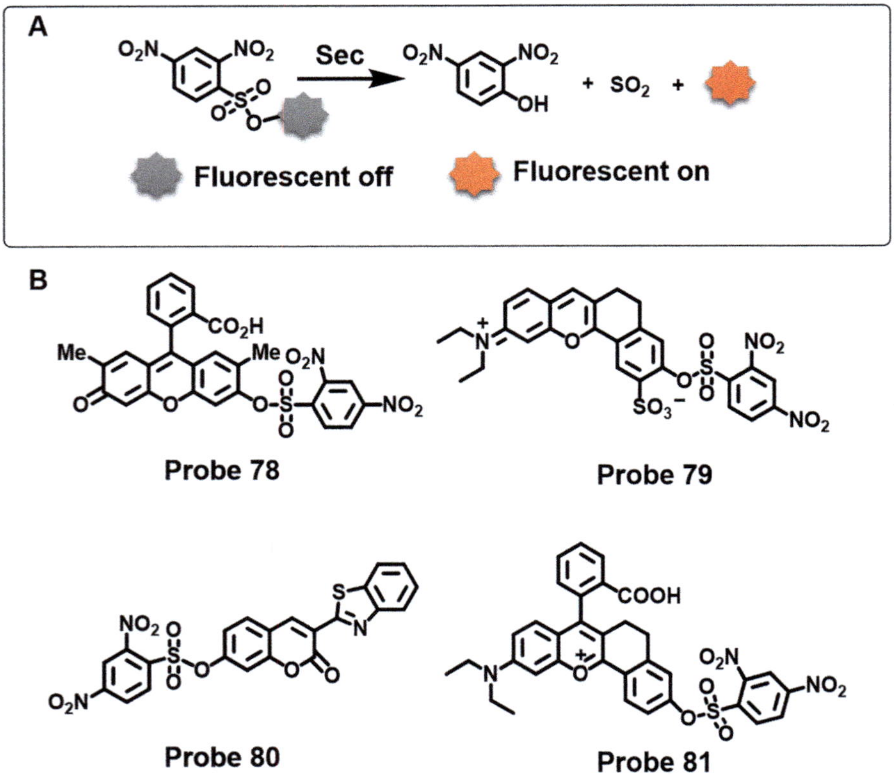

Figure 13.8 (A) Design strategy of fluorescent probes based on nucleophilic substitution reaction of 2,4-dinitrobenzenesulfonate ester. (B) Fluorescent probes 78–81 for the detection of Sec based on nucleophilic substitution reaction of 2,4-dinitrobenzenesulfonate ester.

diseases. Li *et al.* reported a carbon-dot-based fluorescent turn-on probe (probe 83) for selenol detection.[95] The carbon dots (Y-G-CDs) emitted yellow-green fluorescence at 520 nm. The functionalization of the electron-deficient 2,4-dinitrobenzyl moiety onto the surface of Y-G-CDs quenched the fluorescence. The detection of Sec could then cleave the quencher through the nucleophilic substitution process and could recover the fluorescence of probe 83. The detection limit was calculated to be 23 nM. The nanoprobe was capable of imaging exogenous and endogenous selenols in living L929 cells. Qiu *et al.* developed an amphiphilic polymer nano-micelle probe (probe 84) for the selective imaging of Sec.[96] The Sec-responsive micelles could trap the fluorescent doxorubicin (DOX) drug. The selenols activated the response site and generated hydrophilic ends, which destabilized the micelles and released doxorubicin for fluorescence imaging. In the process of detecting Sec, after the probe was excited at 490 nm, the emission at 590 nm was enhanced 2.8 times, and the detection limit was 50 nM. The probe could be used to qualitatively analyze the levels of Sec in HeLa cells and in live cervical neoplasm.

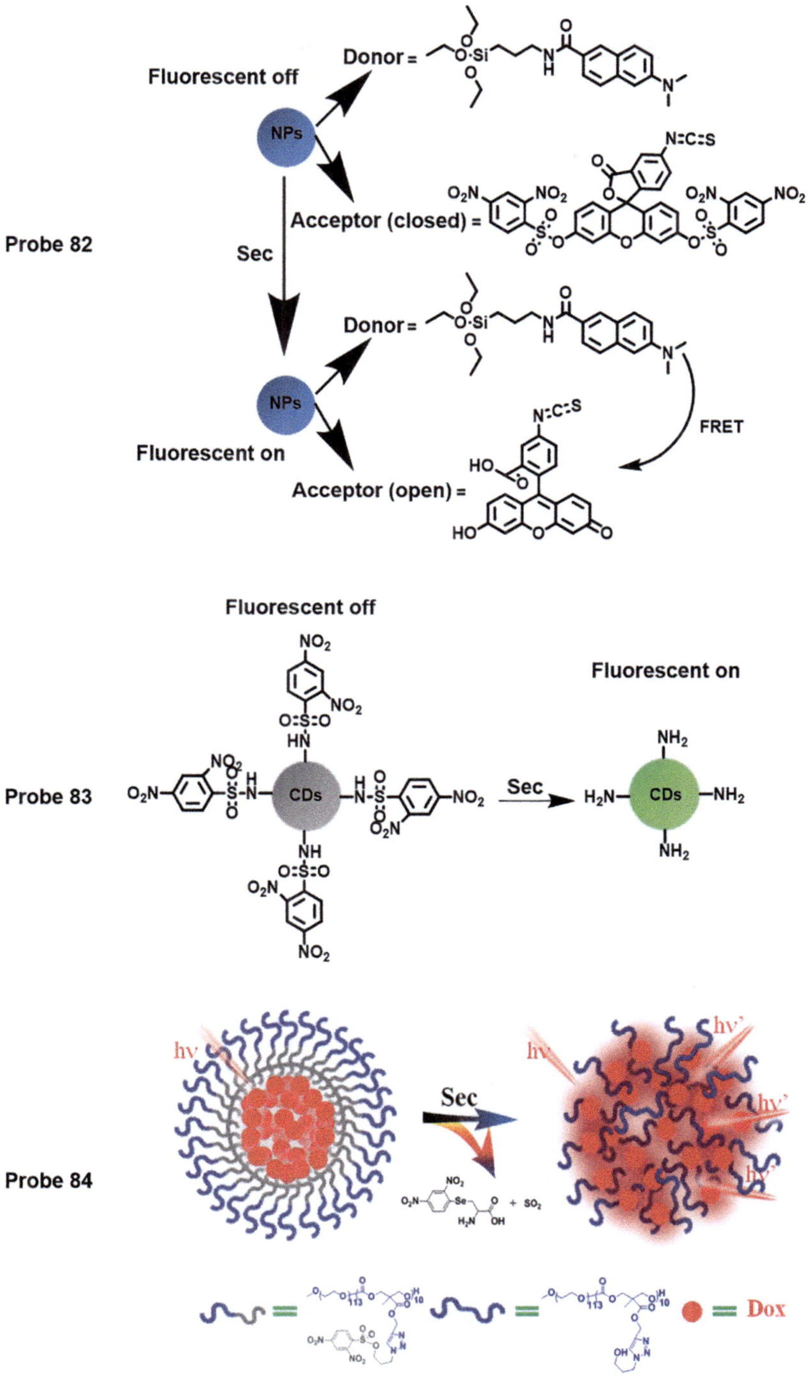

Figure 13.9 Fluorescent nanoprobes constructed by nanoparticles (NPs) (probe 82), carbon dots (CDs) (probe 83), and doxorubicin (DOX) (probe 84) for the detection of Sec based on the nucleophilic substitution reaction of 2,4-dinitrobenzenesulfonate ester. Probe 84 was adapted from ref. 96 with permission from the Royal Society of Chemistry.

13.3.1.2 Nucleophilic Substitution Reaction of 2,4-Dinitrobenzenesulfonamide

Although 2,4-dinitrobenzenesulfonate ester can be used to respond to Sec, the 2,4-dinitrobenzenesulfonamide exhibits weaker electrophilic behaviour, which is a promising candidate for Sec detection, given the stronger nucleophilic ability of selenols than thiols (Figure 13.10A). Zhang *et al.* reported a potential probe (probe 85) for Sec detection.[88] The addition of Sec led to an increase of fluorescence emission centred at 502 nm. The experimental detection limit and theoretical detection limit were 0.5 μM and 62 nM, respectively. Sec and sodium selenite (Na_2SeO_3) could significantly increase the concentration of Sec in HepG2 cells and HeLa cells. The development of fluorescent probes with near-infrared emission is essential for the imaging of living animals. Lin *et al.* reported a near-infrared fluorescent probe

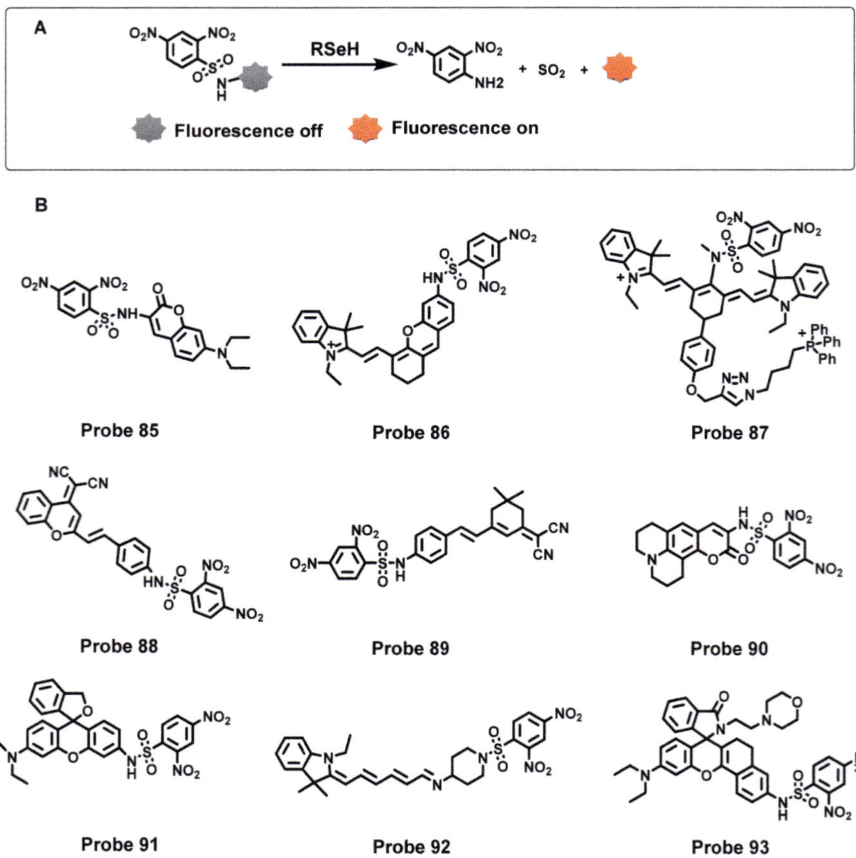

Figure 13.10 (A) Design strategy of fluorescent probes based on nucleophilic substitution reaction of 2,4-dinitrobenzenesulfonamide. (B) Fluorescent probes 85–93 for the detection of Sec based on nucleophilic substitution reaction of 2,4-dinitrobenzenesulfonamide.

(probe 86) for detecting Sec.[97] On reaction with Sec, the probe released a strong fluorescence emission at around 712 nm and the fluorescence quantum yield was 0.53. The probe could be used to detect exogenous Sec in MCF-7 cells and HeLa cells. Moreover, the probe could also image Sec in Kunming mice by intraperitoneal injection of Sec.

In order to detect and selectively quantify changes of endogenous Sec concentration in mitochondria, Yu *et al.* designed and synthesized a mitochondria-targeted ratiometric near-infrared fluorescent probe (probe 87) for evaluating the relationship between CS_2-induced acute/chronic hepatitis and the degree of inflammation.[98] Probe 87 was composed of a responsive group (2,4-dinitrobenzenesulfonamide), a mitochondria-targeting moiety (lipophilic triphenylsulfonium cation), and a fluorophore (heptacyanine). The maximum absorption of the probe was at 790 nm. The reaction with Sec shifted the maximum absorption to 610 nm. At the same time, the fluorescence gradually decreased at 780 nm ($\Phi_{780\ \mathrm{nm}} = 0.03$), and a new emission band increased at 744 nm ($\Phi_{744\ \mathrm{nm}} = 0.15$). The ratiometric images and flow cytometry experiments indicated that the probe could quantitatively detect Sec in HeLa cells, mouse liver cell line (BRL3A), mouse hepatoma cell line (RH-35), human normal liver cell line (HL-7702), human hepatocellular liver carcinoma cell line (HepG2), and human hepatocellular carcinoma cell line (SMMC-7721). This probe was used to sense Sec level fluctuation in CS_2-induced acute hepatitis and chronic inflammation mouse models. Feng *et al.* reported a near-infrared fluorescent turn-on probe (probe 88) for Sec with a large Stokes shift.[99] The fluorophore dicyanomethylene-benzopyran (DCMB) exhibited a high quantum yield of 0.66. The fluorescence emission at 670 nm increased after reaction with Sec within 10 minutes, with a Stokes shift of 192 nm. The calculated detection limit was 60 nM. Probe 88 could be used to detect Sec in fetal bovine serum, in living A549 cells, and in living Balb/c mice. Zhao *et al.* developed a near-infrared fluorescent probe (probe 89) with the ability to observe endogenous Sec.[100] The fluorophore dicyanoisophorone showed a fluorescence emission at 666 nm with fluorescence quantum yield of 0.46. Probe 89 exhibited a Stokes shift of 195 nm. The detection limit was calculated to be 10 nM. The probe was able to detect both exogenous and endogenous Sec in MCF-7 cells, as well as in MCF-7 xenograft tumor Kunming mice. Yang's group reported a probe (probe 90) for Sec detection based on the coumarin fluorophore.[101] The probe exhibited a first-order rate constant of 2.15 min^{-1} and a detection limit of 18 nM. Probe 90 was able to distinguish cancer cells from normal cells *via* detection of intracellular selenols by flow cytometry. The quantum yield of the coumarin fluorophore was 0.3. Probe 90 could detect changes of Sec in HEK293 cells, in A549 cells, and in zebrafish. The level of selenols in cancer cells was significantly higher than that in normal cells. Zhang *et al.* designed a low-background fluorescent probe (probe 91).[102] To avoid the background, the carboxylic acid group of rhodamine dye was reduced to a hydroxymethyl group, ensuring the non-fluorescent spirocyclic form. Probe 91 displayed a fluorescence response at 538 nm, and the detection limit was calculated to be 28 nmol L^{-1}. The probe was used to visualize Sec concentration changes in a lipopolysaccharide-stimulated oxidative stress HepG2 cell model. The selenol levels increased when the cells were resistant to oxidative stress. Li *et al.* designed a fluorescent probe (probe 92) for selenol detection using a pK_a shift strategy.[103] The electron-withdrawing ability of the substituent near the imine group would affect the pK_a value of the imine nitrogen of the imine Schiff base and, therefore, the spectral characteristics of the merocyanine Schiff base. Probe 92 showed two absorptions centred at 423 and 566 nm because of the interchange of two acid–base type forms. Probe 92 emitted at 593 nm, with a calculated detection limit at 68 nM. Probe 92 could be used to selectively sense Sec in SMMC-7721 cells. Zhang's group developed a lysosome-targeting probe (probe 93) for the detection of exogenous and endogenous Sec.[104] The probe exhibited a first-order rate constant (k) of 0.0483 min^{-1}, with a detection limit of 0.38 μM. The probe could be applied for the imaging of endogenous and exogenous Sec in the lysosomes of HeLa cells.

13.3.1.3 Nucleophilic Substitution Reaction of 2,4-Dinitrobenzenoxy Ether

The nucleophilic substitution reaction of 2,4-dinitrobenzenoxy ether was also employed to design fluorescent probes for the detection of Sec (Figure 13.11A). Feng *et al.* developed a colorimetric and near-infrared fluorescent probe (probe 94) for rapidly detecting Sec.[105] The near-infrared fluorophore was dicyanomethylene-benzopyran (DCMB). The addition of Sec shifted the maximum absorption from 430 nm to 560 nm, with a large Stokes shift of 146 nm. The fluorescence emission of the probe was at 706 nm. The experimental detection limit was 0.2 μM and the calculated detection limit was about 62 nM. Probe 94 could quantitatively detect Sec in fetal bovine serum and in HeLa cells. BODIPY derivatives exhibit superior spectral properties, such as high molar extinction coefficients, narrow Gaussian absorption and emission bands, good light stability, *etc.* Song *et al.* reported two probes (probes 95 and 96) for Sec detection based on the BODIPY fluorophore.[106] Probe 96 produced an emission profile centred at 663 nm with a calculated detection limit of 9 nM. Probe 96 could image Sec in living MCF-7 cells and in living animals. Zhou *et al.* reported on a two-photon fluorescent probe (probe 97) for the ratiometric detection of Sec based on the FRET mechanism.[107] The probe's energy donor coumarin moiety exhibited blue fluorescence with an emission peak at 463 nm. Once the probe reacted with Sec, the fluorescence emission shifted to 545 nm, which was the characteristic spectrum of the energy receptor naphthalimide fluorophore. The limit of detection was 7.88 nM. The ratiometric images could be used to detect Sec in living HeLa cells, in frozen liver tissue, and zebrafish. Yin's group described a bioluminescent probe (probe 98) for the detection of Sec.[108] The established strategy was based on the masked 2,4-dinitrophenyl ether moiety at the 6′-hydroxyl of D-luciferin. Probe 98 exhibited an off–on bioluminescence response with a detection limit of 8 nM. The bioluminescent probe could be used for the

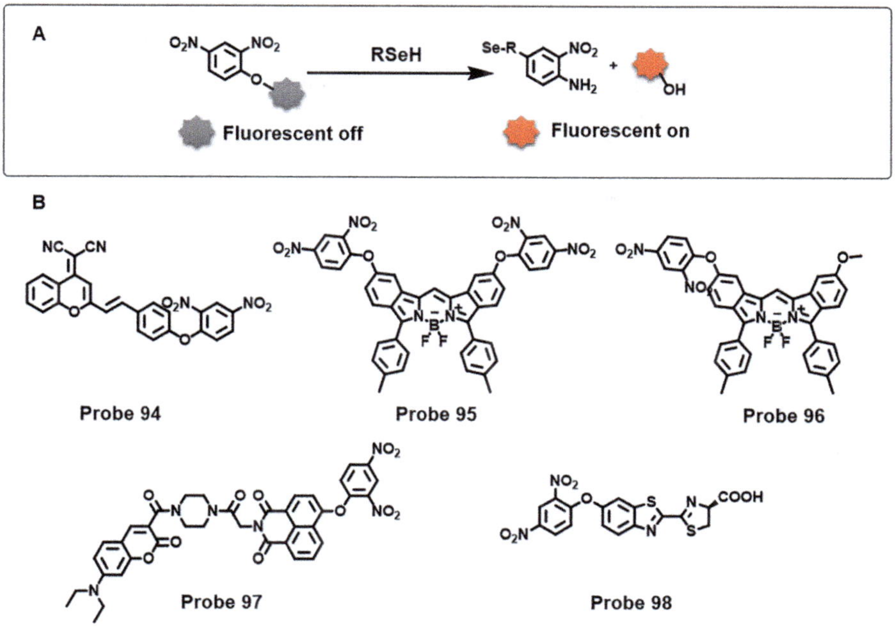

Figure 13.11 (A) Design strategy of fluorescent probes based on nucleophilic substitution reaction of 2,4-dinitrobenzenoxy ether. (B) Fluorescent probes 94–98 for the detection of Sec based on nucleophilic substitution reaction of 2,4-dinitrobenzenoxy ether.

imaging of Sec in living MCF-7-luc cells, in FVB-luc$^+$ mice, and in MCF-7-luc tumor-bearing nude mice.

13.3.1.4 *Other Functional Response Groups*

2,1,3-Benzoselenadiazole contains a Se–N group which could be cleaved by the nucleophilicity of –SeH. Tang *et al.* reported a fluorescent probe (probe 99 in Figure 13.12) integrating the functional moiety into a hemicyanine dye for selenol detection.[109] Sec converted to the Se–N moiety, a diamine group, which was beneficial for fluorescence emission at 580 nm, with a large Stokes shift of 120 nm. The fluorescence quantum yield increased from 0.019 to 0.110. The detection limit was 7.0 nM.

Inspired by the selenium–sulfur bond exchange reaction, Yu *et al.* reported a ratiometric near-infrared fluorescent probe (probe 100) for evaluating the relationship between the redox state of Sec and the process of inflammation in hepatitis.[110] Heptamethine cyanine was selected as the fluorophore, bis(2-hydroxyethyl) disulfide as the response unit, and D-galactose as the liver localization group. After reaction with Sec, the maximum absorption shifted from 782 nm to 610 nm, and the maximum fluorescence wavelength shifted from 800 nm ($\Phi = 0.06$) to 750 nm ($\Phi = 0.11$). The experimental detection limit was determined to be 0.09×10^{-6} M. The probe could qualitatively

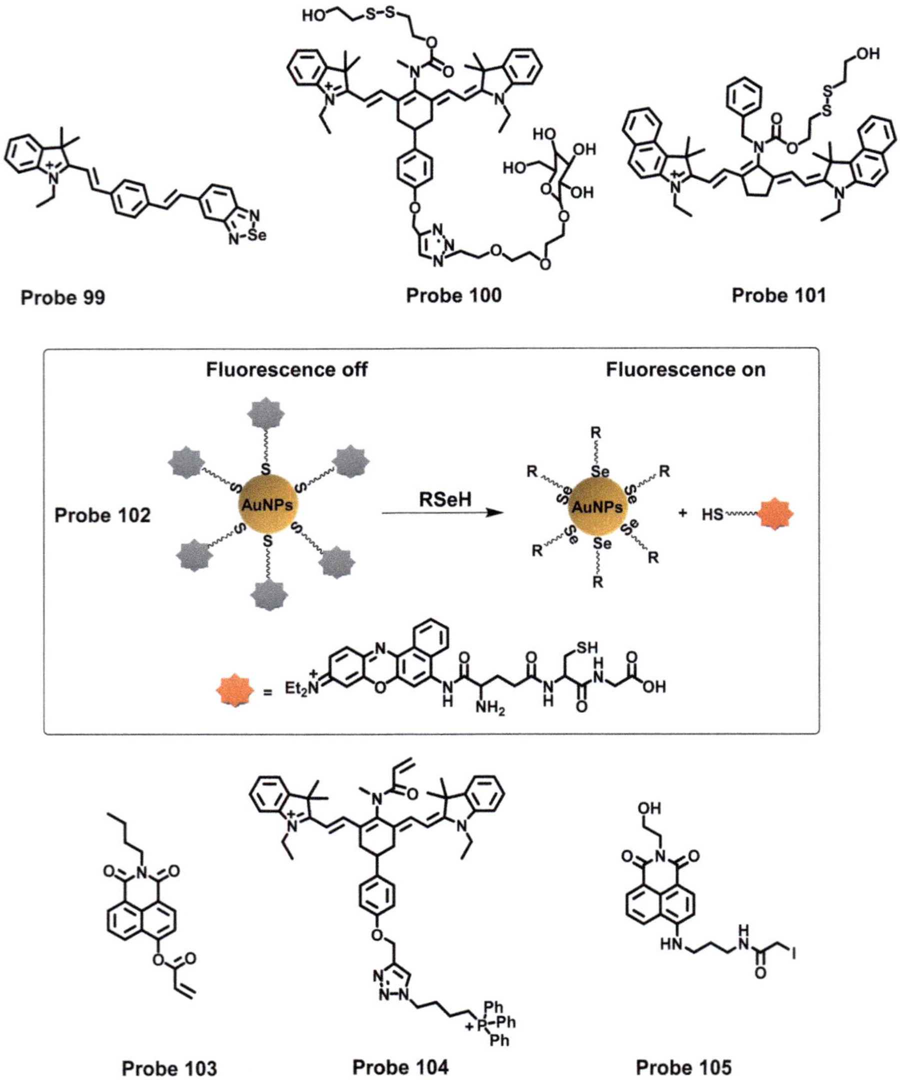

Figure 13.12 Fluorescent probes 99–105 for the detection of Sec based on other functional groups.

and quantitatively measure Sec in HepG2 cells, HL-7702 cells, and primary mouse hepatocytes (female BALB/c mice), with values of $3.09 \pm 0.12 \times 10^{-6}$ M, $4.04 \pm 0.18 \times 10^{-6}$ M, and $4.36 \pm 0.20 \times 10^{-6}$ M, respectively. Probe 100 was used for Sec imaging studies in the BALB/c peritoneal cavity of mice.

Zhao's group developed a selenol-activated proportional photoacoustic imaging probe (probe 101), which was used to visualize the pathological progress of autoimmune hepatitis.[111] The reaction of probe 101 and selenols caused the photoacoustic spectrum to be peak shifted from 860 nm to 690 nm.

Probe 101 was used for photoacoustic imaging of endogenous selenols in HL-7702 cells. Probe 101 could be used to diagnose the development of autoimmune hepatitis (AIH) disease in a Con A-induced AIH murine model by assessing the concentration changes of selenols. Wang *et al.* developed a nanofluorescent probe (probe 102) for selenol detection based on the fluorescence quenching mechanism by gold nanoparticles (AuNPs).[112] The probe integrated GSH into Nile blue, then assembled on AuNPs. The competitive adsorption on the gold surface released the fluorophore to emit fluorescence. The probe can detect exogenous and endogenous selenols in living HepG2 cells, and in selenium-rich rice and tea. Based on the addition reaction of acrylate, Zhang *et al.* developed an ICT-based ratiometric fluorescent probe (probe 103) for Sec detection.[113] The probe was designed by attaching the acrylate moiety to the 4-hydroxynaphthalimide fluorophore. When the probe reacted with Sec, the absorption shifted from 345 nm to 449 nm and exhibited a colour change from blue to yellow. The quantum yield of the probe was 0.077, while that of the fluorophore was 0.32. The calculated detection limit of probe 103 was 12 nM. The probe was used for the quantitative ratiometric imaging of exogenous and endogenous Sec in living HeLa cells. Sec deprotonated to its anion form under physiological conditions, suggesting a stronger nucleophilic capacity than other biothiols. Considering that the amide group is more difficult to hydrolyze than the ester group. Yu *et al.* reported that the acrylamide moiety can be employed as a favourable responsive site for constructing a ratiometric mitochondrial fluorescent probe (probe 104) for Sec detection in living cells and *in vivo*.[114] After reaction of the probe with Sec, the absorption at 784 nm blue shifted to 623 nm. The colour of the probe changed from green to blue. Meanwhile the maximum fluorescence emission at 825 nm decreased, with increase of the fluorescence emission centred at 765 nm. The probe had been applied for investigating the relationship between Sec and oxidative stress in Nthyori3-1 cells and BHT 101 cells, as well as in different thyroid disease mouse models. The probe could behave as a potential chemical tool for the auxiliary diagnosis of thyroid diseases. Zhang *et al.* synthesized a naphthalimide-based fluorescent probe (probe 105) for labelling proteins containing Cys and Sec residues.[115] The alkylation group was iodoacetamide. Probe 105 could quantify serum proteins and hemoglobin in serum. The limit of detection was calculated to be 37.5 nM (2.5 mg L^{-1}). The probe could indicate TrxR in Tan sheep liver and muscle, and could distinguish TrxR 1 and TrxR 2 by SDS–PAGE.

13.3.2 Fluorescent Probe for H_2Se Detection

As a reduced form of selenium, endogenous H_2Se was mainly generated by biothiol-reduced selenite. H_2Se is a crucial intermediate of selenium metabolism and plays a vital role in many physiological and pathological processes. Tang *et al.* developed a H_2Se fluorescent probe (probe 106 in Figure 13.13) by integrating the 2,1,3-benzoselenadiazole moiety into a near-infrared merocyanine fluorophore.[116] The probe exhibited weak fluorescence emission at 735

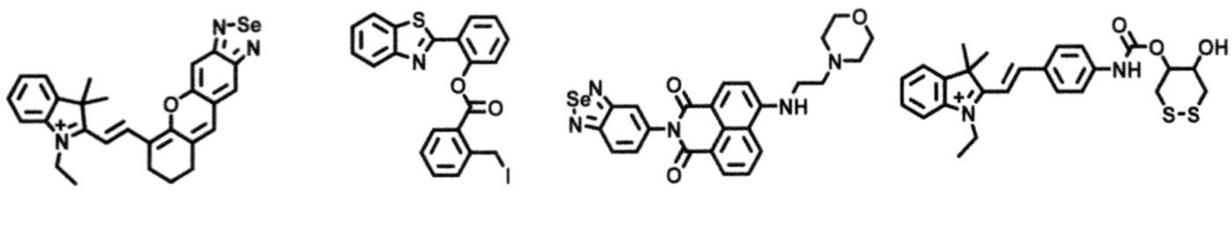

Probe 106 Probe 107 Probe 108 Probe 109

Figure 13.13 Fluorescent probes 106–109 for the detection of H_2Se.

nm because of the heavy atom effect of Se. The probe could be used to image endogenous H_2Se in living HepG2 cells and in H22 tumor-bearing mice.

Zhang *et al.* reported an activatable lysosomal fluorescent probe (probe 107) for H_2Se detection using the intramolecular PeT process.[117] The naphthalimide fluorophore gave its characteristic maximum absorbance at 435 nm. Upon detection of H_2Se, the emission centred at 535 nm increased within 1 min, and the fluorescence quantum yield changed from 0.02 to 0.3. The limit of detection was of 1.2 μM. The probe demonstrated that H_2Se could be generated and gradually accumulated in lysosomes of HepG2 cells under hypoxic conditions. Based on the reaction between H_2Se and the stable disulfide bond in the six-membered ring, Tang's group designed a fluorescent probe (probe 108) for H_2Se detection.[118] The probe was synthesized by integrating the 2-dithiane-4,5-diol response unit into a hemicyanine fluorophore *via* a cleavable ester bond. After treatment with H_2Se, the emission intensity at 535 nm increased and the fluorescence quantum yield changed from 0.023 to 0.081. The detection limit was 6.8×10^{-7} M. The probe was used to visualize H_2Se in living HepG2 cells and in mice bearing subcutaneously implanted hepatoma cell line H22 tumors. Based on a substitution-cyclization cascade reaction, Zhang *et al.* introduced the 2-(iodomethyl) benzoate moiety into the 2-(2-hydroxy-phenyl) benzothiazole fluorophore to generate a fluorescent probe (probe 109) for H_2Se detection.[119] The specific interaction with H_2Se switched on the excited-state intramolecular proton-transfer process and allowed ratiometric detection with the blue emission at 357 nm shifted to an emission band centred at 470 nm. The limit of detection was determined to be 0.2 μM. The probe could be used as a paper-based indicator for gaseous H_2Se detection in environmental pollutants.

13.3.3 Fluorescent Probe for TrxR Detection

The thioredoxin system includes TrxR, thioredoxin (Trx), and nicotinamide adenine dinucleotide phosphate (NADPH), which regulates multiple intracellular redox signals and plays a vital role in maintaining intracellular redox homeostasis.[120,121] Mammalian TrxR is a homodimeric selenoprotein that is involved in special catalytic activity towards selenolthiol/selenenylsulfide. TrxR is commonly overexpressed in many cancer cells and enhancing

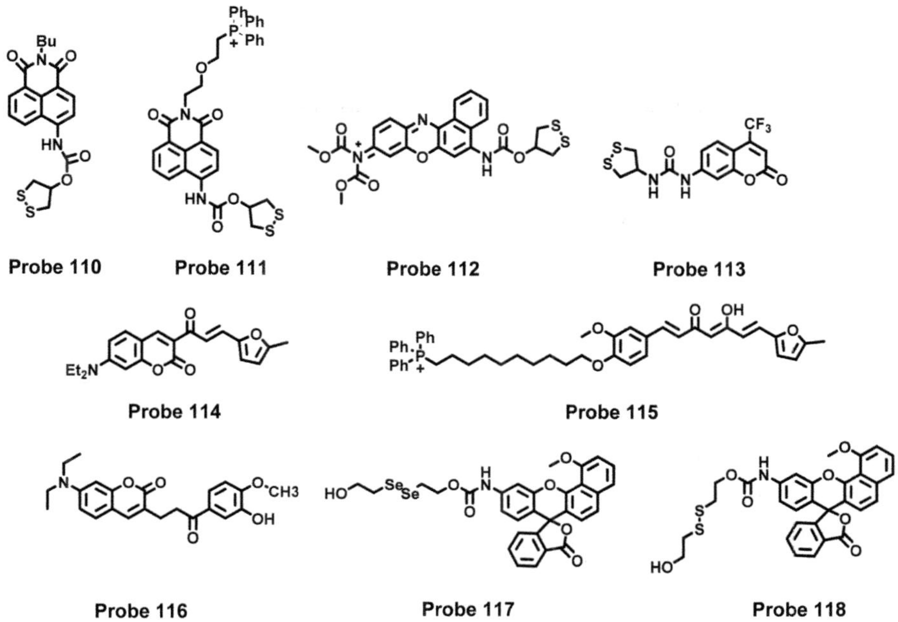

Figure 13.14 Fluorescent probes 110–118 for the detection of TrxR.

enzyme activity of selenoenzyme is considered to be a prospective approach to fight cancer.[120–125]

To detect enzyme activity in living cells, Fang *et al.* reported an off–on fluorescent probe (probe 110 in Figure 13.14) for mammalian TrxR detection.[126] The naphthalimide was equipped with a five-membered cyclic disulfide moiety that could respond to TrxR in the presence of NADPH. When the probe was reduced by TrxR and NADPH, the absorbance was red shifted to 438 nm and accompanied by generation of a green emission band at 538 nm. The probe was useful for imaging TrxR activity in living HepG2 cells. To evaluate the TrxR activity in mitochondria, Fang's group reported another TrxR probe (probe 111).[127] The mitochondrial localization group was triphenylsulfonium. The probe could selectively detect mitochondrial TrxR in HeLa cells. The probe could be used to observe TrxR2 dysfunction in Parkinson's disease model cells. Fang *et al.* also designed a red-emission probe (probe 112) for mammalian TrxR.[128] The probe responded to TrxR, with the maximum emission wavelength at 660 nm. The probe could detect TrxR in the cytosol of HeLa cells. Fang *et al.* improved the cyclization rate of the responsive group to synthesize a naphthalimide-based fast response fluorescent probe (probe 113) for Sec detection within 1 min (λ_{ex} = 345 nm, λ_{em} = 460 nm).[129] The probe could be used to develop a high-throughput screening method to examine inhibitors of TrxR. Based on the Michael reaction with α,β-unsaturated ketone, Bu *et al.* reported a fluorescent probe (probe 114) by conjugating coumarin fluorophore with furanyl acryl moiety for TrxR detection.[130] The α,β-unsaturated ketone moiety could quench the fluorescence of the coumarin, while covalent coupling to TrxR would result in an increase of fluorescence following a blue shift from 470 nm to 440 nm. The fluorescence emission peak appeared at 500 nm, and the detection limit was at the nanomolar level. The probe could be used to detect a high level of TrxR in cisplatin-resistant A549 tumor cells. Bu *et al.* also reported a mitochondria-targeted probe (probe 115) for TrxR.[131] The fluorescent probe 115 was constructed by the linkage of a TrxR inhibitor 1-(4-hydroxy-3-methoxyphenyl)-7-(5-methylfuran-2-yl) hepta-1,6-diene-3,5-dione with the mitochondrial localization group triphenylphosphine *via* a bridging group. With the mitochondrial localization group, the probe could obstruct the function of mitochondria and exhibited a higher inhibitory activity on TrxR in all tested cancer cell lines, including HeLa cells, Bel-7402 cells, and MDA-MB-231 cells. Gao *et al.* evaluated a series of 7-diethylamino-coumarin-chalcone hybrids as fluorescent TrxR inhibitors.[132] The fluorescence agents could visualize the drug delivery and apoptosis-inducing processes in HCT116 cells. The fluorescent agent (probe 116) maximum excitation was at 490 nm, and the emission was at 595 nm. Probe 116 could downregulate the expression levels of TrxR and could accumulate reactive oxygen species to activate the mitochondrial apoptosis pathway. Based on reduction of the disulfide with TrxR, Strongin *et al.* reported two disulfide/diselenide fluorescent probes (probes 117 and 118) based on semi-naphthorhodafluor fluorophore for evaluating TrxR activity.[133] The diselenide fluorescent probe underwent intramolecular nucleophilic addition and faster selenoate attack on the carbonyl carbon than the corresponding thiolate. This probe achieved selective detection of TrxR1 by an order of magnitude of 100 nM against GSH at millimolar concentrations. The binding affinity (K_m) value was 15.89 μM. Both of the probes were used to assess the activity of TrxR1 in lung adenocarcinoma HCC827 cells.

13.4 Conclusions and Perspective

Understanding of the signalling roles and physiological effects involved in redox homeostasis of sulfane sulfur and reactive selenium species are urgently required. Although these dominating biologically reactive species are at nanomolar steady-state concentrations, the contributions of them can trigger signalling pathways by generating and inhibiting redox processes. The recognition of the preferred targets for sulfane sulfur and reactive selenium species, and of their downstream signalling intermediates, are not yet completely understood. While the biofunctional validation of how these species affect individual proteins and regulate intracellular redox flux is still not fully available. The lack of selectivity, sensitivity, accuracy, and readily available chemical tools for quantifying sulfane sulfur or reactive selenium species needs remediation.

The fields of redox biology benefit greatly by investigating the potential formation and decay kinetics and reactivities of sulfane sulfur or reactive selenium species. Appropriate chemical tools that are able to translate subtle variation into a non-invasive, *in situ*, and real-time visual signal are essential for clinical diagnosis of diseases, and therapy. Compared with other biological detection technologies, fluorescent biological imaging can provide researchers with a sensitive, real-time, high-precision method, which can also avoid interference from intracellular endogenous species, providing an opportunity to monitor the concentration changes of reactive biological species more accurately for clinical diagnosis, therapy, and prevention. In this chapter, we have summarized the current significant progress in synthesis and design strategies toward fluorescent probes used to detect sulfane sulfurs and reactive selenium species. The fluorescent probes used to detect sulfane sulfurs include persulfide, polysulfide, and hydrogen polysulfide. The fluorescent probes used to monitor reactive selenium species include Sec, H_2Se, and Trx. All these design strategies are derived from the nucleophilic addition reactions, such as reduction, nucleophilic substitution, Michael addition, thiolysis, and so on.

Although fluorescent probes are currently available for illustrating the vital roles of sulfane sulfurs and reactive selenium species in organisms under physiological conditions, a few problems prevent these detection platforms from being used in clinical application. The fluorescence switch-on response makes the biological function inside the biological system more easily clear, which ultimately offers thorough exposition of the function of these species with good spatial and temporal resolutions. This is one of the key issues for clinical diagnosis. Analyte-specific probes can be beneficial to the accurate detection of the intrinsic behavioural process and trigger discernible detection signals. Moreover, in order to guarantee the response signals emitting from the reaction site under native conditions, an effective organelle-targeting group must be included. The high specificity of molecular-selective labelling must achieve the extended 3D imaging capability. It is important to remember that conceiving a desirable fluorescent probe also requires that the ultimate fluorescence emission occurs at longer wavelengths. Fluorophores with near-infrared excitation/emission wavelengths can facilitate deeper tissue penetration, can minimize photodamage, and can avoid interference from intracellular autoluminescence. We should also note that such fluorescent probes could address current limitations and enable potential clinical applications.

Acknowledgements

This work was supported by the Hainan Provincial Natural Science Foundation (820RC655, 820RC641, 820RC643, 820RC654, 2019RC210, 2019RC220, 2019RC203), National Nature Science Foundation of China (No. 22164009, 21775162, 21864011, 21961010, 21904030, 22064009). Hainan Key Research and Development Project (Grant ZDYF2020133), Nanhai Young-Talent Program of Hainan (20202007, 20202018), and Hundred Talent Program of Hainan (2018).

References

1. H. Sies, *Oxidative Stress*, Academic Press, London & Orlando, 1985.
2. M. T. Lin and M. F. Beal, *Nature*, 2006, **443**, 787–795.
3. C. Cheignon, M. Tomas, D. Bonnefont-Rousselot, P. Faller, C. Hureau and F. Collin, *Redox Biol.*, 2018, **14**, 450–464.
4. E. Monzani, S. Nicolis, S. Dell'Acqua, A. Capucciati, C. Bacchella, F. A. Zucca, E. V. Mosharov, D. Sulzer, L. Zecca and L. Casella, *Angew. Chem., Int. Ed.*, 2019, **58**, 6512–6527.
5. B. C. Dickinson and C. J. Chang, *Nat. Chem. Biol.*, 2011, **7**, 504–511.
6. C. Jacob, W. Maret and B. L. Vallee, *Proc. Natl. Acad. Sci. U. S. A.*, 1999, **96**, 1910–1914.
7. K. Goto, M. Nagahama, T. Mizushima, K. Shimada, T. Kawashima and R. Okazaki, *Org. Lett.*, 2001, **3**, 3569–3572.
8. C. X. Yin, K. M. Xiong, F. J. Huo, J. C. Salamanca and R. M. Strongin, *Angew. Chem., Int. Ed.*, 2017, **56**, 13188–13198.
9. Y. N. Liu, X. H. Feng, Y. N. Yu, Q. Y. Zhao, C. H. Tang and J. M. Zhang, *Anal. Chim. Acta*, 2020, **1110**, 141–150.
10. M. Gao, F. Yu, C. Lv, J. Choo and L. Chen, *Chem. Soc. Rev.*, 2017, **46**, 2237–2271.
11. J. Chan, S. C. Dodani and C. J. Chang, *Nat. Chem.*, 2012, **4**, 973–984.
12. X. Y. Jiao, Y. Li, J. Y. Niu, X. L. Xie, X. Wang and B. Tang, *Anal. Chem.*, 2018, **90**, 533–555.
13. L. Yuan, W. Lin, K. Zheng, L. He and W. Huang, *Chem. Soc. Rev.*, 2013, **42**, 622–661.
14. Y. Yang, Q. Zhao, W. Feng and F. Li, *Chem. Rev.*, 2013, **113**, 192–270.
15. X. Qian and Z. Xu, *Chem. Soc. Rev.*, 2015, **44**, 4487–4493.
16. H. Liu, M. N. Radford, C. T. Yang, W. Chen and M. Xian, *Br. J. Pharmacol.*, 2019, **176**, 616–627.
17. T. Ida, T. Sawa, H. Ihara, Y. Tsuchiya, Y. Watanabe, Y. Kumagai, M. Suematsu, H. Motohashi, S. Fujii, T. Matsunaga, M. Yamamoto, K. Ono, N. O. Devarie-Baez, M. Xian, J. M. Fukuto and T. Akaike, *Proc. Natl. Acad. Sci. U. S. A.*, 2014, **111**, 7606–7611.
18. A. K. Mustafa, M. M. Gadalla, N. Sen, S. Kim, W. Mu, S. K. Gazi, R. K. Barrow, G. Yang, R. Wang and S. H. Snyder, *Sci. Signaling*, 2009, **2**, ra72.
19. M. Gao, R. Wang, F. Yu, J. You and L. Chen, *J. Mater. Chem. B*, 2018, **6**, 2608–2619.
20. X. Han, X. Song, B. Li, F. Yu and L. Chen, *Biomater. Sci.*, 2018, **6**, 672–682.
21. Y. Takano, K. Shimamoto and K. Hanaoka, *J. Clin. Biochem. Nutr.*, 2016, **58**, 7–15.
22. W. Chen, C. Liu, B. Peng, Y. Zhao, A. Pacheco and M. Xian, *Chem. Sci.*, 2013, **4**, 2892–2896.
23. G. Jiang, M. Li, Y. Wen, W. Zeng, Q. Zhao, C. Chen, H. Yuan, C. Liu and C. Liu, *ACS Sens.*, 2019, **4**, 434–440.
24. D. L. Neill, Y.-C. Chang, W. Chen, L. Li and M. Xian, *Sens. Actuators, B*, 2019, **292**, 263–269.

25. M. Gao, R. Wang, F. Yu and L. Chen, *Biomaterials*, 2018, **160**, 1–14.
26. M. Gao, R. Wang, F. Yu, B. Li and L. Chen, *J. Mater. Chem. B*, 2018, **6**, 6637–6645.
27. R. Greiner, Z. Palinkas, K. Basell, D. Becher, H. Antelmann, P. Nagy and T. P. Dick, *Antioxid. Redox Signaling*, 2013, **19**, 1749–1765.
28. Y. Kimura, Y. Mikami, K. Osumi, M. Tsugane, J.-i. Oka and H. Kimura, *Nitric Oxide*, 2014, **39**, S39.
29. C. Liu, W. Chen, W. Shi, B. Peng, Y. Zhao, H. Ma and M. Xian, *J. Am. Chem. Soc.*, 2014, **136**, 7257–7260.
30. F. Yu, M. Gao, M. Li and L. Chen, *Biomaterials*, 2015, **63**, 93–101.
31. M. Gao, F. Yu, H. Chen and L. Chen, *Anal. Chem.*, 2015, **87**, 3631–3638.
32. M. Gao, R. Wang, F. Yu, J. You and L. Chen, *Analyst*, 2015, **140**, 3766–3772.
33. K.-B. Li, F.-Z. Chen, Q.-H. Yin, S. Zhang, W. Shi and D.-M. Han, *Sens. Actuators, B*, 2018, **254**, 222–226.
34. J. Ma, J. Fan, H. Li, Q. Yao, F. Xu, J. Wang and X. Peng, *J. Mater. Chem. B*, 2017, **5**, 2574–2579.
35. L. Zeng, S. Chen, T. Xia, W. Hu, C. Li and Z. Liu, *Anal. Chem.*, 2015, **87**, 3004–3010.
36. H. Shang, H. Chen, Y. Tang, R. Guo and W. Lin, *Sens. Actuators, B*, 2016, **230**, 773–778.
37. L. Yuan, W. Y. Lin, H. Chen, S. S. Zhu and L. W. He, *Angew. Chem., Int. Ed.*, 2013, **52**, 10018–10022.
38. Q. Han, Z. Mou, H. Wang, X. Tang, Z. Dong, L. Wang, X. Dong and W. Liu, *Anal. Chem.*, 2016, **88**, 7206–7212.
39. J. Zhang, X. Y. Zhu, X. X. Hu, H. W. Liu, J. Li, L. L. Feng, X. Yin, X. B. Zhang and W. Tan, *Anal. Chem.*, 2016, **88**, 11892–11899.
40. P. Hou, J. Wang, S. Fu, L. Liu and S. Chen, *Spectrochim. Acta, Part A*, 2019, **213**, 342–346.
41. W. Li, S. Zhou, L. Zhang, Z. Yang, H. Chen, W. Chen, J. Qin, X. Shen and S. Zhao, *Sens. Actuators, B*, 2019, **284**, 30–35.
42. Y. Ren, L. Zhang, Z. Zhou, Y. Luo, S. Wang, S. Yuan, Y. Gu, Y. Xu and X. Zha, *Anal. Chim. Acta*, 2019, **1056**, 117–124.
43. X. Wang, Q. Sun, L. Zhao, S. Gong and L. Xu, *J. Biol. Inorg Chem.*, 2019, **24**, 1077–1085.
44. C. Zhang, Q. Sun, L. Zhao, S. Gong and Z. Liu, *Spectrochim. Acta, Part A*, 2019, **223**, 117295.
45. L. Zhou, G. Yuan and S. Hu, *J. Appl. Spectrosc.*, 2019, **86**, 962–967.
46. Q. Fang, X. Yue, S. Han, B. Wang and X. Song, *Spectrochim. Acta, Part A*, 2020, **224**, 117410.
47. Q. Han, X. Liu, X. Wang, R. Yin, H. Jiang, J. Ru and W. Liu, *Dyes Pigm.*, 2020, **173**, 107877.
48. X. Huang, H. Liu, G. Liu, R. Wang, C. Fan and S. Pu, *J. Photochem. Photobiol., A*, 2020, **391**, 112373.
49. J. Sun, Y. Bai, Q. Ma, H. Zhang, M. Wu, C. Wang and M. Tian, *Spectrochim. Acta, Part A*, 2020, **241**, 118650.

50. C. Wang, J. Xu, Q. Ma, Y. Bai, M. Tian, J. Sun and Z. Zhang, *Spectrochim. Acta, Part A*, 2020, **227**, 117579.
51. F. Zhang, J. Han, J. Wang, X. Li, Y. Wang, B. Wang and X. Song, *Spectrochim. Acta, Part A*, 2020, **242**, 118755.
52. X. Zhang, L. Zhang, M. Gao, Y. Wang and L. Chen, *J. Hazard. Mater.*, 2020, **396**, 122673.
53. X. Zhao, F. He, Y. Dai, F. Ma and Z. Qi, *Dyes Pigm.*, 2020, **172**, 107818.
54. L. Zhang, H. Liu, C. Wu, Y. Zheng, X. Kai and Y. Xue, *Org. Biomol. Chem.*, 2021, **19**, 911–919.
55. Q. Han, L. Yang, Y. Song, J. Ru, H. Zhang, H. Jiang and X. Wang, *Dyes Pigm.*, 2021, **188**, 109190.
56. W. Chen, E. W. Rosser, T. Matsunaga, A. Pacheco, T. Akaike and M. Xian, *Angew. Chem., Int. Ed.*, 2015, **54**, 13961–13965.
57. F. Yang, H. Gao, S. S. Li, R. B. An, X. Y. Sun, B. Kang, J. J. Xu and H. Y. Chen, *Chem. Sci.*, 2018, **9**, 5556–5563.
58. Y. Fang, W. Chen, W. Shi, H. Li, M. Xian and H. Ma, *Chem. Commun.*, 2017, **53**, 8759–8762.
59. A. Anila Hoskere, S. Sreedharan, F. Ali, C. G. Smythe, J. A. Thomas and A. Das, *Chem. Commun.*, 2018, **54**, 3735–3738.
60. W. Meng, W. Shi, Y. Chen, H. Zhang, J. Zhao, Z. Li and K. Xiao, *Sens. Actuators, B*, 2019, **281**, 871–877.
61. H. J. Choi, C. S. Lim, M. K. Cho, J. S. Kang, S. J. Park, S. M. Park and H. M. Kim, *Sens. Actuators, B*, 2019, **283**, 810–819.
62. J.-B. Li, Q. Wang, H.-W. Liu, L. Yuan and X.-B. Zhang, *Chem. Commun.*, 2019, **55**, 4487–4490.
63. Y. Takano, K. Hanaoka, K. Shimamoto, R. Miyamoto, T. Komatsu, T. Ueno, T. Terai, H. Kimura, T. Nagano and Y. Urano, *Chem. Commun.*, 2017, **53**, 1064–1067.
64. W. Chen, A. Pacheco, Y. Takano, J. J. Day, K. Hanaoka and M. Xian, *Angew. Chem., Int. Ed.*, 2016, **55**, 9993–9996.
65. L. Liang, W. Li, J. Zheng, R. Li, H. Chen and Z. Yuan, *Biomater. Sci.*, 2020, **8**, 224–231.
66. Y. Ma, Z. Xu, Q. Sun, L. Wang, H. Liu and F. Yu, *Spectrochim. Acta, Part A*, 2020, 119105.
67. P. Xie, Y. Zhu, Y. Wang, G. Gao, S. Jiang and F. Guo, *Res. Chem. Intermed.*, 2021, **47**, 1123–1138.
68. W. Chen, X. Yue, H. Zhang, W. Li, L. Zhang, Q. Xiao, C. Huang, J. Sheng and X. Song, *Anal. Chem.*, 2017, **89**, 12984–12991.
69. P. Niu, Y. Rong, Y. Wang, H. Ni, M. Zhu, W. Chen, X. Liu, L. Wei and X. Song, *Spectrochim. Acta, Part A*, 2021, **257**, 119789.
70. L. Fu, C. Chen, W. Chen, J. Huang, J. Xiao, L. Yang, J. Sheng and X. Song, *Sens. Actuators, B*, 2020, **304**, 127382.
71. J. Guo, S. Yang, C. Guo, Q. Zeng, Z. Qing, Z. Cao, J. Li and R. Yang, *Anal. Chem.*, 2018, **90**, 881–887.
72. Y. Hou, X.-F. Yang, Y. Zhong and Z. Li, *Sens. Actuators, B*, 2016, **232**, 531–537.

73. W. Chen, L. Fu, C. Chen, J. Xiao, W. Li, L. Zhang, Q. Xiao, C. Huang, J. Sheng and X. Song, *Chem. Commun.*, 2019, **55**, 8130–8133.
74. W. Li, L. Wang, S. Yin, H. Lai, L. Yuan and X. Zhang, *Chem. Sci.*, 2020, **11**, 7991–7999.
75. T. Qiu, H. Wei, D. Cheng, L. Zhang, C. Jiang, S. Luo, L. Yuan and Z. Zeng, *Talanta*, 2017, **164**, 529–533.
76. W. Chen, E. W. Rosser, D. Zhang, W. Shi, Y. Li, W. J. Dong, H. Ma, D. Hu and M. Xian, *Org. Lett.*, 2015, **17**, 2776–2779.
77. W. Chen, X. Yue, J. Sheng, W. Li, L. Zhang, W. Su, C. Huang and X. Song, *Sens. Actuators, B*, 2018, **258**, 125–132.
78. X. Han, F. Yu, X. Song and L. Chen, *Chem. Sci.*, 2016, **7**, 5098–5107.
79. K. Umezawa, M. Kamiya and Y. Urano, *Angew. Chem., Int. Ed.*, 2018, **57**, 9346–9350.
80. R. Kawagoe, I. Takashima, S. Uchinomiya and A. Ojida, *Chem. Sci.*, 2017, **8**, 1134–1140.
81. M. Gao, X. Zhang, Y. Wang, Q. Liu, F. Yu, Y. Huang, C. Ding and L. Chen, *Anal. Chem.*, 2019, **91**, 7774–7781.
82. X. Y. Gong, X. F. Yang, Y. G. Zhong, H. H. Chen and Z. Li, *RSC Adv.*, 2016, **6**, 88519–88525.
83. H. Zhou, J. Tang, L. Sun, J. Zhang, B. Chen, J. Kan, W. Zhang, J. Zhang and J. Zhou, *Sens. Actuators, B*, 2019, **278**, 64–72.
84. J. Zhang, Y. Zhan, Y. Li-Hu, Y. Qi, R. Wang and L. Meng, *Chin. J. Org. Chem.*, 2020, **40**, 1847–1859.
85. J. C. Peeler and E. Weerapana, *Acc. Chem. Res.*, 2019, **52**, 2832–2840.
86. G. V. Kryukov, S. Castellano, S. V. Novoselov, A. V. Lobanov, O. Zehtab, R. Guigo and V. N. Gladyshev, *Science*, 2003, **300**, 1439–1443.
87. K. A. Cupp-Sutton and M. T. Ashby, *Antioxidants*, 2016, **5**, 42.
88. B. Zhang, C. Ge, J. Yao, Y. Liu, H. Xie and J. Fang, *J. Am. Chem. Soc.*, 2015, **137**, 757–769.
89. H. Maeda, K. Katayama, H. Matsuno and T. Uno, *Angew. Chem., Int. Ed.*, 2006, **45**, 1810–1813.
90. S. E. Eriksson, S. Prast-Nielsen, E. Flaberg, L. Szekely and E. S. Arner, *Free Radical Biol. Med.*, 2009, **47**, 1661–1671.
91. P. P. Zhang, Y. Ding, W. M. Liu, G. L. Niu, H. Y. Zhang, J. C. Ge, J. S. Wu, Y. Q. Li and P. F. Wang, *Sens. Actuators, B*, 2018, **264**, 234–239.
92. H. Y. Zhang, M. X. Li, W. Y. Feng and G. Q. Feng, *Dyes Pigm.*, 2018, **149**, 475–480.
93. Z. Wang, H. Zheng, C. Zhang, D. Tang, Q. Wu, W. Dessie and Y. Jiang, *Sensors*, 2020, **20**, 4768.
94. D. Zhang, M. Hu, X. Yuan, Y. Wu, X. Hu, S. Xu, H.-W. Liu, X. Zhang, Y. Liu and W. Tan, *ACS Appl. Mater. Interfaces*, 2019, **11**, 17722–17729.
95. Q. Wang, S. Zhang, Y. Zhong, X. F. Yang, Z. Li and H. Li, *Anal. Chem.*, 2017, **89**, 1734–1741.
96. Y. X. Nan, W. J. Zhao, X. H. Xu, C. T. Au and R. H. Qiu, *RSC Adv.*, 2015, **5**, 69299–69306.
97. H. Chen, B. Dong, Y. Tang and W. Lin, *Chemistry*, 2015, **21**, 11696–11700.

98. X. Han, R. Wang, X. Song, F. Yu and L. Chen, *Anal. Chem.*, 2018, **90**, 8108–8115.
99. W. Feng, M. Li, Y. Sun and G. Feng, *Anal. Chem.*, 2017, **89**, 6106–6112.
100. L. Zhang, X. Kai, Y. Zhang, Y. Zheng, Y. Xue, X. Yin and J. Zhao, *Analyst*, 2018, **143**, 4860–4869.
101. Q. Sun, S. H. Yang, L. Wu, Q. J. Dong, W. C. Yang and G. F. Yang, *Anal. Chem.*, 2016, **88**, 6084–6091.
102. D. Cheng, Y. Pan, B. C. Yin, L. Yuan and X. B. Zhang, *Chin. Chem. Lett.*, 2017, **28**, 1987–1990.
103. S. R. Zhang, Q. Wang, X. W. Liu, J. J. Zhang, X. F. Yang, Z. Li and H. Li, *Anal. Chem.*, 2018, **90**, 4119–4125.
104. Y.-J. Gong, W.-Z. Xu, Y.-P. Zhang, D.-D. Feng, S. Feng and G. Zhang, *Talanta*, 2020, **219**, 121287.
105. M. Li, W. Feng, Q. Zhai and G. Feng, *Biosens. Bioelectron.*, 2017, **87**, 894–900.
106. C. G. Dai, J. L. Wang and Q. H. Song, *J. Mater. Chem. B*, 2016, **4**, 6726–6733.
107. X. Zhao, G. Yuan, H. Ding, L. Zhou and Q. Lin, *J. Hazard. Mater.*, 2020, **381**, 120918.
108. L. Zhang, Y. Shi, Z. Sheng, Y. Zhang, X. Kai, M. Li and X. Yin, *ACS Sens.*, 2019, **4**, 3147–3155.
109. F. P. Kong, B. Hu, Y. Gao, K. H. Xu, X. H. Pan, F. Huang, Q. L. Zheng, H. Chen and B. Tang, *Chem. Commun.*, 2015, **51**, 3102–3105.
110. X. Y. Han, X. Y. Song, F. B. Yu and L. X. Chen, *Adv. Funct. Mater.*, 2017, **27**, 1700769.
111. C. Zhang, Z. Qiu, L. Zhang, Q. Pang, Z. Yang, J.-K. Qin, H. Liang and S. Zhao, *Chem. Sci.*, 2021, **12**, 4883–4888.
112. Y. Guo, Y. Luo, N. Wang, M. Tang, J. Xiao, S.-W. Chen and J. Wang, *Talanta*, 2020, **212**, 120583.
113. Y. Tian, F. Xin, C. Gao, J. Jing and X. Zhang, *J. Mater. Chem. B*, 2017, **5**, 6890–6896.
114. X. Luo, R. Wang, C. Lv, G. Chen, J. You and F. Yu, *Anal. Chem.*, 2020, **92**, 1589–1597.
115. Y. Liu, Y. Yu, Q. Meng, X. Jia, J. Zhu, C. Tang, Q. Zhao, X. Feng and J. Zhang, *Biosensors*, 2021, **11**, 132.
116. F. Kong, L. Ge, X. Pan, K. Xu, X. Liu and B. Tang, *Chem. Sci.*, 2016, **7**, 1051–1056.
117. Y. Tian, F. Xin, J. Jing and X. Zhang, *J. Mater. Chem. B*, 2019, **7**, 2829–2834.
118. F. Kong, Y. Zhao, Z. Liang, X. Liu, X. Pan, D. Luan, K. Xu and B. Tang, *Anal. Chem.*, 2017, **89**, 688–693.
119. F. Xin, Y. Tian and X. Zhang, *Dyes Pigm.*, 2020, **177**, 108274.
120. J. Lu and A. Holmgren, *Antioxid. Redox Signaling*, 2012, **17**, 1738–1747.
121. W. Cai, L. Zhang, Y. Song, B. Wang, B. Zhang, X. Cui, G. Hu, Y. Liu, J. Wu and J. Fang, *Free Radical Biol. Med.*, 2012, **52**, 257–265.
122. E. S. Arner, *Biochim. Biophys. Acta*, 2009, **1790**, 495–526.
123. D. F. Mahmood, A. Abderrazak, K. El Hadri, T. Simmet and M. Rouis, *Antioxid. Redox Signaling*, 2013, **19**, 1266–1303.

124. G. Powis and D. L. Kirkpatrick, *Curr. Opin. Pharmacol.*, 2007, **7**, 392–397.
125. A. Holmgren and J. Lu, *Biochem. Biophys. Res. Commun.*, 2010, **396**, 120–124.
126. L. Zhang, D. Duan, Y. Liu, C. Ge, X. Cui, J. Sun and J. Fang, *J. Am. Chem. Soc.*, 2014, **136**, 226–233.
127. Y. Liu, H. Ma, L. Zhang, Y. Cui, X. Liu and J. Fang, *Chem. Commun.*, 2016, **52**, 2296–2299.
128. H. Ma, J. Zhang, Z. Zhang, Y. Liu and J. Fang, *Chem. Commun.*, 2016, **52**, 12060–12063.
129. X. Li, B. Zhang, C. Yan, J. Li, S. Wang, X. Wei, X. Jiang, P. Zhou and J. Fang, *Nat. Commun.*, 2019, **10**, 2745.
130. L. Huang, Y. Chen, B. Liang, B. Xing, G. Wen, S. Wang, X. Yue, C. Zhu, J. Du and X. Bu, *Chem. Commun.*, 2014, **50**, 6987–6990.
131. B. Liang, W. Shao, C. Zhu, G. Wen, X. Yue, R. Wang, J. Quan, J. Du and X. Bu, *ACS Chem. Biol.*, 2016, **11**, 425–434.
132. Y. Wang, W. Zhang, J. Dong and J. Gao, *Bioorg. Chem.*, 2020, **95**, 103530.
133. T. J. Mafireyi, M. Laws, J. W. Bassett, P. B. Cassidy, J. O. Escobedo and R. M. Strongin, *Angew. Chem., Int. Ed.*, 2020, **59**, 15147–15151.

CHAPTER 14

The Activity-based Sensing of Biological Transition Metals

T. HIRAYAMA*[a]

[a]Gifu Pharmaceutical University, 1-25-4, Daigaku-nishi, Gifu-shi, Gifu, 501-1196, Japan
*E-mail: hirayamat@gifu-pu.ac.jp

14.1 Overview of Activity-based Sensing of Metals

The design strategy for metal-ion probes can be categorized into two types. One is chelation-based sensing and the other is activity-based sensing. The features of each type are summarized in Figure 14.1.

The color, fluorescence intensity, and wavelength of chelation-based probes all change as a result of the formation of metal–chelate complexes. This sensing method has been utilized since the 1980s, and most chelation-based probes are composed of a fluorophore and a metal-chelating site that acts as a fluorescence switch (Figure 14.1). Since the chelation behavior is reversible and stoichiometric, chelation-based probes enable quantitative and time-lapse observations in living cells. Fura-2, a calcium ion fluorescent probe,[1] is a representative chelation-based probe that is routinely used as a fluorescent indicator of intracellular Ca^{2+}. However, the stoichiometric reaction limits the sensitivity for small amounts of metal ions, such as trace elements. The depletion of metal ions is another concern because of the scavenging activity of chelation-based probes, which would unfavorably affect metalloproteins and labile metal ions.

Monographs in Supramolecular Chemistry No. 34
Fluorescent Chemosensors
Edited by Luling Wu, Adam C. Sedgwick, Xiao-Peng He and Tony D. James

Published by the Royal Society of Chemistry, www.rsc.org

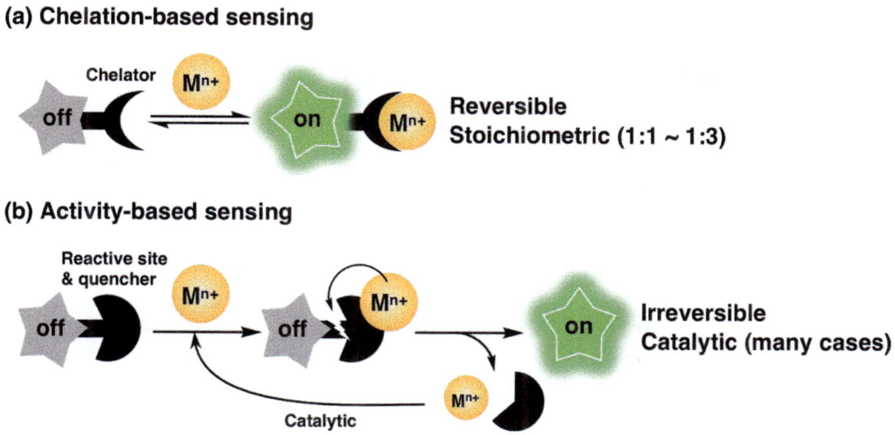

Figure 14.1 Schemes of (a) chelation-based sensing and (b) activity-based sensing.

Activity-based sensing depends on the chemical reactivity of the metal ions. This sensing mode is seen in fluorescent probes for enzymes and reactive oxygen species[2] and has limited application in metal sensing. The chemical reactions used in metal sensing are irreversible and one-way, whereas chelation-based probes show reversible responses. However, there are several advantages gained by avoiding metal-ion depletion and the fluorescence-quenching effect of paramagnetic metal ions, which are the limitations of chelation-based probes. The irreversible detection mode is advantageous for enhancing sensitivity because the fluorescence signal persists once it is in a turn-on state. In addition, most metal-mediated reactions for sensing proceed catalytically, which improves sensitivity and attenuates the artifacts, such as metal depletion in cells. In the last decade, activity-based probes for metal sensing have emerged. In this chapter, the principles, detection mechanism, and features of activity-based fluorescent and related optical probes for metal ions are discussed.

14.2 The Activity-based Sensing of Fe

Fe is the most abundant transition metal species in the human body and stably exists as Fe^{2+} and Fe^{3+} in cells. Most Fe ions are bound to proteins, but a small portion of free or weakly protein-bound Fe species, designated as labile Fe, is also present in living cells.[3] Labile Fe mediates cellular dynamic equilibrium through Fe-regulating proteins, such as ferritin, divalent metal transporter-1 (DMT-1),[4] ferroportin, and poly(γC)-binding protein (PCBP)-1/2,[5,6] and most of these recognize Fe^{2+} as a substrate.[7–9] Thus, the majority of labile Fe is present as Fe^{2+} rather than Fe^{3+}, as reported by Cabantchik *et al.*[10] At the same time, Fe^{2+} is involved in oxidative stress through Fenton chemistry,[11,12] where Fe^{2+} and hydrogen peroxide react to produce highly reactive oxygen species, such as hydroxyl radicals. In this context, sensing Fe^{2+}

is important for studying cellular homeostasis and the pathophysiology of Fe.[13] There are several turn-off-type fluorescent probes with chelation-based designs,[14,15] but Fe^{2+} remains challenging to detect with a turn-on response because of its fluorescence-quenching property.[16] Furthermore, Fe^{2+} is categorized as a medium Lewis acid by the hard–soft acid–base (HSAB) rule and a weak binder on the Irving–Williams series, making selective recognition by chelators difficult. However, an activity-based design can overcome this disadvantage and is introduced here.

14.2.1 *N*-Oxide Deoxygenation

Tertiary amine *N*-oxides are reductively deoxygenated to the corresponding tertiary amines.[17] Hirayama and co-workers found that Fe^{2+} promoted the deoxygenation of *N*-oxides, and thus applied Fe^{2+}-mediated deoxygenation to a turn-on fluorescent probe. The exact reaction mechanism is unclear, but it could be speculated that two Fe^{2+} ions are involved in the two-electron reduction of *N*-oxides *via* $Fe^{4+}{=}O$ or $Fe^{3+}{-}O^{\bullet}$ intermediates (Figure 14.2a).[17] In living cells, abundant glutathione or other reductants, such as ascorbate and NAD(P)H, readily reduce the resultant Fe^{3+} to Fe^{2+}, enabling the catalytic amplification of fluorescence signals.[18] The first *N*-oxide-based fluorescent probe, RhoNox-1, is composed of rhodamine B with an *N*-oxide moiety at one of the diethylamino groups (Figure 14.2b).[19] The fluorescence is attenuated *via* multiple quenching modes, including twisted-internal charge transfer (TICT), photoinduced electron transfer (PeT), and isolation of the *N*-oxide nitrogen atom from the π-conjugation. The Fe^{2+}-induced deoxygenation of RhoNox-1 yields highly fluorescent rhodamine B, resulting in a turn-on response. The high metal selectivity and inertness against biological

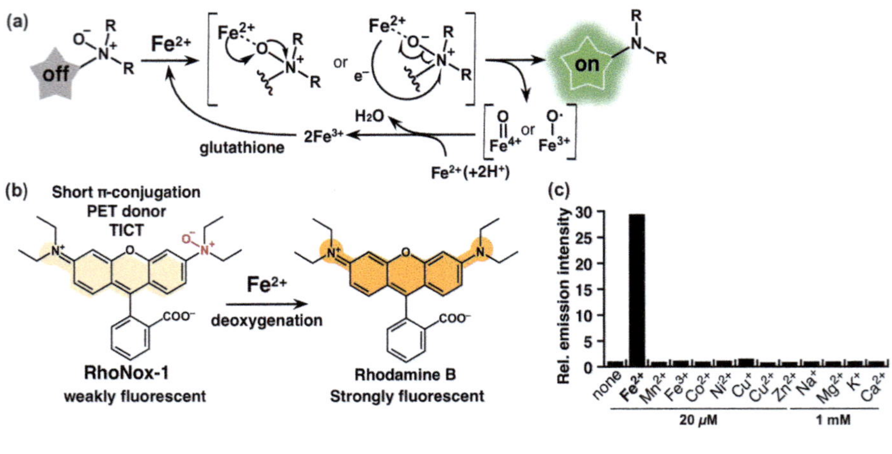

Figure 14.2 (a) The mechanism of Fe^{2+} sensing by an *N*-oxide-based strategy. (b) Structure and reaction of RhoNox-1. (c) Metal selectivity of RhoNox-1 (2 μM) for Na^+, K^+, Ca^{2+}, Mg^{2+} 1 mM; other metal ions, 20 μM. HEPES buffer (50 mM, pH 7.4, 0.2% DMSO). $\lambda_{ex}/\lambda_{em} = 530/575$ nm.

oxidative/reductive species are the specific features of RhoNox-1 (Figure 14.2c), and other *N*-oxide-based fluorescent probes have similar metal selectivities.[20] RhoNox-1 (2 μM) showed a 30-fold turn-on response for Fe^{2+} (20 μM) in HEPES buffer (50 mM, pH 7.4), with a rate constant (k_{obs}) of 1.0×10^{-3} s^{-1}. Furthermore, no side reactions were observed, and water was a theoretical byproduct. The probe was successfully applied to imaging labile Fe^{2+} in living cells and was successfully used to monitor alterations in labile Fe^{2+} levels in disease models.[21–23]

N-Oxidation alters the open/close equilibrium of rhodols and rhodamines between closed spirocyclic (non-fluorescent) and open structures (fluorescent). HMRhoNox-M reportedly favors the spirocycle state when it bears an *N*-oxide unit, while it opens without an *N*-oxide moiety.[24] Thus, *N*-oxides and their deoxygenation could be used to control the spirocycle equilibrium for fluorescence switching. This feature also increases the turn-on contrast 60-fold, which is greater than that of RhoNox-1 (Figure 14.3).

N-Oxide-based strategies are also applicable to other fluorophores with dialkyl aryl amines in their π-conjugated structures.[25] Organelle-targeted probes have also been reported from the same group, which successfully targeted mitochondria,[26] the endoplasmic reticulum (ER),[27] lysosomes,[24,27] Golgi apparatus,[28] and plasma membranes[29] (Figure 14.4); their labile Fe^{2+} levels were also detected. Differences in the excitation and emission wavelengths of Lyso-RhoNox ($\lambda_{ex}/\lambda_{em}$ = 550/575 nm), MtFluNox ($\lambda_{ex}/\lambda_{em}$ = 510/535 nm), and ER-SiRhoNox ($\lambda_{ex}/\lambda_{em}$ = 630/660 nm) enabled their simultaneous use to monitor labile Fe^{2+} levels in their corresponding target organelles.[27] Using this cocktail of organelle-targeted probes, it was revealed that Fe^{2+} levels increase in the lysosomes and ER during the early stages of ferroptosis.

The response of *N*-oxide-based detection has been optimized through the structural screening of the *N*-oxide domain. For example, a highly sensitive probe, RhoNox-4, was developed and applied to high-throughput screening (HTS)[29] (Figure 14.5a), through which a new bioactive compound was discovered that elevates subcellular labile Fe^{2+} levels. In addition to fluorescence imaging probes, *N*-oxides can function as signal switches for ^{19}F magnetic resonance imaging (MRI).[30] F-Nox-1 bears a tetrafluorophenylenediamine *N*-oxide unit that exhibits inequivalent ^{19}F peaks; however, after deoxygenation

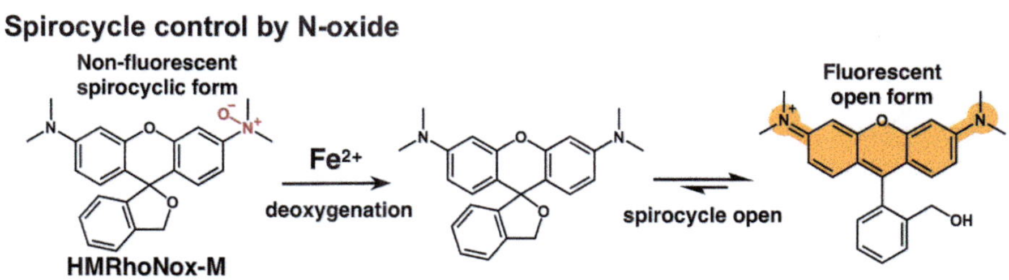

Figure 14.3 The detection mechanism of HMRhoNox-M using spirocycle control induced by an *N*-oxide.

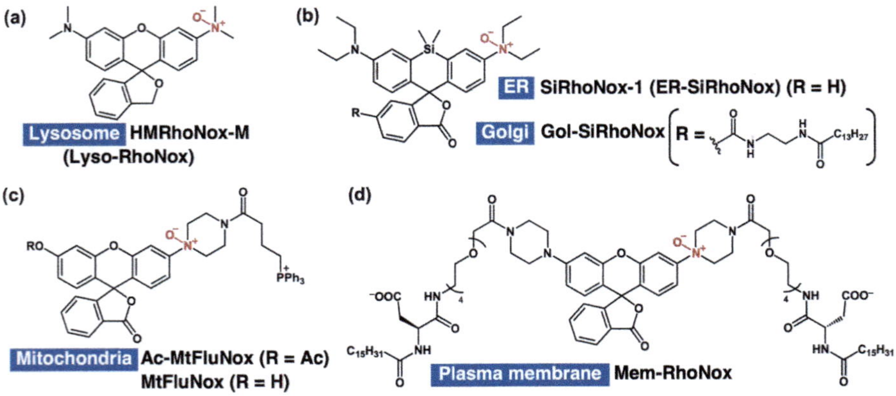

Figure 14.4 (a–d) A chart of *N*-oxide-based organelle-targeted fluorescent probes. (a) HMRhoNox-M (Lyso-RhoNox) (lysosome); (b) SiRhoNox-1 (ER-SiRhoNox) (endoplasmic reticulum, ER) and Gol-SiRhoNox (Golgi apparatus); (c) Ac-MtFluNox and MtFluNox (mitochondria); and (d) Mem-RhoNox (plasma membrane).

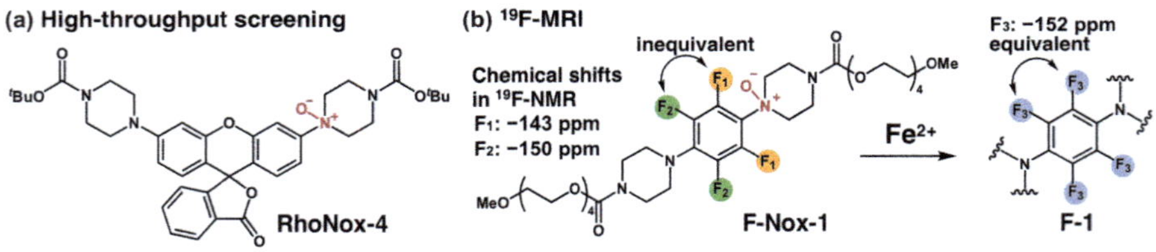

Figure 14.5 The structure of (a) RhoNox-4 and (b) the structure and detection mechanism of F-Nox-1.

by Fe^{2+}, a singlet peak is observed for the four fluorine atoms (Figure 14.5c). Thus, this probe was successfully applied to detect the presence of Fe^{2+} using ^{19}F MRI.

14.2.2 Chelation-assisted Oxidative C–O Bond Cleavage Reaction

Chang *et al.* developed an Fe^{2+} probe, IP-1, that uses a chelate-assisted C–O bond cleavage reaction for detection; this reaction was inspired by the biomimetic chemistry of cytochrome P450.[31] The non-fluorescent, closed, spirocyclic form of the hydroxymethyl-*O*-methylfluorescein moiety of IP-1 is favored due to its attachment to an N4O1 ligand *via* a C–O bond. The ligand coordinates Fe^{2+}, and the Fe–N4O1 complex is activated by molecular oxygen (O_2), resulting in intramolecular C–O bond cleavage (Figure 14.6a and b). This cleavage produces fluorescent hydroxymethyl-*O*-fluorescein (open form), resulting in fluorescence enhancement. The reaction kinetics (k_{obs})

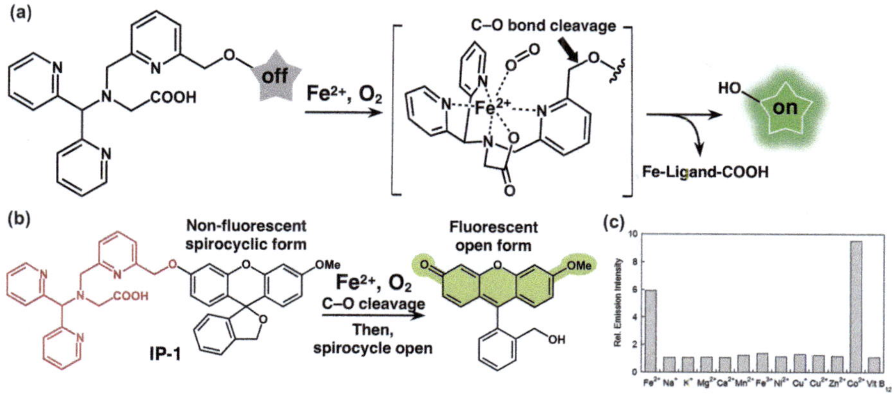

Figure 14.6 (a) The mechanism of the biomimetic C–O cleavage of IP-1 for Fe^{2+} sensing. (b) The structure of IP-1 and its detection scheme. (c) Metal selectivity of IP-1 (1 μM). Na^+, K^+, Ca^{2+}, and Mg^{2+} = 1 mM; other metals = 20 μM. Tris buffer (50 mM, pH 7.6) with 5 mM glutathione. $\lambda_{ex}/\lambda_{em}$ = 470/508 nm. Reproduced from ref. 30 with permission from American Chemical Society, Copyright 2013.

were not reported, but the probe (1 μM) showed a six-fold increase in fluorescence upon incubation with Fe^{2+} (20 μM) for 1 h; although, compared to the enhancements observed for the other Fe^{2+} probes presented here, this increase is relatively low. The probe can distinguish Fe^{2+} from Fe^{3+}, but it also responds to Co^{2+}. However, the authors stated that the effect of ionic Co^{2+} is negligible because the concentration of ionic Co^{2+} is relatively low, and one of the major Co-containing species, cobalamin (vitamin B_{12}), does not trigger fluorescence (Figure 14.6b and c). When applied to imaging, IP-1 enabled the accumulation of Fe^{2+} by hepcidin to be visualized; hepcidin is an iron-regulating hormone that inhibits ferroportin, an iron exporter.

14.2.3 Trioxolane Cleavage

Renslo *et al.* have developed and established an Fe^{2+}-activatable prodrug system, exploiting the Fe^{2+}-induced cleavage of endoperoxide found in antimalarial drugs, such as artemisinin and arterolane.[32] A sterically hindered endoperoxide with an adamantane-type spiro-1,2,4-trioxolane structure, seen in arterolane, has high thermal and chemical stability (Figure 14.7). Fe^{2+} triggers the one-electron reduction of trioxolane to produce Fe^{3+} and oxyl radicals. The oxyl radical intermediate is intramolecularly cleaved to produce carbon radical species and a ketone product. When the ketone counterpart bears a leaving group at its β-position, spontaneous β-elimination occurs to release a free amine (Figure 14.7). The generated Fe^{3+} should be converted to Fe^{2+} after intracellular reduction by glutathione, and the regenerated Fe^{2+} reduces another trioxolane, which repeats catalytically.

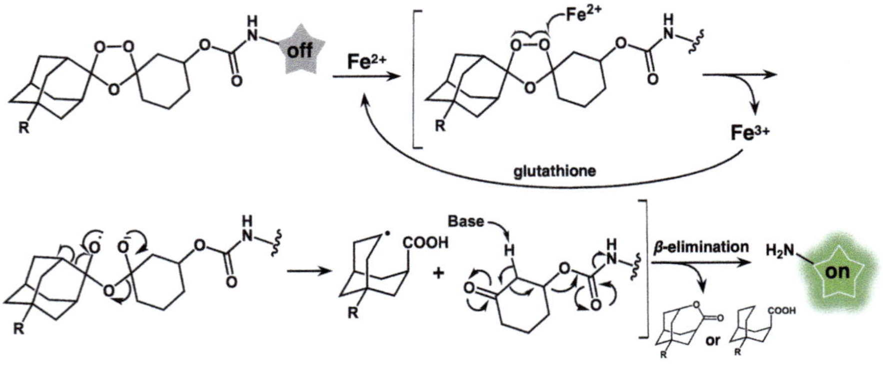

Figure 14.7 The mechanism of Fe^{2+} sensing by the Fe^{2+}-induced cleavage of 1,2,4-trioxolane.

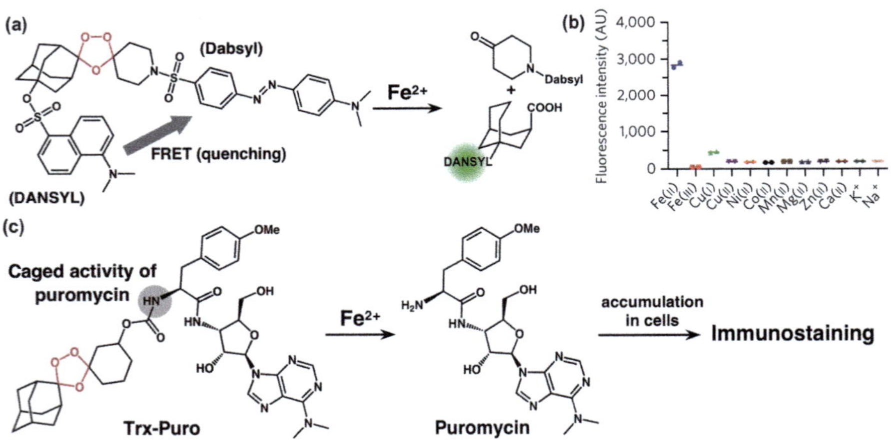

Figure 14.8 (a) The structure of the fluorescent probe using Fe^{2+}-induced 1,2,4-trioxolane cleavage. (b) Metal selectivity of the probe (15 μM) shown in (a). Na^+ = 1 M; other metals = 1.5 mM. 50% DMSO/water. $\lambda_{ex}/\lambda_{em}$ = 337/465 nm. Adapted from ref. 32 with permission from Springer Nature, Copyright 2016. (c) The structure of Trx-Puro and its detection mechanism.

Renslo *et al.* applied this Fe^{2+}-activated prodrug system to activity-based sensing in 2016.[33] As a proof of concept, a fluorophore (DANSYL) and a quencher (Dabsyl) were tethered by the Fe^{2+}-cleavable 1,2,4-trioxolane structure (Figure 14.8a). The reaction with Fe^{2+} triggered fluorescence enhancement upon cleavage of the quencher; this cleavage reaction is highly specific to Fe^{2+} (Figure 14.8b). The kinetics were not reported but the rate (k_{obs}) would likely be approximately 10^{-4} s^{-1} if the concentrations of the probe and Fe^{2+} were 15 μM and 94 μM, respectively. The authors then used puromycin as the substrate to be released upon Fe^{2+}-induced cleavage (Figure 14.8c). It is known that puromycin loses its activity upon acylation of the amino group of its *O*-methyltyrosine site. Trx-Puro, a puromycin caged by the Fe^{2+}-activated system, releases a free puromycin upon Fe^{2+}-induced cleavage of trioxolane, followed by β-elimination (Figure 14.8c). The released puromycin is inserted into the translating terminal ends of proteins and accumulates in cells. The subcellular labile Fe^{2+} levels can be estimated by an immunofluorescence method that quantifies the amount of accumulated puromycin using an anti-puromycin antibody. This Trx-Puro method is not applicable to live-cell imaging because of the requirement for cell fixation; however, it is suitable for a plate-based high-content assay, where it has been used to reveal that the labile Fe^{2+} levels in cancer cell lines (RKO, MDA-MB-231, U-2 OS, and PC-3 cells) are higher than those in normal cell lines (IMR90 and MCF10A cells).

Following the work of Renslo, Chang *et al.* reported a ratiometric fluorescent probe, FIP-1, with two fluorophores, a donor (fluorescein) and an acceptor (Cy5), tethered *via* an Fe^{2+}-cleavable trioxolane linker.[34] FIP-1 shows Cy5 emission upon excitation of fluorescein through Förster resonance energy transfer (FRET), while emission from fluorescein is detected upon reaction with Fe^{2+} due to the release of the acceptor (Figure 14.9). The reaction rate constant was determined to be $1.6 \times 10^{-3}\ s^{-1}$ (1 μM probe + 10 μM Fe^{2+}). Furthermore, measuring the ratio of the emission intensities of fluorescein and Cy5 enables the semi-quantitative imaging of labile Fe^{2+} in cells. This probe was used to measure Fe^{2+} alteration during ferroptosis, where Fe^{2+} levels were found to rise.

A trioxolane-based strategy was applied to bioluminescence imaging. The acylation of aminoluciferin attenuates bioluminescence activity toward luciferase; thus, the amino group was caged by the Fe^{2+}-cleavable trioxolane group. ICL-1, a bioluminescence probe for Fe^{2+}, did not show bioluminescence activity, but Fe^{2+} removed the caged trioxolane to produce active aminoluciferin, which results in bright bioluminescence in the presence of firefly luciferase[35] (Figure 14.10a). *In vivo* imaging using ICL-1 and luciferase-expressing mice demonstrated the accumulation of Fe^{2+} in the mouse bodies

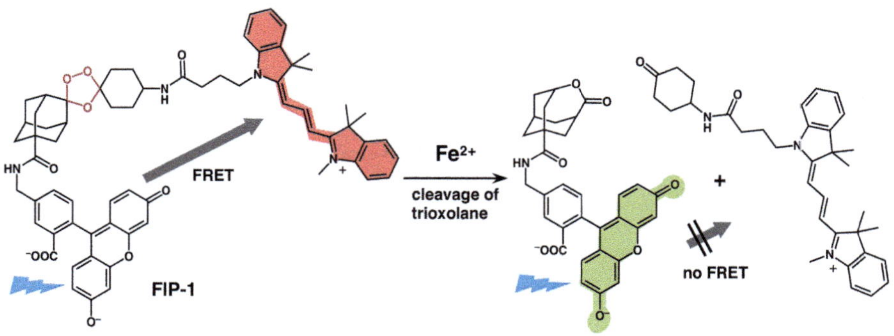

Figure 14.9 The structure of FIP-1 and its detection mechanism.

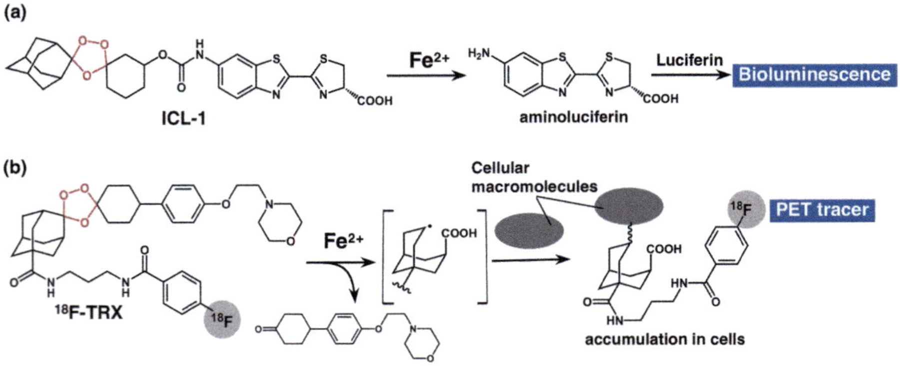

Figure 14.10 (a) The structure of ICL-1 and its detection mechanism. (b) The structure of ^{18}F-TRX, a positron-emitting tomography (PET) imaging probe for Fe^{2+}, and its detection mechanism.

upon infection with *Acinetobacter baumannii*. The success of this *in vivo* imaging and the fact that the probe is inert against the endogenous iron-containing proteins in mammals suggest that the trioxolane-based sensing is suitable for *in vivo* use.

Another *in vivo* application was reported by Renslo *et al.* where they exploited the different aspects of trioxolane chemistry.[36] As shown in Figure 14.8, a carbon radical intermediate appears during the Fe^{2+}-mediated degradation of an adamantane-type spiro-1,2,4-trioxolane. The carbon radical non-specifically modifies biomolecules and accumulates in cells. This cell-trap mechanism enabled the positron-emitting tomography (PET) imaging of Fe^{2+} by installing ^{18}F (a PET tracer) in the carbon radical precursor, the adamantane-spiro-1,2,4-trioxolane unit. The PET imaging using ^{18}F-TRX, a PET probe for Fe^{2+}, of the whole body of a mouse revealed the tissue-specific distribution of Fe^{2+}. Activity-based probes using trioxolane chemistry that is based on the mode of action of antimalarial, drug-inspired, Fe^{2+}-activatable prodrugs are currently the most common design of Fe^{2+}-sensing probes, along with *N*-oxide-based versions, because of their diverse practical applications (see Section 14.2.1).

14.2.4 Nitroxide Radical Reduction

Nitroxide radicals are analogs of the above-mentioned *N*-oxides and act as single-electron acceptors, which enables their use in electron spin resonance (ESR) spectroscopy as a radical trapping agent. The radical species show a fluorescence-quenching effect due to electron spins. Thus, the one-electron reduction of a nitroxide radical by Fe^{2+} (Figure 14.11a), which converts the radical to hydroxylamine, has been employed for the activity-based sensing of Fe^{2+}. The first example of Fe^{2+} sensing using a nitroxide radical is pyrene-TEMPO (Figure 14.11b), as reported by Zhu *et al.*[37] Pyrene-TEMPO (2 μM)

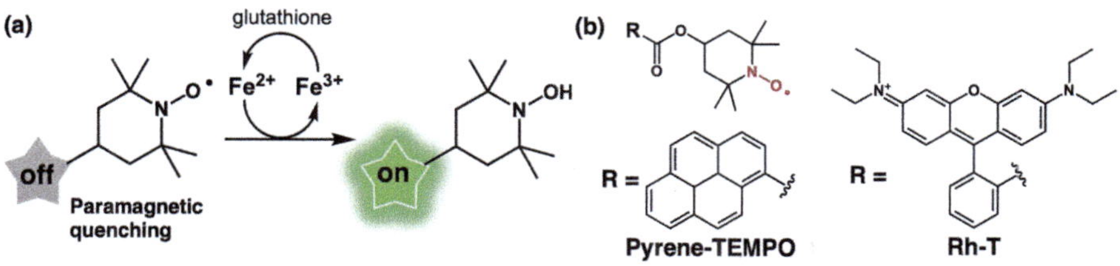

Figure 14.11 (a) The mechanism of Fe^{2+} sensing by nitroxide radical reduction. (b) The structures of pyrene-TEMPO and Rh-T.

is weakly fluorescent, but its fluorescence increases up to three-fold after incubation with Fe^{2+} (3.6 μM) in a 0.2 M H_2SO_4 solution. Furthermore, its response is selective to Fe^{2+} over other metal ions, including Pb^{2+}, Cu^{2+}, Ba^{2+}, Mg^{2+}, Ca^{2+}, Co^{2+}, Cr^{2+}, Ag^{+}, Al^{3+}, Cd^{2+}, Zn^{2+}, Fe^{3+}, Hg^{2+}, Mn^{2+}, and Na^{+}. The fluorescence enhancement is correlated with a decrease in the ESR signals, indicating that the loss of the nitroxide radical contributes to its fluorescence enhancement. Following their success, Guo *et al.* reported an Fe^{2+} probe, RhT, composed of rhodamine B conjugated with TEMPO (Figure 14.11b).[38] The probe (6 μM) showed a two-fold increase in its fluorescence upon reacting with 4 eq. of Fe^{2+} (24 μM). The response was selective to Fe^{2+}, and the existence of other transition metal ions did not interfere with the response. It is speculated that the detection mechanism is based on the reduction of the radical to hydroxylamine, but no product analysis was performed. The stability of Rh-T against reactive oxygen/nitrogen species and NADH was explored, and it was found to be inert against them; however, it should be noted that nitroxide radicals are generally susceptible to ascorbate,[39] which has reportedly triggered modest fluorescence enhancement (1.2-fold). Rh-T selectively targets mitochondria due to its rhodamine B ester structure (a typical mitochondria-targeted motif), and has been successfully applied to the live-cell imaging of Fe^{2+}.

14.2.5 Reduction of Hydroxylamine Acetate

Novak *et al.* reported the Fe^{2+}-mediated conversion of *N*-phenyl-*O*-pivaloyl hydroxylamines to the corresponding anilines.[40] In this reaction, Fe^{2+} promotes the reductive heterolysis of the N–O bond of the *N*-phenyl-*O*-acyl hydroxylamines (Figure 14.12a). The esterification of the hydroxy group of the hydroxylamine is essential for the Fe^{2+}-mediated heterolysis because the coordination of the ester carbonyl oxygen and the nitrogen atom of hydroxylamine to Fe^{2+} triggers the reaction. Wang *et al.* developed Fe-1 as a new type of activity-based probe for Fe^{2+} by converting the amino group of *N*-*n*-butyl-4-aminonaphthalimide to an acetoxyamino group[41] (Figure 14.12b). The fluorescence intensity of Fe-1 was attenuated because of the decrease in the electron-donating property of its nitrogen atom by acetoxylation. The addition of Fe^{2+} (5 μM) to Fe-1 (5 μM) in HEPES buffer (10 mM, pH 7.4)

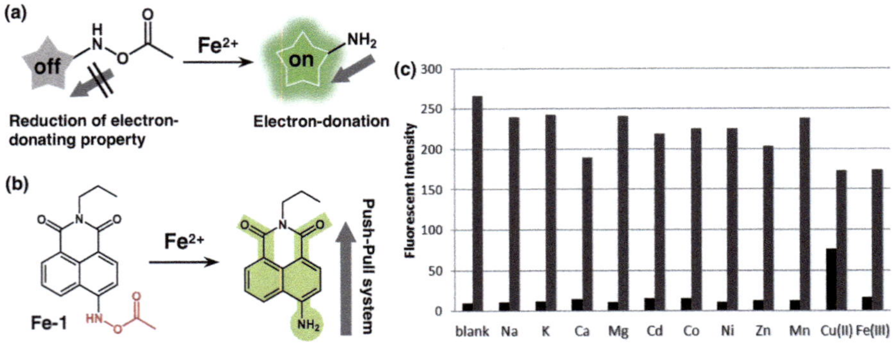

Figure 14.12 (a) The mechanism of Fe^{2+} sensing by the Fe^{2+}-induced reduction of hydroxylamine acetate. (b) The structure of Fe-1. (c) Metal selectivity of Fe-1 (5.0 μM). Black: the indicated metal ions only (50 mM), Gray: Fe^{2+} (5 μM) with the indicated metal ions (50 μM). 10 μM. HEPES buffer (10 mM, pH 7.4, 0.5% DMSO). $\lambda_{ex}/\lambda_{em}$ = 435/544 nm. Adapted from ref. 40 with permission from American Chemical Society, Copyright 2016.

immediately triggered up to a 27-fold increase in fluorescence enhancement within 1 min as a result of the recovery of the push–pull system of the fluorophore. This particularly fast response is a feature of Fe-1 and/or the hydroxylamine acetate reduction system. Fe-1 also showed a modest fluorescence response in the presence of Cu^{2+} with glutathione or ascorbate (Figure 14.12c). The probe was successfully applied to live-cell imaging and tissue imaging, revealing that Zn^{2+} supplementation to brain slices resulted in elevated Fe^{2+} levels. The authors speculated that the Fe^{2+} elevation occurred because of the replacement of the Fe ions held in Fe-binding proteins with Zn^{2+}, but this requires further investigation.

14.2.6 Detelluration

A unique fluorescent probe, Naph-Te, utilizing Fe^{2+}-induced detelluration emerged from the Han group[42] (Figure 14.13a). In this sensing reaction, Fe^{2+} facilitates the detelluration of diaryltelluride and yields a phenolic hydroxy group with concomitant leaving of the phenyltelluro group (Figure 14.13a). Although the detailed mechanism is still under investigation, the Fe^{2+}-mediated detelluration reaction of Naph-Te (5 μM, Figure 14.13b) proceeds efficiently and provokes fluorescence enhancement in a mixed aqueous system of phosphate-buffered saline (PBS)/acetonitrile (1 : 5). Although the turn-on contrast is not described in the literature, it can be estimated to increase at least ten-fold upon the addition of 10 eq. of Fe^{2+}. The presence of acetonitrile is essential for the reaction, which is likely because acetonitrile stabilizes low-valent metal ions such as Cu^{+} and Fe^{2+}; however, this requires further study. The response of Naph-Te is highly selective to Fe^{2+}.

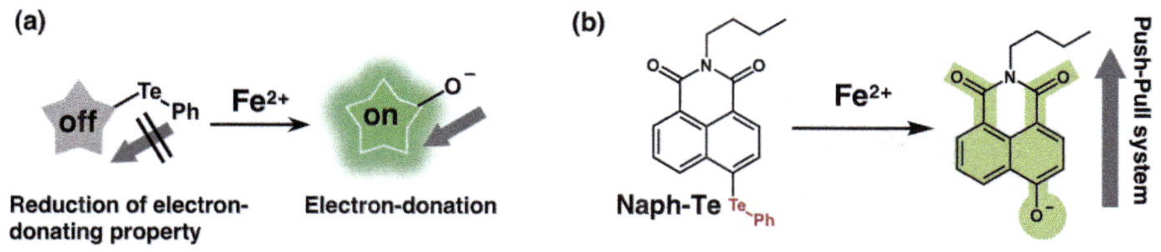

Figure 14.13 (a) The mechanism of Fe^{2+} sensing by Fe^{2+}-induced detelluration. (b) The structure of Naph-Te.

The resultant product was determined to be 4-hydroxy-*N*-butyl-1,8-naphthalimide, indicating that the detelluration proceeded by Fe^{2+}. Additionally, Naph-Te was successfully applied to the live-cell imaging of Fe^{2+}.

14.3 The Activity-based Sensing of Zn^{2+}

Zn^{2+} is the second most abundant biological transition metal ion and acts as a second messenger of cellular signaling.[43] Since the labile Zn^{2+} ranges from nanomolar to millimolar in the human body,[44–46] the choice of a Zn^{2+} probe with a proper detection range for the targets is important for monitoring the alteration of Zn^{2+} concentration in living cells and tissues. The vast majority of conventional probes are chelation-based probes whose fluorescence intensity or wavelength change upon coordination of the probe to Zn^{2+}; this is because the design principles for chelation-based Zn^{2+} fluorescent probes were established in the early 2000s, such as the ZP series[47] and ZnAF series.[48,49] However, chelation by the probes scavenges endogenous Zn^{2+} pools and sometimes protein-bound Zn^{2+}, which is a serious problem for chelation-based probes because the depletion of Zn^{2+} inhibits cell signaling. In addition, the 1 : 1 stoichiometry of the chelation-based probes may limit their detection sensitivity under low Zn^{2+} conditions. In this context, activity-based sensing using a catalytic turn-on reaction is a promising approach for overcoming the drawbacks of chelation-based sensing of Zn^{2+}. In contrast to chelate-based ones, activity-based sensing probes that work catalytically are still rare and limited to one principle based on Zn-assisted hydrolysis, as described below.

14.3.1 Zn-assisted Hydrolysis of β-Lactam

Okuda *et al.* developed Dpa-SoxLC, a new activity-based probe for Zn^{2+} detection, consisting of 2,2′-dipicolylamine (DPA) and umbelliferon bridged by cephem (Figure 14.14).[50] The cephem moiety acts as a Zn^{2+}-degradable linker, where the chelation of the DPA site to Zn^{2+} facilitates the hydrolysis of the β-lactam, followed by the release of a fluorescent umbelliferone anion (pK_a of the phenolic OH ≈ 7). Alkylated umbelliferone at the phenolic hydroxy group is non-fluorescent, and thus a turn-on response is observed.

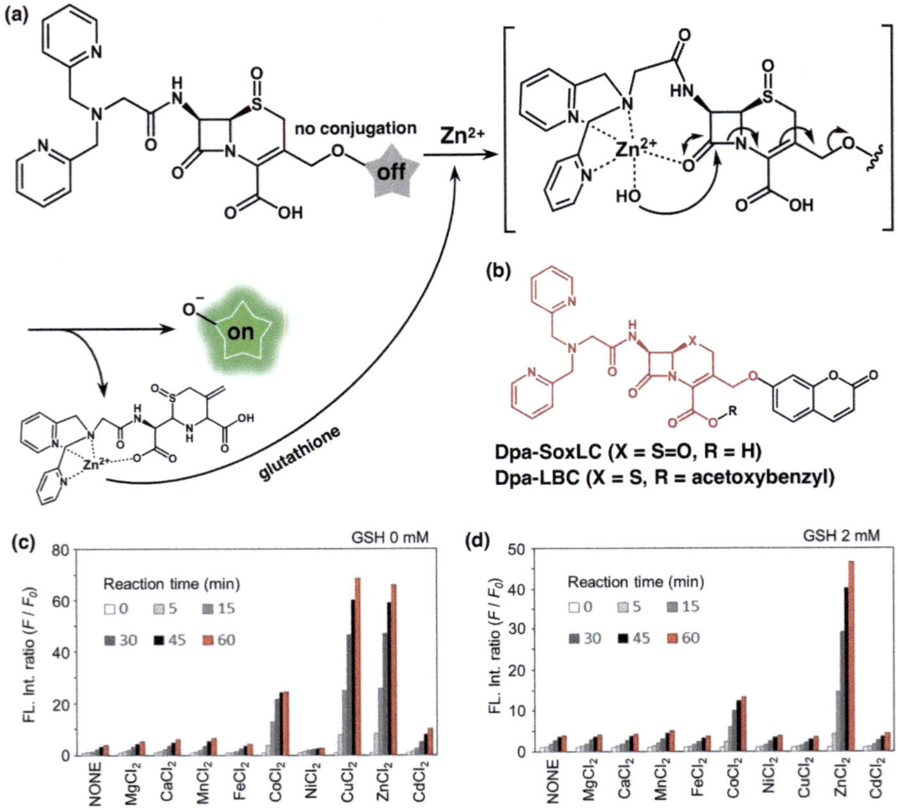

Figure 14.14 (a) The mechanism of Zn^{2+} sensing by the Zn^{2+}-assisted hydrolysis of cephem. (b) The structures of Dpa-SoxLC and Dpa-LBC. (c and d) Fluorescence response of Dpa-SoxLC (10 μM) towards various metal ions in the absence (c) and presence (d) of glutathione (2 mM). Mg^{2+} = 500 μM; Ca^{2+} = 2 mM; others = 10 μM. PIPES buffer (50 mM, pH 7, 100 mM KCl). $\lambda_{ex}/\lambda_{em}$ = 390/470 nm. Reproduced from ref. 49 with permission from the Royal Society of Chemistry.

The chelator counterpart releases Zn^{2+}, which is reused for the hydrolysis of other probe molecules in the presence of glutathione. Therefore, the Zn^{2+}-assisted hydrolysis of the probe proceeds catalytically with a stoichiometry >1:1, and Michaelis–Menten constant (K_m) of 6.73 μM and V_{max} of 8.65 × 10^{-3} M s^{-1}. The presence of glutathione (2 mM) provided a smaller impact on the limits of detection (8 nM/30 nM for glutathione −/+) of Dpa-SoxLC than those (16 nM/191 nM) of ZnAF-2, a chelation-based probe. Live-cell imaging using Dpa-LBC, a cell-compatible analog of Dpa-SoxLC, and ZnAF-2 revealed that Dpa-LBC showed up to a ten-fold signal enhancement in a dose-dependent manner toward Zn^{2+} between 1 and 50 μM. Meanwhile, the response of ZnAF-2 is limited to a two-fold increase at 50 μM addition and no significant increase at 5 μM. The results indicated that the stoichiometric mode and competition with glutathione both attenuated the imaging sensitivity of ZnAF-2. This may be a typical example of catalytic activity-based sensing that can overcome the challenges facing chelation-based detection. The drawback of Dpa-SoxLC is a decrease in the catalytic activity and fluorescence at acidic pH values, which must be improved in the future.

14.4 The Activity-based Sensing of Cu

Cu is the third most abundant transition metal in the human body and has stable redox states of Cu^+ and Cu^{2+}.[51–53] Unlike Cu^{2+}, Cu^+ is a major redox state in subcellular labile Cu pools due to the abundance of glutathione and the reductive cellular environment.[54] Thus, the detection of Cu^+ is more suitable for monitoring and studying labile Cu pools in cells. Cu^+ has an electron configuration of d^{10}, which is the same as that of Zn^{2+}; thus, it has no fluorescence-quenching effect and is considered a soft acid by the HSAB rule. Because Cu^+ is the only soft acid among the first-row transition metal ions, there have been many selective fluorescent probes based on chelation by soft ligands.[16,54,55] In contrast, activity-based sensing probes are still rare and limited to one principle based on the chelate-assisted oxidative C–O cleavage reaction, as described below.

14.4.1 Chelate-assisted Oxidative C–O Cleavage

Taki *et al.* reported an oxidative C–O cleavage reaction mediated by Cu^+ and trispyridylmethlamine (TPA).[56] TPA coordinates to Cu^+, and the resulting Cu-TPA complex is activated in the presence of O_2 to undergo the oxidation of methylene at the *O*-benzyl position (Figure 14.15). The oxidation at the benzyl position yields a hemiacetal or ester, which is readily hydrolyzed to the corresponding aldehyde or carboxylate and phenolate, respectively. The detailed active intermediates are unclear but are speculated to be Cu-OOH (copper hydroperoxo) or Cu-OO• (copper peroxy radical) species.[57,58]

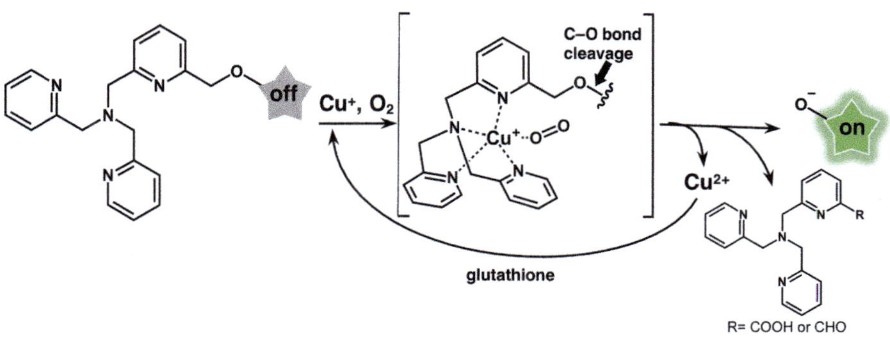

Figure 14.15 The mechanism of Cu^+ sensing by Cu^+-assisted C–O bond cleavage.

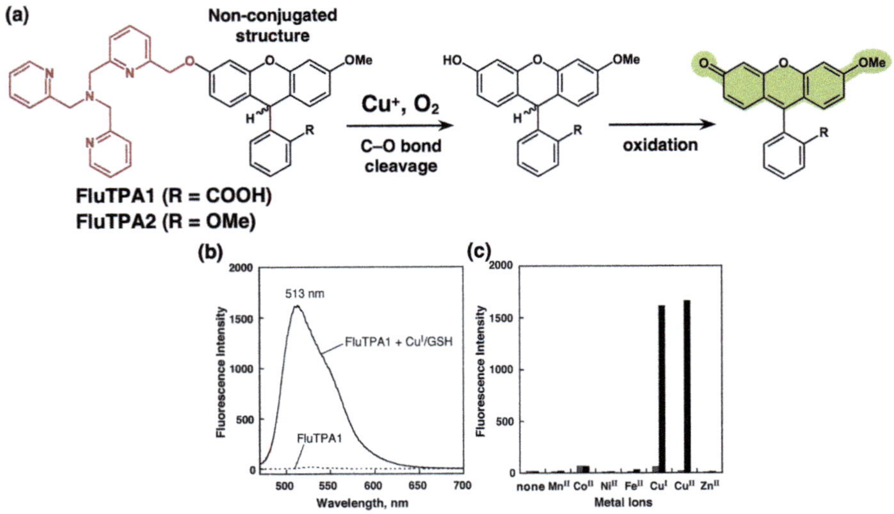

Figure 14.16 (a) The structures and detection mechanisms of FluTPA1 and FluTPA2. (b) Fluorescence response of FluTPA1 (1 μM) before (dotted line) and after (solid line) its reaction with Cu^+ (added as $[Cu(CH_3CN)_4]PF_6$) in HEPES buffer (50 mM, pH 7.2) in the presence of glutathione (2 mM). (c) Metal selectivity of FluTPA1 (1 μM) in the absence (gray) and presence (black) of glutathione (2 mM). Metal ions = 10 μM. HEPES buffer (50 mM, pH 7.2). $\lambda_{ex}/\lambda_{em}$ = 470/530 nm. Reproduced from ref. 55 with permission from American Chemical Society, Copyright 2010.

Taki *et al.* developed the FluTPA1 probe by introducing a TPA group to *O*-methyl-dihydrofluorescein (Figure 14.16a).[56] *O*-Benzylation stabilizes the non-fluorescent dihydrofluorescein structure, and TPA forms a chelate with Cu^+, which triggers the C–O bond cleavage. The resulting *O*-methyl-dihydrofluorescein is readily oxidized to fluorescent *O*-methyl-fluorescein, resulting in fluorescence turn-on. FluTPA1 (1 μM) showed up to a 100-fold increase within 30 min of the addition of Cu^+ (20 μM) in HEPES buffer (50 mM, pH 7.2) in the presence of glutathione (2 mM) (Figure 14.16b). Furthermore, the probe is highly selective towards Cu^+ and responsive to Cu^+ species generated *in situ* by mixing Cu^{2+} and glutathione (Figure 14.16c). The authors analyzed the reaction products *via* LC-MS and observed *O*-methyl-fluorescein and $[Cu\text{-}TPA\text{-}COO^-]^+$, indicating that the expected C–O bond cleavage proceeded upon reaction of the probe with Cu^+. FluTPA2, a cell-permeate analog of FluTPA1, was successfully applied to cell-imaging experiments.

The TPA-based principle for Cu^+ sensing was applied to other fluorescent probes, including those used in mitochondria-targeted imaging (Rdl-TPP), bioluminescence and *in vivo* imaging (CCL-1), and ratiometric imaging (FCP-1) (Figure 14.17). Taki *et al.* developed Rdl-TPP by appending the TPA domain onto a hydroxymethyl-rhodol scaffold with a triphenylphosphonium as a mitochondrial localizer (Figure 14.17a).[59] Subsequent C–O

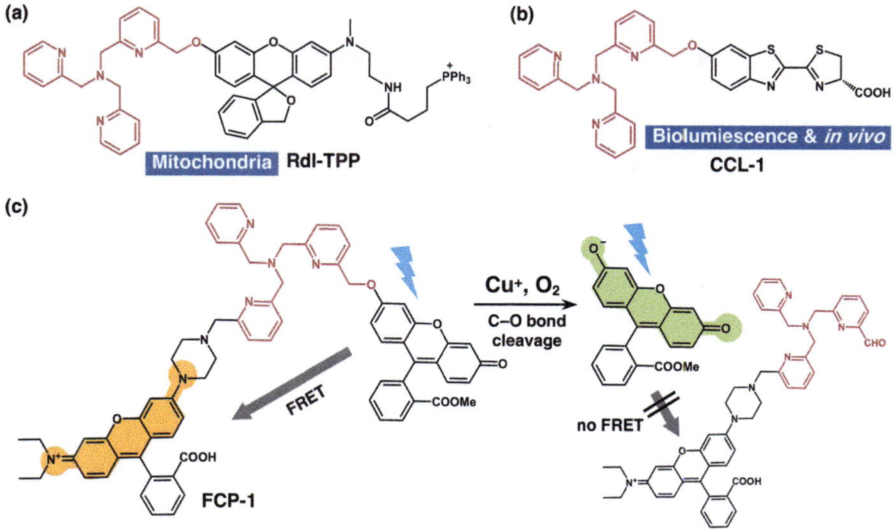

Figure 14.17 (a) The structure of RdI-TPP as a mitochondria-targeted Cu^+ probe. (b) The structure of CCL-1 as a bioluminescent probe for Cu^+. (c) The structure and detection mechanism of FCP-1 as a ratiometric fluorescent probe for Cu^+.

cleavage provides hydroxymethyl-rhodol, which becomes fluorescent *via* the ring-opening of its spirocycle, as illustrated in Figures 14.3 and 14.6b. CCL-1, a bioluminescent probe for Cu^+, was reported by Chang *et al.*[60] *O*-Alkylated luciferin is unaffected by luciferase, and luciferin is released after Cu^+-induced C–O cleavage. The resulting luciferin showed bioluminescence upon its reaction with luciferase. The probe worked in mice and revealed that hepatic Cu deficiency occurred in a mouse model with diet-induced, non-alcoholic fatty liver disease. In addition, Chang *et al.* reported a ratiometric fluorescent probe, FCP-1, composed of a FRET-inducing donor (fluorescein methyl ester) and acceptor (rhodamine) pair tethered by a Cu^+-cleavable linker (Figure 14.17c).[61] An imaging study using the probe revealed a relationship between labile Cu status and glutathione homeostasis that is altered by oncogene-driven phenotypes, which would induce low glutathione levels and a deficiency in labile Cu pools.

14.5 Concluding Remarks and Future Directions

Chelation-based sensing has been the mainstream method for metal sensing for 35 years, and it has made an outstanding contribution to fields of biological and environmental metals. Activity-based sensing has been developed for detecting metal ions, such as paramagnetic metal ions. Activity-based Fe^{2+} sensing probes are typical examples that demonstrate the advantages of the activity-based sensing approach. This chapter aims to highlight the advantages and disadvantages of chelation- and activity-based sensing approaches (Figure 14.1) to provide a guide regarding the selection of the appropriate sensing technique to use. Scientists and researchers are recommended to refer to recent reviews that comprehensively feature chelation-based, activity-based, and protein-based sensing techniques.[16,48,62] The activity-based sensing principles described here are based on previously established biomimetic chemistry and catalytic reactions employed in organic chemistry, which are utilized as the triggers of fluorescence and other signals. Due to the ever-increasing use and development of metal-catalyzed reactions, improvements in sensitivity and selectivity of current metal-detection methods and the production of new sensing techniques for metals that are currently impossible to monitor are expected in the future. Furthermore, these metal-sensing tools need to be further developed to keep up with the rapid progress in the construction of sensing modalities, which will contribute to the future of metal-relevant sciences, including chemistry, biology, medicine, and environmental science.

Acknowledgements

The author thanks Grant-in-Aid Scientific Research (B) (JSPS KAKENHI, Grant Number JP18H02110 to T.H.), Grant-in-Aid for Scientific Research on Innovative Areas (JSPS KAKENHI, "Integrated Metal-bioscience", Grant Number 20H05515 to T.H.).

References

1. G. Grynkiewicz, M. Poenie and R. Y. Tsien, *J. Biol. Chem.*, 1985, **260**, 3440.
2. K. J. Bruemmer, S. W. M. Crossley and C. J. Chang, *Angew. Chem.*, 2020, **59**, 13734.
3. M. Shvartsman and Z. I. Cabantchik, *BioMetals*, 2012, **25**, 711.
4. H. Gunshin, B. Mackenzie, U. V. Berger, Y. Gunshin, M. F. Romero, W. F. Boron, S. Nussberger, J. L. Gollan and M. A. Hediger, *Nature*, 1997, **388**, 482.
5. H. Shi, K. Z. Bencze, T. L. Stemmler and C. C. Philpott, *Science*, 2008, **320**, 1207.
6. I. Yanatori, Y. Yasui, M. Tabuchi and F. Kishi, *Biochem. J.*, 2014, **37**, 25.
7. D. J. R. Lane and D. R. Richardson, *Free Radicals Biol. Med.*, 2014, **75**, 69.
8. W. Breuer, M. Shvartsman and Z. I. Cabantchik, *Int. J. Biochem. Cell Biol.*, 2008, **40**, 350.
9. M. W. Hentze, M. U. Muckenthaler, B. Galy and C. Camaschella, *Cell*, 2010, **142**, 24.
10. O. Kakhlon and Z. I. Cabantchik, *Free Radicals Biol. Med.*, 2002, **33**, 1037.
11. J. A. Imlay, S. M. Chin and S. Linn, *Science*, 1988, **240**, 640.
12. S. Enami, Y. Sakamoto and A. J. Colussi, *Proc. Natl. Acad. Sci. U. S. A.*, 2014, **111**, 623.

13. S. Toyokuni, *Cancer Sci.*, 2009, **100**, 9.
14. S. Epsztejn, O. Kakhlon, H. Glickstein, W. Breuer and I. Cabantchik, *Anal. Biochem.*, 1997, **248**, 31–40.
15. F. Petrat, H. de Groot and U. Rauen, *Arch. Biochem. Biophys.*, 2000, **376**, 74.
16. K. P. Carter, A. M. Young and A. E. Palmer, *Chem. Rev.*, 2014, **114**, 4564.
17. N. Miyata, T. Santa and M. Hirobe, *Chem. Pharm. Bull.*, 1984, **32**, 377.
18. R. C. Hider and X. Kong, *Dalton Trans.*, 2013, **42**, 3220.
19. T. Hirayama, K. Okuda and H. Nagasawa, *Chem. Sci.*, 2013, **4**, 1250.
20. T. Hirayama, *Free Radicals Biol. Med.*, 2019, **133**, 38.
21. T. Mukaide, Y. Hattori, N. Misawa, S. Funahashi, L. Jiang, T. Hirayama, H. Nagasawa and S. Toyokuni, *Free Radical Res.*, 2014, **48**, 990.
22. T. Imamura, T. Hirayama, K. Tsuruma, M. Shimazawa, H. Nagasawa and H. Hara, *Exp. Eye Res.*, 2014, **129**, 24.
23. K. Oshima, Y. Ikeda, Y. Horinouchi, H. Watanabe, H. Hamano, Y. Kihira, S. Kishi, Y. Izawa-Ishizawa, L. Miyamoto, T. Hirayama, H. Nagasawa, K. Ishizawa, K. Tsuchiya and T. Tamaki, *Lab. Invest.*, 2017, **97**, 555.
24. M. Niwa, T. Hirayama, K. Okuda and H. Nagasawa, *Org. Biomol. Chem.*, 2014, **12**, 6590.
25. T. Hirayama, H. Tsuboi, M. Niwa, A. Miki, S. Kadota, Y. Ikeshita, K. Okuda and H. Nagasawa, *Chem. Sci.*, 2017, **8**, 4858.
26. T. Hirayama, S. Kadota, M. Niwa and H. Nagasawa, *Metallomics*, 2018, **10**, 794.
27. T. Hirayama, A. Miki and H. Nagasawa, *Metallomics*, 2019, **11**, 111.
28. T. Hirayama, M. Inden, H. Tsuboi, M. Niwa, Y. Uchida, Y. Naka, I. Hozumi and H. Nagasawa, *Chem. Sci.*, 2019, **10**, 1514.
29. T. Hirayama, M. Niwa, S. Hirosawa and H. Nagasawa, *ACS Sens.*, 2020, **5**, 2950.
30. R. Kakiuchi, T. Hirayama, D. Yanagisawa, I. Tooyama and H. Nagasawa, *Org. Biomol. Chem.*, 2020, **18**, 5843.
31. H. Y. Au-Yeung, J. Chan, T. Chantarojsiri and C. J. Chang, *J. Am. Chem. Soc.*, 2013, **135**, 15165.
32. E. Deu, I. T. Chen, E. M. W. Lauterwasser, J. Valderramos, H. Li, L. E. Edgington, A. R. Renslo and M. Bogyo, *Proc. Natl. Acad. Sci. U. S. A.*, 2013, **110**, 18244.
33. B. Spangler, C. W. Morgan, S. D. Fontaine, M. N. Vander Wal, C. J. Chang, J. A. Wells and A. R. Renslo, *Nat. Chem. Biol.*, 2016, **12**, 680.
34. A. T. Aron, M. O. Loehr, J. Bogena and C. J. Chang, *J. Am. Chem. Soc.*, 2016, **138**, 14338.
35. A. T. Aron, M. C. Heffern, Z. R. Lonergan, M. N. Vander Wal, B. R. Blank, B. Spangler, Y. Zhang, H. M. Park, A. Stahl, A. R. Renslo, E. P. Skaar and C. J. Chang, *Proc. Natl. Acad. Sci. U. S. A.*, 2017, **114**, 12669.
36. R. K. Muir, N. Zhao, J. Wei, Y.-H. Wang, A. Moroz, Y. Huang, Y.-C. Chen, R. Sriram, J. Kurhanewicz, D. Ruggero, A. R. Renslo and M. J. Evans, *ACS Cent. Sci.*, 2019, **5**, 727.

37. J.-L. Chen, S.-J. Zhuo, Y.-Q. Wu, F. Fang, L. Li and C.-Q. Zhu, *Spectrochim. Acta, Part A*, 2006, **63**, 438.
38. S. Maiti, Z. Aydin, Y. Zhang and M. Guo, *Dalton Trans.*, 2015, **44**, 8942.
39. J. J. Warren and J. M. Mayer, *J. Am. Chem. Soc.*, 2010, **132**, 7784.
40. M. Novak and R. K. Lagerman, *J. Org. Chem.*, 1988, **53**, 4762.
41. W. Xuan, R. Pan, Y. Wei, Y. Cao, H. Li, F.-S. Liang, K.-J. Liu and W. Wang, *Bioconjugate Chem.*, 2016, **27**, 302.
42. Z. Qu, P. Li, X. Zhang and K. Han, *J. Mater. Chem. B*, 2016, **4**, 887.
43. S. Yamasaki, K. Sakata-Sogawa, A. Hasegawa, T. Suzuki, K. Kabu, E. Sato, T. Kurosaki, S. Yamashita, M. Tokunaga, K. Nishida and T. Hirano, *J. Cell Biol.*, 2007, **177**, 637.
44. W. Maret, *Metallomics*, 2015, **7**, 202.
45. J. L. Vinkenborg, T. J. Nicolson, E. A. Bellomo, M. S. Koay, G. A. Rutter and M. Merkx, *Nat. Methods*, 2009, **6**, 737.
46. Y. Qin, P. J. Dittmer, J. G. Park, K. B. Jansen and A. E. Palmer, *Proc. Natl. Acad. Sci. U. S. A.*, 2011, **108**, 7351.
47. S. C. Burdette, G. K. Walkup, B. Spingler, R. Y. Tsien and S. J. Lippard, *J. Am. Chem. Soc.*, 2001, **123**, 7831.
48. E. P. S. Pratt, L. J. Damon, K. J. Anson and A. E. Palmer, *Biochim. Biophys. Acta, Mol. Cell Res.*, 2021, **1868**, 118865.
49. T. Hirano, K. Kikuchi, Y. Urano and T. Nagano, *J. Am. Chem. Soc.*, 2002, **124**, 6555.
50. I. Takashima, Y. Inoue, N. Matsumoto, A. Takagi and K. Okuda, *Chem. Commun.*, 2020, **56**, 13327.
51. C. M. Ackerman and C. J. Chang, *J. Biol. Chem.*, 2018, **293**, 4628.
52. F. P. Guengerich, *J. Biol. Chem.*, 2018, **293**, 4603.
53. Y. Li, *IUBMB Life*, 2020, **72**, 1900.
54. R. McRae, M. M. Henary, R. Patel and B. Lai, *Proc. Natl. Acad. Sci. U. S. A.*, 2005, **102**, 11179–11184.
55. L. Zeng, E. W. Miller, A. Pralle, E. Y. Isacoff and C. J. Chang, *J. Am. Chem. Soc.*, 2006, **128**, 10.
56. M. Taki, S. Iyoshi, A. Ojida, I. Hamachi and Y. Yamamoto, *J. Am. Chem. Soc.*, 2010, **132**, 5938.
57. A. Kunishita, M. Kubo, H. Sugimoto, T. Ogura, K. Sato, T. Takui and S. Itoh, *J. Am. Chem. Soc.*, 2009, **131**, 2788.
58. D. Maiti, H. R. Lucas, A. A. N. Sarjeant and K. D. Karlin, *J. Am. Chem. Soc.*, 2007, **129**, 6998.
59. M. Taki, K. Akaoka, K. Mitsui and Y. Yamamoto, *Org. Biomol. Chem.*, 2014, **12**, 4999.
60. M. C. Heffern, H. M. Park, H. Y. Au-Yeung, G. C. Van De Bittner, C. M. Ackerman, A. Stahl and C. J. Chang, *Proc. Natl. Acad. Sci. U. S. A.*, 2016, **113**, 14219.
61. C. Y.-S. Chung, J. M. Posimo, S. Lee, T. Tsang, J. M. Davis, D. C. Brady and C. J. Chang, *Proc. Natl. Acad. Sci. U. S. A.*, 2019, **116**, 18285.
62. T. Hirayama and H. Nagasawa, *J. Clin. Biochem. Nutr.*, 2017, **60**, 39.

CHAPTER 15

Activatable Photodynamic Photosensitizers for Cancer Treatment

E. KILIC[a], M. DIRAK[a] AND S. KOLEMEN*[a]

[a]Department of Chemistry, Koç University, Rumelifeneri Yolu, 34450 Istanbul, Turkey
*E-mail: skolemen@ku.edu.tr

15.1 Photodynamic Therapy: Mechanism of Action

Photodynamic therapy has been considered as a promising alternative to conventional therapies as it offers minimally invasive procedures, opportunities for repeated application without significant side effects and negligible drug resistance.[1–6] In addition, PDT also activates the immune system against cancer cells[2] and is considered as a localized treatment due to the fact that singlet oxygen has a short lifetime (~10–320 ns in aqueous medium), as well as a limited diffusion length (~10–55 nm).[6] PDT action basically requires three key components: a photosensitizer (PS), light and tissue oxygen ($^3O_2$). In a typical singlet oxygen generation mechanism[5,7,8] (Figure 15.1), a PS is first transferred to singlet excited states (S_1 or S_2) from its ground state (S_0) by light irradiation. S_2 decays non-radiatively (IC, internal conversion) to S_1 in a very short time and then excited PS either relaxes back to the ground state by emitting light, which is known as fluorescence, or can undergo intersystem crossing (ISC) to form a long-lived triplet excited state. Thus, fluorescence

Monographs in Supramolecular Chemistry No. 34
Fluorescent Chemosensors
Edited by Luling Wu, Adam C. Sedgwick, Xiao-Peng He and Tony D. James

Published by the Royal Society of Chemistry, www.rsc.org

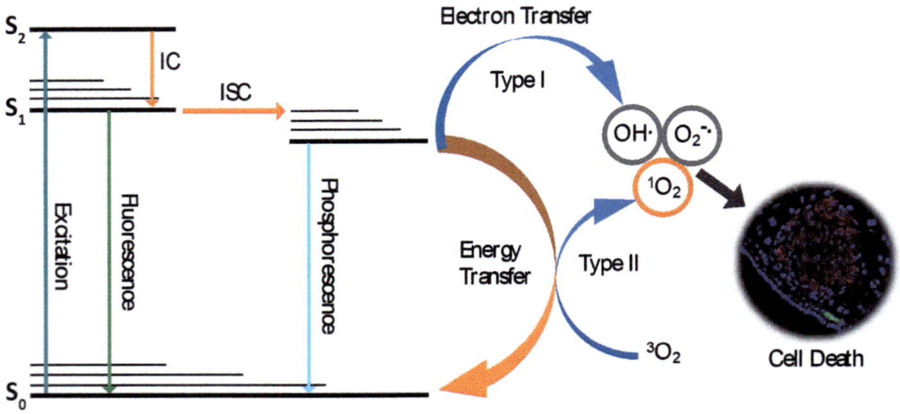

Figure 15.1 Modified Jablonski energy-level diagram for PDT action. IC: internal conversion, ISC: intersystem crossing.

and ISC are two competing pathways, and a good PS should have a high triplet quantum yield for efficient singlet oxygen generation. The triplet excited state energy can be released by phosphorescence or *via* intermolecular energy transfer to the dissolved triplet oxygen (ground state oxygen) in tumor tissues, yielding highly cytotoxic singlet oxygen ($^1O_2$), which is known as a type 2 mechanism (Figure 15.1). The triplet excited state can also be released through type 1 photochemical reactions, which generate ROS, such as hydroxy radical, hydrogen peroxide and superoxide (Figure 15.1). It is generally expected that in most of current PSs, singlet oxygen is predominately responsible for the therapeutic action.[9] During PDT, ROS generation through photosensitization is required since direct excitation of ground state oxygen (spin triplet) to singlet oxygen (spin singlet) is strictly forbidden by selection rules.[8] Upon generation of singlet oxygen, it reacts with vital biomolecules, such as nucleic acids, unsaturated lipids and proteins,[10] which results in cell death through apoptotic or necrotic pathways.[5] The importance of PDT is clearer when it is compared with widely used state-of-the-art treatment methods, such as chemotherapy and radiotherapy, since these therapies have serious side effects, develop drug resistance and they suppress the immune system of patients unlike PDT, which actually activates the immune system against cancer cells.

15.2 Chronic Problems of PDT

Despite its high therapeutic potential and clear advantages, translation of PDT to the clinic has hardly improved in the last decades as evidenced by the limited number of clinical cases in which PDT is actually used as a treatment method. There are currently only seven approved PDT drugs available in the clinic globally,[11] although the first clinical applications were initiated more than a century ago by von Tappeiner.[12] Three long-lasting and inherent problems are at the core of this fact.

1. *Need for external light irradiation.* The penetration depth of the external excitation light through tissues is limited as a result of intrinsic absorption of biomolecules and scattering, which puts most of the tumors that have grown deeply beyond the reach of PDT.[13–16] Additionally, PDT is not an effective treatment for large tumors, as light cannot pass deep into these tumors. In order to address issues related to external light, optical fibers have been employed to deliver the light to deep and large tumors, which is also known as interstitial PDT.[17] However, this technique requires complicated invasive procedures and can only be useful in regions that are accessible through major surgery.

   In recent years, there has been a growing interest towards designing new-generation photosensitizers that can address these chronic problems of PDT.[3,4,18] In an attempt to tackle the light-penetration issue, most popular approaches are to develop near-infrared (NIR) (650–900 nm, therapeutic window) absorbing photosensitizers or to use two-photon excitation or upconversion nanoparticles, as NIR light is known to have maximum penetration.[19–21] Alternatively, in order to bypass the need for external irradiation, self-illuminating PDT approached have been introduced in which a photosensitizer is excited internally (*in situ*) through energy transfer from a bioluminescent or chemiluminescent nanoparticle.[22–24] In typical designs, quantum dots or conjugated polymer-based nanoparticle–luciferase (Fluc) conjugates are employed as internal light sources. Although this approach seems highly promising, its broader acceptance is hampered, either by poor biocompatibility of the employed quantum dots and conjugated polymers or by low therapeutic efficiencies as exhibited by high half-maximal inhibitory concentration (IC_{50}) values. Fenton or Fenton-like reactions that can produce singlet oxygen are also employed to solve the light-penetration problem.[25,26]
2. *Reduced level of tissue oxygen (hypoxia).* Generation of a hypoxic microenvironment is a common feature in most solid tumors. Tumor hypoxia is caused by rapid cell proliferation, as well as insufficient and distorted vasculatures around the tumor region.[27] While hypoxic conditions already reduce the efficacy of the conventional therapies through upregulation of hypoxia-inducible transcription factors and related overexpression of P-glycoprotein (P-gp),[28] the impact on PDT is even more pronounced as singlet oxygen generation through the type 2 mechanism is highly dependent on oxygen.[29] In fact, PDT itself further disrupts the oxygen supply by damaging the tumor vasculature.[30] Thus, in most cases, PDT appears to be a self-limiting treatment.
3. *Lack of cancer cell selectivity.* PDT has an intrinsic selectivity to some extent, arising from the precise routing of the external light onto the tumor region. However, this mode of selectivity is not sufficient

in most cases to eliminate damage to healthy cells and severe post-treatment pain if "always-on" traditional PSs are employed.[11,31] Accordingly, PDT drugs leave patients photosensitive up to months afterwards, forcing them to stay away from sunlight and even indoor light. Thus, additional control mechanisms on PDT action are needed to further improve the selectivity of the photosensitizers. In this direction, one popular approach is to utilize active targeting, in which PSs are selectively attached, either to endogenous surface receptors[32] that are overexpressed in cancer cells, or to artificial chemical receptors that are generated on the cell surface through metabolic glycan labelling of unnatural sugars.[33] Another highly promising strategy is to design activatable or activity-based photosensitizers (aPSs).[34,35] These photosensitizers remain in a passive "off" state (no cytotoxicity) in healthy cells even under light irradiation and are selectively activated at the site where therapy is needed by tumor-associated stimuli. The selectivity is mostly satisfied due to the differential metabolic activities between cancerous and normal cells, which alters the dynamics of cellular microenvironments.

15.3 Activatable Photosensitizers: Design Principles

In a typical aPS design, the PS core is masked with a cage or quencher group, which blocks the $^1O_2$ generation and fluorescence, if any, even under light excitation.[36] Upon removal of the masking unit with a tumor-associated input, such as the acidic microenvironment of cancer cells, elevated levels of biothiols and reactive oxygen species, as well as overexpressed enzymes, several photophysical properties of the PSs change, and consequently both $^1O_2$-induced photocytotoxicity and the emission signal are recovered.[36] Undoubtedly, modulation of the photophysical properties of PSs is at the core of this activation process. This can be realized by judicious molecular engineering through utilizing different design strategies, including photoinduced electron transfer (PeT), intramolecular charge transfer (ICT), Förster resonance energy transfer (FRET) and aggregation-induced emission (AIE) (Figure 15.2).

PeT is a long-known phenomenon, which is heavily used in constructing molecular fluorescent probes.[36–38] It can also be utilized to modulate ISC processes in aPSs. In the design of PeT-based aPSs, an electron-donating group is attached to a PS core with a spacer. Upon excitation of the PS, electron transfer takes place from the donor unit to the core and consequently the excited electron relaxes back to the ground state through non-radiative pathways rather than through the ISC path.[36] This basically blocks ISC, as these two pathways are competing, and the PeT process takes place very rapidly. The ISC path can be reopened for an excited electron when PeT is blocked after an electron-donating unit is involved in a binding interaction with an analyte of interest (Figure 15.2a).

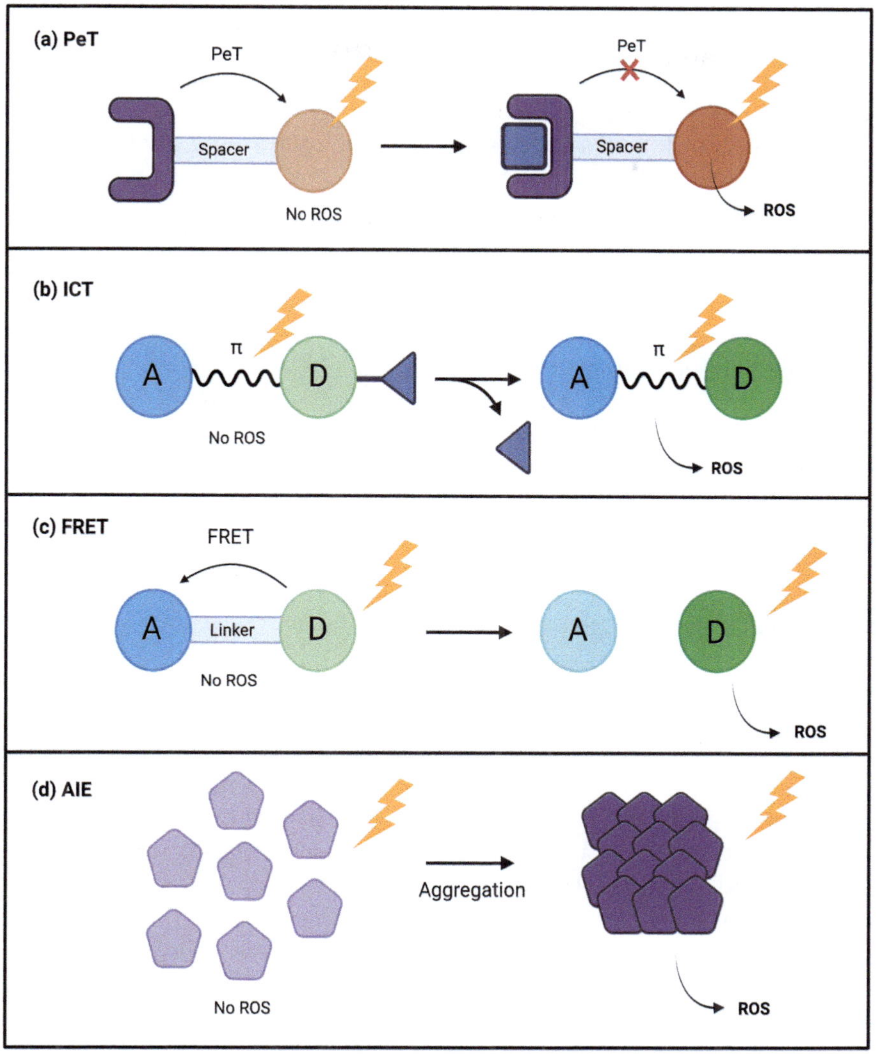

Figure 15.2 Different design strategies for construction of aPSs. (a) PeT, (b) ICT, (c) FRET, and (d) AIE mechanism.

ICT is another well-established mechanism in the design of aPSs. An ICT-based aPS consists of a donor–π-acceptor (D–π-A) structure.[39] Upon excitation, charge transfer occurs from the donor unit, mostly phenolic or amine groups, to the acceptor side, which causes charge redistribution throughout the molecule. If the electron-donating unit is masked with a cleavable cage unit, a dramatic blue shift in the absorption signal of PS is detected due to blocked ICT and charge delocalization.[40–44] Once the cage group is removed selectively by a tumor-associated analyte, ICT turns on and the red-shifted absorption peak is restored. Upon excitation of the PS from the newly apparent red-shifted absorption signal, $^1O_2$ generation is initiated.[45] On the other side, no sign of $^1O_2$ formation can be seen in the caged form, as the external light irradiation, which is adjusted according to the red-shifted signal, cannot excite the PS (Figure 15.2b).

The FRET mechanism is highly attractive as it allows design of highly effective and selective aPSs. In this approach, the PS core acts as a donor and is attached to an acceptor/quencher fluorophore through a cleavable linker. Upon excitation of the PS, energy transfer from the PS to the acceptor unit takes place, thus the ISC path turns out be deactivated. After cleavage of the linker that connects the PS and the acceptor molecule in tumor cells, FRET is broken and excitation of the PS results in effective ISC and consequent $^1O_2$ generation (Figure 15.2c).

AIE arises from the restricted intramolecular rotation (RIR) of fluorophores, known as AIEgens, in the aggregate or solid state.[46,47] On the other side, these types of fluorophores are non-emissive in the molecular state as a result of free molecular rotation-induced, non-radiative relaxation of the excited molecule. AIE is the exact opposite effect of aggregation-caused quenching (ACQ), wherein emission of the hydrophobic fluorophores is quenched because of the π–π stacking in the aggregated state. Propeller-shaped luminogens are mostly known as AIE active due to the RIR effect. However, recent studies also uncovered that there are many different examples of molecules that can act as an AIEgen by utilizing restriction of intramolecular vibration (RIV).[46–49] AIEgens have received widespread attention not only in the field of bioimaging, but also in PDT applications (Figure 15.2d).[50] Most of the AIEgens satisfy ISC by lowering the energy gap between S_1 and T_1 energy states (ΔE_{ST}) thanks to their donor–acceptor-type structures. This is accomplished by separating the HOMO–LUMO distribution, which results in a smaller ΔE_{ST}. For instance, tetraphenyl ethylene (TPE) is one of the most common AIEgen-based PSs. Its ΔE_{ST} is 1.27 eV.[50] With the help of donor–acceptor molecular engineering, it is possible to have smaller ΔE_{ST} values for TPE derivatives, even as small as 0.067 eV.[51] AIEgen-based PSs can also be converted to aPSs by a caging strategy, similarly to conventional PSs.

15.3.1 Enzyme-responsive aPSs

Enzymes are vital catalysts that take roles in a variety of biological functions, and they are indispensable for proper cell function. However, overexpression of certain enzymes is directly associated with cancer, and studies in enzyme biology continue to uncover new enzymes, which also take roles in tumorogenesis.[52] To that end, enzyme-responsive aPSs have attracted great attention in recent decades. These PSs make use of the differential expression levels of enzymes in cancer cells compared to their healthy counterparts to satisfy tumor selectivity. A typical aPS mostly carries a cage unit, which mimics the natural substrate of the enzyme of interest. In the caged form, generation of $^1O_2$ is blocked. Upon removal of the cage, photocytotoxicity is restored. In most cases, selective cleavage of the cage group by the enzyme activates the PS through an ICT, FRET or AIE mechanism. It is also worth mentioning that the PeT mechanism can also be employed in the design of enzyme-activatable PSs. In the current literature, there are numerous enzymes that have been used within the scope of aPS design.

15.3.1.1 β-Galactosidase

β-Galactosidase (β-Gal) is found in a wide range of organisms, such as plants, microorganisms and animals.[53] β-Gal is an enzyme that catalyzes the monosaccharide conversion of lactose by hydrolyzing the β-glycosidic bond.[54,55] It has been reported that β-Gal is overexpressed in a variety of cancer cells, such as primary and metastatic ovarian cancer cell lines.[56]

In 2014, Urano *et al.* introduced an example of a β-Gal-activatable PS (HMDESeR- βgal).[57] The PS is constructed on a selenium-substituted rhodol core, which bears a sugar unit as a cage group. In the caged form HMDESeR-βgal appeared to be colorless due to the intramolecular spirolactone formation. Accordingly, it stays in its off state. Upon selective cleavage of the cage unit, the open xanthene form was recovered, and the singlet oxygen generation ability of the PS was activated, as evidenced from the characteristic $^1O_2$ phosphorescence peak at 1268 nm. HMDESeR-βgal was first tested in HEK cells expressing *lacZ* gene and a dose-dependent drop in cell viability was detected upon irradiation with a light source (510–555 nm, 6 J cm^{-2}). Additionally, larval *Drosophila melanogaster* wing disks were also treated with HMDESeR-βgal and it was shown that the posterior region was expressing β-gal and was the only region affected by the PDT action.

Pu and co-workers devised a β-gal-responsive hemicyanine derivative as a photothermal therapy (PTT) agent (CyGal-P), as shown in Figure 15.3.[58] PTT

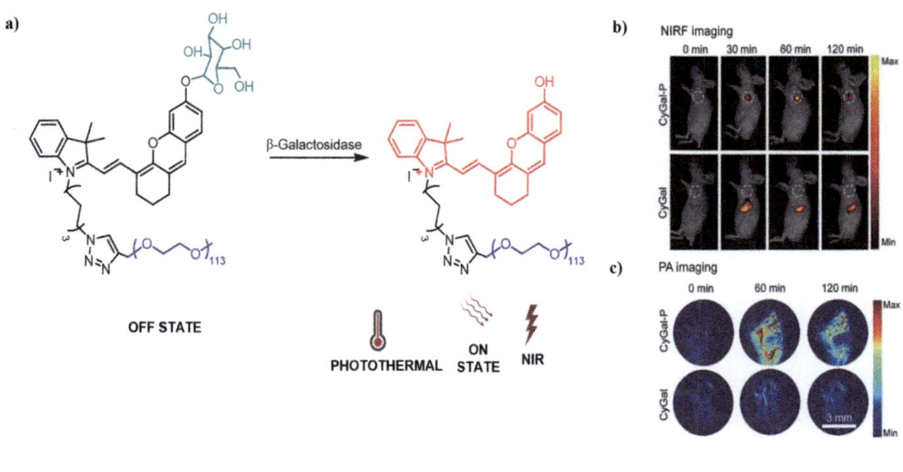

Figure 15.3 (a) Activation of CyGal-P in the presence of β-galactosidase. (b) *In vivo* tumor imaging with CyGal-P. (c) PA images of tumor-bearing mice after injection of CyGal-P. Adapted from ref. 58 with permission from John Wiley & Sons, Copyright © 2018 Wiley-VCH Verlag GmbH & Co. KGaA, Weinheim.

is a type of phototherapy that is based on PSs that can induce local temperature increase upon light irradiation, which results in thermal ablation of the cancer cells. During a PTT action, there is no demand on tissue oxygen, and this is basically the major advantage of PTT over PDT. CyGal-P utilized a typical galactosidase caging unit on the hemicyanine core. Additionally, the agent bears ethylene glycol units to implement water solubility and a passive targeting strategy. In the absence of the enzyme, CyGal-P absorbed around 600 nm and stayed in the off state. Upon cleavage of the cage and activation of the ICT process, the absorption maximum was shifted to 700 nm, with a dramatic increase in the near-IR emission. The PS was also tested in animal models by using SKOV3 tumor-bearing mice. Irradiation of tumors with a 680 nm (0.6 W cm^{-2}) laser decreased the tumor growth rate and initiated NIR fluorescence as well as photoacoustic imaging (Figure 15.3). Thus, CyGal-P acts as a highly effective phototheranostic agent.

15.3.1.2 Nitroreductase

Nitroreductases (NTR) are in the family of proteins that are correlated to evolution. Reduction of compounds substituted with a nitro functional group with the presence of flavin mononuclear cofactor is the characteristic role of NTRs in organisms. Lack of oxygen concentrations in tumors (hypoxia) promotes oxidative stress in the hypoxic cells, which causes the overexpression of nitroreductase along with other intracellular reductases. Therefore, nitroreductase is a suitable marker for cells without sufficient oxygen, such as solid tumors.[59–63]

Song and co-workers designed a lysosome-targeted NTR-activatable theranostic agent on a π-extended fluorescein derivative (compound 2) with a nitrobenzyl caging unit (Figure 15.4).[64] The compound exhibited thermally activated delayed fluorescence (TADF), which makes it a triplet photosensitizer without incorporation of heavy atoms. TADF sensitizers have a very low energy gap between their S1 and T1 states. Thus, both $^1O_2$ generation and fluorescence can be observed with high yields, unlike conventional triplet PSs, which predominantly facilitate ISC. Compound 2 was non-cytotoxic

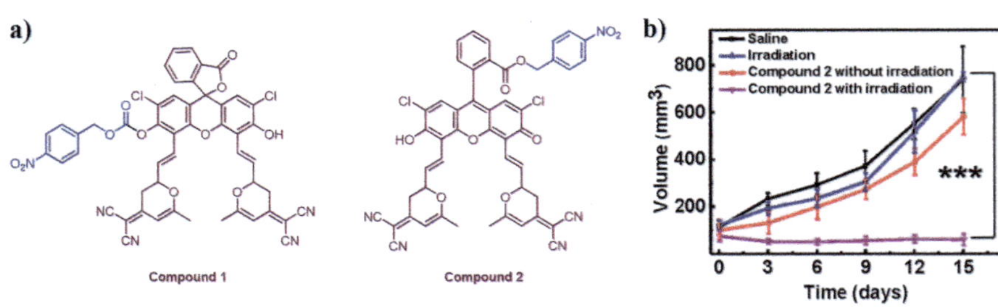

Figure 15.4 (a) TADF fluorescein derivatives. (b) Change in tumor volume with respect to time. Adapted from ref. 64 with permission from American Chemical Society, Copyright 2019.

and non-emissive initially due to an active PeT mechanism. NTR reduces the nitro group to amine and, following 1,6-elimination, cleaves the cage unit and blocks the PeT process. Upon removal of the cage the PS became photocytotoxic and emissive at the same time. Compound 2 was evaluated in HeLa cells under both normoxic (21% O_2) and hypoxic (mild 10% O_2, severe 1% O_2) conditions. Although confocal microscopy images showed that the agent can be activated with a better efficiency under severe hypoxic conditions, PDT action was found to be highly effective under mild hypoxia (10% O_2). The PS was also utilized in *in vivo* experiments successfully upon illumination with a 590 nm LED (16 mW cm^{-2}). Both fluorescence signal and reduction in tumor growth were detected (Figure 15.4).

A mitochondrion-targeted hemicyanine-based PS (ICy-N), which was decorated with the same nitrobenzyl cage unit, was reported by Peng and co-workers in 2019 as another NTR-activatable PDT agent.[65] Reduction of the nitro group cleaved the cage and the resulting ICT mechanism shifted the absorption signal of the PS to 700 nm and restored the emission peak at 716 nm. Irradiation with a 660 nm light source decreased the cell viability (IC_{50} = 0.63 mM) of 4T1 cells in a dose-dependent manner under mild hypoxic conditions (10% O_2). On the other hand, viability remained relatively high under normoxia, suggesting NTR-triggered activation of ICy-N. The efficacy and theranostic nature of the agent were also proven with animal studies using tumor-bearing mouse models.

15.3.1.3 Alkaline Phosphatase

Alkaline phosphatase (ALP) is an omnipresent membrane-bound enzyme that takes part in removal of phosphate groups of small molecules, proteins and nucleic acids *via* hydrolysis. Abnormal levels of ALP in organisms are commonly associated with several diseases, such as Wilson's disease, necrotizing enterocolitis, lymphoma, biliary obstruction, leukemia, osteomalacia, diabetes and osteoblastic bone tumors. Therefore, ALP-activatable photosensitizers have been an object for several studies, since it is involved with various physiological or pathological processes.[66–69]

Li *et al.* employed ALP as an activator in their aPS design. They first synthesized heavy atom-decorated selenium-substituted dicyanomethylene-4*H*-chromene (DCM) derivatives by placing bromine (PS_{Se-Br}) and iodine (PS_{Se-I}) atoms onto the parent core as new PS cores (Figure 15.5). PS_{Se-I} exhibited a higher singlet oxygen quantum yield in solution as expected due to improved spin–orbit coupling-mediated ISC, thanks to the heavier iodine atoms.[40] Later, PS_{Se-I} was converted to an aPS (^{ALP}PS) by masking the electron donor phenolic moiety with an ALP-responsive phosphate cage group (Figure 15.5). Upon phosphorylation, the D–π-A system was broken, and ICT was blocked, which resulted in a blue shift in the absorption maximum (450 nm). Selective removal of the cage unit by ALP activity recovers the dicyanomethylene-4*H*-chromene absorption signal at 600 nm as ICT switched on. Irradiation of HeLa cells incubated with ^{ALP}PS with a Xe lamp (490–700

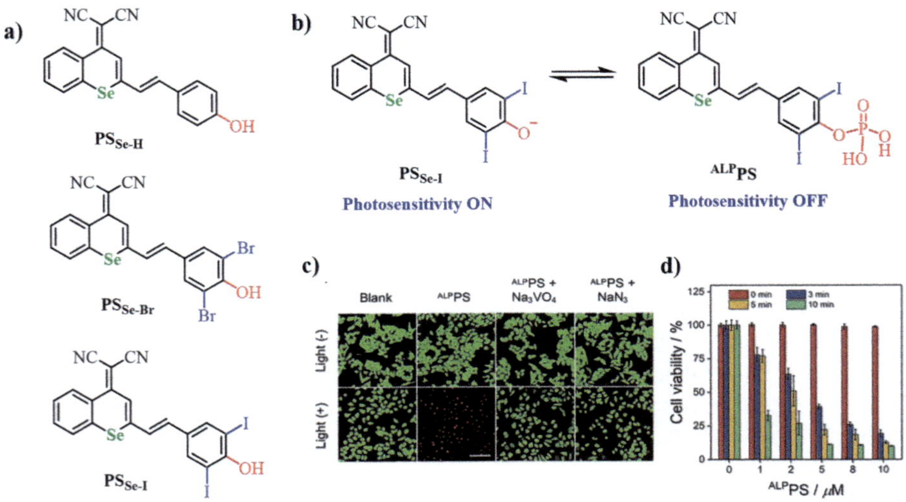

Figure 15.5 (a) D–π-A-type Se-DCM-based photosensitizers with and without heavy atoms. (b) Activation strategy of $PS_{Se\text{-}I}$ with alkaline phosphotase enzyme. (c) Live/dead cell staining of HeLa cells. (d) Cell viability of HeLa cells treated with ^{ALP}PS as determined by MTT assay. Adapted from ref. 40 with permission from John Wiley & Sons, Copyright © 2019 Wiley-VCH Verlag GmbH & Co. KGaA, Weinheim.

nm, 10 mW cm^{-2}) for 10 minutes decreased the cell viability in a dose-dependent manner with no dark toxicity (Figure 15.5). When the cells were pretreated with sodium orthovanadate, a known ALP inhibitor, PDT action was suppressed, indicating that ALP is responsible for activation of the PS and resulting photocytotoxicity.

15.3.1.4 Monoamine Oxidase

Monoamine oxidase (MAO) acts as a catalyst in the reactions of oxidative deamination of various monoamines.[70] MAO-A and MAO-B are the two isoforms of MAO that exist in mammalian tissues. They can be differentiated from each other by their preferences in substrate and inhibitor types, distribution in cell and tissue, and properties in the immune system.[71] Overproduction of MAO has been mostly linked to various neurodegenerative and psychiatric disorders over the years. Recent studies have also shown that abnormal expression of MAO triggers tumor progression and metastasis, which makes MAO an attractive target for cancer research. A high level of MAO in cells is known to cause oxidative stress by triggering redox imbalance. Therefore, it has also attracted great interest in fluorescent probe design towards selective imaging of cancer cells.[72–75]

In 2020, our group introduced the first ever example of an MAO-activatable PS (R1).[76] R1 is an iodinated resorufin that bears an MAO-cleavable propylamine cage group and marks the first resorufin-based PS (Figure 15.6). The

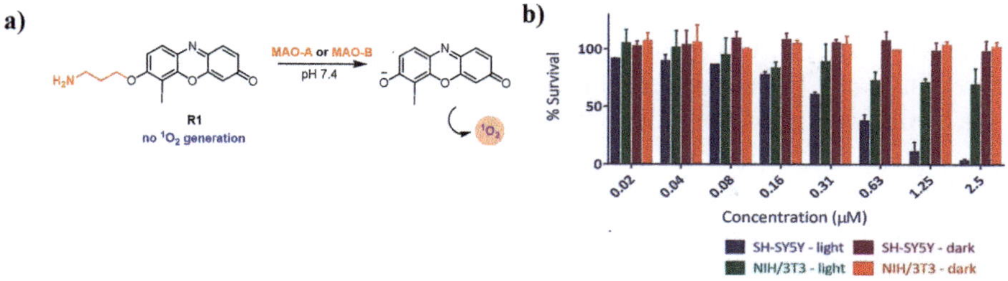

Figure 15.6 (a) Activation of R1 with MAO isoforms. (b) *In vitro* cell viability studies of SH-SY5Y and NIH/3T3 cells upon incubating with different concentrations of R1 in light/dark conditions. Adapted from ref. 76 with permission from American Chemical Society, Copyright 2020.

masked hydroxyl group on R1 blocked the ICT process and R1 showed an absorption peak at 500 nm with a quenched fluorescence signal. Remarkable red shift, with the appearance of a new absorption signal around 600 nm, was detected after removal of the cage unit. Slight increase in emission signal was also observed upon activation. R1 showed highly promising properties as a PS, such as high singlet oxygen generation yield in aqueous solutions (up to 37%), red-shifted absorption signal, photostability and negligible dark toxicity. R1 was then tested in SH-SY5Y neuroblastoma cell lines, which are known to have high MAO activity. Upon excitation of the R1-incubated cells with a 595 nm LED light, a dramatic decrease in cell viability (IC_{50} = 0.42 μM) was reported. On the other hand, when R1 was tested on normal fibroblast NIH-3T3 cells, no remarkable cell death was observed, suggesting selective photocytotoxicity of R1 towards MAO overexpressing cancer cells (Figure 15.6). Intracellular ROS generation was further confirmed in SH-SY5Y cells with a ROS sensor, 2′,7′-dichlorofluorescein diacetate (DCFDA). Use of clorgyline, an MAO inhibitor, extensively blocked $^1O_2$ generation, clearly showing the role of MAO enzyme in triggering the PDT action.

15.3.1.5 *Tyrosinase*

Copper-containing oxidase, known as tyrosinase (TYR), is found in a variety of microorganisms, plants and animals.[77] It is responsible for catalyzing the first two steps of mammalian melanogenesis, which causes hyperpigmentation in the skin of humans. It plays a role in the formation of melanin from tyrosine as a rate-limiting catalyst. As an outcome of melanogenesis, with a combination of chemical reactions catalyzed by enzymes, dark macromolecular pigment, known as melanin, forms.[78] Aggressive malignant melanoma is known for its fast growth and early metastasis.[79] However, early diagnosis of the melanoma is crucial since the survival rate is 97% after surgery for the non-metastatic phenotype.[80] Expression levels in melanoma are correlated with degree of malignancy.[77,81] Therefore, tyrosinase-activatable fluorescent probes and photosensitizers have attracted attention in recent years.

Methylene blue (MB) is a long-known PS, which possesses high singlet oxygen quantum yield in aqueous solutions without losing its emissive character. Liang and Hu converted "always-on" MB to an aPS (MB1) by modifying the amine group on the reduced MB tyrosinase-responsive 4-(2-aminoethyl) phenol cage group.[82] MB1 was colorless and non-fluorescent as a result of broken π-conjugation. After addition of TYR (5 U mL^{-1}), the cage group was oxidized to the quinone derivative and, following intramolecular cyclization, released the urea bond. Upon removal of the cage unit, free amine on the parent core was oxidized under physiological conditions to yield colorful and highly emissive (706 nm) MB. MB1 was evaluated in two different cancer cell lines, B16F10 (melanoma, high TYR activity) and HeLa (cervical, low TYR activity). MB1 induced high phototoxicity in B16F10 cells upon endogenous TYR activation along with a remarkable fluorescence turn-on response, while no activity was detected in HeLa cells as the PS cannot be activated.

15.3.1.6 γ-Glutamyltranspeptidase

γ-Glutamyltranspeptidase (GGT) is a cell-surface enzyme that presents in a wide range of life forms, from bacteria to human.[83] Its role in the cell is selective cleavage of the γ-glutamyl bond of extracellular glutathione (GSH) into cysteinyl-glycine (Cys-Gly). The remaining Cys-Gly goes through dipeptidase catalyzation in the metabolism and breaks down Cys-Gly into cysteine and glycine.[84] GGT is involved in different physiological and pathological processes, such as asthma, diabetes and, most importantly, cancer. It is overexpressed in a variety of cancer cells, such as lung,[85] prostate,[85] liver[85,89] and ovarian.[85,88] Therefore, in the tumor cells, sufficient amounts of cysteine can be provided as the outcome of its regular function and that promote the metastasis, survival and defense mechanisms against the therapy of malignancy.[85–87] Hence, an activatable photosensitizer that is triggered *via* GGT has become a target for many studies.

A selenium-substituted rhodamine derivative (gGlu-HMSeR) was employed as a GGT-responsive PS by caging the rhodamine core with GGT substrate glutamic acid.[90] In the cage form, the PS formed intramolecular spirolactone and consequently appeared to be colorless. Thus, it stayed inactive in normal tissues with low GGT activity. In the presence of GGT enzyme, the amide bond between glutamate and the PS core was selectively cleaved, PS turned to its open state and induced PDT action in cancer cells with high GGT activity upon excitation with a LED light source (50 mW cm^{-2}, 510–550 nm). Studies on tumor spheroids also proved the selectivity of the PS, as the tumor side was ablated during the PDT action, while healthy tissues remained unaffected.

In their recent study, Hua and co-workers introduced a diketopyrrolopyrrole (DPP)-based PS (DPP-GGT), bearing a 4-bromomethylphenyl glutamic acid as a GGT-responsive cage unit (Scheme 15.1).[91] DPP-GGT underwent to a blue shift from 532 nm to 490 nm upon GGT-induced cleavage of the cage due to the ICT mechanism. A similar shift to low wavelengths was observed in the emission signal, and active PS emitted at 548 nm after GGT activation. DPP-GGT showed a remarkable ratiometric response in fluorescence

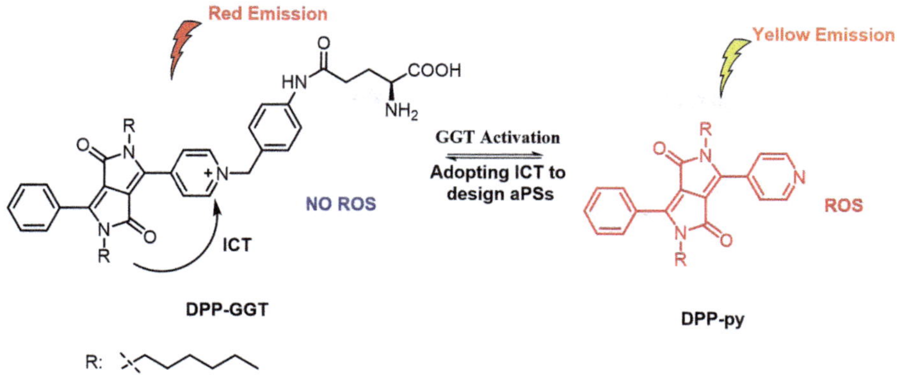

Scheme 15.1 Activation mechanism of γ-GGT activatable DPP-GGT photosensitizer.

and high photocytotoxicity in HepG2 cells, as GGT is known to be overexpressed in this cell type. On the other hand, in low GGT-expressing MCF-7 (breast cancer) and normal LO2 cells, only a red-shifted emission signal was observed, without detectable cell death. DPP-GGT was also used to image *in vivo* HepG2 tumors in animal studies.

15.3.1.7 Cathepsin B

Cathepsin B is a lysosomal cysteine protease that engages in a variety of oncogenic and pathological processes. It plays a role in the control of angiogenesis and invasion throughout the advancement of cancer.[92] Also, it is correlated with metastatic types of cancer. Ordinarily, it is linked with lysosome-participating immune response and autophagy; however, abnormal expression is correlated with cancers.[93]

In 2017, Yoon *et al.* prepared a cathepsin B-responsive aPS (CyA-P-CyB) based on a FRET mechanism.[94] The agent bears two cyanine derivatives, which are linked through a cathepsin B-cleavable peptide (Gly-Phe-Leu-Gly) linker. CyA acts as a donor, while CyB is an energy acceptor and at the same time a quencher. Excitation of CyA-P-CyB at 650 nm resulted in no fluorescence as a result of effective FRET taking place from CyA to non-emissive CyB. Cleavage of the peptide linker *via* cathepsin B enzyme blocked the FRET mechanism and released the NIR-emitting fluorophore and a highly phototoxic PS (Cy-S-Ph-NH_2). Irradiation of the released fluorophore initiated NIR emission at 720 nm, while, on the other hand, excitation of the PS with a 808 nm laser induced $^1O_2$ generation. *In vitro* cytotoxicity of the PS was tested in five different cancer cells overexpressing endogenous cathepsin B enzyme, and in each case decrease in cell viability was noted. When the same experiments were done in normal cells, cell viability remained higher, indicating differential cytotoxicity of the PS towards cancer cells. CyA-P-CyB was also employed successfully in *in vivo* settings. Both NIR tumor imaging and reduction in tumor growth were reported upon laser irradiation.

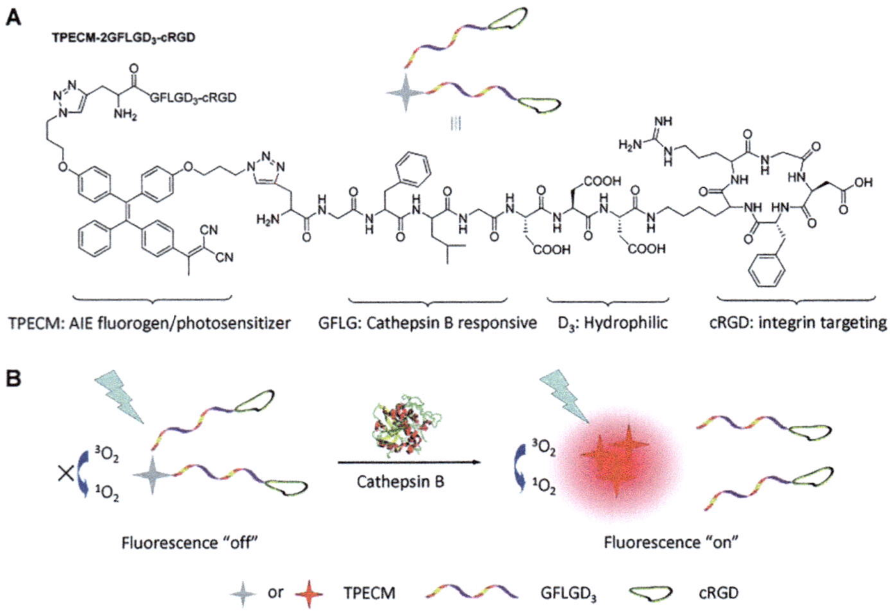

Figure 15.7 (a) Molecular structure and design principles of TPECM-2GFLGD$_3$-cRGD. (b) Activation mechanism of photosensitizer with cathepsin B enzyme. Reproduced from ref. 95 with permission from John Wiley & Sons, Copyright © 2015 WILEY-VCH Verlag GmbH & Co. KGaA, Weinheim.

An AIE-based cathepsin B-activatable PS (TPECM-2GFLGD$_3$-cRGD) was designed by Liu and co-workers.[95] The AIE core (TPECM-2N$_3$) was constructed with a cathepsin B-responsive tetrapeptide linker –Gly-Phe-Leu-Gly– (GFLG), along with a hydrophilic (D$_3$) moiety and $\alpha v\beta_3$ integrin targeting unit (cRGD) (Figure 15.7). TPECM-2GFLGD$_3$-cRGD exhibited almost no fluorescence in water because of the free intramolecular motion, which was sustained by the hydrophilic unit. However, upon cleavage of the cage group it became AIE active and recovered its emission signal at 615 nm upon excitation at 405 nm. ROS generation was detected by using two different trap molecules, 2′,7′-dichlorodihydrofluorescein diacetate (DCFDA) and 1,3-diphenylisobenzo-furan (DPBF). In cell culture studies, no notable cytotoxicity was observed in MDA-MB-231, MCF-7 and 293T cells under dark conditions. Upon light irradiation, a dose-dependent cytotoxicity was observed in MDA-MD231 cells, while cytotoxicity remained negligible for MCF-7 and 293T cells.

15.3.1.8 Aminopeptidase N

Aminopeptidase N (APN), also known as CD13, is a zinc-dependent metalloproteinase that plays a significant role in several pathological and physiological processes, such as adhesion, blood pressure adjustment, coronavirus orientation process, regulation of immune system, sperm motility and signal transduction. This cell surface-located peptidase is known for its characteristic cleavage ability of N-terminal amino acids of various proteins and peptides.[96–98] In comparison with normal cells, APN is overexpressed in plenty of tumor cells due to its participation in numerous processes in tumor cells, such as metastasis, proliferation, invasion and angiogenesis.[99–102] Therefore, it has been a potential target in the field of activatable photosensitizers.

An example of an APN-responsive mitochondria-targeted PS (APN-CyI) was published in 2020 by Peng *et al.*[103] They utilized iodinated hemicyanine core as a PS, which was masked with an APN substrate alanine moiety. ICT was blocked in the case of caged PS, resulting in a blue shift in the absorption signal and quenched fluorescence signal. Removal of the cage recovered the NIR absorption signal around 690 nm, with a contaminant emission peak at 717 nm. *In vitro* imaging studies revealed that APN-CyI exhibited a turn-on response in APN-expressing HepG2 cells, while it remained almost non-fluorescent in normal LO2 cells. Similarly, the PS induced selective photocytotoxicity towards HepG2 cancer cells, as proven using cell viability assays. Additionally, APN-CyI allowed *in vivo* tumor imaging and tumor growth suppression under 660 nm light irradiation.

15.3.1.9 NAD(P)H:Quinone Oxidoreductase

NAD(P)H:quinone oxidoreductase 1 (NQO1) is a cytosolic, intracellular flavoenzyme that catalyzes two-electron reduction of quinones and a range of other compounds along with cofactors NADH and NADPH. In this process, the enzyme and cofactors almost have an equivalency in terms of concentration, which is very uncommon for such co-operations. Ribosyldihydronicotinamide quinone oxidoreductase 2 (NQO2), which has a similar structure to NQO1, also exists in humans but it uses ribosyldihydronicotinamide as a cofactor instead of NAD(P)H.[104] Studies related to activatable photosensitizers have been mostly focused on NQO1 due to its high expression in cancer cells, such as pancreatic, colon, stomach, neck, breast, lung and head cancers, and cholangiocarcinoma.[105–108]

Beharry's group recently reported an NQO1-responsive activatable PS by utilizing the PeT mechanism.[109] They employed a highly efficient phenalenone (PN) core as a theranostic scaffold. PN's cytotoxicity and fluorescence are quenched with a quinone derivative due to active PeT from the PS core to the quinone cage (Figure 15.8). In the caged form, $^1O_2$ and fluorescence quantum yields of the PS were calculated to be 0.28 and 0.0036, respectively. When the PS was treated with hNQO1 enzyme, reduction of the quinone group was followed by intramolecular cyclization, which released the cage and restored the fluorescence and $^1O_2$ capacity of the PN core. After activation, $^1O_2$ and fluorescence quantum yields increased to 0.82 and 0.02, respectively. The PS was evaluated in lung cancer (A549) due to high NQO1 activity, along with a control normal lung cell line, MRC9. A dose-dependent decrease in cell viability was detected when PS-treated A549 cells were irradiated, with

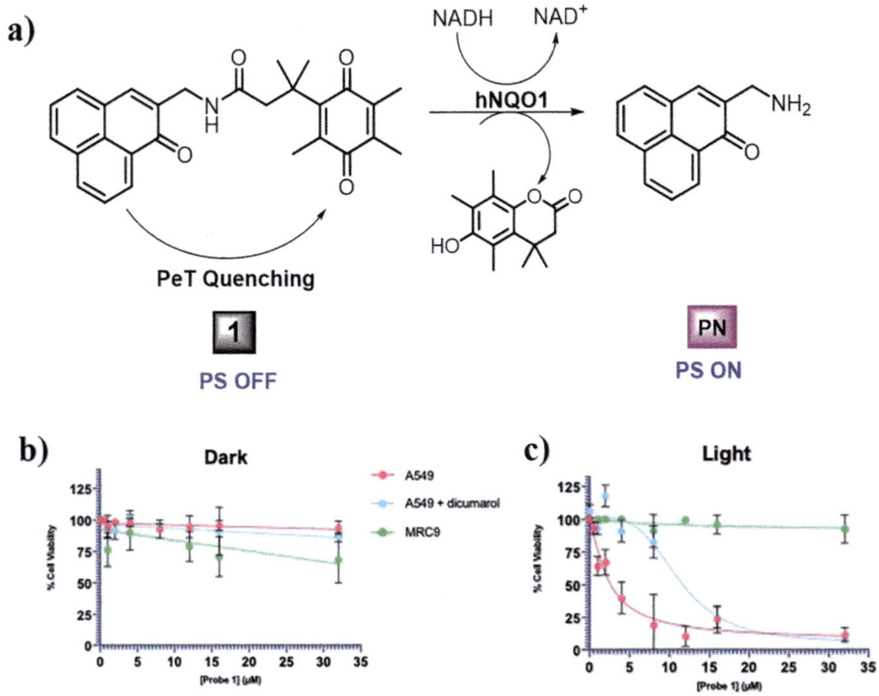

Figure 15.8 (a) Activation of PS (1) *via* reduction of the quinone to quinol in the presence of NQO1. (b) Cell viability assays in (1)-treated A549 (with/without inhibitor) and MRC9 cells under dark and (c) upon irradiation. Adapted from ref. 109 with permission from John Wiley & Sons, Copyright © 2019 Wiley-VCH Verlag GmbH & Co. KGaA, Weinheim.

a negligible dark toxicity (Figure 15.8). As expected, normal lung cells survive, even under light irradiation, as the PS cannot be activated in those cells.

15.3.2 pH-Responsive aPSs

Tumor microenvironments exhibit a slightly acidic pH (6.0–6.5) compared to normal tissues because of anaerobic glycolysis (formation of ATP and pyruvate from glucose *via* pyruvate kinase) of cancer cells, which is activated under hypoxic conditions and after mitochondrial damage.[110] Accordingly, abnormalities in energy metabolism and ion transport processes emerge, which stimulate lactic acid and pyruvate accumulation in tumor cells.[111] Thus, pH-responsive PSs have been designed with precise molecular engineering to develop cancer cell-selective PDT agents. In these studies, PeT is the major mechanism to modulate photocytotoxicity, and BODIPY (dipyrromethеneboron difluoride) is the most popular PS core.

Siegwart and co-workers made use of the acidic pH tumor microenvironment to construct an aPS by utilizing two iodinated BODIPY-based PSs (PEG2k5c-I and PEG2k5c-OMe-I).[112] In their design, a diethylaminophenyl

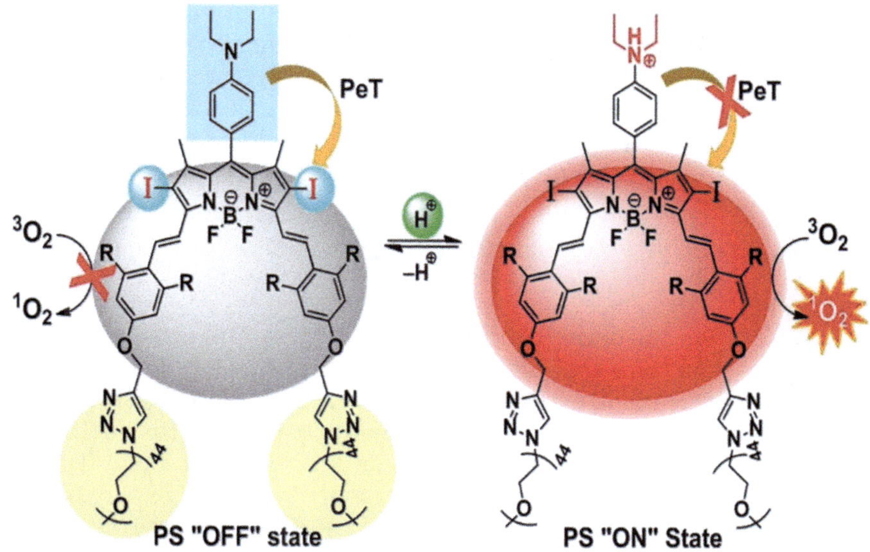

Figure 15.9 Schematic illustration of the working principle of pH-activatable BODIPY-based PSs. Reproduced from ref. 112 with permission from American Chemical Society, Copyright 2018.

moiety was attached to the *meso*-position of the BODIPY cores to control PS activity *via* the PeT mechanism (Figure 15.9). Additionally, π-conjugation of the core was extended through di-styryl ethylene glycolic units to obtain NIR-absorbing/emitting PSs. In addition to that, styryl units improved the water solubility of the hydrophobic BODIPY scaffold. At pH 7.4, both PSs were non-emissive and inactive in terms of PDT action due to the active PeT process. When the agents were tested at pH 5.0, both NIR emission and $^1O_2$ generation were activated in accordance with the design principle. *In vitro* photocytotoxicity of both PDT agents were analyzed in MDA-MB-231 cancer cells under 600–700 nm (0.2 W cm^{-2}, 10 min) light irradiation. Cell viability results indicated that PSs could be activated in cancer cells and induced PDT action by effective $^1O_2$ generation. Two PSs also suppressed the tumor growth in mouse models and could be used for tumor-selective NIR fluorescence imaging at the same time. A similar approach was followed in another study, in which a di-bromophenol unit was attached to an iodinated BODIPY core as a PeT donor to stop ISC at alkaline pH. Under acidic conditions protonation of the phenol quenched the PeT process and $^1O_2$ generation was activated and detected by chemical-trap experiments. A notable cell viability decrease was observed when PS-treated HeLa cells were irradiated with a 532 nm light source for 30 minutes. Alongside this, the PS was emissive enough to allow *in vitro* fluorescence imaging of HeLa cells.

Dong *et al.* introduced an aza-BODIPY derivative (NAB) carrying a dimethylaminophenyl unit as a pH-sensitive PeT donor.[113] The hydrophobic PS was encapsulated with amphiphilic DSPE-mPEG$_{2000}$ in order to prepare the

PS for *in vivo* experiments. NAB had a strong absorption peak in the NIR region at 792 nm and an emission signal at 850 nm at pH 5.0. It exhibited dual PDT and PTT effects when accumulated to lysosomes (pH 5.0) after cellular internalization. *In vivo* experiments showed that the PS accumulated in the tumor, induced synergistic phototherapy and showed pH-dependent photoacoustic imaging upon xenon lamp excitation.

15.3.3 Reactive Oxygen and Nitrogen Species Responsive aPSs

Reactive oxygen and nitrogen species (ROS/RNS) are neutral, anionic and free radical molecules that play essential roles in many different signaling pathways, and are crucial for intracellular redox balance. ROS can be divided into subcategories, such as hydrogen peroxide (H_2O_2), superoxide ($O_2^{\bullet-}$), singlet oxygen ($^1O_2$), hydroxyl radical ($^{\bullet}OH$) and hypochlorous acid (HOCl).[114] ROS are primarily produced in mitochondria, since most of the oxygen is consumed in this organelle to produce ATP.[115] Molecular oxygen (O_2) can only generate one electron at a time and in the reduction of O_2 to H_2O, different ROS are produced as by-products.[116,117] RNS are mainly derived from nitric oxide (NO) and are known as oxidative nitrogen-containing molecules.[114] Peroxynitrite ($ONOO^-$) and nitroxyl (HNO) are among the important RNS. Although these species take important roles in physiological processes, abnormal levels of ROS/RNS lead to oxidative stress and consequently impair the redox cellular redox state and cause cellular dysfunction.[118] Thus, high levels of these reactive species are directly correlated with pathological states, including neurodegenerative and cardiovascular diseases and diabetes, as well as cancer.

15.3.3.1 *Hydrogen Peroxide (H_2O_2)*

Among all the ROS, H_2O_2 is the most stable one. However, the presence of an O–O bond in the structure imparts it with an ambiphilic reactivity. Thus, it can serve as an electrophilic oxidant and nucleophile at the same time. H_2O_2 may trigger several signaling pathways for cell proliferation, migration and differentiation. A high concentration of H_2O_2 could initiate protein function loss and conclude in oxidative cell damage. Along with that, H_2O_2 oxidizes cysteinyl thiol and forms disulfide bonds and sulfenic acids.[119] The residual Cys initiates formation of GSH or sulfoxidation to give methionine. Considering the roles of H_2O_2 in tumorigenesis, it has proven an attractive target in bioimaging and PDT studies. The main problem with first-generation H_2O_2-targeted PDT agents or fluorophores is the selectivity of H_2O_2 over superoxide, nitric oxide and hydroxy radical. To overcome this challenge, Chang's group established a new H_2O_2-mediated transformation of the boronic ester moiety to phenols by making use of the ambiphilic character of H_2O_2. In the boronate structure, the boron atom contains an empty p orbital that can simply be attacked by a nucleophile.

By undergoing Baeyer–Villiger oxidation-type rearrangement and hydrolysis, the phenolic hydroxyl group is formed.[120] In the design of aPSs the boronic ester-based cage group is the primary choice and in most cases the ICT mechanism is utilized to control the ISC pathway.

Recently, we introduced the first ever example of a resorufin-based photosensitizer (RR-1) that can be activated by a hydrogen peroxide (H_2O_2) selectively in cancer cells, as well as the first resorufin-based phototheranostic agent.[121] RR-1 was constructed on an iodinated resorufin core and bears an H_2O_2-cleavable phenylboronic acid pinacol ester cage group. Thus, it serves as an activatable "smart" PS. RR-1 had an absorption maximum at 488 nm prior to H_2O_2 treatment due to the suppressed ICT process. Addition of H_2O_2 activated the ICT mechanism and a remarkable red shift in the absorption signal was detected, with a turn-on response in fluorescence peak, at 597 nm. RR-1 was shown to induce differential photocytotoxicity in MDA-MB-231 and HCT-116 cancer cells with high ROS levels, while staying deactivated in normal NIH-3T3 cells. Addition of *N*-acetyl cysteine and NaN_3, as H_2O_2 and $^1O_2$ scavengers, dramatically increased the cell viability, and clearly indicated H_2O_2-triggered activation and $^1O_2$-induced cell death. Additionally, it can be used to image cancer cells selectively under confocal microscopy. Intracellular ROS generation was further detected by using the cell-permeable ROS sensor. Remarkably, thanks to versatile resorufin chemistry all of these properties are realized by a compound that was synthesized in only two steps.

A similar approach was also reported by Xu and co-workers in which a boronate ester cage group was attached to an iodinated BODIPY (ProBODIPY-2I) derivative to stop $^1O_2$ generation capacity of the PS core as a result of the active PeT mechanism.[122] H_2O_2-induced oxidation and hydrolysis of the cage group not only activated ROS generation, but also converted the PS to an emissive agent. In PDT experiments, dose-dependent photocytotoxicity was obtained in ProBODIPY-2I-treated HCT-116 cancer cells. On the other hand, cell viability was found to be relatively high in normal NCM460 cells, especially at low PS concentrations.

Zhang *et al.* designed a dual therapeutic agent (PNPS) by linking a bromo-hemicyanine-based PDT agent with a chemotherapy drug, 5′-deoxy-5-fluorouridine (5′-DFUR), through a H_2O_2-responsive boronate unit (Figure 15.10).[123] In the absence of H_2O_2, both the PS and the chemotherapy drug were inactive. After cleavage of the boronate linker, active 5′-DFUR and brominated hemicyanine were released. Irradiation of the PS initiates PDT action, in addition to the therapeutic effect that was satisfied by 5′-DFUR. Additionally, the ICT mechanism enabled a turn-on response in the PS NIR emission, which is critical for imaging studies. *In vitro* cytotoxicity of PNPS was tested in cancerous (HeLa and HepG2) and normal (HL7702) cells. As expected, decrease in cell viability was only detected in cancer cells, which correlates well with the design principles (Figure 15.10). Encouraged by the cell culture studies, the fluorescent imaging potential of PNPS was also tested successfully on HCT-116 tumor-bearing mice (Figure 15.10f).

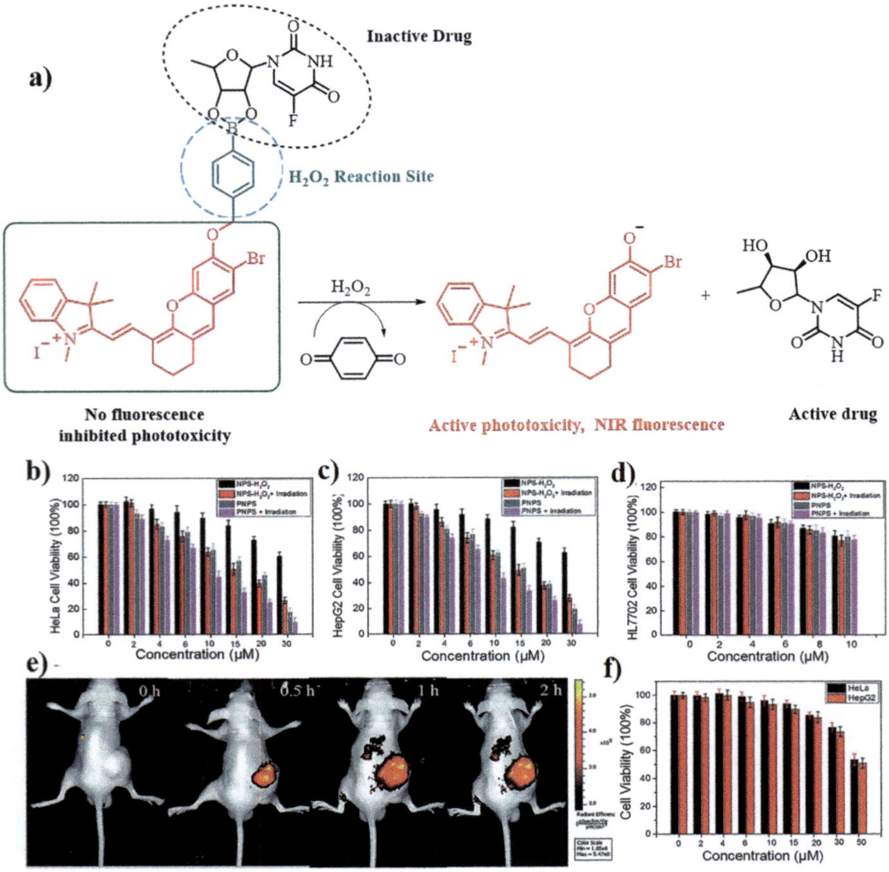

Figure 15.10 (a) Release of active drug 5′-DFUR and NIR PDT agent in the presence of H_2O_2. (b) Cell viability of PNPS-treated HeLa, (c) HepG2 and (d) HL7702 cells. (e) *In vivo* tumor imaging with PNPS. (f) Viability of HeLa and HepG2 cells upon incubating with different concentrations of active drug. Adapted from ref. 123 with permission from the Royal Society of Chemistry.

15.3.3.2 Peroxynitrite ($ONOO^-$)

Peroxynitrite is generated *via* reaction between $O_2^{\bullet-}$ and NO and, due to its reactive nature, causes damage to critical biomolecules, such as lipids, proteins and DNA.[124] Similar to H_2O_2, excessive production of $ONOO^-$ is associated with inflammation, neurodegenerative diseases and cancer.[125]

Li *et al.* expanded their di-iodinated selenium-substituted dicyanomethylene-4*H*-chromene derivative ($^{\text{mito-PN}}$PS) to design a peroxynitrite-responsive aPS.[40] They utilized a boronate ester cage, which can be cleaved by peroxynitrite. Additionally, the core structure was decorated with a cationic triphenylphosphonium unit to target mitochondria. As in the case of their ALP-responsive agent, removal of the cage unit restored the ICT and activated the photocytotoxicity, as evidenced from the cell viability drop in activated (pretreatment with lipopolysaccharide (LPS/interferon-γ)) RAW 264.7 cells.

Peroxynitrite can also be utilized as a highly cytotoxic therapeutic outcome during a PDT action. In this direction, Sortino *et al.* very recently developed a new-generation PS (BPT-NO) based on *N*-nitroso-substituted benzophenothiazine, which can generate $O_2^{\bullet-}$ and release NO at the same time upon red-light irradiation.[126] Released superoxide and nitric oxide are immediately combined to produce cytotoxic $ONOO^-$ (Figure 15.11). Peroxynitrite formation after light irradiation was confirmed by using an $ONOO^-$-selective fluorescein-boronate fluorescent sensor, and then cell viability analyses were performed. When BPT-NO-incubated MDA-MB-231 triple-negative breast cancer cells were irradiated with a red light (1 J cm^{-2}), notable cell death was detected, showing the success of $ONOO^-$-based PDT action. In addition to a high mortality rate of cancer cells, BPT-NO allowed *in vitro* imaging of cancer cells thanks to its emissive nature.

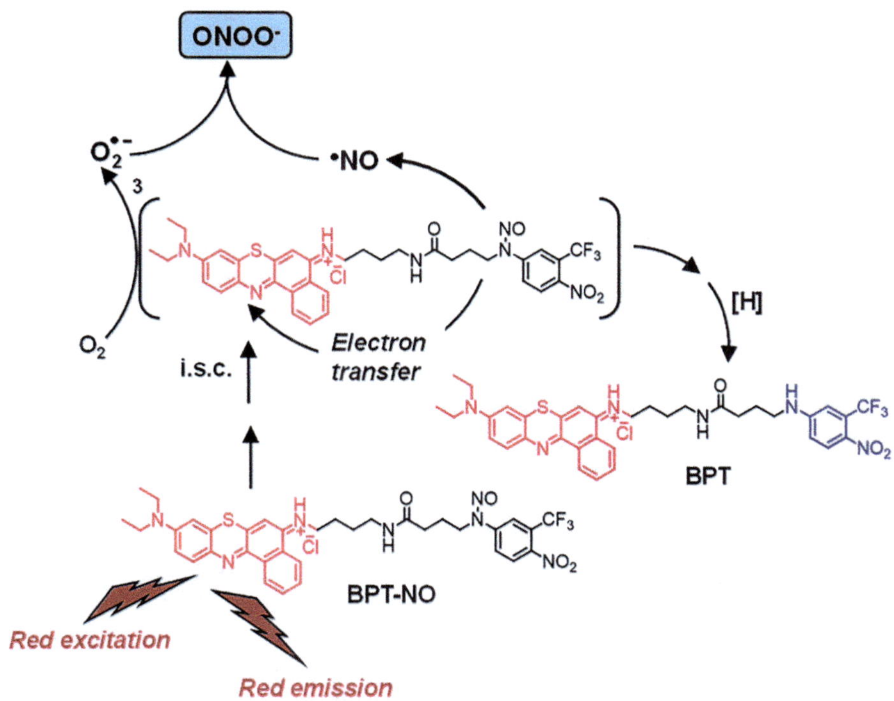

Figure 15.11 Mechanism of cytotoxic $ONOO^-$ generation from BPT-NO under red light. Reproduced from ref. 126 with permission from the Royal Society of Chemistry.

15.3.3.3 Nitric Oxide (NO)

NO is an intracellular gaseous messenger that plays crucial roles in physiological processes. However, high levels of NO are associated with diseased states, including autoimmune disorders, inflammatory diseases and cancer.[127] To that end, it appears to be a promising candidate as a therapeutic target.

DBB-NO is an example of an NO-activatable PS, which was constructed by coupling *o*-phenylenediamine as an NO-responsive unit with a bis(phenylethynyl)benzene derivative.[128] Reaction of NO with the PS cleaves the amide bonds and blocked the PeT mechanism, which in turn increased the two-photon cross-section. Accordingly, the active PS exhibited two-photon PDT and fluorescence, which is highly beneficial in terms of deep tissue penetration. DBB-NO was evaluated in mice, which were treated with LPS to trigger inflammation. Selective cell death was observed in activated macrophages, which overproduce NO, while normal tissues stayed unaffected, even under 750 nm two-photon excitation.

15.3.4 Biothiol-responsive aPSs

Biological thiols, particularly glutathione (GSH), are intracellular reducing agents, which play critical roles in quenching free radicals and maintaining redox homeostasis.[129,130] Thus, these thiols are crucial to protect cells from oxidative stress. However, abnormalities in biothiol concentrations are linked to pathological states, including cancer.[131]

15.3.4.1 Glutathione (GSH)

Glutathione (GSH, γ-glutamine-cysteine-glycine) is a natural tripeptide that is synthesized by γ-glutamylcysteine synthetase and glutathione synthetase.[130,132] In a reduced form GSH is a strong antioxidant and takes a role in the reduction of ROS, and it is oxidized to its disulfide form (GSSG). Later, glutathione reductase enzyme regenerates GSH. There is a balance between GSSG and GSH in healthy cells, which is critical for the maintenance of redox balance. Intracellular GSH level is in the range of millimolar concentrations.[133–135] In a cancer cell, the GSH level is elevated up in order to decrease the oxidative stress initiated by high levels of ROS formation.[136–138] Additionally, detailed studies have uncovered that cancer-promoting transcriptional factors, such as MYC, stimulate glutamine uptake for GSH synthesis.[139]

Akkaya *et al.* reported a GSH-activatable PS based on their heavy atom-free orthogonal BODIPY dimer (Figure 15.12).[140] In a regular setting, symmetric orthogonal BODIPY dimers exhibit a unique doubly substituted tetra radical (DS-TR) singlet excited state, which allows facile ISC crossing to generate $^1O_2$ efficiently. However, the DS-TR state can be broken when asymmetrical BODIPYs are used to build up a dimer structure. In this direction, they modified one of the BODIPY cores on the dimer with a methylpyridinium unit

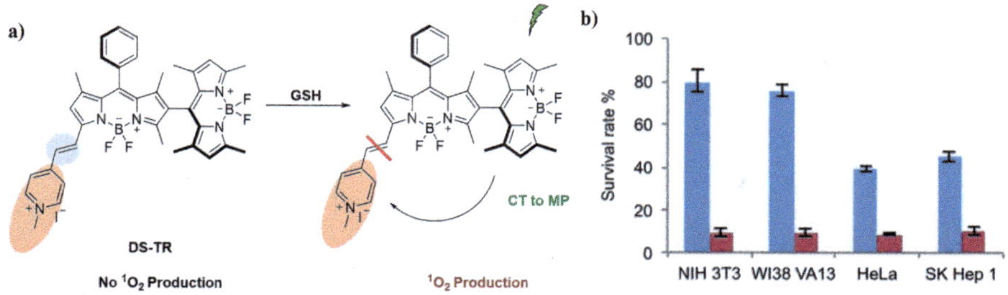

Figure 15.12 (a) Activation mechanism of dimer-BODIPY with GSH. (b) Phototoxic effect of dimer-BODIPY on different cells upon light irradiation. Adapted from ref. 140 with permission from John Wiley & Sons, Copyright © 2015 WILEY-VCH Verlag GmbH & Co. KGaA, Weinheim.

through a styryl bond, and the DS-TR state disappeared, which stopped $^1O_2$ generation. Upon addition of the nucleophilic cysteine unit of GSH with the styryl unit the π-conjugation was broken and isolated the methylpyridinium from the dimer. Irradiation of the GSH-added dimer with a green light triggered charge transfer from the orthogonally separated BODIPY unit and methylpyridinium moiety and initiated charge transfer-mediated ISC (Figure 15.12). Consequently, $^1O_2$ generation and related photocytotoxicity were activated. The PS induced PDT action in HeLa and SK Hep1 cancer cells due to high GSH concentration, while cell viability remained relatively high in normal cell lines (NIH-3T3 and WI38 VA13). The activated PS also enabled *in vitro* fluorescence imaging of cancer cells, indicating the theranostic nature of the agent.

In 2020, Kim, Cao and Lan introduced an AIE-based GSH-activatable PS (TPEPY-S-Fc) with the aim of delivering image-guided PDT.[141] TPEPY-S-Fc bears a GSH-activatable disulfide bond, which is located between ferrocene- and pyridinium-substituted TPE (Scheme 15.2). Fluorescence was quenched as the ferrocene unit activates the PeT process. After addition of GSH, the disulfide bond was selectively cleaved, and PeT was blocked. Therefore, the fluorescence and singlet oxygen generation ability of the remaining TPEPY-SH species were switched on. Red shift in the absorption spectrum around 600 to 620 nm, and increase in fluorescence intensity over time, were attributed to cleavage of the disulfide bond in the presence of GSH. A coherent relation between fluorescence intensity and GSH concentration was found when the GSH concentration was in the range 0–200 mM. Cleavage of the disulfide bond was satisfied by GSH, since it is the most abundant in cells, even though the probe itself was also sensitive to Cys and Hcy. Fluorescence intensity was enhanced in the pH range from 6 to 8, which is desirable considering physiological pH conditions. The singlet oxygen production capacity of the PS was evaluated with a singlet oxygen sensor green (SOSG). The $^1O_2$ quantum yield was measured as 2.4% when GSH was absent. However, it increased to 21.3% when the PS was activated. Strong fluorescence intensity was observed in TPEPY-S-Fc-incubated CT-26 cells, while it remained weak in

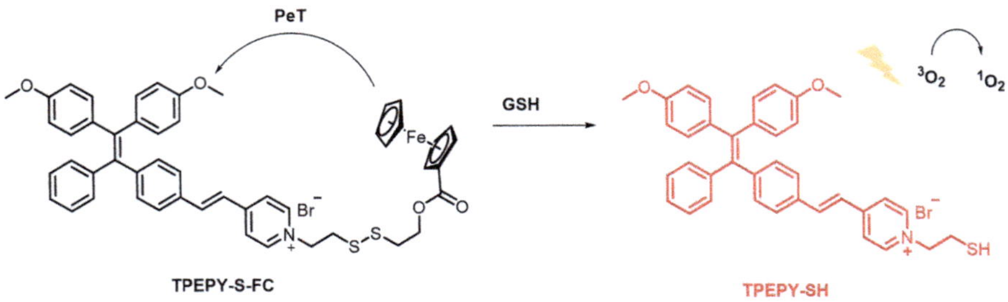

Scheme 15.2 Activation mechanism of TPEPY-S-FC in the presence of GSH.

the *N*-ethylmaleimide (NEM)-treated, a thiol scavenger, group of CT-26 cells. Additionally, mitotracker green dye revealed that the PS was accumulated in mitochondria due to the cationic pyridinium group. No notable cytotoxicity of TPEPY-S-Fc was observed in the dark, yet dose-dependent cytotoxic activity was observed upon light irradiation in CT-26 cells. Even with 10 min of light irradiation, the cell death was increased by 90% in the presence of 25 mM TPEPY-S-Fc.

15.3.5 Dual Analyte-responsive aPSs

Activatable photosensitizers that are responsive to more than one analyte have attracted enormous attention during the last decade as they can further boost the selectivity of the PS compared to single input activatable PDT agents.[142–144] In the design of such aPSs, two of the biological activators (enzymes, pH, thiols and ROS) are combined in most cases, and additional selectivity is satisfied as the PS switches on its photocytotoxicity only in the presence of the two inputs, mimicking the AND logic gate operation.

Torres *et al.* presented a pH- and thiol-responsive silicon(IV) phthalocyanine (PC)-based aPS.[145] The PC core was modified with a pyrene fluorophore through a thiol-cleavable disulfide linker. On the other side, an amino group with two methoxy(triethylenoxy) axial chains was placed, as a pH-sensitive unit. The amino group induced PeT to Si-PC and quenched the ISC path. Additionally, pyrene emission was quenched due to effective energy transfer to the PC scaffold. Under reducing conditions, which were satisfied by addition of dithiothreitol (DTT), the disulfide linker was broken and consequently the energy transfer was stopped. At slightly acidic pH, which is another tumor-associated condition, protonation of the amino group reactivated the ISC and $^1O_2$ generation. In addition to therapeutic outcomes, in the presence of two inputs, both fluorescence of pyrene at 377 nm and the emission signal of PC at 683 nm were restored. Thus, under reducing conditions and a slightly acidic environment, the PS was activated. Another Si-PC-based pH- and thiol-activatable PS was reported by the Ng group, in which the PS core was linked to two quencher ferrocene units through an acid-sensitive hydrazone and thiol-cleavable disulfide linker.[142] Removal of the ferrocene units under slightly acidic conditions with biothiols generated highly photocytotoxic Si-PC, which possessed selective PDT action towards cancer cells.

Wang and co-workers designed a GSH and H_2O_2 dual-activatable benzothiadiazole derivative (HO-NB-Bpin) as a highly selective PDT agent.[146] The thiol group on the PS was protected with an H_2O_2-responsive boronate ester cage unit. This modification not only implemented an additional activation site, but also blocked the reaction of O-NB-Bpin with GSH. Upon removal of the cage group, after addition of H_2O_2 and following 1,6-elimination, HO-NB-SH was generated, which was then converted to cytotoxic and emissive HO–NH_2ArSO_3H *via* H_2O_2 and GSH. HO–NH_2ArSO_3H could induce both type I and type II PDT action as detected by chemical-trap experiments (Figure 15.13). In the same study, a mitochondria-targeted version of HO-NB-Bpin, namely ^{mito}aPS, was also synthesized and tested in HeLa cells (Figure 15.13). A dose-dependent cell death was observed under both normoxic and hypoxic conditions upon light irradiation (Figure 15.13). The success

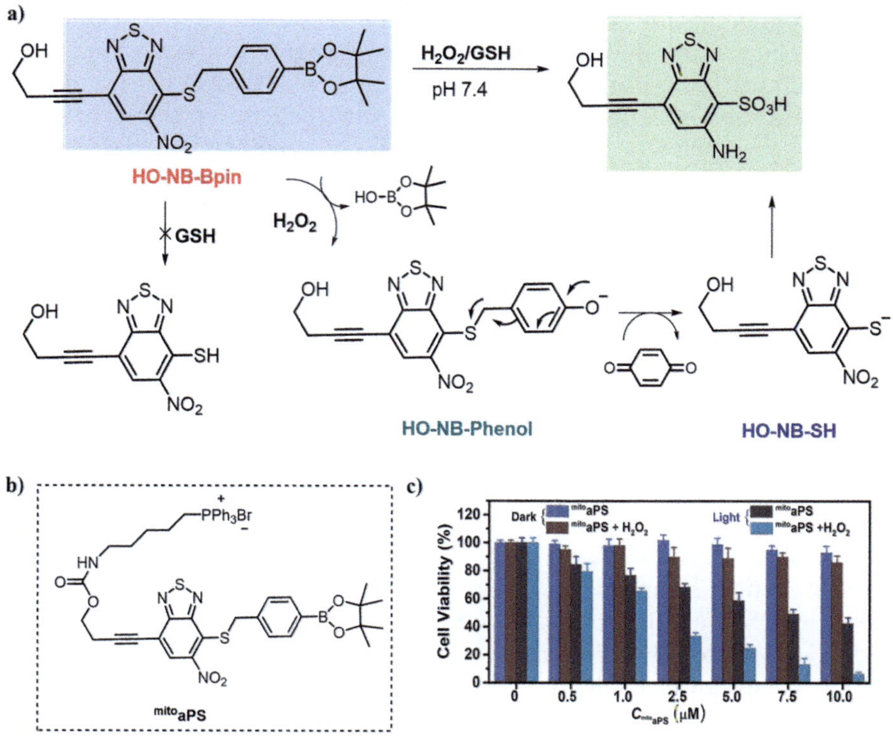

Figure 15.13 (a) Schematic illustration of H_2O_2/GSH-activatable HO-NB-Bpin in the presence of dual analytes: H_2O_2 and GSH. (b) Molecular structure of mitochondria-targeted ^{mito}aPS. (c) Cell viabilities of ^{mito}aPS-treated HeLa cells upon LED light irradiation. Adapted from ref. 146 with permission from John Wiley & Sons, Copyright © 2020 Wiley-VCH Verlag GmbH & Co. KGaA, Weinheim.

of the agent under hypoxia was mostly attributed to its capacity for triggering oxygen-independent type 1 PDT action. ^{mito}aPS can also be used in the scope of two-photon PDT, which is highly beneficial for further biological investigations.

15.4 Conclusion and Future Outlook

The field of photodynamic therapy is growing at a very fast rate. There is no doubt that PSs are at the core of PDT action, and invention of new PS cores is critical for widening the useful PS arsenal. However, current trends in the design of new-generation PSs is forcing researchers to come up with single molecules that can address all the chronic problems of this treatment modality, such as limited light-penetration depth, low efficacy under hypoxia, as well as lack of cancer cell selectivity. Additionally, advancements in tools to modulate excited-state dynamics enable the development of theranostic agents with two or more modes of therapeutic outcome. Thus, more attention should be paid to construct "all-in-one" PSs. There has been an incredible improvement in solving the issues of hypoxia and selectivity during the last few decades; however, the external light irradiation is still a big challenge, restricting the broader acceptance of PDT in clinical practice even in the case of NIR-absorbing PSs. We anticipate that research aiming to design tumor-selective "smart" PSs that do not need any external light irradiation to generate ROS or RNS will be highly popular in the coming years. It is also clear from the current perspective that detailed *in vivo* experiments focusing on the pharmacokinetics and pharmacodynamics of photosensitizers are needed to speed up the clinical translation of promising PDT agents.

List of Abbreviations

PDT	photodynamic therapy
PTT	photothermal therapy
ROS	reactive oxygen species
RNS	reactive nitrogen species
PS	photosensitizer
NIR	near-infrared
aPS	activatable photosensitizer
PeT	photoinduced electron transfer
ICT	intramolecular charge transfer
FRET	Förster-type resonance energy transfer
ISC	intersystem crossing
D–π-A	donor–π-acceptor
AIE	aggregation-induced emission
ACQ	aggregation-caused quenching
β-gal	β-galactosidase
NTR	nitroreductase

hNQO1	human NAD(P)H:quinone oxidoreductase 1
ALP	alkaline phosphatase
APN	aminopeptidase N
MAO	monoamino oxidase
TYR	tyrosinase
GSH	glutathione
PC	phthalocyanine
NO	nitric oxide
HNO	nitroxyl
RIR	restricted intramolecular rotation
RIV	restriction of intramolecular vibration
DCFDA	2′,7′-dichlorodihydrofluorescein diacetate
DPBF	1,3-diphenylisobenzo-furan
BODIPY	boron dipyrromethene
SOSG	singlet oxygen sensor green

Acknowledgements

S.K. thanks Koç University and BAGEP Award of the Science Academy for financial support. Figure 15.2 was created by BioRender.com.

References

1. D. E. Dolmans, D. Fukumura and R. K. Jain, *Nat. Rev. Cancer*, 2003, **3**, 380–387.
2. A. P. Castano, P. Mroz and M. R. Hamblin, *Nat. Rev. Cancer*, 2006, **6**, 535–545.
3. X. Li, S. Lee and J. Yoon, *Chem. Soc. Rev.*, 2018, **47**, 1174–1188.
4. A. A. Beharry, *Biochemistry*, 2018, **57**, 173–174.
5. J. P. Celli, B. Q. Spring, I. Rizvi, C. L. Evans, K. S. Samkoe, S. Verma, B. W. Pogue and T. Hasan, *Chem. Rev.*, 2010, **110**, 2795–2838.
6. P. Agostinis, K. Berg, K. A. Cengel, T. H. Foster, A. W. Girotti, S. O. Gollnick, S. M. Hahn, M. R. Hamblin, A. Juzeniene and D. Kessel, *Ca-Cancer J. Clin.*, 2011, **61**, 250–281.
7. T. J. Dougherty, C. J. Gomer, B. W. Henderson, G. Jori, D. Kessel, M. Korbelik, J. Moan and Q. Peng, *J. Natl. Cancer Inst.*, 1998, **90**, 889–905.
8. J.-M. Aubry, C. Pierlot, J. Rigaudy and R. Schmidt, *Acc. Chem. Res.*, 2003, **36**, 668–675.
9. W. Fan, P. Huang and X. Chen, *Chem. Soc. Rev.*, 2016, **45**, 6488–6519.
10. P. Di Mascio, G. R. Martinez, S. Miyamoto, G. E. Ronsein, M. H. Medeiros and J. Cadet, *Chem. Rev.*, 2019, **119**, 2043–2086.
11. R. Baskaran, J. Lee and S.-G. Yang, *Biomater. Res.*, 2018, **22**, 1–8.
12. H. von Tappeiner and A. Jodlbauer, *Die sensibilisierende Wirkung fluorescierender Substanzen: gesammelte Untersuchungen über die photodynamische Erscheinung*, Vogel, 1907.

13. R. R. Allison, *Future Oncol.*, 2014, **10**, 123–124.
14. P. R. Ogilby, *Photochem. Photobiol. Sci.*, 2010, **9**, 1543–1560.
15. M. J. Garland, C. M. Cassidy, D. Woolfson and R. F. Donnelly, *Future Med. Chem.*, 2009, **1**, 667–691.
16. S. Stolik, J. Delgado, A. Perez and L. Anasagasti, *J. Photochem. Photobiol., B*, 2000, **57**, 90–93.
17. G. Shafirstein, D. Bellnier, E. Oakley, S. Hamilton, M. Potasek, K. Beeson and E. Parilov, *Cancers*, 2017, **9**, 12.
18. J. Zhao, W. Wu, J. Sun and S. Guo, *Chem. Soc. Rev.*, 2013, **42**, 5323–5351.
19. K. Deng, C. Li, S. Huang, B. Xing, D. Jin, Q. Zeng, Z. Hou and J. Lin, *Small*, 2017, **13**, 1702299.
20. J. He, Y. Wang, M. A. Missinato, E. Onuoha, L. A. Perkins, S. C. Watkins, C. M. St Croix, M. Tsang and M. P. Bruchez, *Nat. Methods*, 2016, **13**, 263–268.
21. N. M. Idris, M. K. Gnanasammandhan, J. Zhang, P. C. Ho, R. Mahendran and Y. Zhang, *Nat. Med.*, 2012, **18**, 1580–1585.
22. H. Yuan, H. Chong, B. Wang, C. Zhu, L. Liu, Q. Yang, F. Lv and S. Wang, *J. Am. Chem. Soc.*, 2012, **134**, 13184–13187.
23. Y. R. Kim, S. Kim, J. W. Choi, S. Y. Choi, S.-H. Lee, H. Kim, S. K. Hahn, G. Y. Koh and S. H. Yun, *Theranostics*, 2015, **5**, 805.
24. C.-Y. Hsu, C.-W. Chen, H.-P. Yu, Y.-F. Lin and P.-S. Lai, *Biomaterials*, 2013, **34**, 1204–1212.
25. L. S. Lin, J. Song, L. Song, K. Ke, Y. Liu, Z. Zhou, Z. Shen, J. Li, Z. Yang and W. Tang, *Angew. Chem.*, 2018, **130**, 4902–4906.
26. Z. Zhou, J. Song, R. Tian, Z. Yang, G. Yu, L. Lin, G. Zhang, W. Fan, F. Zhang and G. Niu, *Angew. Chem., Int. Ed.*, 2017, **56**, 6492–6496.
27. W. R. Wilson and M. P. Hay, *Nat. Rev. Cancer*, 2011, **11**, 393–410.
28. D. M. Gilkes, G. L. Semenza and D. Wirtz, *Nat. Rev. Cancer*, 2014, **14**, 430–439.
29. H.-I. Hung, O. J. Klein, S. W. Peterson, S. R. Rokosh, S. Osseiran, N. H. Nowell and C. L. Evans, *Sci. Rep.*, 2016, **6**, 1–13.
30. B. W. Henderson and T. J. Dougherty, *Photochem. Photobiol.*, 1992, **55**, 145–157.
31. I. J. Macdonald and T. J. Dougherty, *J. Porphyrins Phthalocyanines*, 2001, **5**, 105–129.
32. N. Shirasu, S. O. Nam and M. Kuroki, *Anticancer Res.*, 2013, **33**, 2823–2831.
33. S. T. Laughlin and C. R. Bertozzi, *Nat. Protoc.*, 2007, **2**, 2930–2944.
34. N. L. Oleinick and H. H. Evans, *Radiat. Res.*, 1998, **150**, S146–S156.
35. H. Li, H. Kim, J. Han, V. N. Nguyen, X. Peng and J. Yoon, *Aggregate*, 2021, e51.
36. J. F. Lovell, T. W. Liu, J. Chen and G. Zheng, *Chem. Soc. Rev.*, 2010, **110**, 2839–2857.
37. M. A. Filatov, *Org. Biomol. Chem.*, 2020, **18**, 10–27.
38. D. R. Prudhomme, Z. Wang and C. J. Rizzo, *J. Org. Chem.*, 1997, **62**, 8257–8260.

39. W. Hu, M. Liu, X.-F. Zhang, M. Shi, M. Jia, X. Hu, L. Liu and T. Wang, *J. Phys. Chem. C*, 2020, **124**, 23558–23566.
40. W. Zhai, Y. Zhang, M. Liu, H. Zhang, J. Zhang and C. Li, *Angew. Chem., Int. Ed.*, 2019, **58**, 16601–16609.
41. X. Wang, P. Li, Q. Ding, C. Wu, W. Zhang and B. Tang, *J. Am. Chem. Soc.*, 2019, **141**, 2061–2068.
42. X. Wu, X. Sun, Z. Guo, J. Tang, Y. Shen, T. D. James, H. Tian and W. Zhu, *J. Am. Chem. Soc.*, 2014, **136**, 3579–3588.
43. K. Gu, Y. Xu, H. Li, Z. Guo, S. Zhu, S. Zhu, P. Shi, T. D. James, H. Tian and W.-H. Zhu, *J. Am. Chem. Soc.*, 2016, **138**, 5334–5340.
44. W. Sun, J. Fan, C. Hu, J. Cao, H. Zhang, X. Xiong, J. Wang, S. Cui, S. Sun and X. Peng, *Chem. Commun.*, 2013, **49**, 3890–3892.
45. A. Atilgan, E. Tanriverdi Eçik, R. Guliyev, T. B. Uyar, S. Erbas-Cakmak and E. U. Akkaya, *Angew. Chem., Int. Ed.*, 2014, **53**, 10678–10681.
46. J. Mei, Y. Hong, J. W. Lam, A. Qin, Y. Tang and B. Z. Tang, *Adv. Mater.*, 2014, **26**, 5429–5479.
47. Y. Hong, J. W. Lam and B. Z. Tang, *Chem. Commun.*, 2009, 4332–4353.
48. J. Chen, C. C. Law, J. W. Lam, Y. Dong, S. M. Lo, I. D. Williams, D. Zhu and B. Z. Tang, *Chem. Mater.*, 2003, **15**, 1535–1546.
49. D. Ding, K. Li, B. Liu and B. Z. Tang, *Acc. Chem. Res.*, 2013, **46**, 2441–2453.
50. S. Liu, G. Feng, B. Z. Tang and B. Liu, *Chem. Sci.*, 2021, **12**, 6488–6506.
51. W. Wu, D. Mao, F. Hu, S. Xu, C. Chen, C. J. Zhang, X. Cheng, Y. Yuan, D. Ding and D. Kong, *Adv. Mater.*, 2017, **29**, 1700548.
52. M. H. Baig, M. Adil, R. Khan, S. Dhadi, K. Ahmad, G. Rabbani, T. Bashir, M. A. Imran, F. M. Husain, E. J. Lee, M. A. Kamal and I. Choi, *Semin. Cancer Biol.*, 2019, **56**, 1–11.
53. S. S. Asraf and P. Gunasekaran, *Current Research, Technology and Education Topics in Applied Microbiology and Microbial Biotechnology*, Microbiology book series Formatex Research Center, Spain, 2010, pp. 880–890.
54. D. H. Juers, B. W. Matthews and R. E. Huber, *Protein Sci.*, 2012, **21**, 1792–1807.
55. O. Ramos and F. Malcata, *Compr. Biotechnol.*, 2011, 555–569.
56. D. Asanuma, M. Sakabe, M. Kamiya, K. Yamamoto, J. Hiratake, M. Ogawa, N. Kosaka, P. L. Choyke, T. Nagano and H. Kobayashi, *Nat. Commun.*, 2015, **6**, 1–7.
57. Y. Ichikawa, M. Kamiya, F. Obata, M. Miura, T. Terai, T. Komatsu, T. Ueno, K. Hanaoka, T. Nagano and Y. Urano, *Angew. Chem., Int. Ed.*, 2014, **53**, 6772–6775.
58. X. Zhen, J. Zhang, J. Huang, C. Xie, Q. Miao and K. Pu, *Angew. Chem., Int. Ed.*, 2018, **57**, 7804–7808.
59. Z. Liu, F. Song, W. Shi, G. Gurzadyan, H. Yin, B. Song, R. Liang and X. Peng, *ACS Appl. Mater. Interfaces*, 2019, **11**, 15426–15435.
60. X. You, L. Li, X. Li, H. Ma, G. Zhang and D. Zhang, *Chem. – Asian J.*, 2016, **11**, 2918–2923.

61. J. Sun, Z. Hu, R. Wang, S. Zhang and X. Zhang, *Anal. Chem.*, 2018, **91**, 1384–1390.
62. I. M. de Oliveira, J. A. P. Henriques and D. Bonatto, *Biochem. Biophys. Res. Commun.*, 2007, **355**, 919–925.
63. J. Yuan, Y.-Q. Xu, N.-N. Zhou, R. Wang, X.-H. Qian and Y.-F. Xu, *RSC Adv.*, 2014, **4**, 56207–56210.
64. Z. Liu, F. Song, W. Shi, G. Gurzadyan, H. Yin, B. Song, R. Liang and X. Peng, *ACS Appl. Mater. Interfaces*, 2019, **11**, 15426–15435.
65. F. Xu, H. Li, Q. Yao, H. Ge, J. Fan, W. Sun, J. Wang and X. Peng, *Chem. Sci.*, 2019, **10**, 10586–10594.
66. Y. Han, J. Chen, Z. Li, H. Chen and H. Qiu, *Biosens. Bioelectron.*, 2020, **148**, 111811.
67. S. Liu, S. Pang, W. Na and X. Su, *Biosens. Bioelectron.*, 2014, **55**, 249–254.
68. U. Sharma, D. Pal and R. Prasad, *Indian J. Clin. Biochem.*, 2014, **29**, 269–278.
69. W. G. Cox and V. L. Singer, *J. Histochem. Cytochem.*, 1999, **47**, 1443–1455.
70. M. B. Youdim, D. Edmondson and K. F. Tipton, *Nat. Rev. Neurosci.*, 2006, **7**, 295–309.
71. K. Chen and J. C. Shih, in *Advances in Pharmacology*, Elsevier, 1997, vol. 42, pp. 292–296.
72. K. Matsumoto, Y. Akao, H. Yi, M. Shamoto-Nagai, W. Maruyama and M. Naoi, *J. Neural Transm.*, 2006, **113**, 125–135.
73. F. Liu, L. Hu, Y. Ma, B. Huang, Z. Xiu, P. Zhang, K. Zhou and X. Tang, *Oncol. Lett.*, 2018, **15**, 3245–3251.
74. A. Ugun-Klusek, T. S. Theodosi, J. C. Fitzgerald, F. Burté, C. Ufer, D. J. Boocock, P. Yu-Wai-Man, L. Bedford and E. E. Billett, *Redox Biol.*, 2019, **20**, 167–181.
75. Y.-C. Lin, Y.-T. Chang, M. Campbell, T.-P. Lin, C.-C. Pan, H.-C. Lee, J. C. Shih and P.-C. Chang, *Sci. Rep.*, 2017, **7**, 1–17.
76. T. Almammadov, G. Atakan, O. Leylek, G. Ozcan, G. Gunbas and S. Kolemen, *ACS Med. Chem. Lett.*, 2020, **11**, 2491–2496.
77. C. M. Kumar, U. Sathisha, S. Dharmesh, A. A. Rao and S. A. Singh, *Biochimie*, 2011, **93**, 562–569.
78. T.-S. Chang, *Int. J. Mol. Sci.*, 2009, **10**, 2440–2475.
79. L. Chin, L. A. Garraway and D. E. Fisher, *Genes Dev.*, 2006, **20**, 2149–2182.
80. Melanoma Survival Statistics, https://www.cancer.org/cancer/melanoma-skin-cancer/detection-diagnosis-staging/survival-rates-for-melanoma-skin-cancer-by-stage.html, accessed August 2021.
81. A. Korner and J. Pawelek, *Science*, 1982, **217**, 1163–1165.
82. Z. Li, Y.-F. Wang, C. Zeng, L. Hu and X.-J. Liang, *Anal. Chem.*, 2018, **90**, 3666–3669.
83. M. Saini, A. Kashyap, S. Bindal, K. Saini and R. Gupta, *Front. Microbiol.*, 2021, **12**, 441.
84. M. H. Hanigan and W. A. Ricketts, *Biochemistry*, 1993, **32**, 6302–6306.
85. A. Pompella, V. De Tata, A. Paolicchi and F. Zunino, *Biochem. Pharmacol.*, 2006, **71**, 231–238.

86. S. Daubeuf, P. Leroy, A. Paolicchi, A. Pompella, M. Wellman, M. M. Galteau and A. Visvikis, *Biochem. Pharmacol.*, 2002, **64**, 207–216.
87. B. Shi, Z. Zhang, Q. Jin, Z. Wang, J. Tang, G. Xu, T. Zhu, X. Gong, X. Tang and C. Zhao, *J. Mater. Chem. B*, 2018, **6**, 7439–7443.
88. M. H. Hanigan, H. F. Frierson, J. E. Brown, M. A. Lovell and P. T. Taylor, *Cancer Res.*, 1994, **54**, 286–290.
89. M. Tsutsumi, D. Sakamuro, A. Takada, S. Zang, T. Furukawa and N. Taniguchi, *Hepatology*, 1996, **23**, 1093–1097.
90. M. Chiba, Y. Ichikawa, M. Kamiya, T. Komatsu, T. Ueno, K. Hanaoka, T. Nagano, N. Lange and Y. Urano, *Angew. Chem., Int. Ed.*, 2017, **56**, 10418–10422.
91. Z. Yang, W. Xu, J. Wang, L. Liu, Y. Chu, Y. Wang, Y. Hu, T. Yi and J. Hua, *J. Mater. Chem. C*, 2020, **8**, 8183–8190.
92. B. F. Sloane, S. Yan, I. Podgorski, B. E. Linebaugh, M. L. Cher, J. Mai, D. Cavallo-Medved, M. Sameni, J. Dosescu and K. Moin, *Semin. Cancer Biol.*, 2005, **15**, 149–157.
93. C. S. Gondi and J. S. Rao, *Expert Opin. Ther. Targets*, 2013, **17**, 281–291.
94. X. Chen, D. Lee, S. Yu, G. Kim, S. Lee, Y. Cho, H. Jeong, K. T. Nam and J. Yoon, *Biomaterials*, 2017, **122**, 130–140.
95. Y. Yuan, C. J. Zhang, M. Gao, R. Zhang, B. Z. Tang and B. Liu, *Angew. Chem., Int. Ed.*, 2015, **54**, 1780–1786.
96. J. S. Shim, J. H. Kim, H. Y. Cho, Y. N. Yum, S. H. Kim, H.-J. Park, B. S. Shim, S. H. Choi and H. J. Kwon, *Chem. Biol.*, 2003, **10**, 695–704.
97. S.-X. Cui, X.-J. Qu, Z.-H. Gao, Y.-S. Zhang, X.-F. Zhang, C.-R. Zhao, W.-F. Xu, Q.-B. Li and J.-X. Han, *Cancer Lett.*, 2010, **292**, 153–162.
98. L. Chen, Y.-L. Lin, G. Peng and F. Li, *Proc. Natl. Acad. Sci. U. S. A.*, 2012, **109**, 17966–17971.
99. Y. Saitoh, K. Koizumi, T. Minami, K. Sekine, H. Sakurai and I. Saiki, *Biol. Pharm. Bull.*, 2006, **29**, 709–712.
100. Y. Aozuka, K. Koizumi, Y. Saitoh, Y. Ueda, H. Sakurai and I. Saiki, *Cancer Lett.*, 2004, **216**, 35–42.
101. R. Hata, H. Nonaka, Y. Takakusagi, K. Ichikawa and S. Sando, *Angew. Chem., Int. Ed.*, 2016, **55**, 1765–1768.
102. M. Terauchi, H. Kajiyama, K. Shibata, K. Ino, A. Nawa, S. Mizutani and F. Kikkawa, *BMC Cancer*, 2007, **7**, 1–12.
103. X. Zhou, H. Li, C. Shi, F. Xu, Z. Zhang, Q. Yao, H. Ma, W. Sun, K. Shao, J. Du, S. Long, J. Fan, J. Wang and X. Peng, *Biomaterials*, 2020, **253**, 120089.
104. A. L. Pey, C. F. Megarity and D. J. Timson, *Biosci. Rep.*, 2019, **39**, BSR20180459.
105. K. Zhang, D. Chen, K. Ma, X. Wu, H. Hao and S. Jiang, *J. Med. Chem.*, 2018, **61**, 6983–7003.
106. Y. Dong, E. A. Bey, L.-S. Li, W. Kabbani, J. Yan, X.-J. Xie, J.-T. Hsieh, J. Gao and D. A. Boothman, *Cancer Res.*, 2010, **70**, 8088–8096.
107. M. Ough, A. Lewis, E. A. Bey, J. Gao, J. M. Ritchie, W. Bornmann, D. A. Boothman, L. W. Oberley and J. J. Cullen, *Cancer Biol. Ther.*, 2005, **4**, 102–109.

108. A. Marin, A. L. De Cerain, E. Hamilton, A. Lewis, J. Martinez-Penuela, M. Idoate and J. Bello, *Br. J. Cancer*, 1997, **76**, 923–929.
109. E. M. Digby, O. Sadovski and A. A. Beharry, *Chem. – Eur. J.*, 2020, **26**, 2713–2718.
110. E. Boedtkjer and S. F. Pedersen, *Annu. Rev. Physiol.*, 2020, **82**, 103–126.
111. M. K. Jha, I.-K. Lee and K. Suk, *Neurosci. Biobehav. Rev.*, 2016, **68**, 1–19.
112. H. Xiong, K. Zhou, Y. Yan, J. B. Miller and D. J. Siegwart, *ACS Appl. Mater. Interfaces*, 2018, **10**, 16335–16343.
113. Q. Tang, W. Xiao, C. Huang, W. Si, J. Shao, W. Huang, P. Chen, Q. Zhang and X. Dong, *Chem. Mater.*, 2017, **29**, 5216–5224.
114. R. Li, Z. Jia and M. A. Trush, *React. Oxygen Species*, 2016, **1**, 9.
115. L. C. Gomes and L. Scorrano, *Biochim. Biophys. Acta, Mol. Cell Res.*, 2013, **1833**, 205–212.
116. R. J. Mailloux, *Redox Biol.*, 2015, **4**, 381–398.
117. R. J. Mailloux and M.-E. Harper, *Trends Endocrinol. Metab.*, 2012, **23**, 451–458.
118. S. Di Meo, T. T. Reed, P. Venditti and V. M. Victor, *Oxid. Med. Cell. Longevity*, 2016, **2016**, 1245049.
119. D. Ortiz de Orué Lucana, I. Wedderhoff and M. R. Groves, *J. Signal Transduction*, 2012, 2012.
120. M. C. Chang, A. Pralle, E. Y. Isacoff and C. J. Chang, *J. Am. Chem. Soc.*, 2004, **126**, 15392–15393.
121. T. Almammadov and S. Kolemen, *Dyes Pigm.*, 2021, 109499.
122. B. Yuan, H. Wang, J.-F. Xu and X. Zhang, *ACS Appl. Mater. Interfaces*, 2020, **12**, 26982–26990.
123. H.-W. Liu, X.-X. Hu, K. Li, Y. Liu, Q. Rong, L. Zhu, L. Yuan, F.-L. Qu, X.-B. Zhang and W. Tan, *Chem. Sci.*, 2017, **8**, 7689–7695.
124. R. Radi, *Proc. Natl. Acad. Sci. U. S. A.*, 2018, **115**, 5839–5848.
125. P. Pacher, J. S. Beckman and L. Liaudet, *Physiol. Rev.*, 2007, **87**, 315–424.
126. C. Parisi, M. Failla, A. Fraix, L. Menilli, F. Moret, E. Reddi, B. Rolando, F. Spyrakis, L. Lazzarato, R. Fruttero, A. Gasco and S. Sortino, *Chem. Sci.*, 2021, **12**, 4740–4746.
127. N. Tuteja, M. Chandra, R. Tuteja and M. K. Misra, *J. Biomed. Biotechnol.*, 2004, **2004**, 227–237.
128. W. Hu, M. Xie, H. Zhao, Y. Tang, S. Yao, T. He, C. Ye, Q. Wang, X. Lu and W. Huang, *Chem. Sci.*, 2018, **9**, 999–1005.
129. B. L. Tan, M. E. Norhaizan, W.-P.-P. Liew and H. Sulaiman Rahman, *Front. Pharmacol.*, 2018, **9**, 1162.
130. S. Lee, J. Li, X. Zhou, J. Yin and J. Yoon, *Coord. Chem. Rev.*, 2018, **366**, 29–68.
131. S. Wang, Y. Huang and X. Guan, *Molecules*, 2021, **26**, 3575.
132. P. Wexler, B. D. Anderson, S. C. Gad, P. B. Hakkinen, M. Kamrin, A. De Peyster, B. Locey, C. Pope, H. M. Mehendale and L. R. Shugart, *Encyclopedia of Toxicology*, Academic Press, 2005.
133. D. Montero, C. Tachibana, J. R. Winther and C. Appenzeller-Herzog, *Redox Biol.*, 2013, **1**, 508–513.

134. X. Guo, Y. Cheng, X. Zhao, Y. Luo, J. Chen and W.-E. Yuan, *J. Nanobiotechnol.*, 2018, **16**, 1–10.
135. M. Diehn, R. W. Cho, N. A. Lobo, T. Kalisky, M. J. Dorie, A. N. Kulp, D. Qian, J. S. Lam, L. E. Ailles and M. Wong, *Nature*, 2009, **458**, 780–783.
136. G.-Y. Liou and P. Storz, *Free Radical Res.*, 2010, **44**, 479–496.
137. B. Perillo, M. Di Donato, A. Pezone, E. Di Zazzo, P. Giovannelli, G. Galasso, G. Castoria and A. Migliaccio, *Exp. Mol. Med.*, 2020, **52**, 192–203.
138. L. Kennedy, J. K. Sandhu, M.-E. Harper and M. Cuperlovic-Culf, *Biomolecules*, 2020, **10**, 1429.
139. Y. Dong, R. Tu, H. Liu and G. Qing, *Signal Transduction Targeted Ther.*, 2020, **5**, 1–11.
140. S. Kolemen, M. Işık, G. M. Kim, D. Kim, H. Geng, M. Buyuktemiz, T. Karatas, X.-F. Zhang, Y. Dede, J. Yoon and E. U. Akkaya, *Angew. Chem., Int. Ed.*, 2015, **54**, 5340–5344.
141. Y.-H. Zhang, X. Li, L. Huang, H. S. Kim, J. An, M. Lan, Q.-Y. Cao and J. S. Kim, *Chem. Commun.*, 2020, **56**, 10317–10320.
142. J. T. Lau, P.-C. Lo, X.-J. Jiang, Q. Wang and D. K. Ng, *J. Med. Chem.*, 2014, **57**, 4088–4097.
143. Y. Yuan, S. Xu, C.-J. Zhang, R. Zhang and B. Liu, *J. Mater. Chem. B*, 2016, **4**, 169–176.
144. X. Tian, K.-H. Baek and I. Shin, *Chem. Sci.*, 2013, **4**, 947–956.
145. E. van de Winckel, R. J. Schneider, A. de la Escosura and T. Torres, *Chem. - Eur. J.*, 2015, **21**, 18551–18556.
146. J. Sun, K. Du, J. Diao, X. Cai, F. Feng and S. Wang, *Angew. Chem.*, 2020, **59**, 12122–12128.

CHAPTER 16

Fluorescent Platforms for Environmental Sensing

A. A. BOWYER[a] AND E. J. NEW*[a,b]

[a]School of Chemistry, The University of Sydney, NSW, 2006, Australia; [b]The University of Sydney Nano Institute (Sydney Nano), The University of Sydney, NSW 2006, Australia
*E-mail: elizabeth.new@sydney.edu.au

16.1 Chemical Pollutants in the Environment

Many substances, both natural and artificial, are of great environmental concern and therefore must be monitored carefully. This includes essential substances needed for a healthy ecosystem, as well as polluting substances that are not beneficial to the environment in any capacity. The vast majority of fluorescent sensing for environmental monitoring focuses on the detection of pollutants, although there has also been some work towards monitoring delicate balances of naturally occurring species, such as nitrogen/phosphorus nutrient overload observed in eutrophication.[1] Common classes of chemical pollutants include heavy metals, agrochemicals and industrial chemicals and waste products, which will be briefly described here.

Monographs in Supramolecular Chemistry No. 34
Fluorescent Chemosensors
Edited by Luling Wu, Adam C. Sedgwick, Xiao-Peng He and Tony D. James

Published by the Royal Society of Chemistry, www.rsc.org

16.1.1 Toxic Heavy Metals

Heavy metals are metallic elements that have high atomic weights and large densities, commonly classified as those that have densities greater than that of water.[2] They occur naturally in the Earth's crust, but discoveries of their technological, medical, industrial and agricultural applications have led to a high abundance throughout the environment. As a result, heavy metals are most prevalent in source areas, such as mines, smelters and metal-related industries,[3] and are present in solid, dissolved aqueous and gaseous states.

While some heavy metals, such as the first-row transition metals, are essential for life, others are extremely toxic. The toxic heavy metals of greatest human and environmental health concern include lead (Pb), cadmium (Cd), mercury (Hg) and the metalloid, arsenic (As).

In humans, exposure to these high-concern, toxic metals is related to conditions such as cancer, impaired motor and neuronal function and multiple organ damage, even in very small quantities.[4] The biological effect of these toxic heavy metals is reliant on several factors such as the chemical nature, dose, duration and route of exposure. Organic forms of heavy metals, such as organolead and organomercury, are significantly more toxic than the ionic salts of these metals. For example, the lowest lethal dose (LD_{Lo}) of tetraethyllead has been reported as 24 mg kg^{-1} (oral, rat),[5] while the lead nitrate salt has a much higher reported LD_{Lo} of 500 mg kg^{-1} (oral, guinea pig).[6] Despite these differences, any level of toxic heavy metal exposure is of great concern, warranting the need for reliable detection methods in the environment.

16.1.2 Agrochemicals

The two main classes of chemicals used in the agricultural industry are fertilisers and pesticides. Fertilisers can be defined as additives to enhance the productivity, or fertility, of a crop. Pesticides are substances used to destroy pests, such as insects, fungi, animals and plants, that are harmful to agricultural resources. Fertilisers focus on providing crops with three essential nutrients, nitrogen, phosphorus and potassium. As these nutrients all occur naturally in the environment, monitoring of fertilisers often concentrates more on determining nitrogen/phosphorus balances for crop vitality, rather than on detecting harmful chemicals.[7] However, it is important to note that run-off from fertiliser-enriched agricultural sites can have detrimental impacts on the surrounding natural environment.[8,9]

Although naturally derived pesticides exist, most of those used in the global agricultural industry are artificial, introducing foreign chemicals into farming areas and their surroundings. The three main classes of artificial pesticide are organochlorines, organophosphates (OPs) and carbamates, with OPs being the most widely used.[10] The chemical structures of some major pesticides from each category are shown in Figure 16.1. Despite their widespread use, OPs have detrimental impacts upon the environment and human health. Excess pesticides in soil cause acidification, a decrease in

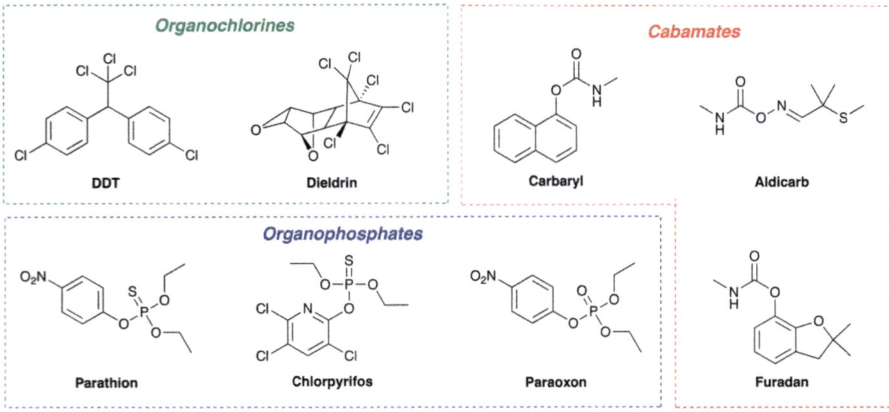

Figure 16.1 Chemical structures of some commonly used pesticides.

microbial biodiversity and water pollution.[10] OPs are largely toxic to humans *via* the inhibition of the essential cholinesterase enzymes.[11] These enzymes moderate central nervous system function, and overexposure to OPs can therefore cause a variety of symptoms, such as weakness, paralysis, respiratory and myocardial defects, and even death. As they can accumulate over time,[12] the presence of OPs in the natural environment and consumer goods, such as fresh produce, must be carefully monitored.

16.1.3 Industrial Chemicals and Waste Products

This final class of important environmental analytes spans multiple industries and chemical species, and includes direct waste products of industrial processes, as well as chemicals that can enter the environment through consumer product use. Examples include exhaust gases, volatile organic compounds, phenolic compounds and explosives.

Some common industrial gases of environmental concern are carbon dioxide (CO_2), carbon monoxide (CO), sulfur dioxide (SO_2), nitrogen dioxide (NO_2), ammonia (NH_3), hydrogen chloride (HCl) and formaldehyde (FA). The release of gaseous waste products into the atmosphere can cause a number of adverse consequences, such as enhancing the greenhouse effect, acidifying oceans and exposing society to toxins, depending on their chemical properties. For example, the burning of fossil fuels, such as coal and natural gas, releases elevated levels of CO_2, CO and nitrous oxides into the atmosphere.[13,14] Alternatively, the greatest source of ammonia emissions is from the agricultural industry, from processes such as fertiliser production and animal husbandry.[15]

Volatile organic compounds (VOCs) comprise a large set of low molecular weight organic chemicals that evaporate easily at room temperature.[16] These include small alkanes, alcohols, carbonyl compounds, ethers, aromatic hydrocarbons and acids. These can be introduced into the environment through a variety of industrial processes, such as textile and furniture manufacturing, burning solid fuels and motor vehicle manufacture and use. Prolonged exposure to VOCs has been linked to the development of chronic illnesses such as asthma, neurological damage, kidney failure and cancer.[17,18]

Phenolic compounds are a general class of small organic molecules that have a hydroxy group directly bonded to a benzene ring. Some common artificial phenolic compounds present in the environment include 2,4-dichlorophenol, 2,4,6-trichlorophenol, pentachlorophenol and bisphenol A (BPA) (Figure 16.2).[19] These molecules are extensively used in agricultural, plastic, pharmaceutical, petrochemical and textile industries in various capacities, but are also generated as unwanted by-products.[10] For example, 2,4-dichlorophenol is a widely used chemical intermediate for the industrial manufacture of miticides, antiseptics, seed disinfectants and wood preservatives,[20] while BPA is used in plastics manufacturing for the production of epoxy, polysulfone, polycarbonate and some polyester resins.[20] Phenolic compounds are generally not biodegradable and therefore accumulate in aquatic and terrestrial ecosystems. Additionally, many phenolic compounds are toxic. BPA is an endocrine disruptor,[21] and chlorophenols are genotoxic, mutagenic and carcinogenic.[22] Their toxicity and persistence in the environment mean that phenolic compounds must be monitored closely.

Residual explosive substances, particularly nitroaromatics and nitroamines, can contaminate the environment *via* soil penetration and ground water seepage from military sites.[23] The most prevalent explosive, 2,4,6-trinitrotoluene (TNT), is poisonous and can be absorbed through the skin, causing irritation.[24] Prolonged exposure has been linked to organ damage, and TNT is classified as a possible human carcinogen.[25,26]

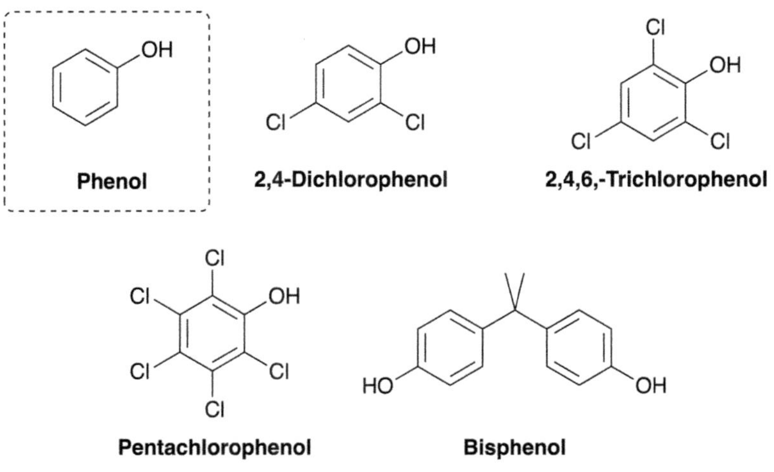

Figure 16.2 Chemical structures of some toxic phenolic compounds of environmental concern.

16.2 Analytical Methods of Environmental Monitoring

Environmental pollutants, such as toxic metals, agrochemicals and industrial chemicals, need to be detected and monitored effectively due to their detrimental effects on the natural environment and society. For effective monitoring of these pollutants, both the analysis matrix (or environment) and the analytical technique must be considered, both of which will be discussed in the following section.

16.2.1 The Scope of Environmental Sensing

The natural environment consists of the atmosphere, hydrosphere, lithosphere and biosphere. "Environmental sensing" is concerned with detecting analytes in the first three of these spheres, while "biological sensing" is only concerned with monitoring analytes in the biosphere. Each environmental domain has its own subclassifications. For example, the hydrosphere can be further divided into reservoirs, such as oceans, rivers, lakes, ground water and frozen water stores. The scope of environmental sensing is therefore very large and diverse, and this is one of the great challenges of environmental sensing.

Some important environments in which pollutants must be sensed are air, oceans, freshwater bodies and soil. Each of these environments has its own challenges for sensing, which will be discussed here briefly.

Air is a complex mixture of gases, liquid droplets and small solid particles.[27] Perhaps the most challenging aspect of sensing pollutants in the air is that gaseous substances are diffuse and so concentrations can fluctuate during measurement. As a result, guidelines and sensing protocols for air sensing are often given with a measure of time, such as 1 h or 24 h exposure increments, or less strictly defined chronic exposure.[13] Another challenge in gas sensing is the typically low analyte concentrations. Unlike liquid matrices, samples cannot be easily concentrated for signal amplification. Therefore, air sensors need to have suitably low levels of detection aligned with guideline values for air pollutants.

The main distinguishing feature of oceans, compared to other aqueous reservoirs, is a high concentration of dissolved salts. Ocean salinity is typically 35 g L^{-1} (600 mM),[28] compared to the micromolar concentrations of important analytes. This is a challenge for pollutant monitoring because high salt levels can interfere with some sensing platforms. In contrast, freshwater reservoirs are smaller than oceans and are therefore susceptible to greater fluctuation in conditions, such as temperature, dissolved oxygen,[29] pH[30] and salinity.[31,32] A greatly fluctuating analysis matrix is a challenge for environmental sensing, as changing conditions often require recalibration of sensing systems.

The development of novel environmental sensors for the lithosphere is less common in the scientific literature. This is because solid samples, such as soil and rock materials, are incompatible with some sensing techniques because they have limited water solubility or low light transmissibility/reflectivity.[33] Analysis of lithospheric samples often requires additional sample preparation, such as dissolution with harsh chemicals or the use of additives. For example, it is generally not possible to determine the pH of soil samples with universal indicator alone because the coloured indicator output will not be visible through the dark-coloured soil. As a result, it has been standard practice for many years to add white, chemically inert barium sulfate to soil samples to visualise the coloured indicator output.[34]

16.2.2 Analytical Techniques for Environmental Sensing

Each facet of the environment has its own challenges for sensing. Consequently, there is a vast array of analytical techniques currently used for sensing. Table 16.1 provides a summary of some of the most common techniques for environmental sensing.

16.2.3 Fluorescence as an Analytical Technique for the Environment

While most analytical techniques for environmental monitoring provide sensitive analyte detection in the parts per billion (ppb) range, many have limitations such as high cost and tedious sample preparation. An attractive alternative to the above techniques is fluorescent sensing. Fluorescent sensing is an optical technique that involves applying an excitation light source to a sample and recording output parameters of the emitted light, such as intensity and wavelength (colour). While some environmental analytes are inherently fluorescent, in most cases artificial fluorescent sensors need to be designed and added to the sample to interact with and report on the analyte of interest. Subsequent changes in the fluorescence properties of added sensors can then be measured to determine the concentration of an analyte of interest. Advantages of fluorescent sensing include high sensitivity (ppb), rapid response times, facile sample preparation, simple instrumentation and low costs.

Approaches to fluorescent sensing are extremely diverse, as sensing platforms can be easily tuned to a variety of environmental sensing applications. Some platforms, such as water-soluble dyes or nanoparticles (NPs),[41] are useful for detecting analytes in aqueous systems, while more sophisticated fluorescent nanomaterial films and solid substrates can be employed for gas sensing.[42] Another attractive feature of fluorescent sensing platforms is that they can be adapted to array-based sensing, allowing for multiple environmental analytes to be detected and discriminated from each other simultaneously.[43] Finally, recent developments in fluorescent sensing platforms include the development of portable devices for on-site testing.[44] Many devices only require simple excitation light sources such as LEDs, and fluorescence detection and analysis can be performed by smartphone cameras

Table 16.1 Summary of commonly used analytical techniques for environmental sensing.

Technique	Sample type	Analytes	Limit of detection	Details
Graphite furnace atomic absorption spectroscopy (GFAAS)	Liquid	Chemical elements	ppb[35]	Inexpensive Small sample volume (100 µL) No screening ability
Inductively coupled plasma–mass spectrometry (ICP–MS)	Liquid	Chemical elements	ppt[35]	Most commonly used for trace elements analysis across water, air, soil and plant analysis[35] Expensive Moderately small sample volume (>1 mL) Auto-sampling available Simultaneous analysis of multiple elements in sample Can differentiate isotopes Samples need to be completely dissolved, often with acid digestion
Anodic stripping voltammetry (ASV)	Liquid	Dissolved metals	ppb[35]	Inexpensive Robust to high levels of salinity
X-ray fluorescence (XRF)	Liquid and solid	Chemical elements	ppm[35]	Inexpensive Minimal sample preparation Applied to portable devices[36]
Inductively coupled plasma–optical emission spectroscopy (ICP–OES)	Liquid	Chemical elements	ppb[35]	Expensive Auto-sampling available Simultaneous analysis of multiple elements in sample Can differentiate isotopes Samples need to be completely dissolved, often with acid digestion
Gas chromatography–mass spectrometry (GC–MS)	Liquid and dissolved gases	Volatile organic compounds and gases	ppb[37]	Inexpensive Applied to portable devices[38]
Liquid chromatography–mass spectrometry (LC–MS)	Liquid	Polar molecules	ppb[39]	Readily modified for specific applications Ability for high throughput screening Inexpensive Tedious sample preparation
Fourier-transformed infrared spectroscopy (FTIR)	Gas and liquid	Organic functional groups	ppm[40]	Inexpensive Data can have low signal-to-noise ratios Highly susceptible to environmental interferences

and purpose-built smartphone applications, respectively. This is perhaps one of the most notable advantages of fluorescent sensing platforms over other analytical techniques, as techniques such as AAS and ICP–MS require large instrumentation that cannot be adapted for *in situ* analysis.

16.3 Considerations for Environmental Fluorescent Sensing Platforms

Environmental chemical pollutants span a broad range of chemical classes and therefore different strategies must be employed to create fluorescent sensors for each class. However, there are also criteria that are common to the design of sensing platforms, regardless of the analyte class. These are sensitivity, selectivity and compatibility to the sensing environment.

16.3.1 Sensitivity

Many environmental pollutants are harmful at very low concentrations. It is therefore important to be able to detect these pollutants at concentrations lower than specified dangerous levels. Table 16.2 outlines guideline thresholds for some important environmental pollutants. To achieve detection

Table 16.2 Recommended guideline values for safe exposure to some environmental pollutants and toxicants.

Pollutant class	Chemical pollutant	Environment	Guideline
Toxic metals	Lead (Pb)	Drinking water	0.01 mg L^{-1} [45]
	Mercury (Hg)	Drinking water	0.006 mg L^{-1} [45]
	Cadmium (Cd)	Drinking water	0.003 mg L^{-1} [45]
	Arsenic (As)	Drinking water	0.01 mg L^{-1} [45]
Herbicides	Pentachlorophenol	Drinking water	0.009 mg L^{-1} [45]
	2,4,5-Trichlorophenoxy-acetic acid	Drinking water	0.009 mg L^{-1} [45]
Insecticides	Dichlorodiphenyltrichloroethane (DDT)	Drinking water	0.001 mg L^{-1} [45]
Pesticides	Various organophosphates	Various fruits and vegetables	0.15–8.2 ppm[46]
Industrial gases	Carbon monoxide	Indoor air	7 mg m^{-3} (24 h)[13]
	Formaldehyde	Indoor air	0.1 mg m^{-3} (30 min)[13]
	Nitrogen dioxide	Indoor air	0.2 mg m^{-3} (1 h)[13]
	Hydrochloric acid	Air	1.8 ppm (1 h)[47]
	Hydrogen sulfide	Air	150 μg m^{-3} (24 h)[48]
	Hydrogen sulfide	Drinking water	0.05–0.1 mg L^{-1} [48]
	Sulfur dioxide	Air	20 μg m^{-3} [49]
Phenolic compounds	2,4,6-Trichlorophenol	Drinking water	0.2 mg L^{-1} [45]
Industrial chemicals	Cyanide	Drinking water	0.07 mg L^{-1} [50]

limits lower than guideline values, fluorescent sensing platforms must exhibit high binding constants and/or fast reaction kinetics accompanied by a significant change in fluorescence response.

16.3.2 Selectivity

It is essential that fluorescent sensing platforms do not respond to any interferents that may be present in the sample. Interferents can include naturally occurring chemical species in the environment or other polluting analytes of interest. Furthermore, it is also important to ensure that interferents will not decrease or mask the positive response of a target analyte. This is especially important for the development of sensors to be used in the natural environment, where more variation is expected than in biological systems. For example, it is acceptable for cellular sensors of essential metals to show some response to toxic metals, as toxic metals are not expected to be significantly present in biological systems. In contrast, an essential metal sensor for the environment should not show any response to toxic metals because they could be present in the analysis matrix.

16.3.3 Environmental Compatibility

Fluorescent sensing platforms can be used to detect pollutants in the air and water, and occasionally in solid substrates such as soil. In designing sensors, it is important to consider the nature of the target sensing environment. Sensors for gaseous pollutants in the air should be robust to oxygen, humidity and temperature fluctuations and ambient light, while sensors for aqueous environments should be water soluble and function in a broad range of pH environments. Furthermore, incompatibility between the sensor and the environment can result in sensor degradation and reduction of sensing reliability and efficiency. Environmental compatibility is often assessed by "spiking" environmental samples with a known concentration of the analyte, quantifying the analyte with the sensing platform, then calculating a "recovery value"—the percentage of the measured to actual concentration of the added analyte.

Besides the characteristics mentioned above, other useful features of fluorescent sensing platforms for the environment include low construction and operation costs, broad accessibility, portability and non-toxicity. It is important to consider this set of criteria in critically evaluating past and future sensors for the environment.

16.4 Examples of Fluorescent Platforms for Environmental Sensing

Fluorescence is a widespread phenomenon, exhibited by a range of molecular substances and materials. These include small-molecule organic dyes,[51,52] oligonucleotides,[53] polymers,[54] gels,[55] peptides,[56] proteins,[57] supramolecules,[58] metal complexes,[59] nanoparticles (NPs)[60,61] and nanoscale materials,[42] and metal–organic frameworks (MOFs).[62] The favourable and tuneable fluorescence properties of these platforms has led to the development of an abundance of sensing systems for the environment. The following section will provide key examples of environmental sensing systems based on different fluorescent platforms, giving emphasis to examples that have successfully detected analytes in real-world environmental samples.

16.4.1 Small-molecule Organic Dyes

Fluorescent chemosensors based on small organic molecules generally contain an organic fluorophore tethered to a sensing group (Figure 16.3). Interaction of the sensor with the analyte of interest can lead to a fluorescence change by one of a broad array of mechanisms.[51] Some probes are reaction based, while others utilise a receptor group to bind analytes (Figure 16.3). Such systems have been extensively reported in the literature for environmental sensing applications owing to their many favourable qualities, including facile tuneability and high brightness. The number of small-molecule sensors developed for environmental analysis is vast; here we present a few representative examples.

Metal ions are amongst the most common analytes of small-molecule probes in the literature. Common receptor groups for metal ions include crown ethers or aza-crown ethers, aromatic amines, thioethers and carboxylic acid derivatives (Figure 16.4).[41,63] Metal selectivity can be tuned by choice of heteroatoms in the receptor group.

Many toxic metal analytes produce a quenching response to fluorescent dyes due to the heavy atom effect.[64] A common approach for turn-on sensing

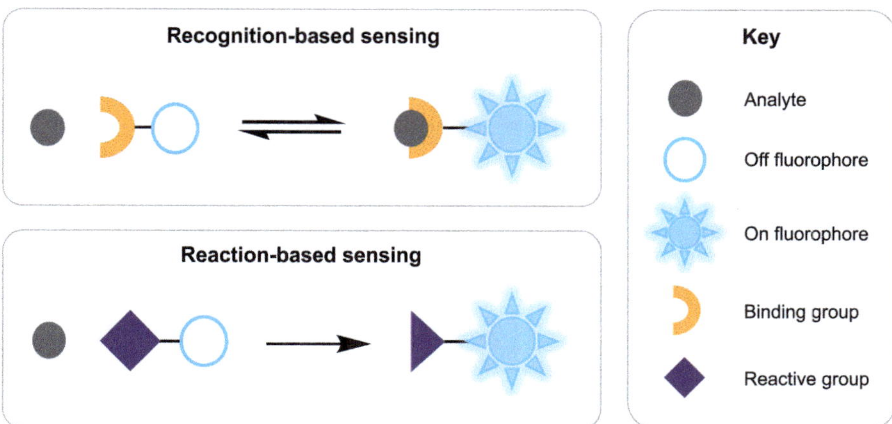

Figure 16.3 Schematic representation of recognition *versus* reaction-based sensing. Recognition-based sensing is a change in fluorescence output upon analyte binding, while in reaction-based sensing the breaking or making of chemical bonds leads to fluorescence changes.

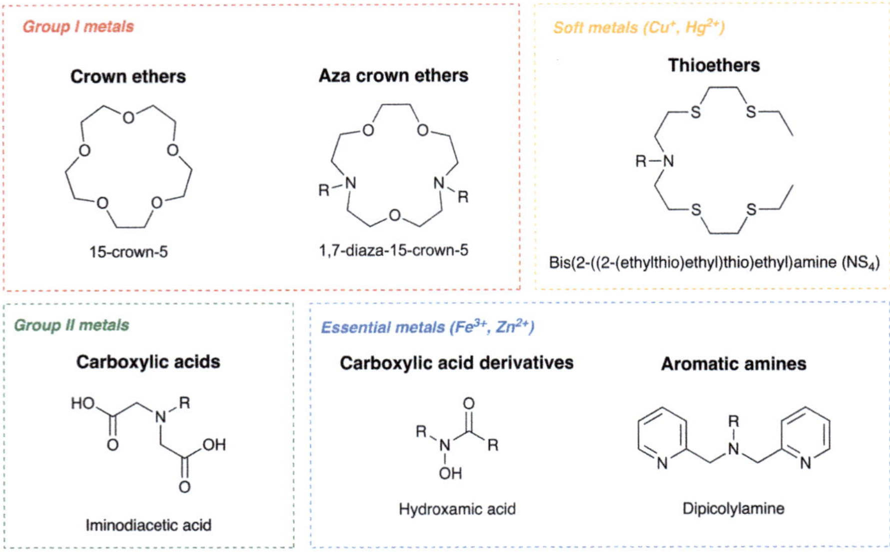

Figure 16.4 Classes of molecular receptor groups for metal ions with chemical structures provided for a key example from each category.

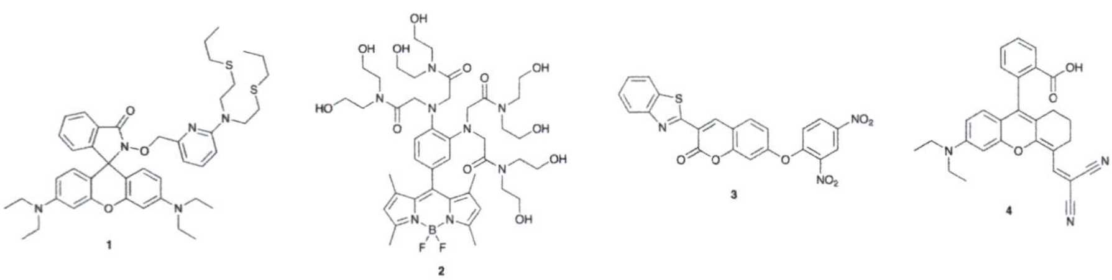

Figure 16.5 Structures of organic small-molecule fluorescent probes for environmental sensing. **1** is a turn-on Hg^{2+} sensor,[67] **2** is a turn-on Pb^{2+} sensor,[68] **3** is a fluorogenic, reaction-based thiol probe[69] and **4** is a dual-responsive probe for hydrazine and cyanide.[70]

of toxic metals is the use of rhodamine B derivatives because the spirolactam ring-opening fluorescence mechanism produces a fluorescence increase.[65,66] For example, Li and co-workers developed a reversible, turn-on rhodamine-based fluorescent sensor for Hg^{2+}(**1**) (Figure 16.5), which also functions as a colorimetric sensor for Cu^{2+}.[67] The metal recognition site of **1** includes a thioether-rich moiety with a pyridine to make six-coordinate sensing complexes. The sensor showed good selectivity for Hg^{2+} and produced reliable calibration plots for quantifying micromolar concentrations of Hg^{2+} in three natural water sources.

Although there are many fluorescent small-molecule probes reported in the literature for toxic heavy metal analysis, the vast majority require organic solvents,[66,71] making them less useful for environmental sensing. This shortfall was addressed by Liu and co-workers, who rationally designed a selective Pb^{2+} probe (**2**) that functions in purely aqueous solvent.[68] The probe consists of a BODIPY florescent core and an amide and alcohol-rich receptor group (Figure 16.5). In the absence of Pb^{2+}, **2** is non-fluorescent due to photoinduced electron transfer (PET) quenching, but when Pb^{2+} is added, a green (λ_{em} = 510 nm) fluorescence band appears in the emission spectrum as Pb^{2+} coordination disables the PET quenching mechanism. The probe demonstrated a 276% increase in fluorescence intensity with the addition of 15 ppb of Pb^{2+}, showing good potential as an environmental probe for trace Pb^{2+} concentrations in aqueous samples.

Small-molecule fluorescent sensors can also be employed for environmental detection of industrial chemicals. Aromatic thiols, such as thiophenol, are used in pesticide and pharmaceutical production but are highly toxic.[72] Lin and co-workers designed a reaction-based small-molecule fluorescent probe (**3**) for environmental thiophenol detection.[69] Probe **3** comprises a coumarin derivative with a reactive nitrophenol ether that undergoes a selective thiolysis in the presence of thiophenol in water with dimethyl formamide (DMF) used as a co-solvent (Figure 16.5). The thiolysis generates a green-fluorescent product, giving a turn-on response. The authors successfully applied this sensor to detect thiophenol as a vapour in air and in spiked river water and soil samples, displaying a high capacity for environmental sensing.

In another example of a small-molecule fluorescent sensor for industrially derived chemical contaminants, Mu *et al.* reported a dual-responsive probe (**4**) for hydrazine and cyanide.[70] Both of these analytes are used in the agrochemical and plastics industries and can pollute the environment in spill events.[73] Probe **3** is a rhodamine B analogue with a dicyano group for hydrazine and cyanide sensing (Figure 16.5). The probe undergoes a nucleophilic addition reaction at the alkene in the presence of cyanide, generating a blue-emissive fluorescent product, while hydrazine transforms the dicyano group into a yellow-emitting hydrazone product. Dual-responsive probes, such as **4**, are useful because fewer sensors are required for the detection of more analytes. However, it is important to note that in systems that have only one reactive group for two analytes, such as this, it is not always possible to detect both analytes simultaneously. In the case of **4**, the hydrazine reaction is preferred over the cyanide reaction, disabling cyanide sensing in the presence of hydrazine. Nonetheless, **4** was able to separately detect hydrazine and cyanide in real tap water and lake water samples with high recovery values, demonstrating its potential utility for environmental sensing.

16.4.2 Nanoparticles and Nanomaterials

Nanomaterials are defined as materials with at least one exterior dimension measuring between 1 and 100 nm.[74] This broad definition means that a large number of substances and materials fall into the nanomaterial category. Common examples of nanomaterials used for sensing applications include metallic nanoparticles (NPs), carbon quantum dots (CQDs) and carbon nanostructures.[75,76] Each class of nanomaterial has its own advantages for sensing, but most nanomaterials have porous structures, large surface areas, high loading capacities, high biocompatibility and specific analyte interactions, which assist with sensing applications.[76]

Metallic nanomaterials, such as AgNPs and AuNPs, are commonly "capped" or functionalised on their surfaces to make sensors for specific applications. Some common coatings include polymers, thiols, amines and acids.[77] AuNPs are often employed as quenchers in Förster resonance energy transfer (FRET)-based fluorescence sensing.[77] For example, Huang and co-workers synthesised AuNPs functionalised with rhodamine B and thiol molecules.[78] In the presence of Hg^{2+}, rhodamine B is displaced from the AuNPs, restoring fluorescence (Figure 16.6). The environmental sensing capacity of their system was confirmed in studies that involved accurately quantifying spiked Hg^{2+} in pond water and estimating Hg^{2+} concentrations in Hg-cathode batteries that were in good agreement with the values derived from ICP–MS analysis.

Another class of nanomaterials frequently employed for environmental sensing are CQDs. CQDs are inherently fluorescent and have additional advantages, such as facile and economical synthesis, water solubility and low toxicity.[79] Yan *et al.* realised the benefits of CQDs in their sensing platform for organophosphorus (OP) pesticides.[80] They combined fluorescent CQDs with a manganese dioxide (MnO_2) nanosheet quencher and treated the resulting system with butyrylcholinesterase (BChE). In the absence of OPs, BChE contributes to MnO_2 decomposition, which restores the CQD fluorescence. In contrast, in the presence of OPs, BChE is inhibited, and the fluorescence of the sensing platform remains quenched, allowing for OP quantification. This CQD sensing platform was successfully applied to OP detection in agricultural samples, including tap water, river water, rice and

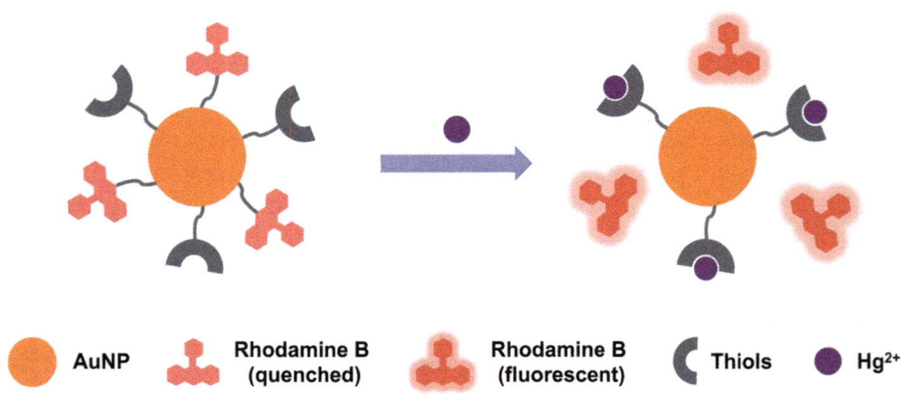

Figure 16.6 Simplified schematic representation of Hg^{2+} sensing with AuNPs as fluorescence quenchers.[78] In the absence of Hg^{2+}, rhodamine fluorescence is quenched by the AuNP. Upon Hg^{2+} binding, rhodamine molecules are liberated and fluorescence is restored.

cabbage. However, the food samples required extensive preparation involving grinding, spiking with OP standards and sonicating in acetonitrile. After centrifugation, filtration and evaporation, OP detection was performed using the MnO_2 nanosheets, with GC–MS analysis also carried out for comparison.

The large porosity and surface area of some nanomaterials, such as nanofibres, makes them ideal candidates for gas and VOC detection.[81] A useful technique to produce one-dimensional nanofibres is electrospinning. In this process, an electric field is applied to a high-voltage stream of polymer solution to generate a mat of nanoscale threads.[82] Fluorescent electrospun fibres have been used in a wide range of environmental sensing applications.[42,82] For example, Petropoulou and co-workers created a fluorescent nanoscale electrospun film for ammonia gas sensing.[83] The film was synthesised with ferrous oxide NP cores with silica shells, functionalised with rhodamine B and integrated with a cellulose acetate polymer. When exposed to ammonia, the spirolactam ring of rhodamine B closes, quenching fluorescence intensity. In this way the concentration of ammonia gas could be detected, showing promise for real-world monitoring in industrial settings.

In another application of fluorescent electrospun fibres, Sun *et al.* devised a platform for explosives detection in aqueous solutions.[84] Their pyrene-polyethersulfone nanofibre films demonstrated good selectivity to nitro-explosives, such as TNT, picric acid and 1,3,5-trinitroperhydro-1,3,5-triazine, with nanomolar limits of detection. In the presence of these analytes, the fluorescence of the pyrene is dramatically quenched by PET. Notably, this system could be regenerated by simply washing with water and drying in air, and therefore has potential as a reusable environmental sensor.

16.4.3 Oligonucleotides

Oligonucleotides (ONs) are short sequences of DNA or RNA, which have a wide variety of applications, including in sensing. Two classes of ONs used as biosensors are aptamers and DNAzymes, which are often classified as a subset of aptamers. Aptamers are ONs capable of binding target molecules,[85] while DNAzymes are larger, tertiary ON structures consisting of enzyme and substate domains that can perform complex reactions, including reaction-based sensing.[10,86] ONs are not inherently fluorescent, but can be modified and combined with other fluorescent platforms, such as small molecules and NPs, for sensing applications. ON sensors have several favourable features for environmental sensing applications, including thermal stability, simple immobilisation, high affinity to analytes (especially metal ions) and facile synthetic modification.[41]

Aptamers are excellent recognition motifs for toxic metal ions, as some nitrogenous bases form strong coordination bonds. For example, Hg^{2+} induces thymine–Hg^{2+}–thymine (T–Hg^{2+}–T) base pair mismatches with T-rich aptamers, while cytosine (C) aptamers can form C–Ag^{+}–C mismatches.[41] Pb^{2+} is also capable of forming guanidine (G) quadruplexes and hence G-rich ONs have selective interactions with Pb^{2+}.[87] There is a vast array of aptamer-based sensors for metals in the literature.[41,85,88] In one example, Zhang and Wei

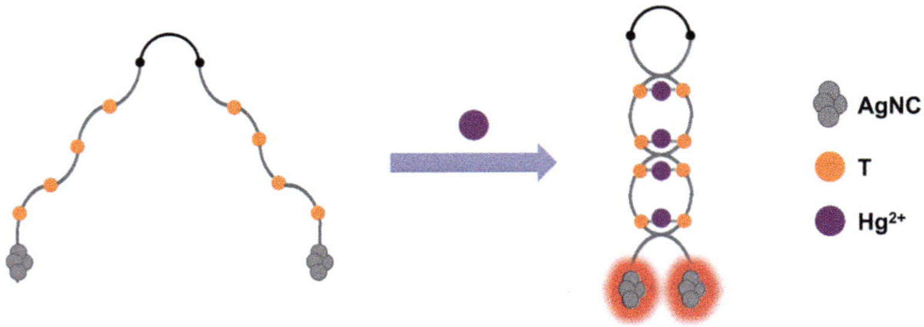

Figure 16.7 Simplified schematic representation of Hg^{2+} sensing with an ON-AgNC fluorescent sensing platform.[89] In the presence of Hg^{2+}, the ON adopts a hairpin structure, bringing AgNCs together and generating a fluorescence signal.

combined silver nanoclusters (AgNCs) with a T-rich aptamer for selective Hg^{2+} sensing.[89] In the presence of Hg^{2+}, the single-strand aptamer adopts a hairpin structure, bringing the two terminal AgNC regions together, accompanied by a turn-on in fluorescence with emission at 620 nm (Figure 16.7). An impressive limit of detection of 0.25 nM was achieved with this sensing platform, which was also applied to real tap water and lake water samples, with accurate Hg^{2+} recoveries when spiked.

Unlike single-stranded aptamers, DNAzymes are larger and can involve more sophisticated design strategies, capable of sensing multiple analytes simultaneously in an amplified, catalytic fashion.[90] For example, Li and co-workers designed a fluorescent DNAzyme sensing platform for simultaneous Pb^{2+} and Cu^{2+} detection in water.[91] The sensor consists of a AuNP functionalised with two types of substrate DNA (S-DNA) strands: 6-carboxyflorescin-labelled S-DNA for Pb^{2+} sensing; and Cy5-labelled S-DNA for Cu^{2+} sensing. A partially hybridised two-strand pincer-type enzyme DNA (E-DNA) is then incorporated onto the AuNP S-DNA structure to create a dual-functional DNAzyme. In the absence of metal analytes, the fluorophores on the DNAzyme are quenched by the AuNP. In the presence of Pb^{2+} or Cu^{2+}, there is a turn-on in fluorescence emission at 520 nm (6-carboxyflorescin) or 657 nm (Cy5), respectively, as the S-DNA is cleaved from the AuNP by the E-DNA. This sensing platform showed minimal interference from other metal ions and was able to quantify Pb^{2+} and Cu^{2+} concentrations accurately in spiked river water. Furthermore, the DNAzyme platform is a useful dual-responsive probe because when both Pb^{2+} and Cu^{2+} are present, the fluorescent responses of each fluorophore are comparable to results obtained from samples of Pb^{2+} or Cu^{2+} alone. This is attributed to the use of separate S-DNA strands for each metal ion, a clever design feature of this DNAzyme.

16.4.4 Metal–Organic Frameworks

MOFs are crystalline coordination polymers built from metal ions or clusters with bridging organic ligand linkers. They have a number of useful sensing properties, such as large surface areas, tuneable pore sizes and excellent chemical stability.[92] As a result, the use of MOFs as chemical sensors is a growing field of research. Early examples of MOF-based sensing platforms use switching mechanisms based on magnetism and ferroelectricity.[93] However, optical MOF sensors are rapidly gaining popularity. MOFs can have luminescent properties as a result of several mechanisms including: (1) fluorescence of π-conjugated organic ligands; (2) emissive metal ions, such as lanthanide ions, d^{10} transition metals and metal clusters; (3) charge transfer luminescence based on excited state to ground state transitions (*e.g.* metal-to-ligand charge transfer); and (4) emission of fluorescent guest chemicals in MOF pores.[93]

MOFs have shown excellent promise as fluorescent sensors for metallic pollutants in water.[94] Zhang *et al.* developed a UiO-66 MOF analogue, **UiO-66-PSM**, for selective Hg^{2+} sensing in water.[95] The authors used a post-synthetic modification strategy to functionalise the MOF with Hg^{2+}-binding triazoles. In the absence of Hg^{2+}, **UiO-66-PSM** has a strong fluorescence emission band at 544 nm due to the π–π* electron transition of organic linkers. Hg^{2+} binding to the triazole recognition sites causes rapid and dramatic fluorescence quenching, but shows no response to other metal ion interferents. The authors also tested **UiO-66-PSM** in spiked tap water and lake water samples, achieving accurate recovery values for micromolar concentrations of Hg^{2+}.

There are numerous examples of MOF-based gas and VOC sensors in the literature.[93] MOFs are efficient gas adsorbers, due to their large pores and high surface-area-to-volume ratios. Hydrogen sulfide (H_2S) is a biologically essential signalling molecule,[96] but is also a toxic industrial gaseous pollutant. Several optically active MOFs have been devised and tested as environmental H_2S sensors. In a multifaceted approach, Zhang and co-workers combined an aluminium-based nano-MOF with a polyvinylidene fluoride polymer to create a mixed-matrix membrane (MMM) sensor.[97] The MMM has exposed electron-withdrawing nitro groups that can be reduced by H_2S, creating a turn-on fluorescence response. The MMM was able to detect H_2S in spiked tap water and lake water samples at micromolar concentrations, demonstrating its capability as an environmental sensor.

In a different approach, Ma *et al.* created an H_2S sensor from an aluminium-copper mixed-metal MOF with Cu^{2+}-metalated porphyrin linkers.[98] The porphyrin fluorescence is quenched by the Cu^{2+} centres, but when treated with H_2S, the porphyrins are demetalated and the fluorescence is restored, generating a significant turn-on response. These two contrasting examples of MOF sensors for H_2S demonstrate the incredible flexibility and diversity of MOFs for sensing applications.

Another gas of concern is sulfur dioxide (SO_2) because it is a potent respiratory irritant and a leading cause of acid rain.[93,99] Wang and co-workers created an amino-functionalised MOF, **MOF-5-NH$_2$**, for gaseous SO_2 sensing.[100] Without SO_2 or sulfite, **MOF-5-NH$_2$** has a low fluorescence output because the amine groups contribute to a charge transfer to the coordinated metal ions (LMCT). However, when an SO_2 or sulfite guest interacts with the amines, LMCT is inhibited, resulting in a fluorescence turn-on. For gaseous sensing

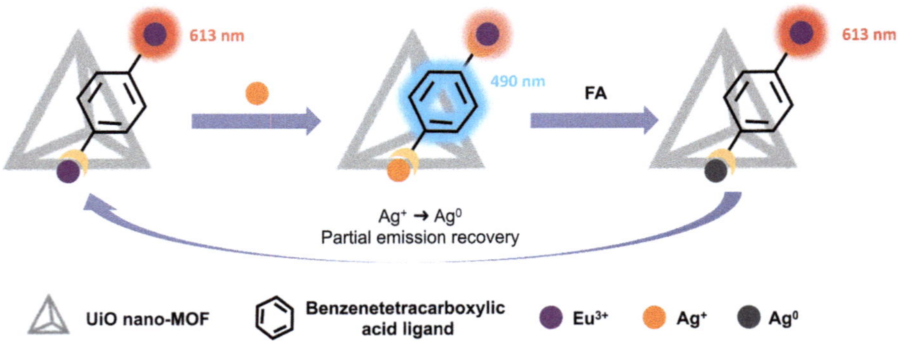

Figure 16.8 Simplified schematic representation of a MOF-based environmental sensor, **1e**, for formaldehyde (FA).[103] The Ag^+-doped MOF quenches Eu^{3+} fluorescence and enhances ligand fluorescence. FA reduces Ag^+, causing a ratiometric change in MOF fluorescence.

applications, test papers were prepared by submerging neutral filter paper strips in a suspension of **MOF-5-NH$_2$**. The paper exhibited bright blue fluorescence under a UV lamp (λ_{ex} = 365 nm) when exposed to SO_2 (2 ppm), but not other common gases, such as air, NH_3, NO_2 or H_2S.

Formaldehyde (FA) is a harmful VOC, causing severe health issues to those exposed to excessive quantities in indoor spaces.[101] For this reason, many fluorescent probes for FA have been developed.[102] Hao and Yan devised the first ratiometric fluorescent MOF sensor for FA.[103] The UiO nano-MOF, **1e**, was functionalised with Eu^{3+} and Ag^+. It exhibits fluorescence from both the π–π* electron transition of the benzenetetracarboxylic acid ligands and from the encapsulated Eu^{3+} ions. Upon FA exposure, the fluorescence peak ascribed to the ligands (490 nm) decreases, while the Eu^{3+} emission peaks (570–700 nm) increase due to the reduction of Ag^+ to Ag^0 by FA (Figure 16.8) The ratio of the emission intensity of the ligand peak (490 nm) to that of the most intense Eu^{3+} peak (613 nm) demonstrated a good linear fit against FA concentration, allowing for accurate quantification of FA in air.

16.4.5 Fluorescent Sensing Arrays

The most common approach to fluorescent environmental sensing, as exemplified in the previous sections, is the use of selective sensors, *i.e.* designing a sensor for a single analyte such that it is non-responsive to other interferents in the analysis matrix. However, creating selective sensors can be challenging, as many analytes of interest have very similar chemical and physical characteristics. An alternative approach to selective sensing is array-based sensing. In this method, multiple sensors are employed within a single sensing platform to provide simultaneous detection and discrimination of more than one analyte. For arrays, cross-reactive sensors are preferred over selective sensors because more analytes can be detected.

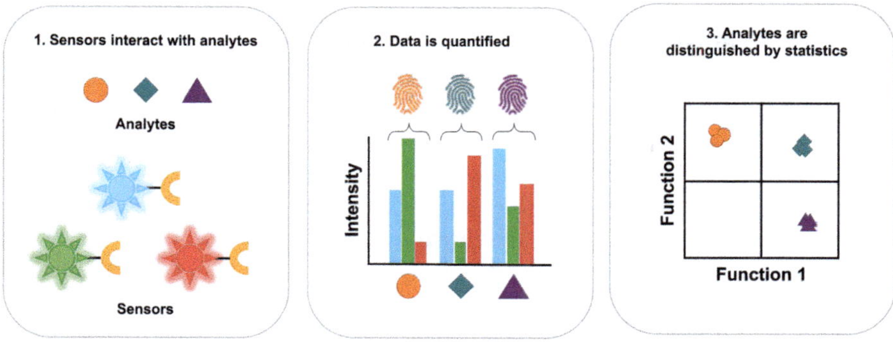

Figure 16.9 Schematic representation of array-based sensing: (1) multiple cross-reactive sensors are tested with numerous analytes; (2) the fingerprint responses of each analyte are quantified with fluorescence spectroscopy; and (3) the quantified responses are analysed with multivariate statistical techniques.

The general method of array-based sensing requires three key steps (Figure 16.9): (1) multiple cross-reactive sensors are tested with numerous analytes and the combination of fluorescence responses across all sensors for each analyte creates a unique "fingerprint" response; (2) the fingerprint responses are quantified with fluorescence spectroscopy, creating a very large data set comprising replicates for each combination of sensor and analyte; and (3) the quantified responses are analysed with multivariate statistical techniques, such as linear discriminant analysis (LDA), to objectively discriminate between each analyte. This general method allows for fewer sensors to be used for a large number of analytes, saving time and costs associated with sensor fabrication and analysis.

Fluorescent small-molecule organic probes have been used extensively in cross-reactive sensing arrays for cations and anions.[43] Many of these arrays have been applied to environmental sensing, particularly for toxic metals. For example, our group has developed several fluorescent small-molecule sensing arrays for metal ions using thiophene,[104] coumarin[105] and thiocoumarin[106] fluorescent scaffolds. One such array was derived from the commercial coumarin-based calcein blue, and five analogues.[105] The analogues were rationally designed to include a variety of hard and soft Lewis bases as metal receptors, and were all prepared in three synthetic steps or fewer. The array was able to accurately discriminate between nine metal ions, including toxic heavy metals. The six-sensor array could also be reduced to three sensors to achieve comparable discrimination accuracy in buffered pure water and pond water.

Oligonucleotides can also be employed in array-based sensing to discriminate between multiple analytes in environmental samples. Kwon *et al.* devised a fluorescent DNA-based sensing array for the detection of 14 common pesticides.[107] Their curated array consisted of ten DNA-like oligomers, where nitrogenous bases were replaced with several fluorophores (pyrene, hydroxyquinoline, perylene) and metal chelators (terpyridine). The array of these ten oligodeoxyfluorosides could discriminate between 100 μM concentrations of pesticides, and quantify lower concentrations of some pesticides with discriminant analysis. The array also demonstrated good discriminating ability in a local lake water sample that was affected by agricultural run-off.

Polymers and other macromolecules have been widely used as sensors in fluorescent arrays.[108] Progress in controlled polymer synthesis has allowed for facile functionalisation with useful sensing elements, such as analyte receptor groups and fluorescent molecules. For example, Tropp and co-workers designed a three-polymer sensing array that demonstrated distinctive fluorescent fingerprint responses for 12 chemically similar azo dyes of environmental concern.[109] The differential outputs were achieved by exploiting the inner filter effect and diverse spectral overlaps, which causes fluorescence quenching of dyes based on interfering absorbance of excited or emitted light, and the data was analysed with LDA. The authors demonstrated that their polymer array significantly improved discrimination of the azo dyes in river and wastewater samples, compared to using just the spectral properties of the dyes alone. Other impressive features of the array were that it achieved accurate discrimination of nanomolar analyte concentrations and functioned well in artificial seawater, demonstrating robustness to high salt concentrations.

16.4.6 Portable Devices

A desirable quality of any environmental sensor is portability, to allow for on-site testing and analysis. Such devices are commonly described as providing "point-of-need testing" because the testing and analysis can be carried out *in situ*, at the location of interest. Sensing strategies such as fluorescence techniques have excellent potential for on-site sensing platforms. This is because many fluorescent platforms can be excited by visible or UV light sources. Furthermore, emission can often be detected by the naked eye, or by widely available smartphone cameras. Portable fluorescent sensors for the environment can be composed of a combination of fluorescent platforms, such as organic dyes, carbon dots and DNA sensors, and are often analysed with smartphone devices. The following section will describe some innovative examples of portable fluorescent sensors for the environment.

Toxic metal ions are a leading target for on-site environmental monitoring. Lee and co-workers have developed a portable sensing system for Hg^{2+} in water samples.[110] The system contains two crucial elements, a "smartphone fluorescence microscope" (SFM) and a biosensor cartridge. The SFM hardware was built from a smartphone camera, emission filter, eye piece, mirror, dichroic mirror, laser (462 nm), objective stage and white LED (Figure 16.10). The cartridge consists of a T–T mismatched DNA Hg^{2+} probe intercalated with the green-fluorescent dye, Sybr Green I (SGI) immobilised on a nano-metallised substrate, and a reaction solution with DNA-conjugated red-fluorescent

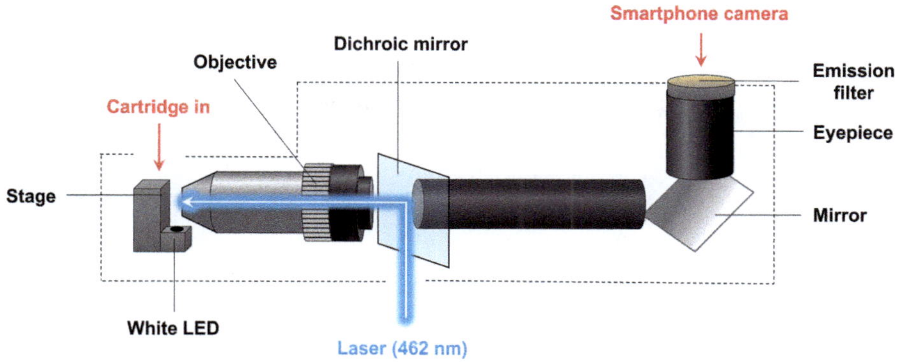

Figure 16.10 Schematic representation of the SFM developed by Lee and co-workers in their portable Hg^{2+} sensor.[110] The SFM consists of a smartphone camera, emission filter, eye piece, mirror, dichroic mirror, laser (462 nm), objective, stage and white LED.

CQDs. In the presence of Hg^{2+}, the DNA probe dehybridises and then rehybridises to the DNA-functionalised CQDs through T–Hg^{2+}–T interactions. This causes a decrease in green SGI fluorescence and an increase in red CQD fluorescence, allowing for ratiometric Hg^{2+} detection using the SFM to measure red and green fluorescence channels. The authors applied their portable sensor to detect nanomolar concentrations of Hg^{2+} in wastewater samples, with comparable results to those determined by *ex situ* AAS analysis.

Array-based sensing can also be applied to portable devices for environmental metal detection. Xiao *et al.* designed a paper-based microarray based on fluorescent CQDs, which can be analysed by a smartphone-based portable reader (SBR).[111] The array contained three types of CQDs, with different emission wavelengths used to simultaneously detect three metal ions, Hg^{2+}, Pb^{2+} and Cu^{2+}, in water. All three CQDs exhibit fluorescence quenching in the presence of these metals, allowing for quantification. The microarray provided good micromolar metal ion concentration recoveries in river water when compared to ICP–MS analysis. The authors emphasise that their sensing platform requires no sample pre-treatment and that sample images can be acquired within 5 s by their SBR for analysis by their purpose-built smartphone application.

Portable fluorescent sensors have also been employed to monitor the health of aquatic ecosystems. Sarwar *et al.* created a portable, paper-based sensing system for inorganic phosphate to monitor for aquatic "dead zones" caused by algal blooms.[112] The sensing apparatus consists of a coumarin dye, MDCC, and an *Escherichia coli* phosphate-binding protein. The fluorescence output of MDCC is sensitive to conformational change to the protein. Phosphate binding causes an increase in the fluorescence emission of MDCC, allowing for phosphate quantification. The sensing system was made portable by incorporating the protein-coumarin duplex onto a paper substrate, and fluorescence was detected by a hand-held reflectance-mode fluorescent

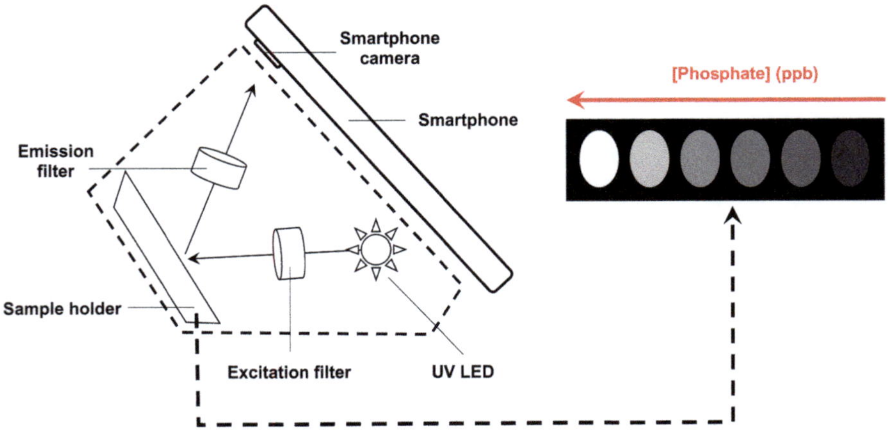

Figure 16.11 Simplified diagram of hand-held reflectance-mode fluorescent imaging device using a paper test for phosphate sensing as described by Sarwar *et al.*[112]

imaging device (Figure 16.11). The device was tested in samples from the flooded-grassland Everglades biome, with test samples showing good agreement to concentrations measured for a pre-prepared linear calibration curve, demonstrating good potential for point-of-need testing.

In addition to molecular analytes, some portable fluorescence devices have also been shown to be able to detect and differentiate phytoplankton organisms.[44] The abundance and diversity of phytoplankton organisms, such as cyanobacteria and green algae, is closely correlated to the health of an aquatic ecosystem.[113] Shin and co-workers developed a hand-held fluorescent device with three excitation light sources to detect green algae and cyanobacteria.[114] Their sensing platform utilised the inherently fluorescent properties of some pigments in phytoplankton. While chlorophyll *a* is present in both green algae and cyanobacteria, chlorophyll *b* and phytocyanin are only present in green algae and cyanobacteria, respectively. In this way, the emission from 680 nm light could be used to detect green algae, while the emission from 645 nm light could be used to detect cyanobacteria. Employment of a multivariate algorithm allowed for accurate biomass quantification of the two phytoplankton species in complex mixtures.

While it is often crucial to detect specific chemical pollutants in the environment, it can sometimes be equally important to analyse all pollution within a broader class, such as dissolved organic matter (DOM). DOM is classified as any organic matter that can be dissolved in water and is smaller than 0.2 μm (a commonly used filtration pore size).[44] Therefore, DOM is a mixture of organic compounds from hydrocarbons to aromatic molecules, which have distinctive spectrophotometric properties that can be analysed with absorbance and fluorescence spectroscopy. Brandl *et al.* devised a portable fluorescence sensing device for DOM using three excitation sources from

LEDs.[115] The fluorescence signals were detected by a photomultiplier tube and DOM concentration could be estimated by calculating the ratios of the different fluorescence emission intensities. The authors tested their device against real water samples from two rivers and recorded DOM concentrations that were within ±10% of those determined using a benchtop, laboratory-based spectrophotometer.

16.5 Conclusion

In this chapter, we have presented examples of fluorescent sensing platforms for common environmental analytes. We have demonstrated how environmental fluorescent sensors can be as simple as a single small-molecule organic dye, or as complex as a novel portable device containing multiple components. It is clear that fluorescent sensing platforms can be very useful for environmental monitoring, evidenced by the large number of sensors reported in the literature that have been tested with real environmental samples. It is also apparent that work in the field of environmental sensing systems is undergoing a shift from laboratory-based to on-site analysis, and that fluorescent sensing platforms have a large role to play in this ongoing field of research.

List of Abbreviations

AgNCs	Silver nanoclusters
ASV	Anodic stripping voltammetry
BChE	Butyrylcholinesterase
BODIPY	Boron dipyrromethene (4,4-difluoro-4-bora-3*a*,4*a*-diaza-*s*-indacene)
BPA	Bisphenol A
C	Cytosine
CQD	Carbon quantum dot
DDT	Dichlorodiphenyltrichloroethane
DMF	Dimethylformamide
DNA	Deoxyribonucleic acid
DOM	Dissolved organic matter
E-DNA	Enzyme DNA
FA	Formaldehyde
FRET	Förster resonance energy transfer
FTIR	Fourier-transform infrared spectroscopy
G	Guanidine
GC–MS	Gas chromatography–mass spectrometry
GFAAS	Graphite furnace atomic absorption spectroscopy
ICP–MS	Inductively coupled plasma–mass spectrometry
ICP–OES	Inductively coupled plasma–optical emission spectroscopy
LC–MS	Liquid chromatography–mass spectrometry

LDA	Linear discriminant analysis
LD_{Lo}	Lowest lethal dose
LED	Light-emitting diode
LMCT	Ligand-to-metal charge transfer
MDCC	*N*-[2-(1-maleimidyl)ethyl]-7-(diethylamino)-coumarin-3-carboxamide
MMM	Mixed-matrix membrane
MOF	Metal–organic framework
NP	Nanoparticle
ON	Oligonucleotide
OP	Organophosphate
PET	Photoinduced electron transfer
RNA	Ribonucleic acid
S-DNA	Substrate DNA
SFM	Smartphone fluorescence microscope
SGI	Sybr Green I
T	Thymine
TNT	2,4,6-trinitrotoluene
UV	Ultraviolet
VOC	Volatile organic compound
XRF	X-ray fluorescence

Acknowledgements

The authors would like to acknowledge the Australian government for a Research Training Program Scholarship (AAB), and the Henry Bertie and Florence Mabel Gritton Medallist Award (AAB).

References

1. M. Li, J. Dong, Y. Zhang, H. Yang, L. Van Zwieten, H. Lu, A. Alshameri, Z. Zhan, X. Chen and X. Jiang, *Water*, 2021, **13**, 225.
2. P. B. Tchounwou, C. G. Yedjou, A. K. Patlolla and D. J. Sutton, in *Molecular, Clinical and Environmental Toxicology: Volume 3: Environmental Toxicology*, ed. A. Luch, Springer Basel, Basel, 2012, pp. 133–164.
3. *Heavy Metals in the Environment: Origin, Interaction and Remediation*, ed. H. Bradl, Elsevier Science, 2005.
4. *The Heavy Elements: Chemistry, Environmental Impact, and Health Effects*, ed. J. E. Fergusson, Pergamon Press, 1990.
5. F. Springman, E. Bingham and K. L. Stemmer, *Arch. Environ. Health*, 1963, **6**, 469–472.
6. G. Tartler, *Arch. Hyg. Bakteriol.*, 1941, 273–292.
7. D. A. Basterrechea, L. Parra, M. Botella-Campos, J. Lloret and P. V. Mauri, *Appl. Sci.*, 2020, **10**, 7222.
8. D. Tilman, K. G. Cassman, P. A. Matson, R. Naylor and S. Polasky, *Nature*, 2002, **418**, 671–677.

9. J. Chen, S. Lü, Z. Zhang, X. Zhao, X. Li, P. Ning and M. Liu, *Sci. Total Environ.*, 2018, **613–614**, 829–839.
10. M. I. Gaviria-Arroyave, J. B. Cano and G. A. Peñuela, *Talanta Open*, 2020, **2**, 100006.
11. G. K. Sidhu, S. Singh, V. Kumar, D. S. Dhanjal, S. Datta and J. Singh, *Crit. Rev. Environ. Sci. Technol.*, 2019, **49**, 1135–1187.
12. C. S. Pundir and N. Chauhan, *Anal. Biochem.*, 2012, **429**, 19–31.
13. World Health Organization, *WHO Guidelines for Indoor Air Quality: Selected Pollutants*, Bonn, 2010.
14. International Energy Agency, *Global Energy Review 2021*, Paris, 2021.
15. S. N. Behera, M. Sharma, V. P. Aneja and R. Balasubramanian, *Environ. Sci. Pollut. Res.*, 2013, **20**, 8092–8131.
16. T. Adamová, J. Hradecký and M. Pánek, *Polymers*, 2020, **12**, 2289.
17. J. Shuai, S. Kim, H. Ryu, J. Park, C. K. Lee, G.-B. Kim, V. U. Ultra and W. Yang, *BMC Public Health*, 2018, **18**, 528.
18. L. Mølhave, B. Bach and O. F. Pedersen, *Environ. Int.*, 1986, **12**, 167–175.
19. M. Bilal, M. Adeel, T. Rasheed, Y. Zhao and H. M. N. Iqbal, *Environ. Int.*, 2019, **124**, 336–353.
20. R. P. Pohanish, *Sittig's Handbook of Toxic and Hazardous Chemicals and Carcinogens*, 4th edn, William Andrew, 2017.
21. S.-H. Rajabnejad, H. Badibostan, A. Verdian, G. R. Karimi, E. Fooladi and J. Feizy, *Microchem. J.*, 2020, **155**, 104722.
22. E. O. Igbinosa, E. E. Odjadjare, V. N. Chigor, I. H. Igbinosa, A. O. Emoghene, F. O. Ekhaise, N. O. Igiehon and O. G. Idemudia, *Sci. World J.*, 2013, **2013**, 460215.
23. P. T. Charles, A. A. Adams, P. B. Howell, S. A. Trammell, J. R. Deschamps and A. W. Kusterbeck, *Sensors*, 2010, **10**, 876–889.
24. K. Boonkitpatarakul, Y. Yodta, N. Niamnont and M. Sukwattanasinitt, *RSC Adv.*, 2015, **5**, 33306–33311.
25. K. Li, C. Sherman, J. Beaumont and M. Sandy, *Evidence on the Carcinogenicity of 2,4,6-trinitrotoluene*, Office of Environmental Health Hazard Assessment's (OEHHA). Reproductive and Cancer Hazard Assessment Branc, California, 2010.
26. H. M. Bolt, G. H. Degen, S. B. Dorn, S. Plöttner and V. Harth, *Rev. Environ. Health*, 2006, **21**, 217–228.
27. D. Fowler, K. Pilegaard, M. A. Sutton, P. Ambus, M. Raivonen, J. Duyzer, D. Simpson, H. Fagerli, S. Fuzzi, J. K. Schjoerring, C. Granier, A. Neftel, I. S. A. Isaksen, P. Laj, M. Maione, P. S. Monks, J. Burkhardt, U. Daemmgen, J. Neirynck, E. Personne, R. Wichink-Kruit, K. Butterbach-Bahl, C. Flechard, J. P. Tuovinen, M. Coyle, G. Gerosa, B. Loubet, N. Altimir, L. Gruenhage, C. Ammann, S. Cieslik, E. Paoletti, T. N. Mikkelsen, H. Ro-Poulsen, P. Cellier, J. N. Cape, L. Horváth, F. Loreto, Ü. Niinemets, P. I. Palmer, J. Rinne, P. Misztal, E. Nemitz, D. Nilsson, S. Pryor, M. W. Gallagher, T. Vesala, U. Skiba, N. Brüggemann, S. Zechmeister-Boltenstern, J. Williams, C. O'Dowd, M. C. Facchini, G. de Leeuw, A. Flossman, N. Chaumerliac and J. W. Erisman, *Atmos. Environ.*, 2009, **43**, 5193–5267.

28. J. Cullum, D. P. Stevens and M. M. Joshi, *Proc. Natl. Acad. Sci. U. S. A.*, 2016, **113**, 4278–4283.
29. S. F. Jane, G. J. A. Hansen, B. M. Kraemer, P. R. Leavitt, J. L. Mincer, R. L. North, R. M. Pilla, J. T. Stetler, C. E. Williamson, R. I. Woolway, L. Arvola, S. Chandra, C. L. DeGasperi, L. Diemer, J. Dunalska, O. Erina, G. Flaim, H.-P. Grossart, K. D. Hambright, C. Hein, J. Hejzlar, L. L. Janus, J.-P. Jenny, J. R. Jones, L. B. Knoll, B. Leoni, E. Mackay, S.-I. S. Matsuzaki, C. McBride, D. C. Müller-Navarra, A. M. Paterson, D. Pierson, M. Rogora, J. A. Rusak, S. Sadro, E. Saulnier-Talbot, M. Schmid, R. Sommaruga, W. Thiery, P. Verburg, K. C. Weathers, G. A. Weyhenmeyer, K. Yokota and K. C. Rose, *Nature*, 2021, **594**, 66–70.
30. D. W. Sutcliffe and T. R. Carrick, *Freshwater Biol.*, 1973, **3**, 437–462.
31. W. B. Langbein, *Salinity and Hydrology of Closed Lakes: A Study of the Long-term Balance Between Input and Loss of Salts in Closed Lakes*, U.S. Government Print Office, 1961.
32. S. Ueda, H. Kawabata, H. Hasegawa and K. Kondo, *Limnol*, 2000, **1**, 57–62.
33. N. J. McKenzie, B. Henderson and W. McDonald, *Monitoring Soil Change – Principles and Practices for Australian Conditions*, CSIRO Land and Water, 2002.
34. L. D. Baver and C. J. Rehling, *Ind. Eng. Chem., Anal. Ed.*, 1930, **2**, 338.
35. M. Hossain, D. Karmakar, S. N. Begum, S. Y. Ali and P. K. Patra, *Microchem. J.*, 2021, **165**, 106086.
36. R. Ravansari, S. C. Wilson and M. Tighe, *Environ. Int.*, 2020, **134**, 105250.
37. D. Barceló, *Trends Anal. Chem.*, 1991, **10**, 323–329.
38. B. A. Eckenrode, *J. Am. Soc. Mass Spectrom.*, 2001, **12**, 683–693.
39. D. Barceló and M. Petrovic, *Trends Anal. Chem.*, 2007, **26**, 2–11.
40. C. M. Simonescu, in *Advanced Aspects of Spectroscopy*, ed. M. A. Farrukh, IntechOpen, 2012.
41. N. De Acha, E. Cesar, J. Corres and F. Arregui, *Sensors*, 2019, **19**, 599.
42. M. Yin and C. Ji, *Nanoscaled Fluorescent Films and Layers for Detection of Environmental Pollutants*, IntechOpen, 2017.
43. D. Smith, I. Topolnicki, V. Zwicker and K. Jolliffe, *Analyst*, 2017, **142**, 3549–3563.
44. Y.-H. Shin, M. T. Gutierrez-Wing and J.-W. Choi, *J. Electrochem. Soc.*, 2021, **168**, 017502.
45. World Health Organization, *Guidelines for Drinking-water Quality*, 4th edn, Geneva, 2017.
46. E. A. Jara and C. K. Winter, *Int. J. Food Contam.*, 2019, **6**, 6.
47. National Research Council, *Acute Exposure Guideline Levels for Selected Airborne Chemicals: Volume 4*, The National Academies Press, Washington, DC, 2004.
48. C. H. S. J. Chou, *Hydrogen Sulfide: Human Health Aspects*, World Health Organization, Geneva, Switzerland, 2003.
49. World Health Organization, *Air Quality Guidelines: Global Update 2005: Particulate Matter, Ozone, Nitrogen Dioxide, and Sulfur Dioxide*, World Health Organization, Copenhagen, Denmark, 2006.

50. World Health Organization, *Cyanide in drinking-water. Background Document for Preparation of WHO Guidelines for Drinking-water Quality*, Geneva, 2003.
51. A. P. De Silva, H. N. Gunaratne, T. Gunnlaugsson, A. J. Huxley, C. P. McCoy, J. T. Rademacher and T. E. Rice, *Chem. Rev.*, 1997, **97**, 1515–1566.
52. J. V. Jun, D. M. Chenoweth and E. J. Petersson, *Org. Biomol. Chem.*, 2020, **18**, 5747–5763.
53. I. Y. Jung, E. H. Lee, A. Y. Suh, S. J. Lee and H. Lee, *Anal. Bioanal. Chem.*, 2016, **408**, 2383–2406.
54. H. N. Kim, Z. Guo, W. Zhu, J. Yoon and H. Tian, *Chem. Soc. Rev.*, 2011, **40**, 79–93.
55. Y. Li, D. J. Young and X. J. Loh, *Mater. Chem. Front.*, 2019, **3**, 1489–1502.
56. L. Choulier and K. Enander, *Sensors*, 2010, **10**, 3126–3144.
57. D. M. Chudakov, M. V. Matz, S. Lukyanov and K. A. Lukyanov, *Physiol. Rev.*, 2010, **90**, 1103–1163.
58. M. Levine, *Front. Chem.*, 2021, **9**, 616815.
59. Y. Y. Chia and M. G. Tay, *Dalton Trans.*, 2014, **43**, 13159–13168.
60. S. M. Ng, M. Koneswaran and R. Narayanaswamy, *RSC Adv.*, 2016, **6**, 21624–21661.
61. O. S. Wolfbeis, *Chem. Soc. Rev.*, 2015, **44**, 4743–4768.
62. J. Dong, D. Zhao, Y. Lu and W.-Y. Sun, *J. Mater. Chem. A*, 2019, **7**, 22744–22767.
63. K. P. Carter, A. M. Young and A. E. Palmer, *Chem. Rev.*, 2014, **114**, 4564–4601.
64. D. S. McClure, *J. Chem. Phys.*, 1952, **20**, 682–686.
65. J. Lian, Q. Xu, Y. Wang and F. Meng, *Front. Chem.*, 2020, **8**, 593291.
66. T. Rasheed, M. Bilal, F. Nabeel, H. M. N. Iqbal, C. L. Li and Y. F. Zhou, *Sci. Total Environ.*, 2018, **615**, 476–485.
67. M. Li, Y. Sun, L. Dong, Q.-C. Feng, H. Xu, S.-Q. Zang and T. C. W. Mak, *Sens. Actuators, B*, 2016, **226**, 332–341.
68. J. Liu, K. Wu, S. Li, T. Song, Y. Han and X. Li, *Dalton Trans.*, 2013, **42**, 3854–3859.
69. W. Lin, L. Long and W. Tan, *Chem. Commun.*, 2010, **46**, 1503–1505.
70. S. Mu, H. Gao, C. Li, S. Li, Y. Wang, Y. Zhang, C. Ma, H. Zhang and X. Liu, *Talanta*, 2021, **221**, 121606.
71. T. Samanta and R. Shunmugam, *Mater. Adv.*, 2021, **2**, 64–95.
72. United States Environmental Protection Agency, *Provisional Peer-Reviewed Toxicity Values for Benzenethiol (CASRN 108-98-5)*, Cincinnati, OH 45268, 2011.
73. F. Wang, L. Wang, X. Chen and J. Yoon, *Chem. Soc. Rev.*, 2014, **43**, 4312–4324.
74. B. D. Malhotra and M. A. Ali, in *Nanomaterials for Biosensors*, ed. B. D. Malhotra and M. A. Ali, William Andrew Publishing, 2018, pp. 1–74.
75. M. Holzinger, A. Le Goff and S. Cosnier, *Front. Chem.*, 2014, **2**, 63.
76. L. Walekar, T. Dutta, P. Kumar, Y. S. Ok, S. Pawar, A. Deep and K.-H. Kim, *Trends Anal. Chem.*, 2017, **97**, 458–467.

77. K. Saha, S. S. Agasti, C. Kim, X. Li and V. M. Rotello, *Chem. Rev.*, 2012, **112**, 2739–2779.
78. C.-C. Huang and H.-T. Chang, *Anal. Chem.*, 2006, **78**, 8332–8338.
79. M. J. Molaei, *Anal. Methods*, 2020, **12**, 1266–1287.
80. X. Yan, Y. Song, C. Zhu, H. Li, D. Du, X. Su and Y. Lin, *Anal. Chem.*, 2018, **90**, 2618–2624.
81. R. K. Ibrahim, M. Hayyan, M. A. AlSaadi, A. Hayyan and S. Ibrahim, *Environ. Sci. Pollut. Res.*, 2016, **23**, 13754–13788.
82. S.-X. Wang, C. C. Yap, J. He, C. Chen, S. Y. Wong and X. Li, *Nanotechnol. Rev.*, 2016, **5**, 51–73.
83. A. Petropoulou, S. Kralj, X. Karagiorgis, I. Savva, E. Loizides, M. Panagi, T. Krasia-Christoforou and C. Riziotis, *Sci. Rep.*, 2020, **10**, 367.
84. X. Sun, Y. Liu, G. Shaw, A. Carrier, S. Dey, J. Zhao and Y. Lei, *ACS Appl. Mater. Interfaces*, 2015, **7**, 13189–13197.
85. E. M. McConnell, J. Nguyen and Y. Li, *Front. Chem.*, 2020, **8**, 434.
86. R. J. Lake, Z. Yang, J. Zhang and Y. Lu, *Acc. Chem. Res.*, 2019, **52**, 3275–3286.
87. Z. Lin, Y. Chen, X. Li and W. Fang, *Analyst*, 2011, **136**, 2367–2372.
88. V. Kumar and P. Guleria, *Curr. Pollut. Rep.*, 2020, DOI: 10.1007/s40726-020-00165-1.
89. B. Zhang and C. Wei, *RSC Adv.*, 2017, **7**, 56289–56295.
90. C. E. McGhee, K. Y. Loh and Y. Lu, *Curr. Opin. Biotechnol.*, 2017, **45**, 191–201.
91. X. Li, W. Yun, W. Guo, W. Zhang and L. Yang, *Sens. Actuators, B*, 2019, **301**, 127170.
92. P. Kumar, A. Deep and K.-H. Kim, *Trends Anal. Chem.*, 2015, **73**, 39–53.
93. H.-Y. Li, S.-N. Zhao, S.-Q. Zang and J. Li, *Chem. Soc. Rev.*, 2020, **49**, 6364–6401.
94. P. Samanta, S. Let, W. Mandal, S. Dutta and S. K. Ghosh, *Inorg. Chem. Front.*, 2020, **7**, 1801–1821.
95. X. Zhang, T. Xia, K. Jiang, Y. Cui, Y. Yang and G. Qian, *J. Solid State Chem.*, 2017, **253**, 277–281.
96. R. Wang, *Antioxid. Redox Signaling*, 2010, **12**, 1061–1064.
97. X. Zhang, Q. Zhang, D. Yue, J. Zhang, J. Wang, B. Li, Y. Yang, Y. Cui and G. Qian, *Small*, 2018, **14**, 1801563.
98. Y. Ma, H. Su, X. Kuang, X. Li, T. Zhang and B. Tang, *Anal. Chem.*, 2014, **86**, 11459–11463.
99. E. Martínez-Ahumada, A. López-Olvera, V. Jancik, J. E. Sánchez-Bautista, E. González-Zamora, V. Martis, D. R. Williams and I. A. Ibarra, *Organometallics*, 2020, **39**, 883–915.
100. M. Wang, L. Guo and D. Cao, *Anal. Chem.*, 2018, **90**, 3608–3614.
101. K.-H. Kim, S. A. Jahan and J.-T. Lee, *J. Environ. Sci. Health, Part C*, 2011, **29**, 277–299.
102. A. Bi, S. Yang, M. Liu, X. Wang, W. Liao and W. Zeng, *RSC Adv.*, 2017, **7**, 36421–36432.

103. J.-N. Hao and B. Yan, *Nanoscale*, 2016, **8**, 12047–12053.
104. D. G. Smith, N. Sajid, S. Rehn, R. Chandramohan, I. J. Carney, M. A. Khan and E. J. New, *Analyst*, 2016, **141**, 4608–4613.
105. A. A. Bowyer, C. Shen and E. J. New, *Analyst*, 2020, **145**, 1195–1201.
106. D. G. Smith, L. Mitchell and E. J. New, *Analyst*, 2019, **144**, 230–236.
107. H. Kwon, K. M. Chan and E. T. Kool, *Org. Biomol. Chem.*, 2017, **15**, 1801–1809.
108. L. Mitchell, E. J. New and C. S. Mahon, *ACS Appl. Polym. Mater.*, 2021, **3**, 506–530.
109. J. Tropp, M. H. Ihde, E. R. Crater, N. C. Bell, R. Bhatta, I. C. Johnson, M. Bonizzoni and J. D. Azoulay, *ACS Sens.*, 2020, **5**, 1541–1547.
110. W.-I. Lee, Y. Park, J. Park, S. Shrivastava, Y.-M. Son, H.-J. Choi, J. Lee, B. Jeon, H. Lee and N.-E. Lee, *Lab Chip*, 2019, **19**, 1502–1511.
111. M. Xiao, Z. Liu, N. Xu, L. Jiang, M. Yang and C. Yi, *ACS Sens.*, 2020, **5**, 870–878.
112. M. Sarwar, J. Leichner, G. M. Naja and C.-Z. Li, *Microsyst. Nanoeng.*, 2019, **5**, 56.
113. S. Spatharis and G. Tsirtsis, *Ecol. Indic.*, 2010, **10**, 840–847.
114. Y.-H. Shin, J. Z. Barnett, M. T. Gutierrez-Wing, K. A. Rusch and J.-W. Choi, *Sens. Actuators, B*, 2018, **262**, 938–946.
115. M. Brandl, T. Posnicek, R. Preuer and G. Weigelhofer, *IEEE Sens. J.*, 2020, **20**, 9132–9139.

CHAPTER 17

Time-delayed Lanthanide Luminescent Sensors and Probes

SAMUEL J. BRADBERRY[a], BRUNO D'AGOSTINO[a], DAVID F. CAFFREY[a], CIDÁLIA M. G. DOS SANTOS[a], OXANA KOTOVA[a] AND THORFINNUR GUNNLAUGSSON*[a]

[a]School of Chemistry, Trinity College Dublin, The University of Dublin, Dublin 2, Ireland
*E-mail: gunnlaut@tcd.ie

17.1 Introduction

In recent times, developments in molecular photochemistry and photophysics have progressed from the study and application of discrete molecules and molecular materials to larger structures, and modular assemblies of interest to supramolecular chemistry.[1] In the same way, the development of molecular materials for sensing, such as chemosensors, remains a highly active area of research. Sensors and probes consist of two main components: a signalling unit and a receptor unit. For a system to function as a chemosensor the recognition must be reversible. Hence, systems where chemical transformation occurs cannot be defined as sensors and the terminology 'probes' is used to describe such systems. Conventionally, optical sensors or probes are either colorimetric or fluorescent in their properties, where modulation of the various photophysical properties of the sensors occur

Monographs in Supramolecular Chemistry No. 34
Fluorescent Chemosensors
Edited by Luling Wu, Adam C. Sedgwick, Xiao-Peng He and Tony D. James

Published by the Royal Society of Chemistry, www.rsc.org

upon recognition of a particular analyte.[2–5] Many elegant examples of fluorescent sensors, in particular, have been developed to date, as this book is testimony to. However, often such sensors (emitting from a single excited state or charge transfer state) suffer from having short fluorescent lifetimes, emit at short wavelengths and have small Stokes shifts, which can hamper their applications. To overcome that, various examples of luminescent systems have been developed. This includes the use of d-metal ions, such as Ru(II), which possess microsecond lifetimes, emitting with metal-to-ligand charge transfer (MLCT) centred transitions (from a triplet state) at long wavelengths, when complexed to ligands such a polypyridyl,[6,7] or complexes possessing ions such as Ir(II), Pt(I), *etc.* However, the problem remains that such structures have broad emission spectra and relatively small Stokes shifts. So, while some of the aforementioned drawbacks are being addressed, other issues still remain.

In the search for new systems that can overcome such drawbacks, luminescent sensors/probes based on the use of the rare-earth metals (f-metal ions), or the lanthanides (Ln(III)), have emerged as attractive alternatives to such complexes in recent times, stemming from their ability to address many of the shortcomings of traditional fluorescent sensors.[8,9] The objective of this chapter is to give some background to the use of lanthanide ion-based sensors/probes for use in chemical and biochemical sensing, and where select examples from both the established and more recent literature will be featured. The aim is to provide an explanation to new researchers within this field, where we demonstrate the versatility in design, properties and applications of lanthanide-based systems. For those who are interested in diving deeper into the subject, we refer the reader to some of the references given below for further and more comprehensive discussion.

17.2 Why Should We Consider the Lanthanides When Designing Luminescent Sensors/Probes?

It is important to point out that not all the lanthanides are emissive. The luminescence of some is short-lived and can be referred as fluorescence, while these most used for sensing/probing purposes emit long-lived luminescence light, referred to as phosphorescence. The lanthanide emission occurs from metal-centred states and therefore is often referred to as '*lanthanide emission*'. For instance, the Ln(III) ions possess long-lived excited states, in the millisecond (ms) range, allowing for monitoring their emission using time delay. On this time-scale, the light scattering and short-lived autofluorescence (nanosecond range, ns) of the biological background is dramatically reduced, which results in an enhanced signal-to-noise ratio for luminescent Ln(III) complexes relative to their fluorescent counterparts, making them invaluable for sensing in biological media,[10–13] or as luminescent imaging agents, an area that is fast growing, and which we and others have recently reviewed.[14,15] Furthermore, the Ln emission can occur within the visible as well as in the near-infrared region of the electromagnetic spectrum, which is often difficult to achieve using all-organic structures.

The lanthanides are remarkable as a series of elements in their broadly similar chemical properties throughout the period, showing little systematic diversity in oxidation state, coordination behaviour or reactivity.[16] These notable phenomena arise from the electronic structure of the Ln(III) ions and present electronic behaviour that is entirely different from that which is familiar from transition metal chemistry or all-organic structures.[17] The trivalent cations of the lanthanide elements have a common electronic configurations [Xe]4f^n where $n = 0–14$. As a result of the energy separation of over 32 000 cm^{-1} between the [Xe]4f^n and [Xe]4f^{n-1}5d^1 electron configurations,[16] transitions from f → d occur only at high energy and are rarely observed.[18] Therefore, the main accessible electronic transitions are magnetic dipole and electric dipole f–f transition operations. Unlike the potential f → d transitions, which are parity allowed, unfortunately, the f → f transitions are, conversely, parity forbidden (referred to as Laporte forbidden).[19] The f–f transitions therefore absorb directly very weakly, and their absorption spectra are characterised by narrow bands with absorbance $\varepsilon < 4$ M^{-1} cm^{-1}. This fundamental selection rule also underpins the emission spectroscopy of Ln(III) ions and is considered advantageous for a number of applications across many disciplines. The theory behind this has been well described in many reviews, some of which are cited above. Consequently, we will not provide in-depth discussion of this phenomenon here. Nevertheless, it is important to state, and in the interests of those new to the area of Ln(III) luminescence, that the unique electronic structure of the Ln(III) ions gives rise to characteristic features of their luminescence spectra, and, as such, some introduction to the photophysical properties of the lanthanides is necessary.[20] Hence, we begin with a short summary of the main properties that have to be considered in the design of luminescent lanthanide sensors and probes.

The various energy gaps, between excited- and ground-state configurations results in the emission of UV light for Gd(III) ($^6P_{7/2} \rightarrow {}^8S_{7/2}$, $\Delta E = 32\,100$ cm^{-1}) through to visible light emission from Eu(III) ($^5D_0 \rightarrow {}^7F_J$, $J = 0–6$, $\Delta E =$ 12 300 cm^{-1}), Tb(III) ($^5D_4 \rightarrow {}^7F_J$, $J = 6–0$, $\Delta E = 14\,800$ cm^{-1}), Sm(III) ($^4G_{5/2} \rightarrow {}^6H_J$, $J = 5/2–13/2$, $\Delta E = 7400$ cm^{-1}) and Dy(III) ($^4F_{9/2} \rightarrow {}^6H_J$, $J = 15/2–9/2$, $\Delta E = 7850$ cm^{-1}), to near-infrared (NIR) for Nd(III) ($^4F_{3/2} \rightarrow {}^4I_J$, $J = 9/2–13/2$, $\Delta E = 5400$ cm^{-1}), Yb(III) ($^2F_{5/2} \rightarrow {}^2F_{7/2}$, $\Delta E = 10\,250$ cm^{-1}) and Er(III) ($^4I_{13/2} \rightarrow {}^4I_{15/2}$, $\Delta E =$ 6500 cm^{-1}).[16,17] These spectra are characterised by sharp line-like emission bands (with small bandwidths), as a result of the aforementioned forbidden nature of the transitions, giving *high purity of the emission colour*. Some example spectra[21] of lanthanides emissive in the visible range and images of solutions of emissive complexes[22] are shown in Figure 17.1a and b. Of the ions currently employed in Ln-luminescent sensing, Eu(III) and Tb(III) are the most frequently used, and give pure red and green emissions, respectively. The line-like emission spectra give the Ln ions predictable and structured spectra that can reveal differences in solid and solution structure, or changes

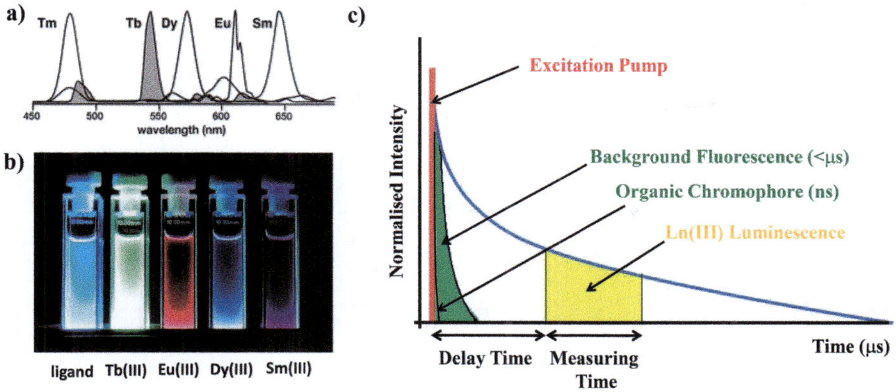

Figure 17.1 Luminescence features of selected emissive Ln(III) ions. (a) Luminescence spectra of UV- and visible-light-emitting Ln ions. Tb-centred emission spectrum highlighted (green). Adapted from ref. 21 with permission from John Wiley & Sons, Copyright © 2013 WILEY-VCH Verlag GmbH & Co. KGaA, Weinheim. (b) Observed emission from complexes under λ_{exc} = 365 nm by Raymond *et al.* Reproduced from ref. 22 with permission from American Chemical Society, Copyright 2009. (c) Schematic representation of time-resolved detection (TRD), the delay time following excitation prior to measurement allows for Ln-emission detection free from ns and >μs fluorescence emissions.

in the coordination environment of the Ln ions, which is reflected in the splitting or relative intensities of the transitions.[23–26]

For the same physical reasons, the Ln ions display much longer luminescence lifetimes compared to the nanosecond (ns) lifetimes of organic dyes and biological molecules. The lifetimes are found in the millisecond (ms) or microsecond (μs) range. These long-lived excited states enable the use of Ln-centred luminescence in time-resolved detection in biological media, where the Ln emission can be distinguished from background autofluorescence by imposing a time delay after the excitation pump and before measurement, as shown schematically in Figure 17.1c. The fluorescence emissions with ns and submicrosecond lifetimes can decay prior to such measurement and, therefore, the detected Ln-centred signal has an improved signal-to-noise ratio, which is highly attractive for luminescent (biomedical) sensing and imaging.[27] Some of the recent work in this area has been elegantly reviewed by Parker and co-workers.[28]

Despite the inefficiency of direct excitation of f–f electronic transitions, emission from Ln-centred excited states is observed as sensitized emission. Sensitized emissions are those that arise from radiative deactivation of metal-centred excited states that were initially populated indirectly by an energy transfer mechanism from a donor species, rather than by direct absorption by the metal ion centre. In brief, the general principles and component processes involved in indirect sensitization, or the "antenna effect", are shown

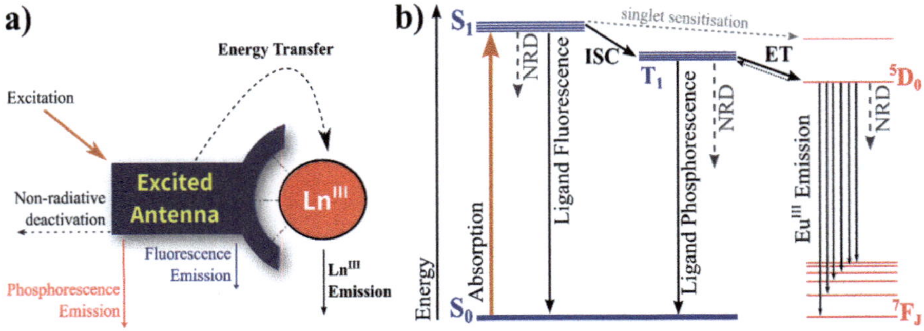

Figure 17.2 (a) Schematic representation of the "antenna effect" showing various deactivation pathways of the antenna excited-state in Ln complexes. (b) Summary Jablonski energy level diagram demonstrating the sensitisation of Eu($^5D_0$) *via* excitation of the ligand from the ground state (S_0) to the first excited state (S_1) and subsequent ligand-centred intersystem crossing (ISC) to the triplet excited-state (T_1) and energy transfer (ET) to the lanthanide ion; also shown are the corresponding non-radiative (NRD) pathways seen in panel (a).

schematically in Figure 17.2a. An emissive Ln(III) ion (such as Eu, Tb, *etc.*) coordinates ligand species that contain chromophores that absorb within the UV and visible regions, these ligand units are excited. The ligand-centred excited states may deactivate *via* various processes that do not facilitate Ln-centred emissions, including internal or external non-radiative processes (*e.g.* vibrational relaxation or vibronic coupling to solvent, respectively) or ligand-centred emissions. However, when the chromophores are held in close proximity (*i.e.* first coordination sphere) to the Ln(III) ion, deactivation of the ligand-centred excited state can result in population of that of the coordinated ion. It is this energy transfer route that constitutes sensitization.

The sensitization through antenna is tuneable to different situations and its efficiency depends upon the competition of these radiative and non-radiative processes, which is also influenced by the balance of various factors in ligand design, chromophore moieties, target lanthanide and conditions of use. This in turn is highly attractive for sensing, as receptors can be synthetically incorporated as part of the antenna design, which allows for interactions with targeted analytes, as demonstrated schematically in Figure 17.3, where the receptor and the chromophore (the Ln-emitting moiety) are either integrated or spaced. The way in which the receptor (recognition unit) and the chromophore (signalling unit) are connected depends on the photophysical property used. As there is either electronic or electron communication between these two parts, it is then the interaction with the analyte at the receptor site that modulates the overall photophysical properties of the receptor (which can be the antennae), or at the lanthanide centre directly (as outlined in Figure 17.2b), which facilitates the direct observation or monitoring of analyte concentrations, species and/or processes in real time.

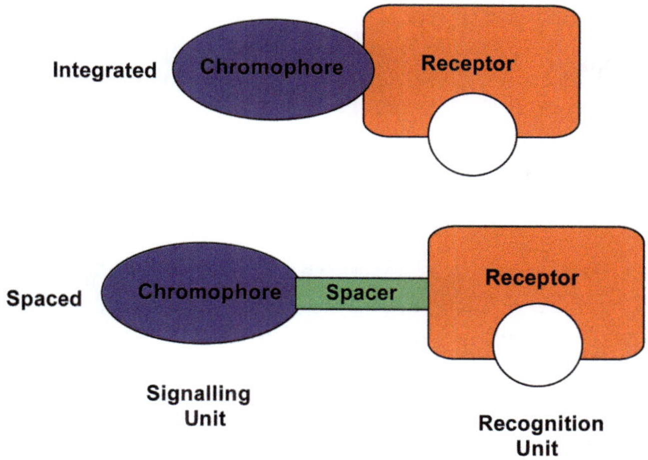

Figure 17.3 Schematic representation of the basic components of a chemosensor or a probe, where the signaling unit, a chromophore (*e.g.* a Ln center) and a receptor are either integrated within a single structure or are separated with a spacer. For both systems, the photophysical properties of the chemosensor or the probe are modulated upon analyte recognition.

The intramolecular energy transfer, from the triplet state of the antenna to the lanthanide centre ($^3\pi\pi^* \rightarrow$ Ln(III)*) was hypothesised and modelled to occur *via* two possible mechanisms. The Förster energy transfer mechanism[29] and the Dexter mechanism.[30] The Förster mechanism involves the overlap of energy levels and it is described in terms of an interaction between the transition dipole moments (a dipolar mechanism). The energy transfer occurs through space and is dependent on r^{-6}, where r is the distance between the metal ion and the excited chromophore/antenna. Consequently, the energy transfer process can be made more efficient by minimizing the distance between the lanthanide ion and the antenna. In the Dexter mechanism, the energy transfer occurs as a result of an electron exchange between the excited chromophore and the metal ion in a through-bond interaction with an e^{-r} dependency. Again, the process can be made more efficient by minimizing the distance *r*.[31]

The sensitization process is highly dependent on the choice of chromophore, which in turn is generally dictated by the energy of the emissive state of the metal. The sensitizer must possess an excited state with energy similar or higher to that of the metal emissive state in order to favour the efficiency of the sensitization process. The use of an antenna possessing too small an energy gap (<1700 cm^{-1}) is disadvantageous, since thermally activated back energy transfer from the metal (Ln(III)*) to the sensitizer ($^3\pi\pi^*$) can then compete with emission.[12,32] Among the lanthanide ions, the two most used visible emitters, Eu(III) and Tb(III), have the largest energy gaps between their emissive and ground states, therefore being the least sensitive to vibrational quenching by water molecules (O–H oscillators) or by N–H and C–H oscillators of other solvents. Their emissive states, $^5D_0$ and $^5D_4$, lie at 17 200 cm^{-1} and 20 500 cm^{-1} for Eu(III) and Tb(III), respectively.

As mentioned above, the issue of O–H oscillator quenching is obviously a concern for the use of these ions in aqueous solution. However, this vibrational energy quenching (through O–H oscillators) can be minimized by shielding the ion from the solvent through coordination to ligands (one or more that work in synergy) that can fulfil the high coordination requirements of the Ln ions. An understanding of this environment can be achieved by employing an empirical method that allows quantification of the number of coordinated water molecules directly attached to the Ln(III) metal centre. This is often called the *q value*, a method that was originally developed by Horrocks *et al.* for a range of Eu(III) and Tb(III) complexes by measuring Ln(III)-centred emission decay in D_2O and H_2O. The O–D oscillators contribute only minimally to the deactivation process of the Ln(III) excited state and all other quenching processes are assumed the same in H_2O and D_2O.[33] The q value or the hydration state (indicating the number of water molecules bound to the metal centre) is a parameter of the complexes that can be determined by the difference in the luminescence decay rates of the Ln(III) in H_2O and D_2O. The original formula was modified by Parker *et al.* by taking into account the deactivation by N–H oscillators in addition to O–H oscillators, to give eqn (17.1) for Eu(III) and eqn (17.2) for Tb(III), where an additional correction factor of 0.075 ms^{-1} should be made for each carbonyl-bound amide NH oscillator with Eu(III):[34,35]

$$q_{\mathrm{Eu(III)}} = 1.2 \times [(1/\tau_{\mathrm{H2O}}) - (1/\tau_{\mathrm{D2O}}) - 0.25] \qquad (17.1)$$

$$q_{\mathrm{Tb(III)}} = 5 \times [(1/\tau_{\mathrm{H2O}}) - (1/\tau_{\mathrm{D2O}}) - 0.06] \qquad (17.2)$$

Efficient sensitization of Ln(III)-centred emission is thus achieved through optimisation of the energy transfer process (ligand to metal) and also by protection and prevention of surrounding solvent vibrating oscillators from entering the first coordination sphere and binding to the Ln(III) ion. To achieve this, design of the ligand has to satisfy several conditions, the most important of which is that it has to fully occupy the coordination environment of the Ln(III) ion and 'shield' it from deactivation. In the following section examples of such lanthanide complexes for sensing ions and molecules, or probes to monitor biological function, *etc.* will be presented and their mode of action discussed. Several reviews have been published, which give a good account of the development of lanthanide complexes for sensing purposes in general, including anions, cations and neutral molecules, and we direct the reader to these reviews for further information.[36–40]

17.3 Examples of Lanthanide Luminescent Sensors/ Probes for Ions, Small Molecules and Biological Molecules

The way to use Ln(III) for *in vivo* sensing is by encapsulation with thermodynamically and kinetically stable ligands, which in turn form stable complexes. This can be achieved by devising ligands that are suitable for encapsulation of Ln(III) ions and can satisfy their high coordination requirements, and so shield them from any solvent coordination that may cause quenching of Ln(III) emission. The hard Lewis acid character of the Ln series has led to the use of corresponding hard oxygen donors and heterocyclic nitrogen donors in a variety of different ligand classes.[41] The coordination to Ln(III) tends to be demanding as they possess high and variable coordination numbers. For example, both Eu(III) and Tb(III) have coordination numbers of 9–10; indeed, this requirement can range from 6 to 12. Hence, this demanding coordination requires polydentate macrocyclic or chelating ligands to saturate the metal ion and create complexes showing thermodynamic stability and kinetic inertness. The macrocyclic chelates, such as 1,4,5,10-tetraazacyclododecane (cyclen) and 1,4,7-triazacyclononane-1,4,7-triacetic acid (NOTA), functionalized with coordinating pendant arms for ion saturation and, where appropriate, antennae, have been commonly employed.

Examples of Ln(III) cyclen-based systems for the detection of ions include those depicted below. Compounds **1a** and **1b** were developed by Butler *et al.* for the detection of fluoride in water.[42] Fluoride is a challenging anion to detect in competitive media and here the authors took advantage of the coordination environment of the Ln complex, which has one metal-bound water molecule that can be exchanged with the anion, due to the Lewis acid character of the lanthanide. The probes are based on a C_2-symmetric monocationic europium complex with two bound quinoline chromophores in the *trans* position and with one water molecule coordinated. The two water-soluble luminescent sensors are able to bind fluoride with minimal interference from the other anions. Upon binding of fluoride, the water is displaced, resulting in enhancement of the Eu(III)-centred emission. Indeed, such direct coordination can modulate the emission properties significantly. Related to this, even fluctuation in pH can cause changes in the emission spectra of the Ln(III) probe, as demonstrated by Plush and co-workers.[43] For this purpose cyclen ligands **2–8** were developed and their complexes with Eu(III) formed. Analysis of their photophysical properties demonstrated that significant changes were observed in all of the Eu(III)-centred transitions. As these structures consist of pendant arms that have aromatic or just aliphatic moieties, the changes could be assigned to variation in the local coordination environment of the Eu(III) ion as a function of pH. This was confirmed by carrying out potentiometric measurements, which showed that the determined pK_a values were due to deprotonation of both metal-bound water molecules and of the amide pendant arms (for those amides with labile protons).

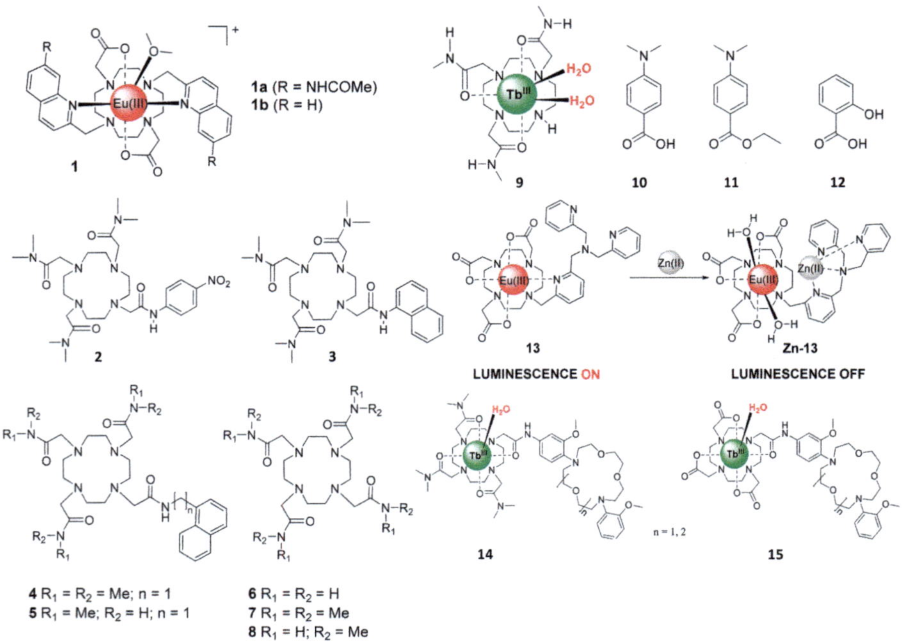

The idea of being able to modulate the lanthanide emission by directly changing the coordination environment of the ion in such cyclen complexes was first proposed by Parker *et al.* who elegantly employed this method for the recognition of anions,[9] and this idea has been further developed by Butler *et al.* for monitoring biological processes.[44] To explore this further, the cyclen complex **9** was developed by Gunnlaugsson *et al.*[45] It has no antenna moiety bound to the Tb(III) ion directly, but the aromatic analytes **10** and **12** work as external antennae, bound in a non-covalent fashion to the Tb(III) core. Complex **9** can be described as photophysically silent, as it lacks the sensitization antenna and has two metal-bound water molecules that quench the emission through O–H deactivation. However, upon addition of one equivalent of the carboxylate external antenna, two molecules of water are displaced, forming a ternary complex. The Tb(III) luminescence can now switch on because of efficient population of the excited states *via* indirect excitation from the antenna. The q value was determined and was found to decrease from two to zero, which indicates a displacement of the water molecules from the metal centre by substitution with the carboxylate groups of the antennae. In contrast, compound **11** does not bind to the Tb(III) ion in the ester-protected form, demonstrating that the carboxylate is the main chelating unit for these analytes, which, due to their aromatic nature, also function as antennae to populate the Tb(III) excited state.

Detection of metal ions such as Zn(II) has been growing in interest in the last few years because of the importance of this metal in many biological processes, such as human growth, and it has been shown to be connected with degenerative conditions such as Alzheimer's disease.

Different sensors have been developed, and one example of such a sensor was reported by Pope *et al.*[46] and is a cyclen-based Eu(III) complex (**13**). This is an on/off luminescent system, where the presence of the Zn(II) ion can quench emission of the Eu(III) centre. Compound **13** consists of a functionalized cyclen–Eu(III) complex incorporating a bis-picolyl unit that binds only Zn(II) ions over other biologically relevant metal ions, such as Mg(II) and Ca(II). The pyridyl unit serves as one of the arms of the cyclen unit and can act as an antenna and sensitize the metal centre emission. Once the Zn(II) is added, the pyridyl unit is displaced from the Eu(III) in order to bind the metal ion. Without coordination to the pyridyl unit, the Eu(III) has a free coordination site, which can be occupied by a water molecule ($q = 2$), and this results in a strong quenching of the luminescence. Furthermore, the emission band $\Delta J = 4$ undergoes significant changes, suggesting a major change in the coordination environment around the Eu(III) metal centre, proving its ability to detect Zn(II) in biological concentrations (200–300 μM).

Using cyclen complexes that have appended crown ethers has also enabled the detection of group I ions. Compounds **14** and **15** were developed by Gunnlaugsson *et al.*[12,47] who showed that these complexes were able to recognize alkali metal ions, such as Na(I) and K(I). These ions are involved in many conditions including cardiovascular diseases, stroke and hypertension. Both terbium complexes incorporating either a diaza-15-crown-5 ether for recognition of Na(I), and a diaza-18-crown-6 ether for selective recognition of K(I), were developed.

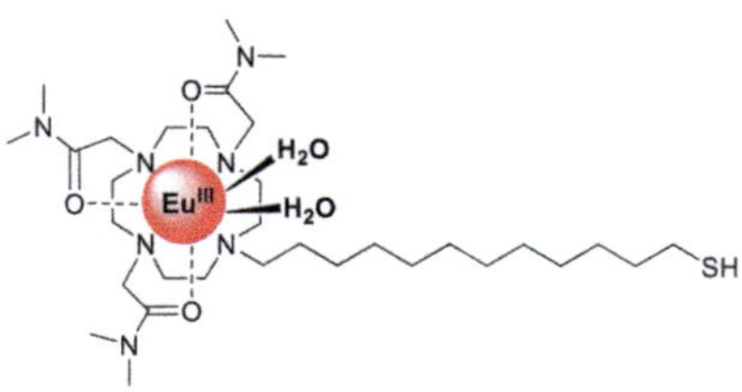

16

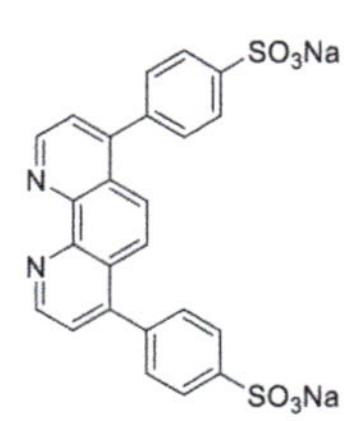

BPS

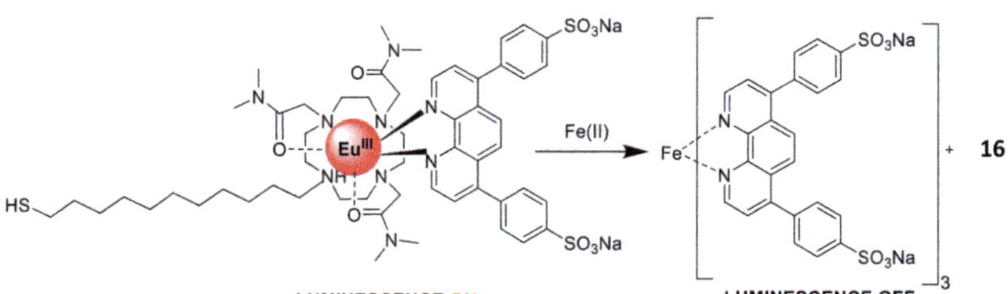

LUMINESCENCE ON LUMINESCENCE OFF

The idea presented for complex **9** for sensing of anions can also be expanded to the sensing of cations, such as d-metal ions. Kotova and co-workers developed such an example, complex **16**, for the recognition of iron in competitive media.[48] The Eu(III) complex **16** has two molecules of water coordinated to the Eu(III) centre. In order to sensitize the metal centre an external antenna, BPS (4,7-diphenyl-1,10-phenanthroline-disulfonate), was added, which is known in the literature to be a good Eu(III) sensitizer and for its selective colorimetric sensing of Fe(II).[49,50] At pH 7.4 formation of the 1:1 complex was observed by monitoring the gradually increasing Eu(III) emission, which reaches a maximum after the addition of 1 equivalent of BPS. Then the photophysical properties of the **16**-BPS complex were evaluated in the presence of different metal ions, such as Ca(II), Cd(II), Co(II), Fe(II), Mg(II), Ni(II) and Zn(II). Upon addition of Fe(II) the emission of Eu(III) strongly decreased, reaching a plateau at *ca.* 0.33 equivalents, indicating displacement of the BPS from **16** and the formation of a new complex, BPS-Fe(II), in a 3:1 stoichiometry. The addition of similar cations, such as Co(II), Cu(II) and Ni(II), caused a decrease in emission of the Eu(III) spectra comparable to that with Fe(II), but only minor changes were observed in the absorption spectra. The complex also proved to be able to sense Fe(II) in the presence of different anions, such as carbonate, lactate, citrate and phosphate. The complex **16-BPS** has been shown to give both good selectivity and effective sensitivity for Fe(II) in competitive media and was the first example of a lanthanide luminescent displacement assay for biologically active d-metal cations. Hence, **16** can be considered as an indicator displacement assay for Fe(II).

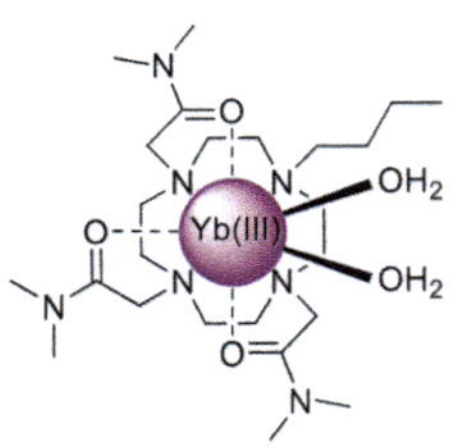

17

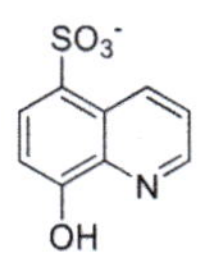

18

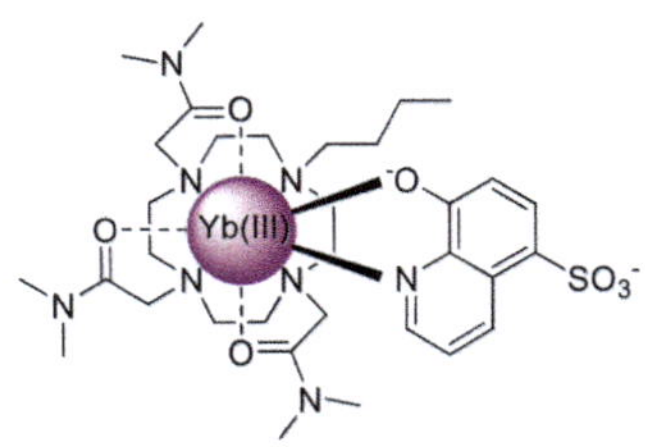

17-18 complex

While the above complexes for d-metal ions have been based on visible-light-emitting Ln ions, the same principle as used for complex **16** can also be extended to NIR-emitting Ln ions. Complex **17** is an example of this approach. The Yb(III) complex was shown to display emission in the NIR range upon binding of the sulfonated 8-hydroxyquinoline sensitizer (**18**) in buffered aqueous solution (pH 7.4).[51] The resulting ternary assembly **17–18** complex was further deployed as a dual-emissive Zn(II) sensor, progressing by the familiar route of antenna displacement, with Zn(II) addition *switching off* NIR Yb(III)-centred emission. Reference to its dual-emissive properties was born from the intense ligand-centred green fluorescence exhibited by the Zn(II)-chelate of **18**. Consequently, two different emission channels, both visible and NIR, could be simultaneously monitored to probe the Zn(II)-binding event.

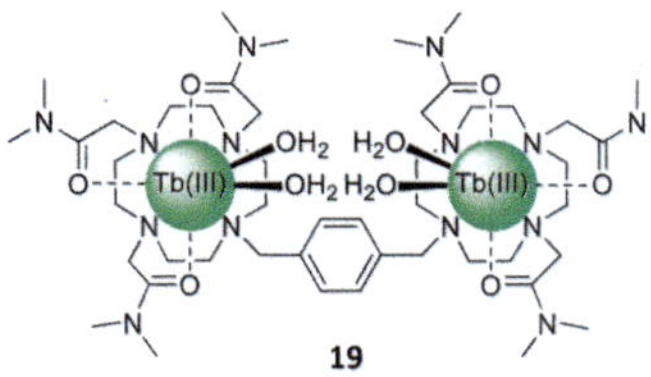

19

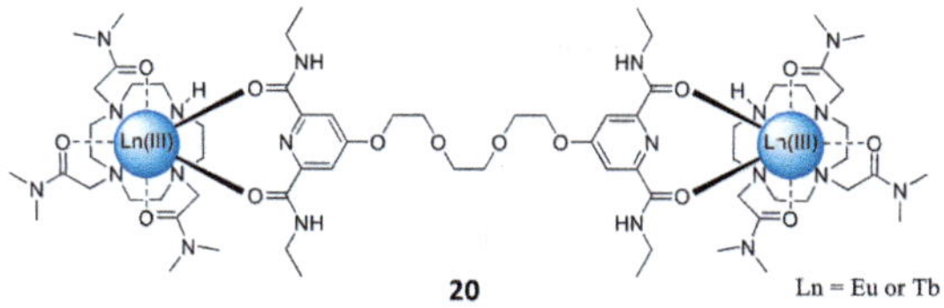

20

The above examples containing coordinatively unsaturated Ln(III) ions were all designed to target simple cations or anions. Sensing of bis-anions is more challenging for such complexes. Two recent examples from the Gunnlaugsson group have demonstrated this. The first example was formed by separating two cyclen ligands by a xylyl-containing linker. Using Tb(III) the bis- lanthanide complex **19** was formed, which was designed for sensing aromatic dicarboxylates.[52] Indeed, this complex was shown to detect the presence of both mono- and bis-aromatic carboxylates, such as terephthalic acid, in buffered aqueous solution (pH 7.4). The mode of sensing in this example again centred around the expulsion of metal-bound water by the antenna, thus leading to significant enhancement of the Tb(III) ion emission. By contrast, the recognition process involving tartaric acid resulted in quenching of the emission. Recently, this design principle has been further developed by Faulkner *et al.* for various bis-carboxylates.[53,54]

The second example from the Gunnlaugsson group is that of complex **20**,[55] where two Ln complexes are bridged by an organic coordination linker, based on the di-picolinic carboxamides, which are well-known antenna for lanthanides. Designed as either Eu(III) or Tb(III) systems, it was predicted that the bridge would function as a coordinating antenna, which could be displaced from the metal core by a more strongly chelating anion. Of these two systems studied, significant changes where observed in the photophysical properties of Tb(III) complexes, where the Tb(III)-centred emission was *switched off* in the sensing of phosphates and nitrates. Other related systems have been developed.

As mentioned earlier, crown ethers have been conjugated to Ln(III) cyclen complexes for sensing of group I ions. However, Ln ions can also complex to crown ethers and Plush and co-workers took advantage of this for sensing of anions by forming homonuclear or mixed-Ln complexes (*i.e.* by using both Eu(III) and Tb(III)), with the latter enabling ratiometric sensing of anions. The former of these, complex **21** (Figure 17.4),[56] was shown to bind small dicarboxylic acids, such as aspartic, malonic, succinic and glutaric acid, in aqueous pH 6.5 media. Nevertheless, only malonic acid gave rise to selective Eu(III) luminescence enhancements, as the emission intensity was reduced for the other acids. The second example, complex **22** (Figure 17.4),[57] was formed by initially forming a Eu(III) complex, followed by careful titration using Tb(III). Alternatively, under thermodynamic control, the homonuclear

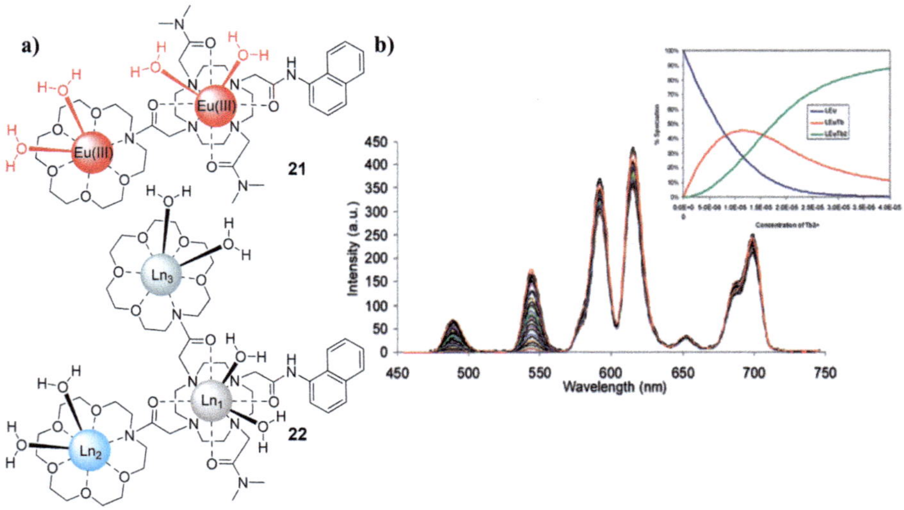

Figure 17.4 (a) The chemical structures of **21** and **22**, developed to form either homonuclear lanthanide complexes (**21**) or heteronuclear complexes (**22**). (b) Changes in Eu(III) and Tb(III) emission upon titration of **22Eu**, formed by mixing **22** (1×10^{-5} mol dm^{-3}) with 1 eq. $Eu(ClO_4)_3$ followed by the titration using $Tb(ClO_4)_3$ ($0 \rightarrow 5 \times 10^{-5}$ mol dm^{-3}). Inset: the speciation diagram for formation of the trimetallic complex Eu_1Tb_2 of **22**.

complex could also be formed using either Tb(III) or Eu(III). The overall emission changes as a function of added equivalents of anions were investigated for all three of these systems (**22Ln**; Ln = Eu_3, Tb_3 and Eu_1Tb_2). The results demonstrated that the Tb(III) emission was quenched upon anion binding to the Eu_1Tb_2 system, while it was enhanced in the case of Tb_3. It was concluded from this investigation that the main binding occurred at the Eu(III)-cyclen site, as the changes in the Tb(III) emission were seen to be significantly smaller than for the Eu(III) $\Delta J = 2$ transition. Indeed, by using the emission changes at 545 nm for the Tb(III) ion as a reference for the changes, a linear relationship was established for the Eu/Tb emission observed for Eu_1Tb_2 with malonic acid.

A series of cationic, zwitterionic and anionic Eu(III) complexes, such as **23–25**, have been developed by Parker *et al.* for the detection of bicarbonate (HCO_3^-).[58,59] An acridone was used as the chromophore, which allowed sensitization of the Eu(III) emission following excitation at 390–410 nm. Each of these complexes was found to selectively bind bicarbonate at physiological pH. The binding process was signalled by changes of the luminescence lifetimes and of the Eu(III) emission intensities, where q changed from 1 to 0. This supported the displacement of the metal-bound inner-sphere water molecule. By monitoring the ratio of the emission intensity of Eu(III) of the 618/588 or 618/702 nm emission bands, the solution concentration of bicarbonate could be assessed in a background of competing anions, such as lactate, citrate and phosphate. The authors then expanded their portfolio of anions towards amino acids, where, in the case of **24**, the binding of phosphorylated amino acids such as Ser-OP and Tyr-OP was achieved; with a Job plot analysis revealing the formation of 1 : 1 adducts. In each case, the obtained data were consistent with formation of mono-aqua species with phosphate anion acting as a monodentate ligand.

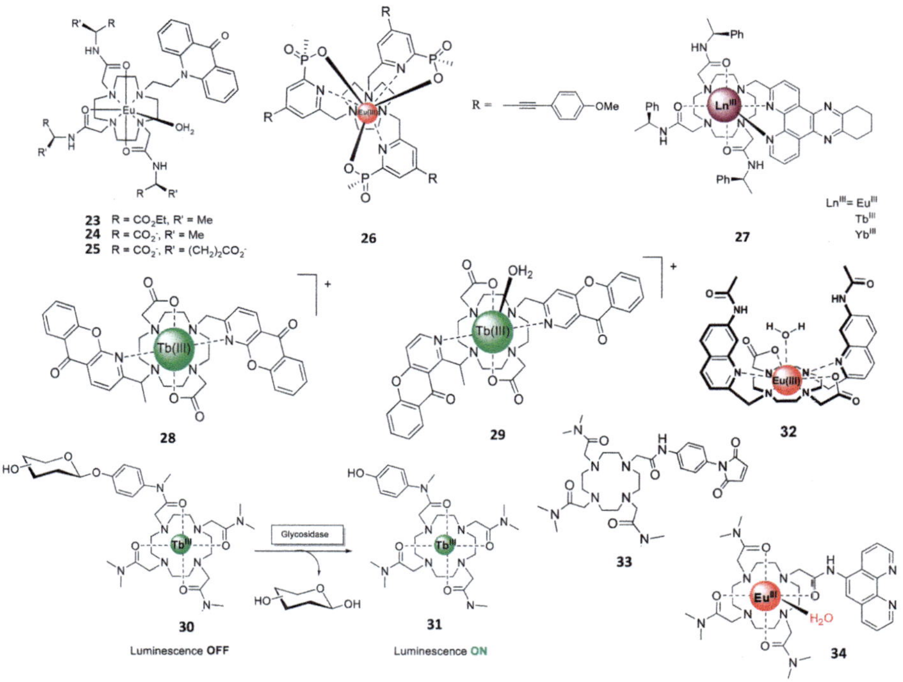

It is worth pointing out at this stage that the application of lanthanide complexes in anion recognition and sensing has been elegantly reviewed by Bodman and Butler.[4]

While most of the cases featured above have had a cyclen ligand as part of their structure, other systems have also emerged in recent times. An example is **26**,[60] developed by Parker and co-workers as an imaging agent, but several other derivatives have also been developed as sensors. The application of such complexes has recently been reviewed by Parker *et al.*[28] Each structure contained three *para*-aryl-alkynyl chromophores absorbing strongly within a biologically beneficial spectral window of 310–340 nm, enabling efficient Eu(III) sensitization. These structures give rise to intensely emissive Eu(III), which has been reported to be an order of magnitude greater than that previously recorded for analogous systems in the literature.

17.4 Lanthanide Complexes with Cyclen Ligands as Biological Probes

The development of lanthanide-based sensors and probes for small organic or biological molecules has been an active area of research. Some of this work has been reviewed recently and the references have been cited herein. Hence, the discussion below will only focus on some select examples. Emissive Ln(III) species constitute an emerging class of metal-based emissive probes that are being rigorously examined for use *in cellulo* for imaging purposes; many recent ones have been developed around the smaller aza-crown ethers. Prior to that, systems were developed based on the cyclen moiety. While being categorized as imaging agents, these species have been designed to target particular organelles or biomolecules, such as DNA, or larger biomolecules, such as complex **27**, which was developed and investigated for its enantioselective ability to bind human serum albumin.[61] Other applicable examples from the Parker group centred around the selective staining of nuclear DNA by treatment with two constitutionally isomeric Tb(III)-cyclen complexes, such as **28** and **29**, possessing *trans*-related azaxanthone sensitizers.[62] Administering the two complexes at low concentrations (<1 μM) to HeLa cell lines revealed, by means of luminescence microscopy, chromosomal DNA staining in less than 10% of the total cell population upon excitation at 340 nm. The observed discriminatory pattern was consistent with preferential labelling of those cells undergoing replication. Cells at each stage of the mitotic cycle were visualised and discerned by real-time monitoring of these Tb(III)-based optical probes, arising from the noted affinity for exposed nuclear DNA.

The use of lanthanide complexes for monitoring biological processes using lanthanide luminescent systems has also attracted significant attention in recent times.[63] Gunnlaugsson and co-workers developed a synthetic chemical probe for the enzymatic detection of glycosidase enzyme activity in real time.[64] The system, a Tb(III)-cyclen-based complex, **30**, possessing a conjugate carbohydrate unit was not luminescent upon excitation of the antenna. However, upon enzymatic hydrolysis of the glycosidic bond the sugar would be cleaved off, resulting in release of the phenolic group, which had the correct excited-state energy to populate the Tb(III) ion in complex **31**. Hence, the emission was *switched on* as a direct consequence of the enzymatic activity. Hence, **30** can be described as a lanthanide probe because the reaction is irreversible upon formation of **31**.

A few lanthanide-based probes have been developed in recent years to monitor the phosphorylation process in solution. This includes the elegant work of Butler and co-workers, where a lanthanide complex **32** was used in an assay to monitor the enzyme reaction for detecting peptide modification by phosphorylation and dephosphorylation.[65] Other related systems that can be employed as probes for monitoring enzymatic processes include both **33** and **34**. In **33**, a cyclen conjugate possessing maleimide functionality was synthesised. The use of Tb(III) showed that upon 1,4-Michael addition of a thiol group to the alkene bond large changes in the Tb(III)-centred emission were displayed. With this in mind the authors demonstrated high selectivity of the Tb(III) complex of **33** for glutathione (GSH), and that, in combination with the enzyme glutathione reductase, the formation of GSH from GSSG can be monitored in real time with high sensitivity.[66]

Similarly, the Eu(III) complex **34** was also employed in observing a biochemical process in real time,[67] the focus being on monitoring the activity of urease by following the changes in pH upon urease-mediated hydrolysis of urea in aqueous solution. The sensor has previously been demonstrated to give highly dependent luminescence responses[68] to changes in pH and behaved here as an *on–off* luminescent switch, where the emission was quenched upon urea being enzymatically converted into ammonia and carbon dioxide. The intention of investigating the potential application of this probe was in detecting the onset of catheter-associated urinary tract infections. This was shown by incorporating the sensors into water-permeable hydrogels, which are commonly employed in the coatings for catheters.

A number of lanthanide-based biomarker probes have been reported using acyclic ligands and have also been developed.[69–71] Recently, Miller *et al.* used Eu(III) and Tb(III) complexes of the multidentate polyaminocarboxylate ligand **35**, with an attached azido- and ester-functionalized quinolinone precursor, to monitor cystathionine γ-lyase activity using time-gated Ln emission (Figure 17.5).[72] The generation of H_2S by the enzymatic activity was found to convert the precursor chromophore (which was not suitable for Ln(III) sensitization) into the 7-amino-4-methyl quinolinone (which was a good sensitizing antenna) and gave rise to the significant lanthanide-centred

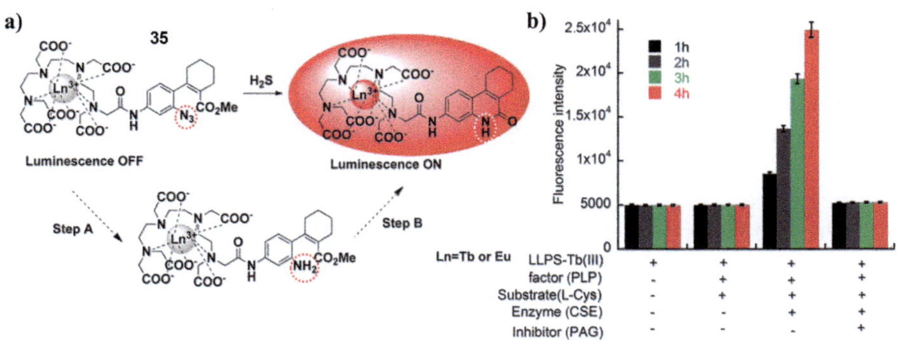

Figure 17.5 (a) Schematic representation of luminescence *switch-on* response of [Ln(**35**)] upon exposure to H_2S generating an active antenna. (b) The fluorescence intensity responses upon exposure to lyase enzyme, substrate and inhibitor showing selective response to active enzyme. Reproduced from ref. 72 with permission from John Wiley & Sons, Copyright © 2017 Wiley-VCH Verlag GmbH & Co. KGaA, Weinheim.

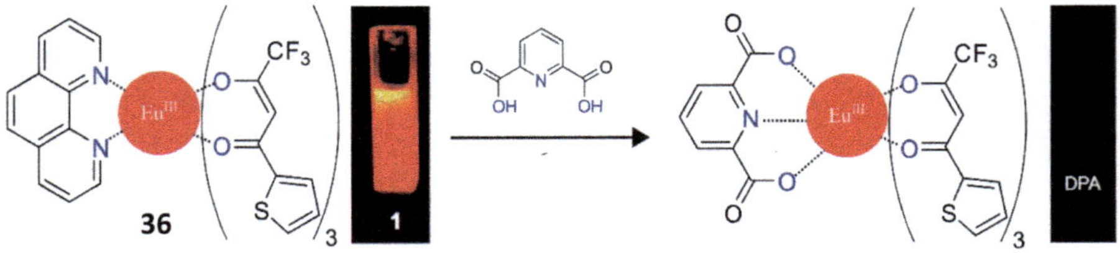

Figure 17.6 *Switch-off* sensor **36** for *B. anthracis* biomarker DPA, showing phen-sensitized emission in the resting state, which becomes displaced upon exposure to DPA. The offset excitation wavelength between phen and DPA resulted in poorer sensitization by DPA at λ_{exc} = 340 nm. Reproduced from ref. 74 with permission from the Royal Society of Chemistry.

emission. Luminescence enhancements of >25-fold were observed with low false-positive rates for the Eu complexes and an overall performance comparable to established commercially available alternatives.

Lu and Blight have independently reported Eu(III)-based biomarker nanoparticle sensors for *Bacillus anthracis* (anthrax), which is, in fact, dipicolinic acid (DPA). The approach of Lu *et al.* was to implement a *switch-on* sensor in which ethylenediaminetetraacetic acid (EDTA)-bound Eu(III) complexes become emissive upon coordination of trace DPA.[73] The probe was shown to be selective over related carboxylic acids, with a limit of detection of 0.2 nM and a response time of 2 minutes.

On the other hand, Blight and co-workers used a *switch-off* approach with complex **36**, shown in Figure 17.6, in which the presence of DPA was shown to compete with and displace the 1,10-phenanthroline (phen) antenna, reducing the sensitization efficiency when irradiated at λ_{exc} = 340 nm. The operation of the probe was shown to be selective for DPA in the presence of coordinating phosphates and water, while also maintaining sensitivity at nanomolar concentrations of the biomarker.[74]

Also, in order to address the consequences of warfare, Albrecht and co-workers have efficiently shown the Eu(III)-based probe **37** to be selective for the explosive marker picric acid over other nitro-aromatic compounds (Figure 17.7).[75] The phen-Schiff base ligand **37** sensitizes the Eu emission, but in the presence of picric acid it is quenched, with a selective efficiency of *ca.* 80%. The same complexes operated as selective anion sensors for nitrite, showing eight-fold selectivity over other anions as a result of discrimination in coordination abilities.

17.5 Lanthanide Luminescent Materials

Some of the above examples demonstrate how soft material can be employed in probing mechanisms and sensing ions and molecules. The use of lanthanide luminescent materials is a fast-growing area and their application in

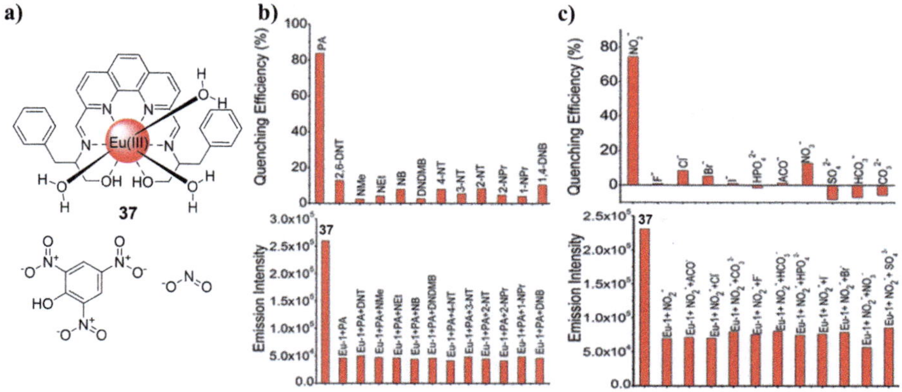

Figure 17.7 (a) Structure of responsive probe [$37 = Eu(L)(H_2O)]^{3+}$ and selective analytes: picric acid and nitrite. (b) Selective response of **37** for picric acid by quenching efficiency over other nitro-aromatic compounds and (c) selective response for nitrite by quenching efficiency over other coordinating anions. Graphs reproduced from ref. 75 with permission from John Wiley & Sons, Copyright © 2016 WILEY-VCH Verlag GmbH & Co. KGaA, Weinheim.

sensing has become apparent. One of the first examples of such a system was **38**, a pH-responsive hydrogel, where the Eu(III)-responsive cyclen complex was doped non-covalently within a matrix of poly(MMA-*co*-HEMA-*co*-EGDMA) *via* direct encapsulation during polymerisation.[76]

The same antenna structure was then employed in the generation of the Eu(III) or Tb(III) luminescent complexes **39**, which had three such ligands attached to a cyclen macrocycle, with the fourth position containing a long alkyl thiol spacer that was used to surface modify gold nanoparticles (**AuNP**), giving rise to the formation of **AuNP-39**. This nanosystem is then a sensor for pH, but also has two water molecules at each Ln centre, which could be exchanged for an appropriate organic-binding molecule, and, as such, could also be used as a molecular logic gate mimic with two inputs. The photophysical properties of **AuNP-39** were investigated and the excitation at the absorption maximum at 318 nm was found to result in the Ln(III)-centered emission. An absorption band with a $\lambda_{max} = 540$ nm was also observed, corresponding to the surface plasmon resonance band of the **AuNPs**, which was red-shifted *ca.* 6 nm compared to the non-functionalized **AuNPs**; confirming the adsorption of the complexes onto the nanoparticle surface. The number of complexes bound to each **AuNP** was evaluated and it was found that *ca.* 50 complexes bound each **AuNP**. The ability of these conjugates to work as probes was evaluated upon the addition of various chemical inputs as pH, pO_2 and ternary-sensitive antennae. Upon pH change, a modulation of the Ln(III)-centered emission was observed, making this system applicable as a *switch-on/switch-off* sensor, and its parameterization for application as a logic gate was also evaluated.[77]

Other examples of **AuNPs** functionalized with lanthanide complexes that have been employed for sensing include systems responsive to the presence of various anions, such as demonstrated for **AuNP-40**, through the use of displacement assays using ligand/antenna **41** and anions, such as **42**.[78] When bound to proteins such as bovine serum albumin this Ln-based "biosensor" could then be used for sensing of two drug candidates that can selectively bind at two different sites on the protein,[79] or as novel NIR-emitting pH sensors[80] and as ratiometric sensors, as demonstrated for **43**, using antennae, such as **44**.[81,82] The latter example (**43**) has also been developed as an MRI contrast agent by exchanging the Eu(III) ions for Gd(III).

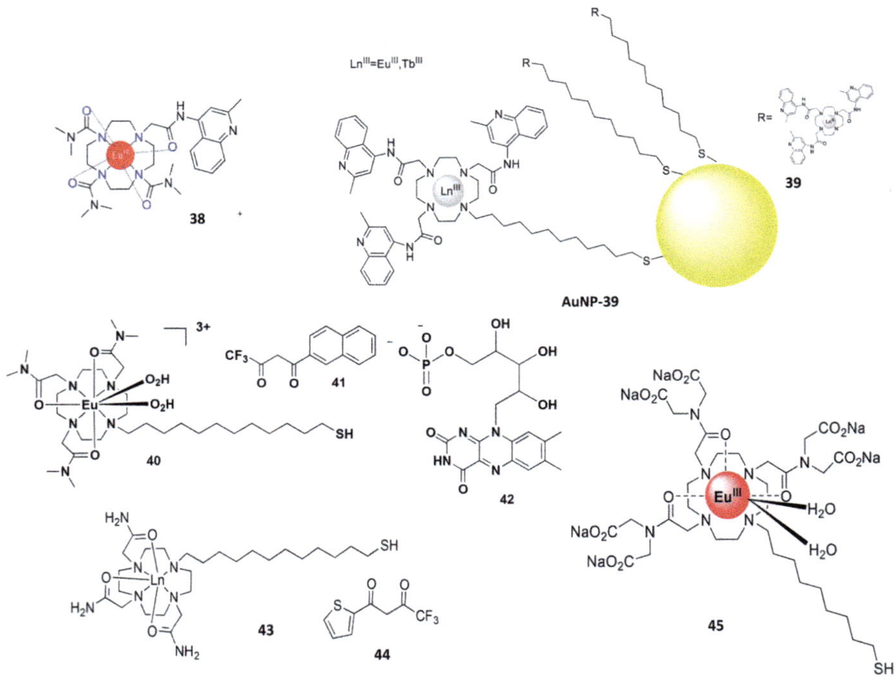

The use of such nanoparticles has significant advantages over the use of single complexes, particularly in the application of systems such as MRI probes or for luminescent imaging. This was demonstrated by Surender and co-workers using the cyclen derivative **45** in combination with the antennae **41**.[83] The ternary complexes of **45** with **41** (gave $q = 0$) were anchored to the surface of 5 nm-sized gold nanoparticles, and the authors showed that the resulting nanoagent could bind to an exposed calcium site (or microcracks) on bone structures, which was then imaged *via* sensitization through antenna **41** using two-photon microscopy. This then generates a three-dimensional map of the damaged bone area by observing the long-wavelength lanthanide-centered emission, opening a new way of imaging such damage, where background emission from the bone structure was minimized or removed

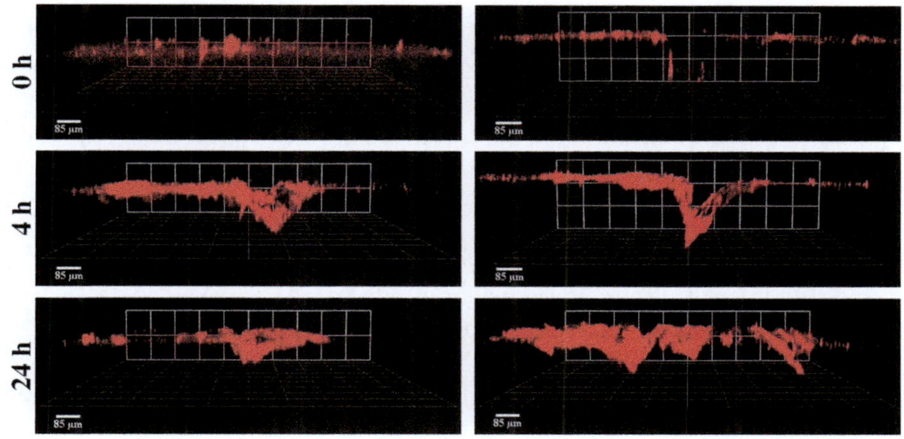

Figure 17.8 The two-dimensional two-photon excitation microscopy of microdamaged bovine bone structure showing the cracks formed on the surface of the bone and the depth of those cracks. The scale on the images corresponds to 85 µm. Reproduced from ref. 83.

altogether (Figure 17.8). This demonstrates that such Ln-based imaging systems have significant advantages over more conventional fluorescent imaging agents within a biological setting.

17.6 Summary

In this chapter some select examples of sensors and probes, as well as imaging agents, based on the use of lanthanide luminescence have been presented. The objective was to give some flavour of the area, which is rapidly growing, and to discuss examples that demonstrate the versatility of the use of such long-lived and long-wavelength emitting agents based on f-metal ions, such as Eu(III), Tb(III), Nd(III) and Yb(III). Some basic background information into the photophysical properties of such systems was included to enable the reader to appreciate some of the conditions that need to be considered in the design of such structures. As mentioned several times throughout the chapter, there have been some excellent reviews published on specific areas for the use of such sensors and we direct interested readers towards those reviews for further information on developments within this fast-growing and highly active area of research.[84]

Acknowledgements

The support from Science Foundation Ireland (SFI), The Irish Research Council (IRC) and Trinity College Dublin, The University of Dublin is gratefully acknowledged.

References

1. S. Erbas-Cakmak, S. Kolemen, A. C. Sedgwick, T. Gunnlaugsson, T. D. James, J. Yoon and E. U. Akkaya, *Chem. Soc. Rev.*, 2018, **47**, 2228.
2. M. V. Ramakrishnam Raju, S. M. Harris and V. C. Pierre, *Chem. Soc. Rev.*, 2020, **49**, 1090.
3. A. B. Aletti, D. M. Gillen and T. Gunnlaugsson, *Coord. Chem. Rev.*, 2018, **354**, 98.
4. S. E. Bodman and S. J. Butler, *Chem. Sci.*, 2021, **12**, 2716.
5. E. Mathieu, A. Sipos, E. Demeyere, D. Phipps, D. Sakaveli and K. E. Borbas, *Chem. Commun.*, 2018, **54**, 10021.
6. S. Estalayo-Adrián, S. Blasco, S. A. Bright, G. J. McManus, G. Orellana, D. C. Williams, J. M. Kelly and T. Gunnlaugsson, *Chem. Commun.*, 2020, **56**, 9332.
7. S. Estalayo-Adrián, S. Blasco, S. A. Bright, G. J. McManus, G. Orellana, D. C. Williams, J. M. Kelly and T. Gunnlaugsson, *ACS Appl. Bio Mater.*, 2021, **4**, 6664.
8. C. M. G. dos Santos, A. J. Harte, S. J. Quinn and T. Gunnlaugsson, *Coord. Chem. Rev.*, 2008, **252**, 2512.
9. S. J. Butler and D. Parker, *Chem. Soc. Rev.*, 2013, **42**, 1652.
10. D. Parker, *Coord. Chem. Rev.*, 2000, **205**, 109.
11. J.-C. G. Bünzli, *Chem. Rev.*, 2010, **110**, 2729.
12. T. Gunnlaugsson and J. P. Leonard, *Chem. Commun.*, 2005, 3114.
13. J. P. Leonard, C. B. Nolan, F. Stomeo, and T. Gunnlaugsson, *Photochemistry and Photophysics of Coordination Compounds: Lanthanides*, Springer, 2007, vol. 281.
14. *Supramolecular Chemistry in Biomedical Imaging, Monographs in Supramolecular Chemistry No. 33*, ed. S. Faulkner, T. Gunnlaugsson and G. M. Ó Máille, The Royal Society of Chemistry, 2022.
15. E. Calatrava-Pérez, E. Surender, L. Truman, G. M. Ó. Máille, A. F. Henwood, E. Scanlan and T. Gunnlaugsson, *Supramolecular Chemistry in Biomedical Imaging, Monographs in Supramolecular Chemistry No. 33*, ed. S. Faulkner, T. Gunnlaugsson and G. M. Ó Máille, The Royal Society of Chemistry 2022, ch. 1, pp. 1–42.
16. J. C. G. Bünzli and C. Piguet, *Chem. Soc. Rev.*, 2005, **34**, 1048.
17. J.-C. G. Bünzli and S. V. Eliseeva, *Lanthanide Luminescence*, Springer Berlin Heidelberg, Berlin, Germany, 2011, vol. 7, p. 1.
18. V. N. Makhov, M. Kirm, G. Stryganyuk, S. Vielhauer, G. Zimmerer, B. Z. Malkin, O. V. Solovyev and S. L. J. Korableva, *J. Lumin.*, 2012, **132**, 418.
19. O. Laporte and W. F. J. Meggers, *Opt. Soc. Am.*, 1925, **11**, 459.
20. D. Parker and J. A. G. Williams, *J. Chem. Soc., Dalton Trans.*, 1996, 3613.
21. E. Pazos and M. E. Vázquez, *Biotechnol. J.*, 2014, **9**, 241.
22. E. G. Moore, A. P. S. Samuel and K. N. Raymond, *Acc. Chem. Res.*, 2009, **42**, 542.
23. K. Binnemans, *Coord. Chem. Rev.*, 2015, **295**, 1.
24. K. Binnemans and C. Gorller-Waldrand, *J. Rare Earths*, 1996, **14**, 173.

25. A.-S. Chauvin, F. Thomas, B. Song, C. D. B. Vandevyver and J.-C. G. Bünzli, *Philos. Trans. R. Soc., A*, 2013, **371**, 20120295.
26. N. M. Shavaleev, S. V. Eliseeva, R. Scopelliti and J.-C. G. Bünzli, *Inorg. Chem.*, 2015, **54**, 9166.
27. A. D'Aléo, F. Pointillart, L. Ouahab, C. Andraud and O. Maury, *Coord. Chem. Rev.*, 2012, **256**, 1604.
28. D. Parker, J. D. Fradgley and K.-L. Wong, *Chem. Soc. Rev.*, 2021, **50**, 8193.
29. T. H. Förster, *Discuss. Faraday Soc.*, 1959, **27**, 7.
30. D. L. Dexter, *J. Chem. Phys.*, 1953, **215**, 836.
31. S. Faulkner and J. L. Matthews, in *Comprehensive Coordination Chemistry*, ed. M. D. Ward, Elsevier, Oxford, 2nd edn, 2004, vol. 9, pp. 1–32.
32. J.-C. G. Bünzli, *J. Alloys Compd.*, 2006, **408–412**, 934.
33. D. Parker, R. S. Dickins, H. Puschmann, C. Crossland and J. A. Howard, *Chem. Rev.*, 2002, **102**, 1977.
34. S. Faulkner, L. S. Natrajan, W. S. Perry and D. Sykes, *Dalton Trans.*, 2009, 3890.
35. A. Beeby, I. Clarkson, R. Dickins, S. Faulkner, D. Parker and L. Royle, *J. Chem. Soc., Perkin Trans.*, 1999, **2**, 493.
36. J.-C. G. Bunzli, *Acc. Chem. Res.*, 2006, **39**, 53.
37. S. Faulkner, S. J. A. Pope and B. P. Burton-Pye, *Appl. Spectrosc. Rev.*, 2005, **40**, 1.
38. D. E. Barry, D. F. Caffrey and T. Gunnlaugsson, *Chem. Soc. Rev.*, 2016, **45**, 3244.
39. J.-C. G. Bünzli, *J. Lumin.*, 2016, **170**, 866.
40. A. de Bettencourt-Dias, P. S. Barber and S. Viswanathan, *Coord. Chem. Rev.*, 2014, **273–274**, 165.
41. J.-C. G. Bünzli, *J. Coord. Chem.*, 2014, **67**, 3706.
42. S. J. Butler, *Chem. Commun.*, 2015, **51**, 10879.
43. S. E. Plush, N. A. Clear, J. P. Leonard, A.-M. Fanning and T. Gunnlaugsson, *Dalton Trans.*, 2010, **39**, 3644.
44. S. H. Hewitt, G. Macey, R. Mailhot, M. R. J. Fernanda Duarte, A. M. Kenwright and S. J. Butler, *Chem. Sci.*, 2020, **11**, 3619.
45. T. Gunnlaugsson, A. J. Harte, J. P. Leonard and M. Nieuwenhuyzen, *Supramol. Chem.*, 2003, **15**, 505.
46. S. J. Pope and R. H. Laye, *Dalton Trans.*, 2006, 3108.
47. T. Gunnlaugsson and J. P. Leonard, *Dalton Trans.*, 2005, 3204.
48. O. Kotova, S. Comby and T. Gunnlaugsson, *Chem. Commun.*, 2011, **47**, 6810.
49. M. Fujiwara, S. Tsukahara and H. Watarai, *Phys. Chem. Chem. Phys.*, 1999, **1**, 2949.
50. E. P. Achterberg, T. W. Holland, A. R. Bowie, R. F. C. Mantoura and P. J. Worsfold, *Anal. Chim. Acta*, 2001, **442**, 1.
51. S. Comby, S. A. Tuck, L. K. Truman, O. Kotova and T. Gunnlaugsson, *Inorg. Chem.*, 2012, **51**, 10158.
52. A. J. Harte, P. Jensen, S. E. Plush, P. E. Kruger and T. Gunnlaugsson, *Inorg. Chem.*, 2006, **45**, 9465.

53. O. A. Blackburn, M. Tropiano, L. S. Natrajan, A. M. Kenwright and S. Faulkner, *Chem. Commun.*, 2016, **52**, 6111.
54. T. J. Sørensen, L. R. Hill and S. Faulkner, *ChemistryOpen*, 2015, **4**, 509–515.
55. D. F. Caffrey and T. Gunnlaugsson, *Dalton Trans.*, 2014, **43**, 17964.
56. S. E. Plush and T. Gunnlaugsson, *Org. Lett.*, 2007, **9**, 1919.
57. S. E. Plush and T. Gunnlaugsson, *Dalton Trans.*, 2008, 3801.
58. Y. Bretonnière, M. J. Cann, D. Parker and R. Slater, *Chem. Commun.*, 2002, 1930.
59. Y. Bretonnière, M. J. Cann, D. Parker and R. Slater, *Org. Biomol. Chem.*, 2004, **2**, 1624.
60. J. W. Walton, A. Bourdolle, S. J. Butler, M. Soulie, M. Delbianco, B. K. McMahon, R. Pal, H. Puschmann, J. M. Zwier, L. Lamarque, O. Maury, C. Andraud and D. Parker, *Chem. Commun.*, 2013, **49**, 1600.
61. R. A. Poole, G. Bobba, M. J. Cann, J.-C. Frias, D. Parker and R. D. Peacock, *Org. Biomol. Chem.*, 2005, **3**, 1013.
62. G.-L. Law, C. Man, D. Parker and J. W. Walton, *Chem. Commun.*, 2010, **46**, 2391.
63. S. H. Hewitt and S. J. Butler, *Chem. Commun.*, 2018, **54**, 6635.
64. H. M. Burke, T. Gunnlaugsson and E. M. Scanlan, *Org. Biomol. Chem.*, 2016, **14**, 9133.
65. S. H. Hewitt, R. Ali, R. Mailhot, C. R. Antonen, C. A. Dodson and S. J. Butler, *Chem. Sci.*, 2019, **10**, 5373.
66. B. K. McMahon and T. Gunnlaugsson, *J. Am. Chem. Soc.*, 2012, **134**, 10725.
67. E. M. Surender, S. J. Bradberry, S. A. Bright, C. P. McCoy, D. C. Williams and T. Gunnlaugsson, *J. Am. Chem. Soc.*, 2017, **139**, 381.
68. T. Gunnlaugsson, A. J. Harte, J. P. Leonard and K. Senegal, *J. Am. Chem. Soc.*, 2003, **125**, 12062.
69. B. H. M. Hussein, G. M. Khairy and R. M. Kamel, *Spectrochim. Acta, Part A*, 2016, **158**, 34.
70. V. Fernandez-Moreira, B. Song, V. Sivagnanam, A.-S. Chauvin, C. D. B. Vandevyver, M. Gijs, I. Hemmilä, H.-A. Lehr and J.-C. G. Bünzli, *Analyst*, 2010, **135**, 42.
71. E. G. Moore, J. Xu, C. J. Jocher, T. M. Corneillie and K. N. Raymond, *Inorg. Chem.*, 2010, **49**, 9928.
72. Y. Yao, C. Kong, L. Yin, A. D. Jain, K. Ratia, G. R. J. Thatcher, T. W. Moore, T. G. Driver and L. W. Miller, *Chem. – Eur. J.*, 2017, **23**, 752.
73. K. Ai, B. Zhang and L. Lu, *Angew. Chem., Int. Ed.*, 2009, **48**, 304.
74. M. A. Shipman, K. J. Ramhit and B. A. Blight, *J. Mater. Chem. B*, 2016, **4**, 3043.
75. J. Sahoo, S. B. Waghmode, P. S. Subramanian and M. Albrecht, *ChemistrySelect*, 2016, **1**, 1943.
76. C. P. McCoy, F. Stomeo, S. E. Plush and T. Gunnlaugsson, *Chem. Mater.*, 2006, **18**, 4336.
77. L. K. Truman, S. J. Bradberry, S. Comby, O. Kotova and T. Gunnlaugsson, *ChemPhysChem*, 2017, **18**, 1746.

78. J. Massue, S. Quinn and T. Gunnlaugsson, *J. Am. Chem. Soc.*, 2008, **130**, 6900.
79. S. Comby and T. Gunnlaugsson, *ACS Nano*, 2011, **5**, 7184.
80. L. K. Truman, S. Comby and T. Gunnlaugsson, *Angew. Chem., Int. Ed.*, 2012, **51**, 9624.
81. J. K. Molloy, A. M. Nonat, J. E. O'Brien, D. F. Brougham and T. Gunnlaugsson, *Supramol. Chem.*, 2020, **32**, 373.
82. S. Comby, E. M. Surender, O. Kotova, L. K. Truman, J. K. Molloy and T. Gunnlaugsson, *Inorg. Chem.*, 2014, **53**, 1867.
83. E. M. Surender, S. Comby, B. Cavanagh, O. Brennan, T. C. Lee and T. Gunnlaugsson, *Chem.*, 2016, **1**, 438.
84. M. C. Heffern, L. M. Matosziuk and T. J. Meade, *Chem. Rev.*, 2014, **114**, 4496.

Subject Index